AF556708

Cellular and Molecular Aspects of Inflammation

NEW HORIZONS IN THERAPEUTICS
Smith Kline & French Laboratories Research Symposia Series

Series Editors: George Poste and Stanley T. Crooke
Smith Kline & French Laboratories, Philadelphia, Pennsylvania

---

CELLULAR AND MOLECULAR ASPECTS OF INFLAMMATION
Edited by George Poste and Stanley T. Crooke

DOPAMINE RECEPTOR AGONISTS
Edited by George Poste and Stanley T. Crooke

MECHANISMS OF RECEPTOR REGULATION
Edited by George Poste and Stanley T. Crooke

NEW FRONTIERS IN THE STUDY OF GENE FUNCTIONS
Edited by George Poste and Stanley T. Crooke

NEW INSIGHTS INTO CELL AND MEMBRANE TRANSPORT PROCESSES
Edited by George Poste and Stanley T. Crooke

Cellular and Molecular Aspects of Inflammation

Edited by

GEORGE POSTE *and*
STANLEY T. CROOKE
Smith Kline & French Laboratories
Philadelphia, Pennsylvania

PLENUM PRESS • NEW YORK AND LONDON

Library of Congress Cataloging in Publication Data

Cellular and molecular aspects of inflammation / edited by George Poste and Stanley T. Crooke.

p. cm.—(New horizons in therapeutics)

Proceedings of the 5th Smith, Kline, and French Research Symposium on New Horizons in Therapeutics, held in Philadelphia in 1987.

Includes bibliographies and index.

ISBN 0-306-42852-0

1. Inflammation—Congresses. 2. Inflammtion—Mediators—Congresses. I. Poste, George. II. Crooke, Stanley T. III. Smith, Kline, and French Research Symposium on New Horizons in Therapeutics (5th: 1987: Philadelphia, Pa.) IV. Series.

RB131.C44 1988

616′.0473—dc19 87-37400

CIP

A Division of Plenum Publishing Corporation

233 Spring Street, New York, N.Y. 10013

Printed in the United States of America

Contributors

Joanna Balcarek, Department of Molecular Pharmacology, Smith Kline & French Laboratories, Philadelphia, Pennsylvania 19101

Vinay S. Bansal, Division of Hematology–Oncology, Departments of Internal Medicine and Biological Chemistry, Washington University School of Medicine, St. Louis, Missouri 63110

Laurent Baud, Howard Hughes Medical Institute and Departments of Medicine and Microbiology–Immunology, University of California Medical Center, San Francisco, California 94143-0724

Robert M. Bell, Department of Biochemistry, Duke University Medical Center, Durham, North Carolina 27710

C. Frank Bennett, Department of Molecular Pharmacology, Smith Kline & French Laboratories, Philadelphia, Pennsylvania 19101

Druie Cavender, The University of Texas Health Science Center, Dallas, Texas 75235-9030

Anthony Cerami, Laboratory of Medical Biochemistry, The Rockefeller University, New York, New York 10021

M. K. William Chan, Department of Pathology, University of Toronto, Toronto, Ontario, Canada M5S 1A8

Mike Clark, Department of Molecular Pharmacology, Smith Kline & French Laboratories, Philadelphia, Pennsylvania 19101

Thomas M. Connolly, Division of Hematology–Oncology, Departments of Internal Medicine and Biological Chemistry, Washington University School of Medicine, St. Louis, Missouri 63110

Stanley T. Crooke, Department of Molecular Pharmacology, Smith Kline & French Laboratories, Philadelphia, Pennsylvania 19101

Pedro Cuatrecasas, Glaxo Research Laboratories, Research Triangle Park, North Carolina 27709

Myron I. Cybulsky, Department of Pathology, University of Toronto, Toronto, Ontario, Canada M5S 1A8

Florence F. Davidson, Department of Chemistry, University of California, San Diego, La Jolla, California 92093

Hans Deckmyn, Division of Hematology–Oncology, Departments of Internal Medicine and Biological Chemistry, Washington University School of Medicine, St. Louis, Missouri 63110

Raymond A. Deems, Department of Chemistry, University of California, San Diego, La Jolla, California 92093

Edward A. Dennis, Department of Chemistry, University of California, San Diego, La Jolla, California 92093

Susan B. Dillon, Howard Hughes Medical Institute and the Division of Rheumatology and Immunology, Department of Medicine, Duke University Medical Center, Durham, North Carolina 27710

John H. Exton, Howard Hughes Medical Institute and Department of Molecular Physiology and Biophysics, Vanderbilt University School of Medicine, Nashville, Tennessee 37232

Brian J. Fitzsimmons, Merck Frosst Canada, Inc., Pointe-Claire-Dorval, Quebec, Canada H9R 4P8

T. D. Geppert, Harold C. Simmons Arthritis Research Center, University of Texas Health Science Center at Dallas, Southwestern Medical School, Dallas, Texas, 75235-9030

Edward J. Goetzl, Howard Hughes Medical Institute and Departments of Medicine and Microbiology–Immunology, University of California Medical Center, San Francisco, California 94143-0724

Daniel W. Goldman, Howard Hughes Medical Institute and Departments of Medicine and Microbiology–Immunology, University of California Medical Center, San Francisco, California 94143–0724

Warner C. Greene, Howard Hughes Medical Institute, Duke University School of Medicine, Durham, North Carolina 27710

Kathleen A. Haines, Department of Pediatrics, New York University Medical Center, New York, New York 10016

Perry V. Halushka, Departments of Cell and Molecular Pharmacology and Experimental Therapeutics and Medicine, Medical University of South Carolina, Charleston, South Carolina 29425

Yusuf A. Hannun, Department of Medicine, Duke University Medical Center, Durham, North Carolina 27710

Jeanne P. Harvey, Howard Hughes Medical Institute and Departments of Medicine and Microbiology–Immunology, University of California Medical Center, San Francisco, California 94143-0724

Dorian Haskard, The University of Texas Health Science Center, Dallas, Texas 75232-9030

Fusao Hirata, Department of Environmental Health Sciences, School of Hygiene and Public Health, The Johns Hopkins University, Baltimore, Maryland 21205-2179

Roger C. Inhorn, Division of Hematology–Oncology, Departments of Internal Medicine and Biological Chemistry, Washington University School of Medicine, St. Louis, Missouri 63110

Catherine H. Koo, Howard Hughes Medical Institute and Departments of Medicine and Microbiology–Immunology, University of California Medical Center, San Francisco, California 94143-0724

Stephen M. Krane, Department of Medicine, Harvard Medical School and the Medical Services (Arthritis Unit), Massachusetts General Hospital, Boston, Massachusetts 02114

P. E. Lipsky, Harold C. Simmons Arthritis Research Center, University of Texas Health Science Center at Dallas, Southwestern Medical School, Dallas, Texas 75235-9030

Carson R. Loomis, Department of Biochemistry, Duke University Medical Center, Durham, North Carolina 27710

Dale E. Mais, Departments of Cell and Molecular Pharmacology and Experimental Therapeutics, Medical University of South Carolina, Charleston, South Carolina 29425

Philip W. Majerus, Division of Hematology–Oncology, Departments of Internal Medicine and Biological Chemistry, Washington University School of Medicine, St. Louis, Missouri 63110

Kirk R. Manogue, Laboratory of Medical Biochemistry, The Rockefeller University, New York, New York 10021

Steven B. Mizel, Department of Microbiology and Immunology, Wake Forest University Medical Center, Winston-Salem, North Carolina 27103

Seymour Mong, Department of Molecular Pharmacology, Smith Kline & French Laboratories, Philadelphia, Pennsylvania 19101

Henry Z. Movat, Departments of Pathology and Immunology, University of Toronto, Toronto, Ontario, Canada M5S 1A8

Philip Needleman, Department of Pharmacology, Washington University School of Medicine, St. Louis, Missouri 63110

Paul G. Polakis, Howard Hughes Medical Institute and the Division of Rheumatology and Immunology, Department of Medicine, Duke University Medical Center, Durham, North Carolina 27710

James W. Putney, Jr., Section of Calcium Regulation, Laboratory of Pharmacology, National Institute of Environmental Health Sciences, Research Triangle Park, North Carolina 27709

Joan Reibman, Department of Medicine, New York University Medical Center, New York, New York 10016

Richard J. Robb, Medical Products Department, Glenolden Laboratory, E. I. du Pont de Nemours & Company, Glenolden, Pennsylvania 19036

Joshua Rokach, Merck Frosst Canada, Inc., Pointe-Claire-Dorval, Quebec, Canada H9R 4P8

Henry Sarau, Department of Molecular Pharmacology, Smith Kline & French Laboratories, Philadelphia, Pennsylvania 19101

David L. Saussy, Jr., Department of Cell and Molecular Pharmacology and Experimental Therapeutics, Medical University of South Carolina, Charleston, South Carolina 29425. *Present address:* Department of Molecular Pharmacology, Smith Kline & French Laboratories, Philadelphia, Pennsylvania 19101.

Jeffrey W. Sherman, Howard Hughes Medical Institute and Departments of Medicine and Microbiology–Immunology, University of California Medical Center, San Francisco, California 94143-0724

Ralph Snyderman, Howard Hughes Medical Institute and the Division of Rheumatology and Immunology, Department of Medicine, Duke University Medical Center, Durham, North Carolina 27710

Stephen M. Spaethe, Department of Pharmacology, Washington University School of Medicine, St. Louis, Missouri 63110

Allen M. Spiegel, Molecular Pathophysiology Section, National Institute of Diabetes, Digestive, and Kidney Diseases, National Institutes of Health, Bethesda, Maryland 20892

Artis P. Truett III, Howard Hughes Medical Institute and the Division of Rheumatology and Immunology, Department of Medicine, Duke University Medical Center, Durham, North Carolina 27710

Ronald J. Uhing, Howard Hughes Medical Institute and the Division of Rheumatology and Immunology, Department of Medicine, Duke University Medical Center, Durham, North Carolina 27710

Raju Vegesna, Department of Molecular Pharmacology, Smith Kline & French Laboratories, Philadelphia, Pennsylvania 19101

Gerald Weissmann, Department of Medicine, New York University Medical Center, New York, New York 10016

James D. Winkler, Department of Molecular Pharmacology, Smith Kline & French Laboratories, Philadelphia, Pennsylvania 19101

Angela Wong, Department of Molecular Pharmacology, Smith Kline & French Laboratories, Philadelphia, Pennsylvania 19101

Morris Ziff, The University of Texas Health Science Center, Dallas, Texas 75235-9030

Preface to the Series

The unprecedented scope and pace of discovery in modern biology and clinical medicine present remarkable opportunities for the development of new therapeutic modalities, many of which would have been unimaginable even a few years ago. This situation reflects the unprecedented progress being made not only in disciplines such as pharmacology, physiology, organic chemistry, and biochemistry that have traditionally made important contributions to drug discovery, but also in new disciplines such as molecular genetics, cell biology and immunology that are now of sufficient maturity to our understanding of the pathogenesis of disease and to the development of novel therapies. Contemporary biomedical research, embracing the entire spectrum of biological organization from the molecular level to whole body function, is on the threshold of an era in which biological processes, including disease, can be analyzed in increasingly precise and mechanistic terms. The transformation of biology from a largely descriptive, phenomenological discipline to one in which the regulatory principles underlying biological organization can be understood and manipulated with ever-increasing predictability brings an entirely new dimension to the study of disease and the search for effective therapeutic modalities. In undergoing this transformation into an increasingly mechanistic discipline, biology and medicine are following the course already charted by the sister disciplines of chemistry and physics, albeit still far behind.

The consequences of these changes for biomedical research are profound: new concepts; new and increasingly powerful analytical techniques; new advances generated at a seemingly ever-rapid pace; an almost unmanageable glut of information dispersed in an increasing number of books and journals; and the task of integrating this information into a realistic experimental framework. Nowhere is the challenge more pronounced than in the pharmaceutical industry. Drug discovery and development have always required the successful coordination of multiple scientific disciplines. The need to assimilate more and more disciplines within the drug discovery process, the extraordinary pace of discovery in all disciplines, and the growing scientific and organizational complexity of coordinating increasingly ultra-specialized and resource-intensive scientific skills in an ever-enlarging framework of collaborative research activities represent formidable challenges for the pharmaceutical industry. These demands are balanced, however, by the excitement and the scale of the potential opportunities for achieving dramatic improvements in health care and the quality of human life over the next twenty years via the development of novel therapeutic modalities for effective treatment of major human and animal diseases.

It is against this background of change and opportunity that the present symposium series, *New Horizons in Therapeutics,* was conceived as a forum for providing critical and up-to-date surveys of important topics in biomedical research in which significant advances were occurring and which offer new approaches to the therapy of disease. Each volume will contain authoritative and topical articles written by investigators who have contributed significantly to their respective research fields. While individual articles will discuss specialized topics, all papers in a single volume will be related to a common theme. The level will be advanced, directed primarily to the needs of the active research investigator and graduate students.

Editorial policy will be to impose as few restrictions as possible on contributors. This is appropriate since each volume is limited to the papers presented at the symposium and no attempt will be made to create a definitive monograph dealing with all aspects of the selected subject. Although each symposium volume will provide a survey of recent research accomplishments, emphasis will also be given to the examination of controversial and conflicting issues, to the presentation of new ideas and hypotheses, to the identification of important unsolved questions and to future directions and possible approaches by which such questions might be answered.

The range of topics for future volumes in the symposium series will be broad and will embrace the full repertoire of scientific disciplines that contribute to modern drug discovery and development. We thus look forward to the publication of what we hope will be viewed as a worthy series of volumes that reflect the excitement and challenge of contemporary biomedical research in defining new horizons in therapeutics.

George Poste
Stanley T. Crooke

Philadelphia

Preface

The characterization of the cellular and molecular mechanisms that mediate inflammation provides a foundation that supports future studies that will define mechanisms more intimately. It encourages substantial optimism about the opportunities to understand the inflammatory process and to use that information to develop novel therapeutic approaches. Recent progress has defined the cells that mediate the inflammatory response, many of the intercellular transmitters, the receptors, signal transduction processes and regulatory mechanisms. Thus, we now have the opportunity to understand inflammation in pharmacologic terms and to attack the key molecular targets to develop new therapeutics.

Among the cells involved in the inflammatory response are the lymphocytes, neutrophils and endothelial cells. Maintenance of homeostasis, response to proinflammatory stimuli and pathophysiologic responses are products of complex interactions between these and other elements of the immune systems. Each of these cells displays a variety of receptors to define the stimuli to which they respond. The receptors displayed that the signal transduction processes and cellular responses are regulated genetically and epigenetically. The critical role of membranes and particularly the phospholipid components of the membranes is emphasized by recent studies.

Several classes of intercellular transmitters are now defined. Peptides such as the interleukins and cachectin are employed to induce a number of activities in several cells. Substantial progress is reported in the characterization of the mediators and their receptors, including identification and characterization of the ligand binding sites on the receptors and mechanisms of genetic regulation of the receptors, synthesis of the mediators and other steps in the process. All the tools are now in place to support studies that will define the interactions between the peptides and their receptors in chemical terms.

Lipid mediators are also involved. Arachidonic acid and metabolites are of critical importance and progress in this area is extraordinary. Critical enzymes in the biosynthetic pathways have been identified and purified. Many phospholipases A_2 and other phospholipases are now purified and being characterized. Moreover, regulatory proteins and other factors involved are identified. The purification of enzymes in the cycloxogenase and lipoxygenase pathways is proceeding and one can expect all the genes of all major enzymes to be cloned and sequenced within the next few years. Moreover, the key products have been identified and substantial advances in understanding their roles are reported. These advances and controversies such as the roles and mechanisms of action of the Lipocortins and the complexity of the enzymology argue persuasively that improved methods of lipid chemistry and enzymologic studies on enzymes active in lipid environments are essential.

Signal transduction processes have been identified for many of the mediators, and a common pattern has emerged. The receptors which, despite their diversity, appear to share many homologous regions, are coupled to a family of guanine nucleotide binding proteins. This family shares considerable homology and continues to grow as new members of the family are identified.

The guanine nucleotide binding proteins then mediate coupling of the receptor stimulus to an enzyme that produces a key intracellular mediator. The list of key enzymes is also growing and includes anenylate cyclase, guanylate cyclase, PI-specific phospholipase C, PC-specific phospholipase C and probably other enzymes. Again the genes for each of the proteins analyzed are either characterized or will be shortly, portending even more extraordinary progress.

The central roles of calcium and inositol phosphates in signal transduction is emphasized by recent progress. Major methodologic advances in calcium measurements and separation of inositides allow a much more detailed understanding of these processes and the enzymes and intercellular receptors involved. Furthermore, substantial advances in understanding genetic and epigenetic regulatory processes for many of the systems is reported, and attention is focused on protein kinase C as a key regulatory molecule.

The advances in each area constitute a basis for excitement. The montage of advances in so many related areas knit an elegant fabric of increasing clarity. The cells, mediators, receptors, signal transduction mechanisms and regulatory processes simplify into common themes that allow us to understand the processes in more detail and set the stage for more effective exploitation of the various targets to create new generations of therapeutics—an exhilarating opportunity.

This volume attempts to place the recent advances in context and provide a foundation that supports the progress to come in areas of science of explosive growth and profound importance.

George Poste
Stanley T. Crooke

Philadelphia

Contents

I. CELLULAR INTERACTIONS IN INFLAMMATORY PROCESSES

Chapter 6

Interleukin-1: Biology and Molecular Biology

Chapter 7

Chapter 8

III. BIOSYNTHESIS AND RELEASE OF LIPID MEDIATORS OF INFLAMMATION

Chapter 9

VI. THE ROLE OF PHOSPHOLIPASES IN INFLAMMATION

Chapter 23

Philip W. Majerus, Thomas M. Connolly, Vinay S. Bansal, Roger C. Inhorn, and Hans Deckmyn

I

CELLULAR INTERACTIONS IN INFLAMMATORY PROCESSES

1

Cellular Interactions Regulating Inflammation

Activation of T Lymphocytes at Inflammatory Sites and Their Role in Perpetuating Chronic Inflammation

T. D. GEPPERT and P. E. LIPSKY

1. Introduction

Chronic inflammation characteristic of a variety of diseases is driven by immunologic processes. The introduction of a foreign antigen initiates a cascade of events that results in various manifestations of chronic inflammation. Central to this process is the T lymphocyte that recognizes antigen and is thereby triggered to undergo a series of morphologic and metabolic changes. Once activated, the T lymphocyte plays a central role in the initiation and maintenance of immunologically mediated inflammation. One mechanism whereby activated T lymphocytes propagate the inflammatory response is the production of lymphokines. Lymphokines are antigen-nonspecific-secreted products that exert effects on a variety of cell types that bear specific receptors for various lymphokines. Lymphokines may affect cells of the same lineage or different cell types both inside and outside the immune system. Moreover, some lymphokines have a variety of effects on several different cell types.

T. D. GEPPERT and P. E. LIPSKY • Harold C. Simmons Arthritis Research Center, University of Texas Health Science Center at Dallas, Southwestern Medical School, Dallas, Texas 75235–9030.

The most extensively studied T cell lymphokine is interleukin-2 (IL-2). IL-2 promotes the expansion of antigen-reactive T cells, the differentiation of antigen-specific cytotoxic T lymphocytes, and the growth and differentiation of B cells into cells that produce antibody (Ruscetti *et al.*, 1977; Gillis, 1983; Jelinek *et al.*, 1986). In addition, IL-2 enhances the capacity of a subset of lymphocytes—natural killer cells—to kill malignant cells in an antigen-nonspecific manner (Dempsey *et al.*, 1982). Another lymphokine released by activated T cells—gamma interferon (IFN-γ)—induces or enhances the expression of class I and II major histocompatibility complex (MHC) encoded gene products by a variety of cells involved in cellular interactions characteristic of an inflammatory response (Steeg *et al.*, 1982; Prober *et al.*, 1983). Macrophages, activated in the presence of IFN-γ, secrete interleukin-1 (Gerrard *et al.*, 1987), which additionally promotes T cell lymphokine production (Mizel, 1982). Interleukin-1 has a number of other activities, including the ability to enhance B cell growth and differentation into cells that produce antibody (Lipsky *et al.*, 1983), the capacity to act as a chemoattractant for polymorphonuclear leukocytes (PMN) and lymphocytes (Moissec *et al.*, 1984; Sauder *et al.*, 1984), the capacity to increase the adhesion of circulating PMN to post capilliary venules (Bevilacqua *et al.*, 1985; Schleimer and Rutledge, 1986), and the capacity to induce a variety of systemic effects, including fever and the production of acute-phase reactants by the liver (Murray *et al.*, 1980; Kushner, 1982). Activated macrophages also have increased phagocytic capabilities (Imanishi *et al.*, 1975) and are responsible for producing several factors that damage tissue directly, including toxic oxygen metabolites and potent hydrolytic enzymes (Nathan *et al.*, 1983; Vogel and Friedman, 1984). Finally, activated macrophages produce the cytokine—tumor necrosis factor (TNF)—that has the capacity to lyse certain malignant cells (Carswell *et al.*, 1975) and has a variety of other physiologic and metabolic effects both locally and systemically (Collins *et al.*, 1985; Shalaby *et al.*, 1985; Beutler and Cerami, 1986; Tsuyimoto *et al.*, 1986; Vilcek *et al.*, 1986). Lymphotoxin is another lymphokine released by activated T cells that shares much of the activity of TNF, since the molecules are homologous at the portion that binds the specific receptor (Gray *et al.*, 1984; Pennica *et al.*, 1984). Finally, T cells produce a variety of lymphokines besides IL-2 that affect B cell growth and differentiation into cells that produce antibody. Several of these factors have only recently been well characterized, including interleukin-4 (Lee *el al.*, 1986), low molecular weight B cell growth factor (Sharma *et al.*, 1987), and interleukin-5 (Kinushi *et al.*, 1986). These factors were initially identified because of their effects on B cell function, but more recent evidence suggests that they have pleiotropic effects altering the function of a variety of cell types. For example, it has recently been demonstrated that IL-4 promotes the growth of the subset of T cells that produces IL-4 and also facilitates the growth of mast cells (Mosmann *et al.*, 1986). Activated T cells also have a variety of effects that cannot be clearly ascribed to the production of lymphokines (Waltenbaugh *et al.*, 1977). Some T cells have the capacity to downregulate B cell antibody production, while others have the capacity to lyse cells bearing specific foreign antigens (Brunner *et al.*, 1970).

T cells can be divided into subsets based on their capacity to carry out the various functions listed above. One subset of helper/inducer T lymphocytes appears to be responsible for promoting B cell growth and differentiation and also the differentiation of suppressor cells (Reinherz *et al.*, 1979b, 1980). These cells can be identified by their expression of a surface molecule termed CD4. The functional subsets of CD4-bearing T cells may be distinguished by the expression of additional phenotypic markers (Morimoto *el al.*, 1985a,b). In addition, it has recently been suggested that the helper subset can be subdivided further into cells that produce IL-2 and IFN-γ and those that produce IL-4 (Mosmann *et al.*, 1987). Another T cell subset is responsible for lysing cells bearing foreign antigen and also functions as an antigen-specific suppressor cell (Reinherz *et al.*, 1979a). Such T cells can be distinguished by the expression of the T cell differentiation antigen, CD8. The suppressor and cytotoxic subsets within the CD8-positive T cell population can be distinguished by the expression of additional T cell markers (Morishita *et al.*, 1986). Both CD4- and CD8-positive subsets are represented at sites of immunologically mediated inflammation (Janossy *et al.*, 1981; Lindbald *et al.*, 1983), suggesting that the end result of that process evolves as a result of the complex interplay of various T cell subsets with diverse functions.

The normal immune response is self-limited. When the inciting agent has been removed the inflammation subsides, leaving the host primed to deal with a second encounter with the same antigen. Occasionally, however, the inflammatory process persists, with the resultant development of chronic inflammation. The mechanism behind the persistence of the immunologic reactivity underlying chronic inflammation is not completely understood. One possibility is that the antigen responsible for the inflammation cannot be eliminated and therefore continues to stimulate primed T cells locally. This could occur because the antigen is part of a living organism that is not effectively killed, or because the antigen cannot be degraded within mononuclear phagocytes. Another possibility is that the regulatory networks that control T cell activation do not function properly and thereby permit persistent immunologic reactivity. Finally, it is possible that the immune response becomes directed at antigens present on normal autologous tissues or cells because they resemble the inciting antigen, or the local inflammation alters host tissues, rendering them immunogenic. Regardless of the explanation for persistent immunologic activity, it is clear that ongoing T cell activation can lead to many of the manifestations of chronic inflammation. Therefore, an understanding of the mechanisms controlling T cell activation should provide insight into the initiation and perpetuation of chronic immunologically mediated inflammation.

2. *Antigen-Induced T Cell Activation*

T lymphocytes are unable to recognize soluble protein antigens directly (Rosenthal and Shevach, 1973). Antigen-induced T cell activation requires the participation of an antigen presenting cell (APC). The APC takes up antigen, degrades it, and presents the relevant antigenic peptides to the T cell in the context of determinants encoded by class II major histocompatibility complex (MHC) genes, the Ia antigens (Rosenthal and Shevach, 1973; Ellner *et al.*, 1977; Weinberger *et al.*, 1981; Ziegler and Unanue, 1981; Scala and Oppenheim, 1983). T cell activation is triggered by an interaction between the T cell antigen receptor and the antigen–Ia complex displayed by an APC. It is therefore necessary for a cell to express Ia molecules to function as an APC.

The expression of Ia antigens, and therefore the capacity to initiate an immune response, is carefully regulated. A few cells express Ia antigens constitutively, including B cells, dendritic cells, Langerhans cells, and monocytes (Mϕ), whereas the vast majority of cells do not (Giles and Capra, 1985). As a result, in the resting state the initiation of an immune response is dependent on the participation of these specialized cells of the immune system. In contrast, a variety of additional cell types have been found to express Ia antigens at inflammatory sites *in vivo* (Barclay and Mason, 1982; Breathnach and Katz, 1983; DeWaal *et al.*, 1983; Lindbald *et al.*, 1983; Volc-Platzen *et al.*, 1984; Sobel and Colvin, 1985). This is thought to relate to the action of a T-cell-derived lymphokine, IFN-γ, that has been shown to induce Ia-antigen expression on a variety of Ia-negative cell types (Prober *et al.*, 1983; Geppert and Lipsky, 1985). For example, endothelial cells (ECs) and fibroblasts (FBs), which are normally Ia-antigen-negative, express Ia at inflammatory sites (DeWaal *et al.*, 1983; Lindbald *et al.*, 1983).

The induction of Ia-antigen expression by ECs and FBs as a result of incubation with IFN-γ *in vitro* has recently been examined (Prober *et al.*, 1983; Geppert and Lipsky, 1985). Whereas control FBs and ECs do not express Ia antigens, nearly 100% of IFN-γ-treated FBs and ECs become Ia-positive. Interestingly, IFN-γ does not induce the expression of all class II MHC antigens comparably. Thus, nearly 100% of IFN-γ-treated FBs and ECs expressed HLA-DR, whereas approximately 50% of each cell type expressed HLA-DP antigens. Neither IFN-γ-treated cell type expressed HLA-DQ antigens. In additional studies, it was demonstrated that the capacity of

IFN-γ to induce Ia-antigen expression resulted from the induction of class II MHC gene transcription (Collins *et al.*, 1984). An important unresolved question regarding the capacity of IFN-γ-treated Ia-positive cells to contribute to immunologic responses relates to their capacity to function as APC.

The capacity of IFN-γ-treated FBs and ECs to function as APCs was examined by culturing control and IFN-γ-treated ECs and FBs with allogeneic T4 cells and assessing T cell proliferation. Highly purified T4 cells prepared from human peripheral blood were used as responder cells because this population includes the major antigen- and alloantigen-reactive T cells (Reinherz *et al.*, 1979a). Despite comparable HLA-DR expression, IFN-γ-treated HLA-DR (+) ECs, but not FBs, stimulated allogeneic T4 cell DNA synthesis (Geppert and Lipsky, 1985). IFN-γ-treated FBs were similarly unable to present soluble antigens to autologous T4 cells. The inability of IFN-γ-treated HLA-DR (+) FBs to function as APCs could not be explained by nonspecific immunosuppression. Moreover, IFN-γ-treated FBs expressed functionally effective HLA-DR antigens as defined by the ability to stimulate alloreactive T cell lines or clones. The data suggest that IFN-γ-treated FBs lack an APC characteristic necessary for activating resting T4 cells but not for stimulating cycling T cells.

A number of other cells have been shown to be inefficient APCs despite expressing Ia antigens. Thus, tissue macrophages of the lung and liver are ineffective APCs for resting T cells despite the expression of Ia antigens (Toews *et al.*, 1984; Lipscomb *et al.*, 1986; Rubinstein and Lipsky, 1986). These results suggest that the interaction between the T cells antigen receptor and the antigen–Ia complex may not be sufficient for the activation of resting T cells. In addition to expressing Ia antigens, APCs may also be required to deliver additional signals to the resting T cells necessary for activation and cell cycle entry.

One explanation for the inability of IFN-γ-treated FBs to function as effective APCs could relate to an inability of IFN-γ-treated FBs to degrade antigen to reveal the immunogenic moiety recognized by T cells. Little is known about the series of physiologic steps involved in this process. It is clear, however, that it involves uptake of the antigen by the APC and passage through an acid compartment, since it is suppressed by inhibitors of lysosomal function and paraformaldehyde fixation (Scala and Oppenheim, 1983). Moreover, it is clear that not all antigens need to be processed prior to recognition, as several can be presented by paraformaldehyde-fixed APCs or artificial membranes containing Ia molecules (Shimonkevitz *et al.*, 1983; Watts *et al.*, 1984). The capacity of IFN-γ-treated FBs to degrade antigen and generate an immunogenic moiety was assessed by examining the capacity of IFN-γ-treated antigen-bearing FBs to stimulate T cell activation when the additional nonspecific signals were provided by a second APC. Since Mφ and ECs had been found to function as effective APCs, it was expected that they might provide the additional signals that do not involve the antigen–Ia complex. It was anticipated that, by coculturing antigen-bearing Ia-positive fibroblasts with such APCs that did not contain antigen and were unable to present antigen because of the expression of inappropriate or no Ia molecules, the capacity of FBs to degrade antigen and generate the immunogenic moiety recognized by T cells could be examined. The recall antigen streptokinase-streptodornase (SKSD) was utilized, since it had been demonstrated that this antigen required processing in order to induce T4 cell proliferation. Thus, paraformaldehyde-fixed APCs could not present SKSD to T cells unless they had been pulsed with SKSD before fixation (Moreno and Lipsky, 1986). To prevent the additional APCs from removing the antigen from the antigen-bearing FBs and presenting it to the T4 cell, the antigen-bearing Ia-positive FBs were fixed with paraformaldehyde before being cultured with the T4 cells.

In the absence of additional APC's, antigen-bearing Mϕ but not control Mϕ stimulated autologous but not allogeneic T4 cell DNA synthesis. In contrast, neither control FBs, IFN-γ-treated FBs, antigen-pulsed control FBs nor antigen-pulsed IFN-γ-treated FBs stimulated signficiant T4 cell DNA synthesis. When additional accessory cells such as Mϕ or ECs were added to the cultures, however, antigen-bearing IFN-γ-treated FBs and antigen-bearing fixed Mϕ stimulated comparable T4 cell DNA synthesis. Neither antigen-bearing Mϕ nor antigen-pulsed IFN-γ-treated FBs presented antigen to T4 cells obtained from HLA-D region mismatched donors with or without the additional APCs. Genetic restriction of antigen presentation was thus apparent. Moreover, the class II MHC phenotype of the additional APCs was irrelevant. Thus, support of T cell activation in this system was not genetically restricted. The model depicted in Fig. 1 indicates that IFN-γ-treated FBs were able to take up antigen and process it effectively but lacked accessory cell properties possessed by ECs and Mϕ that permitted effective T4 cell activation. When ECs or Mϕ were added to the cultures, they provided the nonspecific signals necessary for antigen-bearing IFN-γ-treated FBs to function as effective APCs. Thus, the capacity of an APC to degrade antigen to an immunogenic moiety is a property that is found not only among specialized cells of the immune system but in a variety of cells. By contrast, the data suggest that the capacity of APCs to provide nonspecific signals that do not involve the T cell receptor may be the characteristic common only to a limited number of cells with the capacity to function as APCs. In addition, antigen and Ia unrelated signals necessary for resting T cell activation may be provided by adjacent cells and not only by the cells presenting antigen. The data therefore, suggest that the interaction between the T cell antigen receptor and the antigen–Ia complex is necessary for antigen-induced T cell activa-

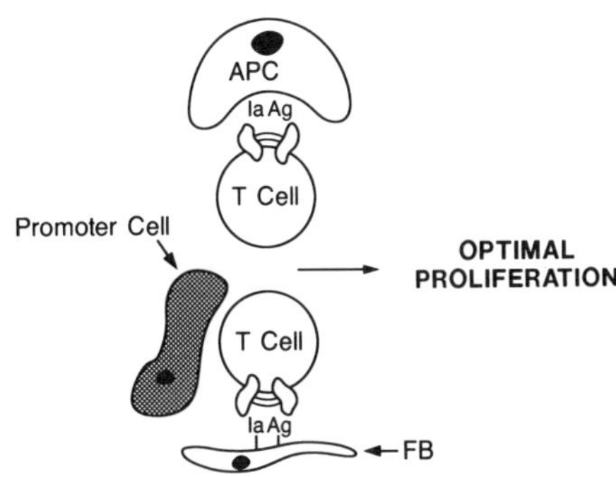

Figure 1. Antigen presentation by gamma-interferon-treated fibroblasts: effect of additional antigen presenting cells. This illustrates the ability of antigen-bearing IFN-γ-treated FBs to function as effective APCs in the presence of ECs or MΦ as promoter cells.

tion, but that it is not sufficient for stimulation of resting T cells. An additional antigen-independent and Ia-unrestricted event is also necessary for the activation of resting T cells. FBs lack this latter capability.

It was possible that the inability of IFN-γ-treated FBs to function as effective APCs related to their inability to promote production of IL-2 by antigen-responsive T4 cells. To examine this possibility, T4 cells were cultured with antigen-bearing Mϕ and IFN-γ-treated antigen-bearing FBs alone or with IL-2. Whereas antigen-bearing IFN-γ-treated FBs stimulated some T4 cell DNA sythesis in the presence of IL-2, they remained markedly less effective than antigen-bearing Mϕ. Thus, the inability of IFN-γ-treated FBs to function as effective APCs could not be explained entirely by their inability to support IL-2 production. Moreover, the data suggest that the interaction between the T cell receptor and the antigen–Ia complex is insufficient to induce T cell reponsiveness to IL-2 in the absence of other signals.

Tp44 (CD28) is a 44-kD molecule expressed on 95% of T4 cells and a subset of T8 cells (Hansen *et al.*, 1980). It has recently been demonstrated that monoclonal antibodies to the CD28 molecule (9.3) deliver an accessory cell-like signal to the T cell. Thus, purified T cells that do not respond to monoclonal antibodies to the CD3 complex conjugated to Sepharose beads will proliferate when costimulated with 9.3 (Martin *et al.*, 1986). 9.3 alone does not stimulate proliferation. It has been suggested that the signal delivered to the T cell when 9.3 binds CD28 mimics a physiologic signal transmitted during accessory cell–T cell interaction (Martin *et al.*, 1986). It was therefore, possible, that the inability of IFN-γ-treated FBs to function as effective APCs related to their inability to deliver a signal via CD28. To examine this possibility, T4 cells were cultured with IFN-γ-treated antigen-bearing FBs in the presence or absence of 9.3. When 9.3 was present, antigen-bearing IFN-γ-treated FBs stimulated autologous T4 cell DNA synthesis, but they remained somewhat less effective than antigen-bearing Mϕ.

When IL-2 was also added to culture, however, antigen-bearing IFN-γ-treated FBs and antigen-bearing Mϕ stimulated comparable T4 cell proliferation. These results suggest a model illustrated in Fig. 2. IFN-γ-treated FBs are ineffective APCs despite their ability to express Ia and generate the relevant immunogenic moieties recognized by T cells because they lack certain APC characteristics possessed by ECs and Mϕ. When ECs or Mϕ are present in the cultures, they can deliver these nonspecific signals and promote activation and proliferation of T4 cells recognizing the Ia–antigen complex present on the FBs. Similarly, antigen-bearing IFN-γ-treated FBs can function as effective APCs when IL-2 and 9.3 are present in the cultures. Thus, IFN-γ-treated FBs differ from Mϕ or ECs in the capacity to function as effective APCs because they are unable to provide the T cell with nonspecific signals mimicked by 9.3 interacting with the CD28 molecule and other as yet unknown interactions.

The data suggest that the role of cells that have been induced to express Ia antigens in the inflammatory response depends on the activation state of the T cell, the capacity of the cell to deliver antigen-independent Ia-unrestricted signals to the T cell, and the capacity of adjacent cells to provide these nonspecific signals. Complete APCs with the capacity to provide the additional nonspecific signals necessary for the activation of resting T cells may function to stimulate newly arriving T cells at the inflammatory site, in addition to promoting the growth of previously activated T cells. Cells without this capacity may still serve a similar function at inflammatory sites, depending on the capacity of adjacent Mϕ and ECs to provide these signals.

3. *Cellular Interactions Involved in the Initiation of T Cell Activation*

The T cell receptor is noncovalently linked to a molecular complex made up of at least five and perhaps seven components, termed CD3 (Bren-

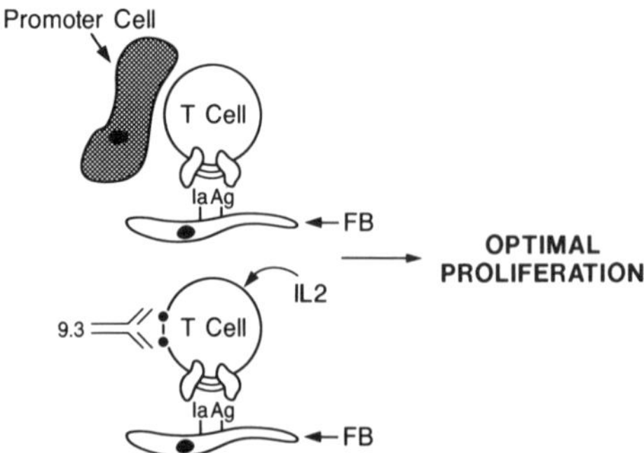

Figure 2. Antigen presentation by gamma-interferon-treated fibroblasts: effect of monoclonal antibodies to Tp44 (CD28) and IL-2. This illustrates the ability of antigen-bearing IFN-γ-treated FBs to function as effective APCs in the presence of 9.3, a monoclonal antibody to Tp44 (CD28) and IL-2.

ner *et al.*, 1985; Weissman *et al.*, 1986). It has been suggested that signals generated as a result of recognition of the antigen–Ia complex by the T cell receptor are transduced to the cells via CD3. Support for this notion comes from the demonstration that the structure of the T cells receptor lacks a number of features of receptor molecules known to be involved in transmembrane signal transduction (Hedrick *et. al.*, 1984; Sim *et al.*, 1984). In addition, CD3 and the T cell receptor are physically associated on the cell surface, although not covalently linked (Brenner *et al.*, 1985). Moreover, monoclonal antibodies to either the T cell receptor or CD3 stimulate T cell activation (Van Wause *et al.*, 1980; Chang *et al.*, 1981; Infante *et al.*, 1982; Meuer *et al.*, 1983). The capacity of anti-CD3 monoclonal antibodies to induce T cell activation provides a system whereby the requirement for recognition of the antigen–Ia complex by the T cell antigen receptor can be obviated and the requirement for other T cell–accessory cell interactions examined. Since anti-CD3-induced T cell activation is accessory-cell-dependent (Kaneoka *et al.*, 1983; Ceuppens *et al.*, 1985; Clement *et al.*, 1985), it provides a model system for examining the antigen- and Ia-independent cellular signaling involved in the activation of T cells. Moreover, since anti-CD3 binds and activates all T cells regardless of their antigen specificity, this system facilitates the study of interactions between the T cell and the accessory cell. In addition, the ability to control the concentration of anti-CD3 and the physical state of the anti-CD3 makes it possible to examine the number of CD3 molecules that must be engaged to trigger T cells, the effect of cross-linking the CD3 complex, and the duration of the engagement of the CD3 complex necessary to deliver an activation signal.

The role of accessory cells in anti-CD3-induced T cell proliferation is not yet understood. Several studies have demonstrated that at least one role of the accessory cell is to provide a stimulatory matrix, binding anti-CD3 by means of Fc receptors (Kaneoka *et al.*, 1983; Looney and Abraham, 1984, Ceuppens *et al.*, 1985; Clement *et al.*, 1985). This suggests that anti-CD3-induced T cell proliferation, like antigen-induced T cell activation, involves an interaction with a ligand that is bound to a solid matrix. In addition to the interaction between the Fc-receptor-bound anti-CD3 and CD3, the accessory cell provides additional signals, as evidenced by the requirement for accessory cells to support proliferation induced by Sepharose-bound anti-CD3 (Manger *et al.*, 1985). The similarity between anti-CD3-induced activation and antigen-induced activation therefore includes not only the requirement for stimulation by a solid-phase ligand but also the need for additional signals that do not involve the T cell receptor–CD3 complex.

It has recently been demonstrated that Fc-receptor-negative ECs can support T cell activation induced by soluble anti-CD3 (Geppert and Lipsky, 1985) and that monocytes can support anti-CD3-induced T cell activation when Fc receptors are blocked by aggregated human immunoglobulin (Geppert and Lipsky, 1986). In both circumstances, stimulation requires large concentrations of the anti-CD3. Finally, it has been demonstrated that individuals that do not possess Fc receptors for mouse IgG1 Mab can support T4 cell activation induced by high concentrations of an IgG1 mouse anti-CD3 monoclonal antibody (Clement *et al.*, 1985). These results suggest the possibility that signals delivered via the CD3 complex by soluble ligands can be sufficient to induce T cell activation when additional accessory cell signals are provided.

In order to understand the T cell–accessory cell interactions involved in promoting anti-CD3-induced T4 cell proliferation more completely, the capacity of monoclonal antibodies directed at various T cell and accessory cell surface proteins to inhibit anti-CD3-induced T cell proliferation was examined. To minimize the possibility that the monoclonal antibodies were inhibiting anti-CD3-induced responses by competing for the accessory cell Fc receptors, experiments were carried out in medium supplemented with human serum containing additional aggregated immunoglobulin. Under these conditions, it was unlikely that the additional monoclonal antibody would significantly increase competition for the accessory cell Fc receptor. Monoclonal antibodies directed at CD11a and class I and II MHC encoded gene products were chosen because it has been suggested that these molecules may be involved in T cell–accessory cell interactions (Beatty *et al.*, 1983; Marrack *et al.*, 1983; Turco *et al.*, 1985). Each of these monoclonal antibodies inhibited Mϕ-supported anti-CD3-induced T4 cell proliferation by 60–95%. Anti-CD11a, but not monoclonal antibodies to class II MHC encoded gene products, inhibited responses supported by Fc-receptor-negative ECs by 70% The inhibition could not be reversed completely by the addition of IL-2. Moreover, the inhibition could not be attributed to a nonspecific inhibitory effect of the monoclonal antibodies on T cell proliferation. The data suggest that each of the monoclonal antibodies inhibited anti-CD3-induced T cell proliferation by inhibiting accessory cell–T cell interactions. The data therefore, suggest that accessory cell–T cell interactions involve class I and II MHC encoded gene products and CD11a. Whether these interactions actually provide signals that promote T cell activation or serve as adhesion molecules facilitating other interactions that deliver these signals could not determined from these studies.

The previous studies demonstrated that the nature of T cell–accessory cell interactions is complex, involving a variety of cell surface molecules in addition to the T cell receptor–CD3 complex. To define these interactions more completely, it was necessary to develop a model of T cell stimulation in which the activation signal and accessory cell signals could be defined more precisely. To accomplish this, monoclonal antibodies directed at various T cell surface proteins were immobilized onto polystyrene plastic microtiter wells in an attempt to provide a sufficiently dense matrix of stimulatory antibodies to induce the activation of T cells in the absence of accessory cell signals. Since anti-CD3 recognizes a well-defined epitope present on all T cells (Chang *et al.*, 1981), it was chosen as the mitogen. Initial studies examined the capacity of immobilized anti-CD3 to induce T4 cell proliferation. T4 cells were cultured with various densities of immobilized anti-CD3 monoclonal antibodies, 64.1 or OKT3, with or without Mϕ or IL-2, and T cell activation was assessed. High-density 64.1 induced nearly all of the T4 cells to proliferate in the absence of accessory cells; the response was not enhanced by IL-2 or Mϕ. In contrast, low-density 64.1 and high- or low-density OKT3 induced only a small minority of the cells to enter the cell cycle. This suboptimal response was enhanced by IL-2 and Mϕ. Responses induced by still lower densities of anti-CD3 were completely dependent on IL-2 or Mϕ. IL-1 also enhanced responses but was much less effective in this regard.

To determine whether the responses were dependent on a few contaminating accessory cells in the purified responding cell population, decreasing numbers of T4 cells were cultured with immobilized anti-CD3. Responses obtained when 1000 T4 cells were cultured in each well were comparable to those obtained when 100,000 T4 cells were employed when normalized for the number of responding T4 cells. If responses were dependent on a small number of contaminating accessory cells, they would have been expected to decrease as the number of cells cultured was decreased and contaminating accessory cells became limiting. Since this was not observed, the results indicate that high-density immobilized anti-CD3 induces T4 cell proliferation in the complete absence of accessory cells.

Using this system, it was possible to examine whether interactions involving the T cell receptor were required throughout the culture period or only for a finite period early in the culture. T cells were cultured with immobilized anti-CD3 for various time periods and then removed from the wells and cultured alone, with IL-2, or with immobilized anti-CD3. Optimal responses were dependent on continued stimulation with anti-CD3. Cells removed from the well before 16 h of stimulation with immobilized anti-CD3 did not respond even when cultured with supplemental IL-2. Cells stimulated for 16–24 h responded suboptimally in the presence of IL-2 but did not respond when cultured alone. Cells stimulated for more than 24 h proliferated when subsequently cultured with or without supplemental IL-2, but responses were never comparable in magnitude to those obtained when the cells were stimulated throughout the culture period with immobilized anti-CD3. Thus, anti-CD3-induced T4 cell proliferation is dependent on the epitope recognized by the anti-CD3 monoclonal antibody, the number of CD3 molecules bound by anti-CD3, and the duration of the CD3–anti-CD3 interaction.

The results of these studies are summarized in Fig. 3. Immobilized anti-CD3 induces T cell activation in the absence of accessory cells when the density of anti-CD3 is sufficient. Although not illustrated in this model, the

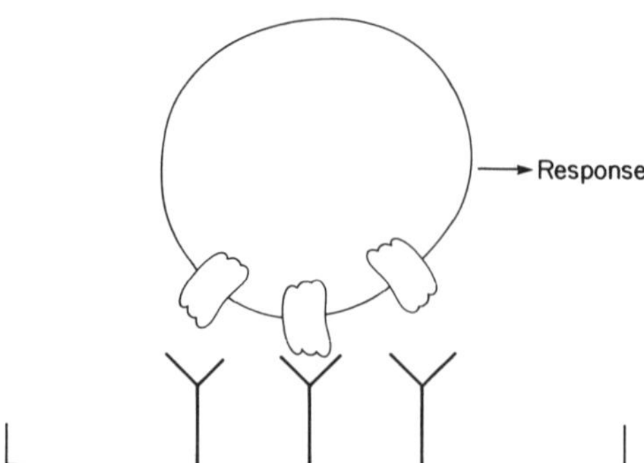

Figure 3. Immobilized anti-CD3 induces T4 cell proliferation in the absence of accessory cells.

epitope recognized by the anti-CD3 monoclonal antibody and the duration of the CD3–anti-CD3 interaction are also important determinants of the capacity of immobilized anti-CD3 to induce T cell activation.

The data suggest that the accessory cell requirements for T cell activation vary inversely with the number of CD3 molecules and therefore with the number of T cell receptors engaged by an antigen–Ia complex. The factors that control the number of T cell receptors that recognize and bind the antigen–Ia complex are the avidity of the T cell receptor for the antigen–Ia complex and the number of antigen–Ia complexes on the APCs. The number of antigen–Ia complexes may relate to the density of Ia antigens on the surface of the APCs and the concentration of antigen. As discussed above, the density of Ia antigens on many cell types is upregulated by lymphokines produced by activated T cells. Thus, for example, IFN-γ upregulates monocyte Ia expression and induces Ia expression on a variety of Ia-negative cell types (Steeg *et al.*, 1982; Pober *et al.*, 1983). TNF appears to enhance the effect of IFN-γ on some cell types (Pfizenmaier *et al.*, 1987). IL-4 upregulates B cell Ia expression (Noelle *et al.*, 1984). At the inflammatory site, the activity of a variety of cytokines may decrease the importance of nonspecific T cell–accessory cell interactions by upregulating Ia-antigen expression. This may facilitate expansion of the antigen-reactive T cells at the inflammatory site.

The data described here suggest a possible explanation for the observation that T cell lines and clones appear to have fewer accessory cell requirements than resting T cells (Inaba and Steinman, 1984; Geppert and Lipsky, 1985). The process of deriving T cell lines and clones may select for T cells with T cell receptors that bind the antigen–Ia complex more avidly. Antigen recognition by high-avidity T cell receptors could generate a stronger activation signal, thereby diminishing the need for additional accessory cell signals. T cells proliferating at inflammatory sites might be enriched for cells that express high-avidity receptors for the antigen responsible for the inflammation. It is possible that these cells also have decreased accessory cell requirements. The decreased requirement for these additional signals may facilitate the capacity of the APCs such as Ia(+) FBs to propagate the local immune response.

4. CD2, CD11a, and HLA-A,B,C Molecules Are Involved in T Cell–Accessory Cell Interactions that Promote T Cell Responses

Antigen-induced resting T cell activation is dependent on antigen-independent Ia-unrestricted events. Therefore, to examine nonspecific accessory cell–T cell interactions that might support antigen-induced T cell activation, a model was developed utilizing the suboptimal activation signal delivered by immobilized OKT3. Additional monoclonal antibodies directed at various T cell surface molecules were then immobilized to the same plastic surface in an attempt to provide the additional signals necessary for immobilized-OKT3-induced responses. The monoclonal antibodies used were directed at CD2, CD11a, and class I MHC determinants. These were chosen because, as suggested above, these antigens appear to be involved in the accessory cell–T cell interaction (Beatty *et al.*, 1983; Marrack *et al.*, 1983; Martin *et al.*, 1983; Sprent and Schaefer, 1985; Turco *et al.*, 1985; Huet *et al.*, 1986; Vollger *et al.*, 1987). Immobilized but not soluble monoclonal antibodies directed at CD2, CD11a, and class I MHC determinants dramatically enhanced immobilized-OKT3-induced T4 cell proliferation. In the presence of these additional immobilized monoclonal antibodies, immobilized OKT3 induced nearly all T4 cells to enter the cell cycle. None of the additional immobilized monoclonal antibodies induced T4 cell proliferation alone or with IL-2.

The effect of immobilized monoclonal antibodies to CD2, CD11a, or HLA-A,B,C on immobilized-OKT3-induced T4 cell proliferation is depicted in Fig. 4. Low-density immobilized anti-CD3, which is insufficient to induce T4 cell activation alone, induces responses when immobilized monoclonal antibodies directed at CD2, CD11a, or class I MHC determinants are also present.

The mechanism whereby immobilized monoclonal antibodies directed at CD2, CD11a, and HLA-A,B,C enhanced immobilized OKT3-induced responses remains to be determined. It is possible that each enhances responsiveness by a different mechanism. One possibility is that the additional immobilized monoclonal antibodies served to stabilize the interaction between the CD3 molecule on the T4 cell and the immobilized anti-CD3. The immobilized monoclonal antibodies directed at additional T4 cell surface proteins besides CD3 might serve to facilitate and prolong the interaction between CD3 and anti-CD3. Although this is still a possibility, the observa-

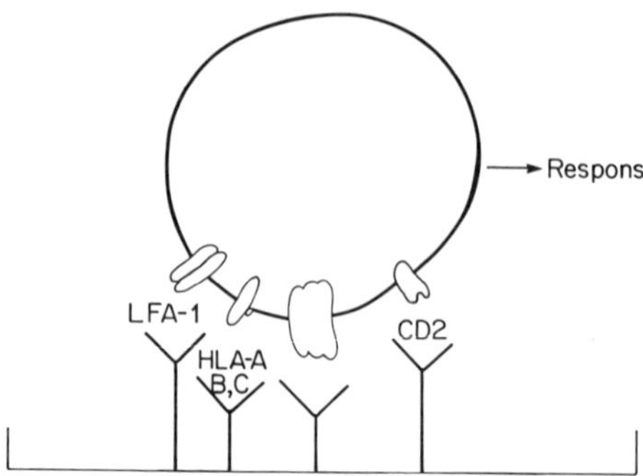

Figure 4. Immobilized anti-CD3-induced T4 cell activation: effect of immobilized monoclonal antibodies to CD2, CD11a (LFA-1), or HLA-A,B,C. The responses induced by low-density immobilized anti-CD3 are dramatically enhanced by immobilized monoclonal antibodies to CD2, CD11a (LFA-1), or HLA-A,B,C.

tion that immobilized 64.1 did not appear to require these additional immobilized monoclonal antibodies to stimulate T4 cells effectively, despite being no more avid an anti-CD3 than OKT3, argues against this explanation. Furthermore, as will be discussed below, not all antibodies were comparably able to enhance immobilized-OKT3-induced responses. Moreover, the ability to enhance responses did not correlate with the density of the surface molecules recognized by the additional monoclonal antibody.

It is more likely that the additional monoclonal antibody promoted anti-CD3-induced T cell activation by crosslinking CD2, CD11a, or class I MHC determinants or intercalating them between CD3 molecules, thereby enhancing the strength of the stimulatory signal. These results are consistent with the conclusion that signals delivered through these molecules may play a role in activating T cells.

The possibility that class I MHC molecules may be involved in accessory cell–T cell interactions has been suggested previously (Sterkers *et al.*, 1983; Sprent and Schaefer, 1985; Turco *et al.*, 1985). In those studies, monoclonal antibodies to class I MHC determinants were found to inhibit class II restricted antigen-specific responses and T cell proliferation induced by nonspecific mitogens. On the other hand, polyclonal antibodies to β_2 microglobulin have been shown to induce accessory-cell-dependent T cell proliferation (Hammarstrom and Smith, 1985). These results suggest that interactions between class I MHC determinants found on T lymphocytes and an accessory-cell-bound ligand may facilitate triggering of T4 cell activation by providing an additional signal that promotes T4 cell proliferation.

CD11a has been found to be involved in a variety of cellular interactions that promote cellular adhesion. Thus, anti-CD11a antibodies have been found to inhibit T and B cell homotypic adhesion, T cell–EC binding, and a variety of accessory-cell-dependent T cell functions, including antigen-, alloantigen-, and mitogen-induced T4 cell proliferation and cytotoxicity

(Beatty *et al.*, 1983; Mentzer *et al.*, 1985; Haskard *et al.*, 1986; Rothlein and Springer, 1986). The capacity of immobilized anti-CD11a to enhance immobilized anti-CD3-induced T4 cell proliferation suggests that the CD11a molecule may function to transmit regulatory signals in addition to its role as a cellular adhesion molecule. A similar role for CD11a was noted previously for B cell activation, when anti-CD11a was found to promote B cell activation (Mishra *et al.*, 1986).

It has also been suggested that CD2 is involved in T cell–accessory cell interactions. More specifically, it has been demonstrated that CD2 on the T cell binds to LFA-3 on the accessory cell (Selvaraj *et al.*, 1987). The importance of this interaction in T cell activation has been suggested by the finding that anti-CD2 antibodies inhibit antigen-, alloantigen-, and mitogen-induced T cell proliferation and cytotoxicity (Martin *et al.*, 1983). The capacity of combinations of anti-CD2 monoclonal antibodies to induce T cell activation suggests that interactions involving CD2 may deliver signals that promote T cell activation (Meuer *et al.*, 1984; Huet *et al.*, 1986). The capacity of immobilized anti-CD2 to promote immobilized anti-CD3-induced T4 cell proliferation supports this concept and suggests that crosslinking may be critical for signal transduction via CD2.

The data support the concept that antigen-bearing APCs stimulate T cell proliferation by delivering several signals to the T cell. Recognition of the Ia–antigen complex by the T cell antigen receptor delivers a necessary but not sufficient signal that dictates the specificity of an immune response. In addition to this interaction, however, various other T cell surface molecules, including CD2, CD11a, and class I MHC encoded gene products, bind accessory cell surface molecules. The accessory cell membrane then provides a matrix that serves to crosslink these molecules and deliver the additional nonspecific signals that are necessary for T cell proliferation.

5. *The Role of CD5 and CD28 in T Cell Activation*

CD5 is a 67-kD molecule present on the surface of all T cells (Martin *et al.*, 1980). Like soluble monoclonal antibodies to CD28, soluble antibodies to CD5 can promote T cell responses (Ledbetter *et al.*, 1986). Since soluble antibodies to CD28 and CD5 enhance responses, the effect of immobilized monoclonal antibodies directed at CD28 (9.3) and CD5 (10.2) on anti-CD3-induced T4 cell proliferation was examined. Immobilized-OKT3-induced responses were enhanced by both immobilized and soluble 9.3 and 10.2. Moreover, soluble antibodies were as effective as immobilized antibodies at promoting anti-CD3-induced responses. Neither soluble nor immobilized 9.3 or 10.2 was as effective as immobilized monoclonal antibodies to CD2,

CD11a, or class I MHC determinants. Thus, monoclonal antibodies directed at certain T cell surface determinants deliver signals that promote suboptimal anti-CD3-induced responses only when immobilized, whereas others deliver enhancing signals in both immobilized and soluble form. These results further support the conclusion that the additional monoclonal antibody is not simply stabilizing the interaction between CD3 and anti-CD3. If this were the case, immobilized 9.3 or 10.2 should be more effective than soluble 9.3 or 10.2. The finding that soluble monoclonal antibodies to CD5 and CD28 enhance responses suggests that the physiologic ligand for these determinants, if one exists, may be a soluble factor. Thus, CD28 and CD5 may be receptors for secreted growth factors.

The effect of soluble monoclonal antibodies to CD5 and CD28 on immobilized anti-CD3-induced T cell activation is illustrated in Fig. 5. Responses induced by low-density anti-CD3 are dependent on 9.3 or 10.2 or, as mentioned above, on intact accessory cells, IL-1 or IL-2.

6. *The Role of CD4 in T Cell Proliferation*

CD4 is a cell surface determinant with a molecular weight of 55,00 expressed by 60–70% of T cells, the helper/inducer subset (Maddon *et al.*, 1985). The vast majority of T cells whose antigen responsiveness is restricted by class II MHC antigens express CD4 (Engleman *et al.*, 1981; Biddison *et al.*, 1982). This observation has suggested that CD4 molecules might interact with nonpolymorphic regions on class II MHC antigens and thereby enhance antigen recognition. It was therefore surmised that the ability of soluble monoclonal antibodies to CD4 to inhibit antigen-, mitogen-, alloantigen-

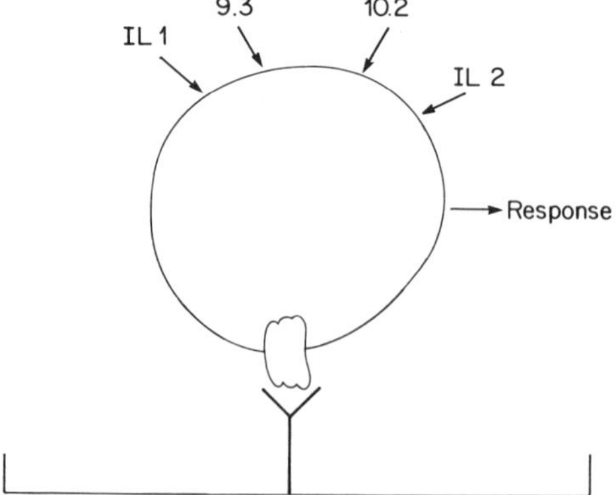

Figure 5. T4 cell proliferation induced by suboptimal stimulation with immobilized anti-CD3 is dependent on additional signals. This illustrates that T4 cell proliferation induced by low-density anti-CD3 requires signals delivered by IL-1, IL-2, 9.3, or 10.2.

induced T cell proliferation and cytotoxicity might be related to their ability to inhibit Ia–CD4 interactions. Support for this notion was obtained when it was reported that the capacity of a class I restricted CD4 (+) cytotoxic T cell clone to lyse target cells expressing the relevant class I MHC antigen could be inhibited by an anti-CD4 monoclonal antibody when the target cell expressed Ia but not when the target cell was Ia-negative (Greenstein *et al.*, 1984).

More recent data have suggested that inhibition of an interaction between CD4 and Ia could not explain entirely the effects of anti-CD4 monoclonal antibodies. Thus, for example, it was demonstrated that soluble anti-CD4-inhibited Ia(-) FBs supported Con-A-induced IL-2 receptor expression (Malek *et al.*, 1985), Ia(-) EC supported OKT3-induced T4 cell proliferation (Geppert and Lipsky, 1986), and anti-CD3 directed cytotoxicity of Ia(-) targets (Fleischer *et al.*, 1986). Moreover, soluble anti-CD4 monoclonal antibodies inhibited immobilized anti-CD3-induced T cell responses in the absence of accessory cells (Geppert and Lipsky, 1987). The degree of inhibition varied inversely with the density of anti-CD3 employed. Thus, it appears that soluble anti-CD4 delivers a negative signal directly to T4 cells. In addition, the effect of this negative signal appears to vary with the strength of the activation signal. Although it is still possible that CD4 interacts with Ia, the data in support of this hypothesis must be reexamined in light of evidence that anti-CD4 directly inhibits T4 cell function. For example, it is possible that soluble anti-CD4 did not inhibit the class I specific cytotoxic T-cell-mediated killing of the Ia-negative target cell because the cytotoxic T cell received a stronger activation signal from the Ia-negative target. Moreover, it is possible that the ability of soluble anti-CD4 to inhibit the responses of some but not all T cell clones also relates to the strength of the activation signal delivered through the T cell receptor–Ia complex. T cell clones with a T cell receptor that has a higher avidity for the antigen–Ia complex might receive a stronger activation signal when it interacts with the antigen-bearing APC than a T cell clone with a lower-avidity antigen receptor. Thus, T cell clones with high-avidity T cell receptors might be inhibited by anti-CD4 antibodies less effectively than T cell clones with low-avidity T cell receptors. Indeed, such a relation between the avidity of the T cell clone and the ability of anti-CD4 to inhibit responses has previously been described (Marrack *et al.*, 1983; Shaw *et al.*, 1985).

It was possible that soluble anti-CD4 delivered a negative signal to T4 cells because they did not crosslink CD4. A similar phenomenon has been observed with monoclonal antibodies to CD3; soluble anti-CD3 has been shown to deliver a negative signal to T cells (Davis *et al.*, 1986). To examine this possibility, the effect of immobilized anti-CD4 on immobilized anti-CD3-induced responses was examined. In contrast to the effects of soluble anti-CD4, immobilized anti-CD4 enhanced responses. The magnitude of the enhancement was considerably less than that obtained with immobilized monoclonal antibodies to CD2, CD11a, or class I MHC antigens, but immobilized anti-CD4 did enhance responses. Therefore, monoclonal antibodies to CD4 determinants can mediate both negative and positive effects on immobilized anti-CD3-induced T4 cell responses. A similar observation regarding the effect of monoclonal antibodies to CD8 on immobilized anti-CD3-induced T8 cell proliferation has recently been described (Emmrich *et al.*, 1986). Crosslinking of CD3 and CD8 was found to promote IL-2 responsiveness of CD8 cells. This was in contrast to previous reports demonstrating that soluble monoclonal antibodies to CD8 inhibited T8 cell responsiveness (Geppert and Lipsky, 1986). In the current studies, immobilized anti-CD4 could facilitate anti-CD3-induced responses in the absence of exgenous IL-2. The results suggest that both CD4 and CD8 molecules may enhance anti-CD3-induced responses when the interacting ligand is bound to a matrix that effectively crosslinks CD4 or CD8

These results have important implications for the role of CD4 in the regulation of T cell activation. If CD4 molecules interact with a soluble ligand, they may serve to downregulate the immune response. Indeed, activation signals delivered in the presence of this signal may tolerize the T cell. A breakdown in this pathway of regulation could lead to a lack of tolerance or autoreactivity. In contrast, if CD4 interacts with a ligand that is linked to a solid matrix, it may transmit a positive signal, thereby enhancing the strength of the activation stimulus.

7. *Proposed Model of the Signals Involved in T Cell Activation*

These results have led to the development of the model depicted in Fig. 6. In this model, the interaction between the Ia–antigen complex and the T cell receptor is replaced by the interaction between an immobilized anti-CD3 and the CD3 molecule. Since most antigen-responsive resting T cells appear to require signals in addition to those delivered via the T cell receptor, it is likely that the signal delivered by immobilized OKT3 or low-denity 64.1 is the most relevant model of antigen-induced responses. This probably reflects the fact that the density of antigen present on APCs is small. T cell determinants such as CD2, CD11a, and class I MHC molecules are likely to interact with accessory-cell-bound ligands, since the monoclonal antibodies that recognize these molecules promote anti-CD3-induced responses only when they are immobilized. In contrast, CD28 and CD5 are likely to receive signals from soluble ligands, since soluble and immobilized monoclonal antibodies directed at CD5 and CD28 are comparably effective promotors of anti-CD3-

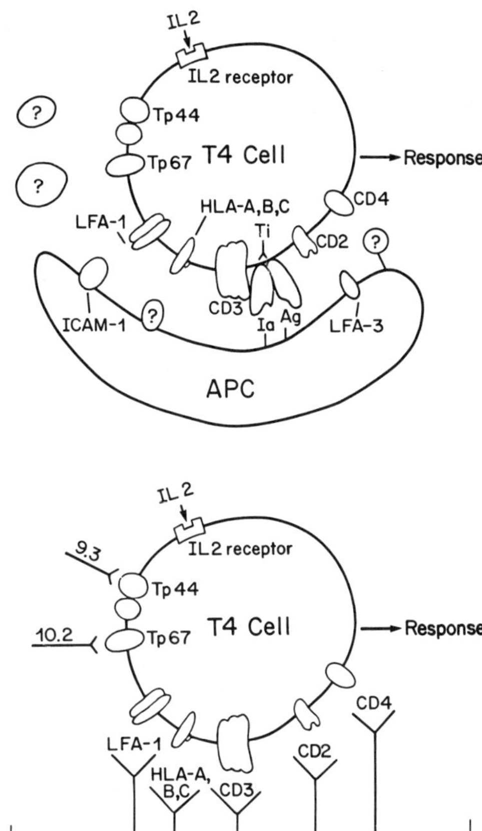

Figure 6. Proposed model of the accessory cell–T4 cell interaction required for antigen or low-density immobilized anti-CD3-induced T4 cell activation.

induced T4 cell proliferation. Since soluble and immobilized monoclonal antibodies to CD4 exert contrasting effects on immobilized anti-CD3 responses, it is not clear whether CD4 in a receptor for a soluble factor that transmits a negative signal or a receptor for a membrane-bound ligand that provides a positive signal, or both.

Interactions involving CD2, CD4, CD5, CD11a, CD28, and HLA-A,B,C that promote T cell activation may play an important role in perpetuating chronic inflammation. It is possible that the underlying immune response is initially stimulated by specific antigen, but that with time it can be maintained by these additional signals in the absence of the initiating antigen. Thus, for example, a T cell that is initially activated by an exogenous antigen may express a T cell receptor that weakly binds an epitope present on normal autologous tissues or cells. This interaction might normally be insufficient to induce T cell activation. However, at inflammatory sites such T cells may be stimulated by the autoantigen because of the presence of an excess of additional nonspecific signals. Defects in the negative regulatory signals such as those delivered by soluble anti-CD4 may also be involved in perpetuating the inflammatory reponse.

8. *Conclusions*

The activation of antigen-specific T cells is a critical determinant of the extent and chronicity of an inflammatory response. Antigen-induced T cell activation is dependent on the participation of APCs. The APC takes up antigen, degrades it to reveal an immunogenic moiety recognized by T cells, and presents it in the context of Ia molecules. The capacities to express Ia and process antigen are properties common to a variety of cell types and are not limited to cells of the immune system. Antigen receptors on the T cell then bind the antigen–Ia complexes and, as a result of this interaction, the T cell receives a signal, the strength of which depends on the avidity of the T cell receptor, the number of T cell receptors that bind the antigen–Ia complex, and the duration of this interaction. This recognition may be sufficient to induce T cell activation under ideal circumstances when antigen–Ia complexes are displayed at a high density on the APCs and the T cells possess high-avidity receptors for this complex. Under most circumstances, however, T cell activation is dependent on additional signals. The ability to provide the additional signals necessary for antigen-induced activation of resting T cells appears to be restricted to a very few cell types, including Mϕ and ECs. The delivery of these additional signals appears to involve a variety of T cell surface molecules including CD2, CD4, CD5, LFA-1 (CD11a), CD28, and class I MHC encoded gene products. Moreover, the data suggest that CD2, CD11a, and class I MHC molecules must be crosslinked to provide the growth-promoting signal, whereas soluble ligands can augment responsiveness via CD28 or CD5. The promotion of T cell activation by these antigen and Ia-unrelated signals may be responsible for the maintenance of chronic inflammation by enhancing the activation of clones of T cells that were initially stimulated by exogenous antigen but express receptors that also recognize epitopes on normal tissue with low avidity.

References

Barclay, A. N., and Mason, D., 1982, Induction of Ia antigen in rat epidermal cells and gut epithelium by immunologic stimuli, *J. Exp. Med.*, **156** 1665–1676.

Beatty, P. G., Ledbetter, J. A., Martin, P. J., Price, T. H., and Hansen, J. A., 1983, Definition of a common leukocyte cell-surface antigen (Lp95-150) associated with diverse cell mediated immune functions, *J. Immunol.* **131:** 2913–2918.

Beutler, B., and Cerami, A., 1986, Cachectin and tumor necrosis factor as two sides of the same biological coin, *Nature* **320:** 584–588.

Bevilacqua, M. P., Pober, J. S., Wheeler, M. E., Cotran, R. S., and Gimbrone, M. A., Jr., 1985, Interleukin 1 acts on cultured human vascular endothelium to increase the adhesion of polymorphonuclear leukocytes, monocytes, and related leukocyte cell lines, *J. Clin. Invest.* **76:**2003–2011.

Biddison, W. E., Rao, P. E., Talle, M. A., Goldstein, G., and Shaw, S., 1982, Possible involvement of the OKT4 molecule in T cell recognition of class II HLA antigens. Evidence from studies of cytotoxic T lymphocyctes specific for SB antigens, *J. Exp. Med.* **156:**1065–1076.

Breathnach, S. M., and Katz, S. I., 1983, Keratinocytes synthesize Ia antigen in acute cutaneous graft-versus-host disease, *J. Immunol.* **131:**2741–2745.

Brenner, M. B., Trowbridge, I. S., and Strominger, J. S., 1985, Cross-linking of human T cell receptor proteins: Association between the T cell idiotype B subunit and the T3 glycoprotein heavy subunit, *Cell* **40:**183–190.

Brunner, K. T., Mauel, J., Roedoif, H., and Chapuis, B., 1970, Studies of allograph immunity in mice. I. Induction development, and *in vitro* assay of cellular immunity, *Immunology* **18:**501–515.

Carswell, E. A., Old, J., Kassel, R. L., Greene, S., Fiorey, N., and Williamson, B., 1975, An endotoxin-induced serum factor that causes necrosis of tumors, *Proc. Natl. Acad. Sci. U.S.A.* **72:**3666–3670.

Ceuppens, J. L., Bloemmen, F. J., and Van Wauwe, J. P., 1985, T cell unresponsiveness to the mitogenic activity of OKT3 antibody results from a deficiency of monocyte Fcγ receptors for murine IgG2a and inability to crosslink the T3–Ti complex, *J. Immunol* **135:**3882–3886.

Chang, T. W., Kung, P. C., Gingras, S. P., and Goldstein, G., 1981, Does OKT3 monoclonal antibody react with an antigen-recognition structure on human T cells?, *Proc. Natl. Acad. Sci. U.S.A.* **78:**1805–1808.

Clement, L. T., Tilden, A. B., and Dunlap, N. E., 1985, Analysis of the monocyte Fc receptors and antibody-mediated cellular interactions required for the induction of T cell proliferation by anti-T3 antibodies, *J. Immunol.* **135:**165–171.

Collins, T., Korman, A. J., Wake, C. T., Boss, J. M., Kappes, D. J., Fiers, W., Ault, K. A., Bimbron, M. A., Strominger, J. O., and Prober, J. S., 1984, Immune interferon activates class II major histocompatibility complex genes and the associated invariant chain gene in human endothelial cells and dermal fibroblasts, *Proc. Natl. Acad. Sci. U.S.A.* **81:**4917–4921.

Collins, T., Lapierre, L. A., Fiers, W., Strominger, J. L., and Prober, J. S., 1985, Recombinant human tumor necrosis factor increases mRNA levels and surface expression of HLA-A,B antigens in vascular endothelial cells and dermal fibroblast in vitro, *Proc. Natl. Acad. Sci. U.S.A.* **83:**446–450.

Davis, L., Vida, R., and Lipsky, P. E., 1986, Regulation of human T lymphocyte mitogenesis by antibodies to CD3, *J. Immunol.* **137:**3758–3767.

Dempsey, R. A., Dinarello, A., Mier, J. W., Rosenwasser, L. J., Allegretta, M., Brown, T. E., and Parkinson, D. R., 1982, The differential effects of human leukocytic pyrogen/lymphocyte activating factor, T cell growth factor and interferon on human natural killer cell activity, *J. Immunol.* **129:**2504–2510.

DeWaal, R. M. W., Bagman, J. J., Maass, C. N., Carnelissen, L. M. H., Tax, W. J. M., and Koene, R. A. P., 1983, Variable expression of Ia antigens on the vascular endothelium of mouse skin allografts, *Nature* **303:**426–429.

Ellner, J., Lipsky, P. E., and Rosenthal, A. S., 1977, Antigen handling by guinea pig macrophages: Further evidence for the sequestration of antigen relevant for activation of primed T lymphocytes, *J. Immunol* **118:**2053–2057.

Emmrich F., Strittmatter, U., and Eichmann, K., 1986, Synergism in the activation of human CD8 T cells by cross-linking the T-cell receptor complex with CD8 differentation antigen, *Proc. Natl. Acad. Sci. U.S.A.* **83:**8298–8302.

Engleman, E. G., Benike, C., Glickman, E., and Evans, R. L., 1981, Activation of human T cell subsets: Helper and suppressor/cytotoxic T cells recognize and respond to distinct histocompatibility antigens, *J. Immunol* **127:**2124–2129.

Fleischer, B. H., Schrezenmeier, H., and Wagner, H., 1986, Function of the CD4 and CD8 molecules on human cytotoxic T lymphocytes: Regulation of T cell triggering, *J. Immunol.* **136:**1625–1628.

Geppert, T. D., and Lipsky, P. E., 1985, Antigen presentation by interferon-γ-treated endothelial cells and fibroblasts: Differential ability to function as antigen presenting cells despite comparable Ia expression, *J. Immunol.* **135:**3750–3763.

Geppert, T. D., and Lipsky, P. E., 1986, Accessory cell–T cell interactions involved in anti-CD3 induced T4 and T8 cell proliferation: Analysis with monoclonal antibodies, *J. Immunol.* **137:**3065–3073.

Geppert, T. D., and Lipsky, P. E., 1987, Accessory cell independent proliferation of human T4 cells stimulated by immobilized antibodies to CD3, *J. Immunol,* **138:**1660–1666.

Gerrand, T. L., Siegel, J. D., Dryer, D. R., Zoon, K., 1987, Differential effects of interferon-α and interferon-γ on interleukin 1 secretion by monocytes, *J. Immunol.* **138:**2535–2540.

Giles, R. C., and Capra, D., 1985, Structure, function and genetics of human class II molecules, *Adv. Immunol.* **37:**1–71.

Gillis, S., 1983, Interleukin-2: Biology and biochemistry, *J. Clin. Immunol.* **3:**1–13.

Gray, P. W., Aggarwal, B. B., Benton, C. V., Bringman, T. S., Henzel, W. J., Jarret, J. A., Leung, D. W., Moffat, B., Ng, P., Svedensky, L. P., Palladino, M. A., and Nedwig, G. E., 1984, Cloning and expression of cDNA for human lymphotoxin, a lymphokine with tumor necrosis activity, *Nature* **312:** 721.–724.

Greenstein, J. L., Kappler, J., Marrack, P., and Burakoff, S. J., 1984, The role of L3T4 in recognition of Ia by a cytotoxic, H-2D^{d}-specific T cell hybridoma, *J. Exp. Med.* **159:**1213–1224.

Hammarstrom, L., and Smith, C. I. E., 1985, Rabbit anti-human B_2 microglobulin antibodies: A T-cell mitogen, *Scand. J. Immunol* **22:**677–681.

Hansen, J. A., Martin, P. J., and Nowinski, R. C., 1980, Monoclonal antibodies identifying a novel T-cell antigen and Ia antigens of human lymphocytes, *Immunogenetics* **10:**247–260.

Haskard, D., Cavender, D., Beatty, P., Springer, T., and Ziff, M., 1986, T lymphocyte adhesion to endothelial cells: Mechanisms demonstrated by anti-LFA-1 monoclonal antibodies, *J. Immunol* **137:**2901–2906.

Hedrick, S. M., Cohen, D. F., Nielsen, E. A., and Davis, M. M., 1984, Isolation of cDNA clones encoding T cell specific membrane associated proteins, *Nature* **308:**149–153.

Huet, S., Wakasugi, H., Sterkers, G., Gilmour, J., Tursz, T., Boumsell, L., and Bernard, A., 1986, T cell activation via CD2 [T,gp50]: The role of accessory cells in activating resting T cells via CD2, *J. Immunol.* **137:**1420–1428.

Imanishi, J., Yokota, Y., Kishida, T., Mukainaka, T., and Matsuo, A., 1975, Phagocytosis-enhancing effect of human leukocyte interferon preparations of human peripheral monocytes in vitro, *Acta Virol.* **79:**52–58.

Inaba, K., and Steinman, R. M., 1984, Resting and sensitized T lymphocytes exhibit distinct stimulatory (antigen presenting cell) requirements for growth and lymphokine release, *J. Exp. Med.* **160:**1717–1735.

Infante, A. J., Infante, P. D., Gillis, S., and Fathman, C. G., 1982, Definition of T cell idiotypes using anti-idiotypic antisera produced by immunization with T cell clones, *J. Exp. Med.* **155:**1100–1107.

Janossy, G., Duke, O., Poulter, C. W., Panayi, G., Bofill, M., Goldstein, G., 1981, Rheumatoid arthritis: A disease of T-lymphocyte/macrophage immunoregulation, *Lancet* **2(825):**839–842.

Jelinek, D. F., Splawski, J. B., and Lipsky, P. E., 1986, The roles of interleukin 2 and interferon-γ in human B cell activation, growth and differentiation, *Eur. J. Immunol* **16:**925–932.

Kaneoka, H., Perez-Rojas, G., Sasasuki, T., Benike, C. J., and Engleman, E. G., 1983, Human T lymphocyte proliferation induced by a pan-T monoclonal antibody (anti-Leu 4): Heterogeneity of response is a function of monocytes, *J. Immunol.* **131:**158–164.

Kinashi, T., Harada, T., Severinson, E., Tanabe, T., Siderus, P., Konishi, M., Azuma, L., Tominaga, A., Bergstedt-Lindquist, S., Takahashi, M., Matsuda, F., Yaoita, Y., Takatsu, K., Honjo, T., 1986, Cloning of complementary DNA encoding T cell replaceing factor and identity with B cell growth factor II, *Nature* **324:**70–72.

Kushner, I., 1982, The phenomenon of the acute phase response, *Ann. N.Y. Acad. Sci.* **339:**39–48.

Ledbetter J. A., June, C. K., Martin, P. J., Spooner, C. E., Hansen, J. A., and Meier, K. E., 1986, Valency of CD3 binding and internalization of the CD3 cell-surface complex control T cell responses to second signals: Distinction between effects on protein kinase C, cytoplasmic free calcium and proliferation, *J. Immunol.* **136-**3945–3952.

Lee, F., Yokota, T., Otsuka, T., Meyerson, P., Valleret, D., Coffman, R., Mosmann, T., Rennick, D., Roehm, N., Smith, C., Zlotnick, A., Arai, K., 1986, Isolation and characterization of a mouse interleukin cDNA clone that expresses B cell stimulatory factor I activities and T-cell and marker cell stimulating activities, *Proc. Natl. Acad. Sci. U.S.A.* **83:**2061–2065.

Lindbal, S., Klareskog, L., Medfors, E., Forsum, U., and Sundstrom, C., 1983, Phenotypic characterization of synovial tissue cells in situ in different types of synovitis, *Arthritis Rheum* **26:**1321–1332.

Lipscomb, M. F., Lyons, C. R., Nunez, G., Ball, E. J., Stastny, P., Vial, W., Lem, V., Weissler, J., Miller, L. M., and Toews, G. B., 1986, Human alveolar macrophages: HLA-DR positive macrophages that are poor stimulators of a primary mixed lymphocyte reaction, *J. Immunol* **136:**497–504.

Lipsky, P. E., Thompson, P. A., Rosenwasser, L. J., and Dinarello, C. A., 1983, The role of interleukin 1 in human B cell activation: Inhibition of B cell proliferation and the generation of immunoglobulin-secreting cells by an antibody against human leukocytin pyrogen, *J. Immunol.* **130:**2708–2714.

Looney, R. F., and Abraham, G. N., 1984, The Fc portion of intact IgG blocks stimulation of human PBMC by anti-T3, *J. Immunol.* **133:**154–156.

Maddon, P. J., Littman, D. R., Godfrey, M., Maddon, D. E., Chess, L. and Axel, R., 1985, The isolation and nucleotide sequence of a cDNA encoding the T cell surface protein, T4: A new member of the immunoglobulin gene family, *Cell* **42:**93–104.

Malek, T. R., Chan, C., Glimcher, L. H., Germain, R. N., and Shevach, E. M., 1985, Influence of accessory cell and T cell surface antigens on mitogen-induced IL 2 receptor expression, *J. Immunol.* **135:**1826–1833.

Manger, B., Weiss, A., Weyland, C., Goronzy, J., and Stobo, J., 1985, T cell activation: Differences in the signals required for IL 2 production by nonactivated and activated T cells, *J. Immunol.* **135:**3669–3673.

Marrack, P., Endres, R., Shimonkevitz, R., Zlotnick, A., Dialynas, D., Fitch, F., and Kappler,

J., 1983, The major histocompatibility complex-restricted antigen receptor on T cells. II. Role of the L3T4 product, *J. Exp. Med.* **158:**1077–1091.

Martin, P. J., Hansen, J. A., Nowinski, R. C., and Brown, M. A., 1980, A new human T cell differentiation antigen: Unexpected expression on chronic lymphocytic leukemia cells, *Immunogenetics* **11**429–439.

Martin, P. C., Longton, Ledbetter, J. A., Newman, W., Braun, M. P., Beatty, P. A., and Hansen, J. A., 1983, Identification and functional characteristics of two distinct epitopes on the human T cell surface protein T, p50, *J. Immunol.* **131:**180–185.

Martin, P. J., Ledbetter, J. A., Morishita, Y., June, C. H., Beatty, P. G., and Hansen, J. A., 1986, A 44 kilodalton cell surface homodimer regulates interleukin 2 production by activated human T lympohcytes, *J. Immunol.* **136:**3282–3287.

Mentzer, S. J., Gromkowski, S. H., Krensky, A. M., Burakoff, S. J., and Martz, E., 1985, LFA-1 membrane molecule in the regulation of homotypic adhesions of human B lymphocytes, *J. Immunol.* **135:**9–11.

Meuer, S. C., Hodgdon, J. C., Hussey, R. E., Protentis, J. P., Schlossman, S. F., and Reinherz, E., 1983, Antigen-like effects of monoclonal antibodies directed at receptors on human T cell clones, *J. Exp. Med.* **158:**988–993.

Meuer, S. C., Hussey, R. E., Fabbi, M., Fox, D., Acuto, O., Fitzerald, K. A., Hodgdon, J. C., Potentis, J. F., Schlossman, S. F., and Reinherz, E. L., 1984, An alternate pathway of T-cell activation: A functional role for the 50 kd T11 sheep erythrocyte receptor protein, *Cell* **36:**897–906.

Mishra, G. C., Berton, M. T., Oliver, K. G., Krammer, P. H., Uhr, J. W., and Vitetta, E. S., 1986, A monoclonal anti-mouse LFA-1α antibody mimics the biological effects of B cell stimulating factor (BSF-1), *J. Immunol.* **137:**1590–1598.

Mizel, S. B., 1982, Interleukin 1 and T cell activation, *Immunol. Rev.* **63:**51–72.

Moissec, P., Yu, C., and Ziff, M., 1984, Lymphocyte chemotactic activity of human interleukin 1, *J. Immunol.* **133:**2007–2011.

Moreno, J., and Lipsky, P. E., 1986, Differential ability of fixed antigen presenting cells to stimulate nominal antigen reaction and alloreactive T4 lymphocytes, *J. Immunol.* **136:**3579–3578.

Morimoto, C., Letvin, N. L., Disaso, J. A., Aldrich, W. R., and Schlossman, S. F., 1985a, The isolation and characterization of the human suppressor inducer T cell subset, *J. Immunol.* **134:**1508–1515.

Morimoto, C., Letvin, N. C., Boyd, A. W., Hagan, M., Brown, H. M., Kornacki, M. M., and Schlossman, S. F., 1985b, The isolation and characterization of the human helper inducer T cell subset, *J. Immunol.* **134:**3762–3769.

Morishita, Y., Martin, P. J., Bean, M. A., Yamada, H., and Hansen, J. A., 1986, Antigen-specific functions of a CD4+ subset of human T lymphocytes with granular morphology, *J. Immunol.* **136:**2095–2102.

Mosmann, T. R., Bond, M. W., Coffman, R. C., Ohara, J., and Baal, W., 1986, T cell and mast cell lines respond to B cell stimulatory factor I, *Proc. Natl. Acad. Sci. U.S.A.* **83:**5654–5662.

Mosmann, T. R., Chevwinski, H., Bond, M., Giedlin, R. C., and Coffman, J., 1987, Two types of murine helper T cell clones: Definition according to profiles of lymphokine activities and secreted position, *J. Immunol.* **136:**2348–3570.

Murray, P. A., Simon, P. L., and Willoughby, W. K., 1980, Endogenous pyrogens made by rabbit peritoneal macrophages are identical with lymphocyte-activating factors made by rabbit alveolar macrophages, *J. Immunol.* **124:**2498–2501.

Nathan, C. F., Murray, H. W., Weibe, M. E., and Rubin, B., 1983, Identification of interferon gamma as the lymphokine that activates human macrophage oxidation metabolism and microbial activity, *J. Exp. Med.* **158:**670–689.

Noelle, R., Krammer, P. H., Ohara, J., Uhr, J. W., and Vitetta, E. S., 1984, Increased expression of Ia antigens on resting B cells: A new role for B cell growth factor, *Proc. Natl. Acad. Sci. U.S.A.* **81:**6149–6153.

Pennica, D., Nedwig, G. E., Hayflick, J. S., Seeburg, P. H., Derynck, R., Palladino, M. A., Kohn, W. J., Aggarwal, B. B., and Goeddel, D. V., 1984, Human lymphotoxin, *Nature* **312:**724–729.

Pfizenmaier, K., Scheurich, P., Schluter, C., and Kronke, M., 1987, Tumor necrosis factor enhances HLA-A,B,C and HLA-DR gene expression in human tumor cells, *J. Immunol.* **138:**975–980.

Pober, J., Collins, T., Gimbrone, M., Jr., Cotran, R., Gitlin, J., Fiers, W., Clayberger, C., Krensky, A., Burakoff, S., and Reiss, C., 1983, Lymphocytes recognize human vascular endothelial and dermal fibroblast Ia antigens induced by recombinant immune interferon, *Nature* **305:**726–729.

Reinherz, E. L., Kung, P. C., Goldstein, G., and Schlossman, S. F., 1979, Separation of functional subsets of human T cells by a monoclonal antibody, *Proc. Natl. Acad. Sci. U.S.A.* **76:**4061–4065.

Reinherz, E. L., Kung, P. C., Goldstein, G., and Schlossman, S. F., 1979b, Further characterization of human inducer T cell subsets defined by monoclonal antibody, *J. Immunol.* **123:**2894–2896.

Reinherz, E. L., Kung, P. C., Breard, J. M., Goldstein, G., and Schlossman, S. F., 1980, T cell requirements for generation of helper factors in man: Analysis of the subsets involved, *J. Immunol.* **124:**1883–1887.

Rosental, A. S., and Shevach, E. M., 1973, Function of macrophages in antigen recognition by guinea pig T lymphocytes, *J. Exp. Med.* **138:**1194–1212.

Rothlein, R., and Springer, T., 1986, The requirement for lymphocyte function-associated antigen one in homotypic leukocyte adhesive stimulated by phorbol ester, *J. Exp. Med.* **163:**1132–1149.

Rubinstein, D., and Lipsky, P. E., 1983, Liver sinusoidal lining cells express class II major histocompatibility complex antigens but are poor stimulators of fresh allogeneic T lymphocytes, *J. Immunol.* **137:**1803–1810.

Ruscetti, F. W., Morgan, D. A., and Gallo, R. C., 1977, Functional and morphologic characterization of human T cells continously grown in vitro, *J. Immunol* **119:**131–138.

Sauder, D. N., Manuessa, N. L., Katz, S. I., Dinarello, C. A., and Gallin, J. I., 1984, Chemotactic cytokines: The role of leukocytic pyrogen and epidermal cell thymocyte-activating factor in neutrophil chemotaxis, *J. Immunol.* **132:**828–832.

Scala, G., Oppenheim, J. J., 1983, Antigen presentation by human monocytes: Evidence for stimulant processing and requirement for interleukin one, *J. Immunol.* **131:**1160–1166.

Schleimer, R. P., and Rutledge, B. K., 1986, Cultured human vascular endothelial cells acquire adhesiveness for neutrophils after stimulation with interleukin one, endotoxin and tumor-promoting phorbol diester, *J. Immunol.* **136:**649–654.

Selvaraj, P., Plunkett, J. L., Sanders, M. E., Dustin, M., Shaw, S., and Springer, T. A., 1987, The T lymphocyte glycoprotein CD2 (LFA-2/T11/E rosette receptor) binds the cell surface ligand LFA-3, *Nature* **326:**400–403.

Shalaby, M. R., Aggarwal, B. B., Rinduknecht, E., Avedeasky, L. P., Finkle, B. S., and Palladino, M. A., Jr., 1985, Activation of human polymorphonuclear neutrophil function by interferon-γ and tumor necrosis factors, *J. Immunol.* **135:**2069–2073.

Sharma, S., Mehta, S., Morgan, J., and Maizel, A., 1987, Molecular cloning and expression of a human B-cell growth factor growth gene in *Escherichia coli, Science* **235:**1489–1492.

Shaw, S., Goldstein, G., Springer, T. A., and Biddison, W. E., 1985, Susceptibility of cytotoxic T lymphocyte (CTL) clones to inhibition by anti-T3 and anti-T4 (but not anti-LFA-1) monoclonal antibodies varies with the "avidity" of the CTL–target interaction, *J. Immunol.* **134:**3019–3026.

Shimonkevitz, R., Kappler, J., Marrack, P., and Grey, H., 1983, Antigen recognition by H-2 restricted cells. I. cell-free antigen processing, *J. Exp. Med.* **158:**303–316.

Sim, G. K., Yague, J., Nelson, J., Marrack, P., Palmer, E., Augustin, A., and Kappler, J., 1984, Primary structure of human T cell receptor α-chain *Nature* **312:**771–775.

Sobel, R. A., and Colvin, R. B., 1985, The immunopathology of experimental allergic encephalomyelitis (EAE). III. Differential in situ expression of strain 13 Ia on endothelial and inflammatory cells of (Strain 2X strain 13) F_1 guinea pigs with EAE, *J. Immunol.* **134:**2333–2337.

Sprent, J., and Schaefer, M., 1985, Properties of purified T cell subsets I *In vitro* responses to class I and class II H-2 allo-antigen, *J. Exp. Med.* **162:**2068–2088.

Steeg, P. S., Moore, R. N., Johnson, H. M., and Oppenheim, J. J., 1982, Regulation of macrophage Ia antigen expression by a lymphokine with immune interferon activity, *J. Exp. Med.* **155:**1780–1793.

Sterkers, G., Henin, Y., Kalill, J., Bagot, M., and Levy, J., 1983, Influence of HLA class I and class II specific monoclonal antibodies on DR-restricted lymphoproliferative responses. I Unseparated populations of effector cells, *J. Immunol.* **131:**2735–2740.

Toews, G. B., Vial, W. C., Dunn, M. M., Guzzetta, P., Nunez, G., Stastny, P., and Lipcomb, M., 1984, The accessory cell function of human alveolar macrophages in specific T cell proliferation, *J. Immunol.* **132:**181–186.

Tsujimoto, M., Yokota, S., Vilcek, J., and Weissman, G., 1986, Tumor necrosis factor provokes superoxide anion generation from neutrophils, *Biochem Biophy Res Commun* **137:**1094–1100.

Turco, M. C., DeFelice, M., Corbo, L., Morone, G., Mertelsmann, R., Ferrone, S., and Venuta, S., 1985, Regulatory role of a monomorphic determinant of HLA class I antigens in T cell proliferation, *J. Immunol.* **135:**2268–2273.

Van Wause, J. P., De Mey, J. R., and Gossens, J. G., 1980, OKT3: A monoclonal anti-human T lymphocyte antibody with potent mitogenic properties, *J. Immunol.* **124:**2708–2713.

Vilcek, J., Palombella, V. J., Henrykson-Destefans, D., Swenson, C., Feinman, R., Hirai, M., and Tsujimoto, M., 1986, Fibroblast growth enhancing activity of tumor necrosis factor and its relationship to other polypeptide growth factors, *J. Exp. Med.* **163:**632–643.

Vogel, S. N., and Friedman, R. M., 1984, Interferons and macrophages: Activation and cell surface changes, in: *Interferons: Volume 2—Interferon and the Immune System* (J. Vilcek and E. DeMaeyer, eds.), Elsevier, Amsterdam.

Volc-Platzen, B., Majdic, O., Knapp, W., Wolff, K., Hinterberger, W., Lechner, K., and Stingl, G., 1984, Evidence of HLA-DR antigen biosythesis by human keratinocytes in disease, *J. Exp. Med.* **159:**1784–1789.

Vollger, L. W., Tuck, D. T., Springer, T. A., Haynes, B. F., and Singer, K. H., 1987, Thymocyte binding to human thymic epithelial cells is inhibited by monoclonal antibodies to CD-2 and LFA-3 antigens, *J. Immunol.* **138:**358–363.

Waltenbaugh, C., Theze, J., Kapp, J. A., and Benacerraf, B., 1977, Immuno suppressive factors specific for L-glutamic acid^{50}-L-tyrosine^{50}. III. Generation of suppressor T cells by a suppressive extract derived from GT primed lymphoid cells, *J. Exp. Med.* **146:**970–985.

Watts, T. M., Brian, A. A., Kappler, J. W., Marrack, P., and McConnell, H. M., 1984, Antigen presentation by supported planar membrances containing I-A^d, *Proc. Natl. Acad. Sci. U.S.A.* **81:**7564–7568.

Weinberger, O., Herrman, S., Mescher, M. F., Benacerraf, B., Burakoff, S. J., 1981, Antigen presenting cell function in induction of helper T cells for cytotoxic T-lymphocyte responses: Evidence for antigen processing, *Proc. Natl. Acad. Sci. U.S.A.* **78:**1796–1799.

Weissman, A. M., Samuelson, L.E., and Klausner, R. D., 1986, A new subunit of the human T cell antigen receptor complex, *Nature* **324:**480–485.

Ziegler, K., and Unanue, E. R., 1981, Identification of macrophage antigen-processing event required for I-region restricted antigen presentation to T lymphocytes, *J. Immunol.* **127:**1869–1875.

2

Triggering and Activation of Human Neutrophils

Two Aspects of the Response to Transmembrane Signals

KATHLEEN A. HAINES, JOAN REIBMAN, and GERALD WEISSMANN

The function of the neutrophil can be summed up as follows: Upon engagement of its surface receptors by immune reactants or chemoattractants, this motile postmitotic cell seeks and destroys microbes by chemotaxis and phagocytosis. After engagement of these membrane receptors by inflammatory ligands such as C5a, f-Met-Leu-Phe, or leukotriene B_4, neutrophils undergo: (1) activation of phospholipases and turnover of membrane phospholipids (Walsh *et al.*, 1981, 1984; Sherman *et al.*, 1983; Korchak *et al.*, 1986; Wynkoop *et al.*, 1986), (2) alterations in ion fluxes and membrane potential (Naccache *et al.*, 1977; Korchak *et al.*, 1978; Horne *et al.*, 1981; Jones *et al.*, 1981; Whitin *et al.*, 1980), (3) increases in cytosolic calcium (Goldstein *et al.*, 1974; O'Flaherty *et al.*, 1978), and (4) phosphorylation of cellular proteins (Schneider *et al.*, 1981; Andrews and Babior, 1984; White *et al.*, 1984). These events regulate the biological functions of aggregation, chemotaxis, degranulation, release of reactive oxygen species, and production of the lipid-derived inflammatory substances acetyl-glyceryl-ether, phosphorylcholine (platelet activating factor, Benveniste *et al.*, 1972), leukotriene B_4 (Borgeat and Samuelsson, 1979), and lipoxin A (Serhan *et al.* 1984).

KATHLEEN A. HAINES • Department of Pediatrics, New York University Medical Center, New York, New York 10016. *JOAN REIBMAN* and *GERALD WEISSMANN* • Department of Medicine, New York University Medical Center, New York, New York 10016.

1. Phospholipid Remodeling and Intracellular Messengers

The *twin signal* hypothesis of Nishizuka (Takai *et al.*, 1980) proposes that engagement of cell surface receptors generates two intracellular signals: increments in cytosolic calcium and activation of the calcium and phospholipid-dependent kinase—protein kinase C. Triggering of the neutrophil seems to proceed according to this scheme. Ligand–receptor interactions result in activation of a phospholipase C that cleaves phosphatidylinositol 4,5-bisphosphate (PIP_2) to inositol 1,4,5-trisphosphate (IP_3) and diacylglycerol (DAG). IP_3 functions to release calcium from intracellular stores, raising cytosolic calcium, $[Ca]_i$, from 50–75 to 500 nM (Korchak *et al.*, 1986) and thus providing a cofactor for calcium-dependent enzymes. DAG interacts with protein kinase C (PKC), resulting in its translocation to the cell membrane, activation, and subsequent phosphorylation of cellular proteins (Schneider *et al.*, 1981; Andrews and Babior, 1984; White *et al.*, 1984). Smith *et al.* (1985) and Okajima *et al.* (1985) have both demonstrated that these events are triggered via activation of a pertussis-toxin-inhibitable, GTP-binding protein (G_{pc}) that is linked to phospholipase C.

Accordingly, cellular IP_3 and DAG concentrations substantially increase within seconds after engagement of the fMLP receptor (Berridge, 1983; Berridge and Irvine, 1984; Cockcroft *et al.*, 1985; Irvine and Moor, 1986). By 5 s, the IP_3 signal begins to decline (Berridge, 1983; Korchak *et al.*, 1984b). In contrast, both DAG and phosphatidic acid (PA) continue to increase over the course of the next 120 s. Although the mechanism by which the neutrophil generates these elevated levels of DAG and PA is not yet understood, we postulate that these compounds play a critical role in maintaining activation of the neutrophil. We suggest that intracellular signaling in the neutrophil is a dual event. In order for functional neutrophil responses to proceed, two signals must be generated: a *triggering* signal, resulting in an immediate increase in intracellular messengers, and an *activation* signal, resulting in the maintenance of these messengers.

Indirect evidence for this hypothesis began with our studies on calcium movements in the neutrophil (Korchak *et al.*, 1984a,b). After treatment with fMLP, the neutrophil increases its levels of cytosolic calcium, reaching a peak by 2–5 s. Over the next 2 min, $[Ca]_i$ slowly decreases and returns to baseline. The peak levels are achieved primarily by IP_3-induced mobilization from intracellular stores, since in the absence of extracellular Ca, similar levels are achieved. Influx of extracellular ^{45}Ca begins at approximately 5 s after Ca has been released from intracellular sites, and while IP_3 levels are decreasing. Although IP_4 may regulate Ca channels (Berridge and Irving, 1984; Irvine and Moor, 1986), we would like to postulate an important role for PA.

Neutrophils break down Phosphatidyl Inositol to form DAG and PA within the first 5 s of engagement of receptors by ligands such as fMLP, immune complexes, or concanevilin A (Korchak *et al.*, 1984a,b; Spat *et al.*, 1986). Suggestions that some molecule in the PI–PA cycle mediates these changes in Ca permeability include influx via PA-activated Ca gates (Michell, 1975) of formation of inositol (1,3,4,5)P_4 from inositol (1,4,5)P_3 by specific kinases (Irvine and Moor, 1986). However, the turnover of IP_3/IP_4 in consequence of specific phosphatases (Michell, 1986) is extremely rapid, while DAG and PA continue to accumulate (30–120 s) after treatment of neutrophils with fMLP. Preliminary evidence from our laboratory reveals that there is a biphasic wave of DAG formation. The first peaks at 2–5 s, which is consistent with the release of the triggering messengers, IP_3 and DAG, by the hydrolysis of polyphosphoinositides. Before this first wave is completed (by 15–30 s after treatment with fMLP, a second wave of DAG formation is seen. The concentration of DAG remains elevated compared to that of resting neutrophils for more than 300 s. This second wave of DAG formation is consistent with the activation phase of neutrophil stimulation, when persistence of intracellular messages is required.

DAG is rapidly phosphorylated by a specific kinase to form PA. Additionally, the *de novo* pathway of DAG formation uses PA as a precursor of DAG. Of all the phospholipid classes, there is now general agreement that PA functions as a *Ca-selective* ionophore *in vitro* (Salmon and Honeyman, 1980; Serhan *et al.*, 1981, 1982; Chauhan and Brockerhoff, 1984; Nayer *et al.*, 1984; Smaal *et al.*, 1985). It is by no means clear, however, that Ca ionophoresis explains the biological effects of PA in neutrophil activation. Although exogenously added PA has relatively few direct effects in the triggering of neutrophils, addition of PA or treatment with phospholipase C does activate a variety of cells (Putney *et al.*, 1980; Philipson and Nashimoto, 1984). In human carcinoma cells, exogenously applied PA or treatment with phospholipase D provoked rises of $[Ca]_i$, promoted Na^+/H^+ exchange, and induced the expression of c-*fos* and c-*myc* proto-oncogenes and DNA synthesis, a *growth-factor-like* effect of PA (Moolenar *et al.*, 1986). We still do not know how PA, which is formed in the plasmalemma by the activation of DAG kinase, can activate cells in a manner similar to that of receptor–ligand interactions or calcium ionophores.

To have a role in generating increments in $[Ca]_i$, PA must be able to assume a nonbilayer phase that permits Ca movements in the direction of a [Ca] gradient. As reviewed by Cullis *et al.* 1985), most phospholipids of living cells assume the L_a or lamellar configuration in bilayers. However, cardiolipin and PA assume the H_{ii} or inverse hexagonal phase upon complexing with calcium in the presence of pH 6 or proton movements. The studies of Yuli and Oplatka (1987) provide the basis by which we postulate that PA can act as a Ca ionophore *in vivo*. They demonstrated that treatment of neutrophils with fMLP or sodium propionate resulted in cytosolic acidification and changes in perpendicular light scattering, a correlate of chemotaxis. After 10 s the cytosol was then alkalinized via plasmalemmal Na^+/H^+ exchange. These proton movements and associated pH changes would provide the ionic requirement for the newly generated PA to form the Hex_{ii} phase. The time course of the acidification/alkalinization of the neutrophil correlates with the increments of PA formed in cells, the kinetics of $_{45}Ca$ influx, and the formation of $[Ca]_i$-dependent poorly permselective Ca channels. These channels, found by means of patch-clamp studies of neutrophil activation, are not opened by IP_3 or ligand (GW36) but may be related to PA in the bilayer.

The source of the activation phase of DAG (and thus PA) formation can only be postulated. Hydrolysis of phosphatidylcholine via specific phospholipases is certainly a likely pathway for DAG formation, since a phosphatidylcholine-specific phospholipase C has recently been described in mammalian cells (Bokkino *et al.*, 1985; Besterman *et al.*, 1986b; Clark *et al.*, 1986; Daniel *et al.*, 1986). Moreover, preliminary data from our laboratory suggest that phorbol myristate acetate is able to provoke the second wave of DAG formation in the presence of cytochalasin B. Phorbol esters do not stimulate phosphoinositide hydrolysis (Labarca *et al.*, 1984; Brock *et al.*, 1985; Besterman *et al.*, 1986a) but do generate phosphocholine and DAG in hepatocytes, HL60 cells, 3T3 fibroblasts, and chick myoblasts. Furthermore, the lipids found in the *sn*-2 position of DAG are similar to those in PC (Grove and Schimmel, 1982).

2. *Phospholipase A Activation in Neutrophils*

In addition to release of IP_3 and DAG, neutrophils respond to phlogistic stimuli by mobilization of arachidonate form the *sn*-2 position of cellular phospholipids (Walsh *et al.*, 1981, 1983; Wynkoop *et al.*, 1986). These phospholipids undergo a deacylation/reacylation cycle as described by Lands (1960), since incubation of cells with free [^{3}H]arachidonate leads to a rapid incorporation of the label. Two pathways have been proposed for mobilization of arachidonate from membrane stores: *directly* via a PLA_2 or *indirectly* via a phospholipase C (PLC), followed by the action of a diacylglycerol lipase on DAG. Phospholipase A_2 is associated with both neutrophil granules and the plasma membrane. It has two pH optima (5.5 and 7.5) and in purified preparations requires high concentrations of calcium (Franson *et al.*, 1977; Victor *et al.*, 1981). Neutrophils contain two phospholipase C, one that acts specifically on PI and a second that acts on PC to yield DAG. Both PLA and PLC appear to be functioning in activated neutrophils. After treatment with calcium ionophore or opsonized zymosan particles (activating neutrophils via the C3b receptor), neutrophils release 20:4 from both phosphatidyl inositol and phosphatidylcholine to an equivalent degree (Walsh *et al.*, 1981, 1983). Our laboratory has shown that treating neutrophils with fMLP results in a pattern of lipid metabolism consistent with activation not only of PLA but also of PLC followed by DAG lipase.

3. *Arachidonate Metabolism*

Once released from its membrane stores, 20:4 can be oxidized nonenzymatically by active oxygen species such as superoxide (Perez *et al.*, 1979) or via two major enzymatic pathways, the cyclooxygenase pathway or the lipoxygenase pathway. Of the circulating blood cells, platelets and monocytes have active cyclooxygenase enzymes (Hamberg *et al.*, 1974; Pawlowski *et al.*, 1983). Neutrophils rarely, if ever, utilize this pathway. However, neutrophils are capable of synthesizing many lipoxygenase products, acting alone or in concert with other cells (Borgeat and Samuelsson, 1979a,b; Marcus *et al.*, 1982). When neutrophils are treated with calcium ionophores, the mobilized arachidonic acid is first metabolized to 5-hydroperoxyeicosatetraenoic acid, or 5-HPETE. Its peroxide moiety either spontaneously forms 5-hydroxyeicosatetraenoic acid (5-HETE) or reacts with the enzyme leukotriene A_4 synthase to form LTA_4, which has an epoxide at the 5,6 position (Hammarstrom, 1983; Samuelsson, 1983). The 5-lipoxygenase enzyme has been purified by Rouzer and Samuelsson (1985). It is an extremely unstable enzyme, requiring Ca and ATP for activity, as well as at least three nondialyzable components (Powell, 1982; Rouzer and Samuelsson, 1985; Rouzer *et al.*, 1986). 5-Lipoxygenase and LTA_4 synthase are probably a single enzyme with two active sites (Rouzer *et al.*, 1986). LTA_4 may then be acted on by LTA_4 hydrolase (LTB_4 synthetase) to form 5S, 12R, 6,14-*cis*, 8,10-*trans*-dihydroxyeicosatetraenoic acid, or LTB_4. Alternatively, LTA_4 can break down nonenzymatically to 5S, 6S, or 5S, 6R-6,8,10-*trans*, 14-*cis*-dihydroxyeicosatetraenoic acid. These nonenzymatic metabolites have, at best, one-tenth the activity of LTB_4 in activating neutrophils (Lindgren *et al.*, 1982; Ford-Hutchinson *et al.*, 1983). In the presence of exogenous arachidonic acid, neutrophils also possess an active 15-lipoxygenase enzyme that, in a parallel manner, produces 15-HPETE and 15-HETE. These products are acted on in turn by an LTA_4-like enzyme to make a 14,15-dihydroxy product (Radmark *et al.*, 1982), and finally, in concert with 5-lipoxygenase, can yield the trihydroxy compounds, lipoxins A and B (Serhan *et al.*, 1984). The link(s) between these biochemical pathways for LTB_4 synthesis and the *in vivo* conditions for activation of these pathways are as yet undefined.

It has long been assumed that the diacyl phospholipids of the cell membrane are the source of arachidonic acid for the lipoxygenase products synthesized upon cell activation. We have recently shown that the lipoxygenase pathway of arachidonate metabolism in neutrophils is only activated by receptor–ligand interactions in the presence of more than 10 μM arachidonic acid. Calcium ionophores provoke synthesis of prodigious quantities of lipoxygenase products, including leukotriene B_4 from neutrophils. This is clearly a paradox: neutrophils can, but do not, synthesize one of the most potent phlogistic agents yet identified when treated with inflammatory ligands.

The *sn*-2 position of triglycerides in chylomicrons and lipoproteins is a major source of 20:4 available to tissues. Lipids circulate as large supramolecular aggregates. Their hydrophobic core of triglyceride and cholesterol esters is surrounded by phospholipids and specific proteins. Receptors for these molecules, particularly the low-density lipoproteins (LDLs), have been found on a variety of cell types, and LDLs can be subsequently modified (acetylated) to a form that allows them to specifically bind to macrophages. This modification is associated with hydrolysis of the LDL–phosphatidyl choline to lyso-PC. The fate of the free fatty acids released by this reaction is unknown. The interactions of neutrophils and plasma proteins have been poorly studied. Neutrophils can metabolize triglyceride from chylomicrons to DAG and free fatty acid (Elsbach and Kayden, 1965). However, it is not known whether neutrophils have receptors for LDLs or acLDLs, or whether neutrophils can hydrolyze fatty acids from circulating LDLs.

References

Andrews, P., and Babior, B., 1q984, Phosphorylation of cytosolic proteins in resting and activated human neutrophils, *Blood* **64:**883–890.

Benveniste, J., Henson, P., and Cochrane, C., 1972, Leukocytes dependent histamine release from rabbit platelets: The role of IGE, basophils, and a platelet-activating factor, *J. Exper. Med.* **136:**1356.

Berridge, M., 1983, Rapid accumulation of inositol trisphosphate reveals that agonist hydrolyse polyphospho-inositides instead of phosphatidylinositol, *Biochem.* . **212:**849–858.

Berridge, M., and Irvine, R., 1984, Inositol trisphosphate, a novel second messenger in cellular signal transduction, *Nature* **312:**315.

Besterman, J., Watson, S., and Cuatrecasas, P. 1986a, Lack of association of epidermal growth factor—Insulin—and Serum-induced Mitogenesis with Stimulation of phosphoinositide degradation in BALB/c 3T3 fibroblalsts, *J. Biol. Chem.* **261:**723–727.

Besterman, J., Duronio, V., and Cuatrecasas, P., 1986b, Rapid formation of diacylglycerol from phosphatidylcholine: A pathway for generation of a second messenger, *Proc. Natl. Acad, Sci. U.S.A.* **83:**6785–6789.

Bokkino, S., Blackmore, P., and Exton, J., 1985, Stimulation of 1,2-diacylglycerol accumulation in hepatocytes by vasopressin, epinephrine, and angiotensin II, *J. Biol. Chem.* **260:**14201–14207.

Borgeat, P., and Samuelsson, B., 1979a, Metabolism of arachidonic acid in polymorphonuclear leukocytes. Structural analysis of novel hydroxylated compounds, *J. Biol. Chem.* **254:**7865–7869.

Borgeat, P. and Samuelsson, B., 1979b, Transformation of arachidonic acid by rabbit polymorphonuclear leukocytes. *J. Biol. Chem.* **254:**2643.

Borgeat, P., and Samuelsson, B., 1979c, Arachidonic acid metabolism in polymorphonuclear leukocytes: Effects of ionophore A23187, *Proc. Natl. Acad. Sci. U.S.A.* **76:**2148–2152.

Brock, T., Rittenhouse, S., Powers, C., Ekstein, L., Gimbrone, W., and Alexander, R., 1985, Phorbol ester and 1-oleoyl-2-acetylglycerol inhibit angiotensin activation of phospholipase C in cultured vascular smooth muscle cells, *J. Biol. Chem.* **260:**14158–14162.

Chauhan, V., and Brockerhoff, H., 1984, Ca (phosphatidate)$_2$ can traverse liposomal bilayers, *Life Sci.* **13:**1395.

Clark, M., Shorr, R., and Bomalaski, J., 1986, Antibodies prepared to *Bacillus cereus* phospholipase C crossreach with a phosphatidylcholine preferring phospholipase C in mammalian cells, *Biochem. Biophys. Res. Commun.* **140:**114–119.

Cockcroft, S., Barrowman, M., and Gomperts, B., 1985, Breakdown and synthesis of polyphosphoinositides in fMET Leu Phe-stimulated neutrophils, *FEBS Lett.* **181:**259.

Cullis, P., Hope, M., de Kruiff, B., Verkley, A., and Tilcock, C., 1985, Structural properties and functional roles of phospholipids in biological membranes, in: *Phospholipids and Cellular Recognition* (J. F. Kuo, ed.), pp. 1–59, CRC Press, Boca Raton, FL.

Daniel, L., Waite, M., and Wykle, R., 1986, A novel mechanism of diglyceride formation, *J. Biol. Chem.* **261:**9128–9132.

Elsbach, P., and Kayden, H., 1965, Chylomicron-lipid-splitting activity in homogenates of rabbit polymorphonuclear leukocytes, *Am. J. Physiol.* **209:**765–769.

Ford-Hutchinson, A., Rackham, A., Zamboni, R., Rokach, J., and Roy, S., 1983, Comparative biological activities of synthetic leukotriene B_4 and its w-oxidation products, *Prostaglandins* **25:**29–37.

Franson, R., Weiss, J., Martin, L., Spitznagel, J., and Elsbach, P., 1977, Phospholipase A activity associated with the granules of human polymorphonuclear leukocytes, *Biochem. J.* **167:**839–841.

Grove, R., and Schimmel, S., 1982, Effects of 12-0 letradecanoylphorbol 13-acetate on glycerolipid metabolism in cultured myoblasts, *Biochem. Biophys. Acta* **711:**272–280.

Goldstein, I., Horn, J., Kaplan, H. and Weissmann, G., 1974, Calcium-induced lysozyme secretion from human polymorphonuclear leukocytes. *Biochem. Biophys. Res. Comm.* **60:**807–812.

Hamberg, M., Svensson, J., and Samuelsson, B., 1974, Prostaglandin endoperoxides. A new concept concerning the mode of action and release of prostaglandins, *Proc. Natl. Acad. Sci. U.S.A.* **79:**3711.

Hammarstrom, S., 1983, Leukotrienes, *Ann. Rev. Biochem.* **52:**355–377.

Horne, W., Norman, N., Schwartz, D., and Simons, E. 1981. Changes in cytoplasmic pH and membrane potential in thrombin-stimulated human platelets. *Eur. J. Biochem.* **120:**295–302.

Irvine, R., and Moor, R., 1986, Micro-injection of inositol 1,3,4,5-tetrakisphosphate activate sea urchin eggs by a mechanism dependent on external CA^{2+}, *Biochem. J.* **240:**917.

Jones, G., VanDyke, K., and Castranova, V., 1981, Transmembrane potential changes associated with superoxide release from human granulocytes. *J. Cell. Physiol.* **106:**75–83.

Korchak, H., and Weissman, G., 1978, Changes in membrane potential of human granulocytes antecede the metabolism responses to surface stimulation. *Proc. Nat. Acad. Sco. U.S.A.* **75:**3813–3822.

Korchak, H., Rutherford, L., and Weissman, G., 1984a, Stimulus response coupling in the human neutrophil: I. Kinetic analysis of changes in calcium permeabiliity, *J. Biol. Chem.* **159:**4070–4075.

Korchak, H., Vienne, K., Rutherford, L., Wilkenfeld, C., Finkelstein, M., and Weissman, G., 1984b, Stimulus response coupling in the human neutrophil: II. Temporal analysis of change in cytosolic calcium and calcium efflux, *J. Biol. Chem.* **259:**4076–4082.

Korchak, H., Vosshall, L., Reibman, J., Rich, A., and Haines, K., 1986, Superoxide anion generation in LTB_4 activated neutrophils: Inadequacy of cytosolic calcium and diacyl glyceros as signals, *J. Cell. Biol.* **103:**504a.

Lands, W., 1960, Metabolism of glycerolipids II. The enzymatic acylation of lysolecithin, *J. Biol. Chem.* **235:**2233–2237.

Labarca, R., Jamowsky, A., Patel, J., and Paul, S. M., 1984, Phorbol esters inhibit agonist-induced [^{3}H] inositol-1-phosphate accumulation in rat hippocampal slices, *Biochem. Biophys. Res. Commun.* **123:**703–709.

Lindgren, J., Hansson, G., Claesson, H., and Samuelsson, B., 1982, Formation of novel biologically active leukotrienes by χ-oxidtion in human leukocyte preparation, in: *Leukotrienes and Other Lipoxygenase Products* (B. Samuelsson and R. Paoletti, eds.), pp. 53–60, Raven Press, New York.

Marcus, A., Broekman, M., Safier, L., Ullman, H., Islam, N., Serhan, C., Rutherford, L., Korchak, H., and Weissman, G., 1982, Formation of leukotrienes and other hydroxy acids during platelet–neutrophil interactions in vitro, *Biochem. Biophys. Res. Commun.* **109:**130–137.

Michell, R., 1975, Inositol phospholipids and cell surface receptor function, *Biochem. Biophys. Acta* **415:**81.

Michell, R., 1986, A seocnd messenger function for inositol tetrakisphosphate, *Nature* **324:**613.

Moolenar, W., Kruijer, W., Tilly, B., Verlaan, I., Bierman, A. J., and de Laat, S., 1986, Growth factor-like action of phosphatidic acid, *Nature* **323:**171.

Naccache, P., Showell, Becker, E., and Sha'afi, R., 1977, Changes in ionic movements across rabbit polymorphonuclear leukycyte membranes during lysosomal enzyme release. Possible ionic basis for lysosomal enzyme release. *J. Cell Biol.* **75:**635–649.

Nayer, R., Mayer, L., Hope, M., and Cullis, P., 1984, Phosphatidic acid as a calcium ionophore in large unilamellar vesicle systems, *Biochem. Biophys. Acta* **777:**343.

O'Flaherty, J., Showell, H., Becker, E. and Ward, P., 1978, Substances which activate neutrophils. Mechanism of action. *Am. J. Pathol.* **92:**155–166.

Okajima, F., Katada T., and Ui, M., 1985, Coupling of the guanine nucleotide regulatory protein to chemotactic peptide receptors in neutrophil membranes and its uncoupling by islet-activating protein, pertussis toxin, *J. Biol. Chem.* **260:**6761–6768.

Pawlowski, N., Kaplan, G., Hamill, A., Cohn, Z., and Scott, W., 1983, Arachidonic acid metabolism by human monocytes. Studies with platelet-depleted cultures, *J. Exp. Med.* **158:**393–412.

Perez, H., Weksler, B., and Goldstein, I., 1979, A new mechanism for the generation of biologically active products from arachidonic acid, *Clin. Res.* **17:**464A.

Philipson, K., and Nashimoto, A., 1984, Interaction of charged amphiphiles with Na^+-Ca^{2+} exchange in cardiac sarcolemmal vesicles, *J. Biol. Chem.* **259:**16

Powell, W., 1982, Rapid extraction of arachidonic acid metabolities from biological samples using octadecylsilyl silica, *Methods Enzymol.* **86:**467–477.

Putney, J., Weiss, S., van de Walle, C., and Haddas, R., 1980, Is phosphatidic acid a calcium ionophore under neurohumoral control?, *Nature* **284:**345.

Radmark, O., Lundberg, U., Jubiz, W., Malmsten, C., and Samuelsson, B., 1982, New group of leukotrienes formed by initial oxygenation at C-15, in: *Leukotrienes and Other Lipoxygenase Products* (B. Samuelsson and R. Paoletti, eds.), Vol. 11, pp.61–70, Raven Press, New York.

Rouzer, C., and Samuelsson, B., 1985, On the nature of the 5-lipoxygenase reaction in human leukocytes: Enzyme purification and requirement for multiple stimulatory factors, *Proc. Natl. Acad. Sci. U.S.A.* **82:**6040–6044.

Rouzer, C., Matsumoto, T., and Samuelsson, B., 1986, On the nature of the 5-lipoxygenase reaction in human leukocytes: Enzyme purification and requirement for mulitple stimulatory factors, *Proc. Natl. Acad. Sci. U.S.A.* **83:**857–861.

Salmon, D., and Honeyman, T., 1980, Proposed mechanism of cholinergic action in smooth muscle, *Nature* **284:**344

Samuelsson, B., 1983, Leukotrienes: Mediators of immediate hypersensitivity reactions and inflammation, *Science* **220:**568–575.

Schneider, C., Zanetti, M., and Romeo, D., 1981, Surface-reactive stimuli selectively increase protein phosphorylation in human neutrophils, *FEBS Lett.* **127:**4–8.

Serhan, C., Anderson, P., Goodman, E., Dunham, P., and Weissmann, G., 1981, Phosphatidate and oxidized fatty acids are calcium ionophores: Studies employing arsenazo III in liposomes, *J. Biol. Chem.* **257:**4746.

Serhan, C., Broekman, M., Korchak, H., Smolen, J., Marcus, A., and Weissman, G., 1983, Changes in phosphatidylinositol and phosphatidic acid in stimulated human neutrophils: Relationship to calcium mobilization, aggregation, and superoxide radical generation. *Biochem. Biophys. Acta* **762:**420–428.

Serhan, C., Fridovich, J., Goetzl, E., Dunham, P., and Weissman, G., 1982, Leukotriene B4 and phosphatidic acid are calcium ionophores: Studies employing arsenazo III in liposomes, *J. Biol. Chem.* **257:**4726.

Serhan, C., Hamberg, M., and Samuelsson, B., 1984, Lipoxins: Novel series by biologically active compounds formed from arachidonic acid in human leukocytes, *Proc. Natl. Acad. Sco. U.S.A.* **81:**5335–5339.

Smaal, E., Mandersloot, J., de Kruiff, B., and de Gier, J., 1985, Essential Adaptation for the calcium influx assay into liposomes with entrapped arsenazo III for studies on the possible calcium translocating properties of acidic phospholipids, *Biochim. Biophys. Acta* **816:**418.

Smith, C., Lane, B., Kusaka, I., Verghese, M., and Snyderman, R., 1985, Chemoattractant receptor induced hydrolysis of phosphatidylinositol 4, 5-bisphosphate in human polymorphonuclear leukocyte membranes. Requirement for a guanine nucleotide regulatory protein, *J. Biol. Chem.* **260:**5875–5878.

Spat, A., Bradford, W., McKinney, J., Rubin, R., and Putney, J., 1986, A saturable receptor for P-inositol-1,4,5-trisphosphate in hepatocytes and neutrophils, *Nature* **319:**514.

Takai, Y., Kishimoto, A., Kikkawa, U., Mori, T., and Nishizuka, Y., 1980, Activation of a calcium and phospholipid-dependent protein kinase by diacylglycerol, its possible relation to phosphatidylinositol turnover, *J. Biol. Chem.* **244:**2273–2276.

Victor, M., Weiss, J., Klempner, M. and Elsbach, P., 1981, Phospholipase A_2 activity in the plasma membrane of human polymorphonuclear leukocytes, *FEBS Lett.* **136:**298–300.

Walsh, C., Waite, M., Thomas, M., and DeChatelet, L., 1981, Release and metabolism of arachidonic acid in human neutrophils, *J. Biol. Chem.* **256:**7228–7234.

Walsh, C., DeChatelet, L., Chilton, F., Wynkle, R., and Waite, M., 1983, Mechanism of arachidonic release in human polymorphonuclear leukocytes, *Biochim. Biophys. Acta* **750:**32–40.

White, J., Huang, C., Hill, J., Naccache, P., Becker, E., and Sha'afi, R., 1984, The effect of 4a-phorbol 12,13-didecanoate on protein phosphorylation and lysosoman enzyme release in rabbit neutrophils, *J. Biol. Chem.* **259:**8605–88610.

Whitin, J., Chapman, C., Simons, E., Chovaniec, M., and Cohen, H., 1980, Correlation between membrane potential changes and superoxide production in human granulocytes stimulated by phorbol myristate acetate. Evidence for defective activation in chronic granulomatous disease, *J. Biol. Chem.* **255:**1874–1878.

Wynkoop, E., Broekman, M., Korchak, H., Marcus, A., and Weissmann, G., 1986, Phospholipid metabolism in f-Met-Leu-Phe activated human neutrophils degranulation is not required for release of arachidonic acid: Studies with neutrophils and neutrophil-derived cytoplasts, *Biochem. J.* **236:**829–837.

Yuli, I., and Oplatka, A., 1987, Cytosolic acidification as an early transductory signal of human neutrophil chemotaxis, *Science* **235:**340.

3

Neutrophil Emigration

Quantitation, Kinetics, and Role of Mediators

MYRON I. CYBULSKY, M. K. WILLIAM CHAN, and HENRY Z. MOVAT

1. Introduction

The emigration of neutrophil leukocytes from the blood of postcapillary and larger venules into tissues and their accumulation at sites of microbial infection, immune complex deposition, or various forms of nonspecific injury, represent the hallmark of an acute inflammatory reaction (Colditz, 1985; Movat, 1985). Other vascular phenomena of inflammation, such as hyperemia, stasis, and increase in vasopermeability with edema formation, are important components of the acute inflammatory reaction, but emigrated neutrophils constitute the single most outstanding feature of an acute inflammatory exudate.

Our present concept of leukocyte emigration evolved gradually, with early studies in the 1840s and considerable advances in the second half of the 19th century, particularly those of Cohnheim, Metchnikoff, and many others. Further important landmarks were the detailed *intra vitam* observations of the Clarks in the 1930s and the ultrastructural observations made primarily by Marchesi and Florey in the early 1960s (reviewed by Movat, 1985). The currently accepted view is that at the site of injury leukocytes adhere to microvascular endothelium and emigrate by active ameboid move-

MYRON I. CYBULSKY and *M. K. WILLIAM CHAN* • Department of Pathology, University of Toronto, Toronto, Ontario, Canada M5S 1A8. *HENRY Z. MOVAT* • Departments of Pathology and Immunology, University of Toronto, Toronto, Ontario, Canada M5S 1A8.

ments between adjacent endothelial cells into the surrounding tissue. Beyond the vessel wall chemotaxis plays an important role; this was recognized 100 years ago, but good progress and considerable insight into the mechanism of chemotaxis were acquired after the development by Boyden (1962) of an *in vitro* assay that permitted quantitation of chemotaxis. To this day some investigators believe that chemotactic substances exert their effect not only as a gradient in the interstitial tissue but also into the wall and lumen of vessels. In the lumen they would be swept away immediately by the flowing blood. This may apply to *in vitro* situations in which leukocytes stimulated with chemotaxins become adherent to cultured endothelium *(vide infra)*. *In vivo*, when chemotaxins gain access to blood, leukocyte aggregation and sequestration is induced (reviewed by Cybulsky and Movat, 1983). Adsorption of chemotaxis to the luminal membrane of endothelial cell seems unlikely in view of desensitization experiments (Colditz and Movat, 1984a). In addition to chemotaxis assays *in vitro*, recent leukocyte–endothelial adhesion assays have contributed to our present understanding of leukocyte emigration, as discussed briefly below.

From a teleological point of view therefore, neutrophil emigration plays a central and pivotal role in an acute inflammatory reaction. Moreover, although neutrophil accumulation has the beneficial effects of eliminating infections and harmful agents, microvascular and tissue injury occurs frequently in this process and this is most frequently mediated by the neutrophils.

2. *Quantitation and Kinetics*

The introduction of reproducible methods for quantitating several components of the acute inflammatory reaction in rabbits led to the elucidation of the kinetics of these components or parameters (Kopaniak *et al.*, 1980). These assays determined the following in an acute inflammatory reaction: increase in vasopermeability ([^{125}I]-albumin), microhemorrhage ([^{59}Fe]-erythrocytes), emigration of neutrophils (^{51}Cr) and of platelets (^{111}In), and blood flow, with the aid of micropheres variously radiolabeled.

Radiolabeling these cells with $Na_2{}^{51}CrO_4$ (Issekutz and Movat, 1980) is of particular interest for quantitation of neutrophil emigration and accumulation into inflammatory sites. This enabled the study of (1) the magnitude of the response to various stimuli, (2) the kinetics of these responses, and (3) the mediators capable of eliciting a response. This method was further improved to enable expression of the results in absolute numbers of neutrophils that have accumulated. It was found useful to relate this to the level of circulating neutrophils, which when altered (neutropenia, neutrophilia) may considerably influence the delivery of the blood neutrophils to the inflammatory site (Cybulsky *et al.*, 1986b).

In our earliest studies, inflammation was induced by injecting formalin killed *E. coli* or chemotaxins into the skin of rabbits (Issekutz and Movat, 1980; Issekutz *et al.*, 1980). With the killed microorganisms, most of the neutrophil influx into the dermal connective tissue occurred within the first 4 h after intradermal (i.d.) injection, which was also observed more recently with live *E. coli* (Fig. 1). The chemotaxin $C5a_{des\ Arg}$ elicited essentially a similar reaction. The chemotactic oligopeptide fMLP (formyl methionyl leucyl phenylalanine) induced a very transient emigration of neutrophils that peaked 1 h after the intradermal injections (Fig. 2) and did not vary with three concentrations of fMLP tested (Colditz and Movat, 1984*b*). Skin sites were injected hourly with three doses of chemotaxin. One hour before sacrifice, [^{51}Cr]neutrophils were injected intravenously. The leukocytes circulated with a half-life between 3.0 and 3.8 h (Issekutz and Movat, 1980; Cybulsky and Movat, 1983) and all the chemotaxin-injected sites were exposed to these cells for 1 h. Both the total neutrophil accumulation over the period of 4 h for each concentration of chemotaxin and the proportion of the total accumulation that occurred during each hour were calculated. Figure 2 (right) demonstrates that, for each concentration of fMLP, a comparable portion of the total neutrophil accumulation occurred each hour. This implies that over a 1000-fold range of fMLP concentration, the kinetics of neutrophil emigration are independent of the concentration of the chemotaxin (Colditz and Movat, 1984b). In addition to the examples presented, the pattern of neutrophil kinetics was found to be similar with a large variety of inflammatory agents tested. It had already become apparent in our early studies that, despite some variations in kinetics, all inflammatory stimuli induced a transient neutrophil influx into tissues, which never appreciably exceeded 4 h after the intradermal injection of the stimulus. This applied to *E. coli*, immune precipitates, several chemotaxins ($C5a_{des\ Arg}$, fMLP, PAF, LTB_4), and endotoxin.

Of all the inflammatory agents tested, endotoxin was about 3–4 log-fold more potent than the chemotaxins (Colditz and Movat, 1984a) (Fig. 3). Since endotoxin is not chemotactic *in vitro* (Issekutz and Bhimji, 1982), its effect(s) in tissues may be mediated by one or more endogenous substances. This led to studies implying that neutrophil emigration induced by endotoxin was mediated primarily by interleukin-1 (IL-1), and further studies that led to the concept that IL-1 and tumor necrosis factor α (TNF-α) acted synergistically to induce neutrophil emigration.

Recent findings on the kinetics of neutrophil emigration are discussed in conjunction with the profiles elicited by various mediators (see Fig. 10).

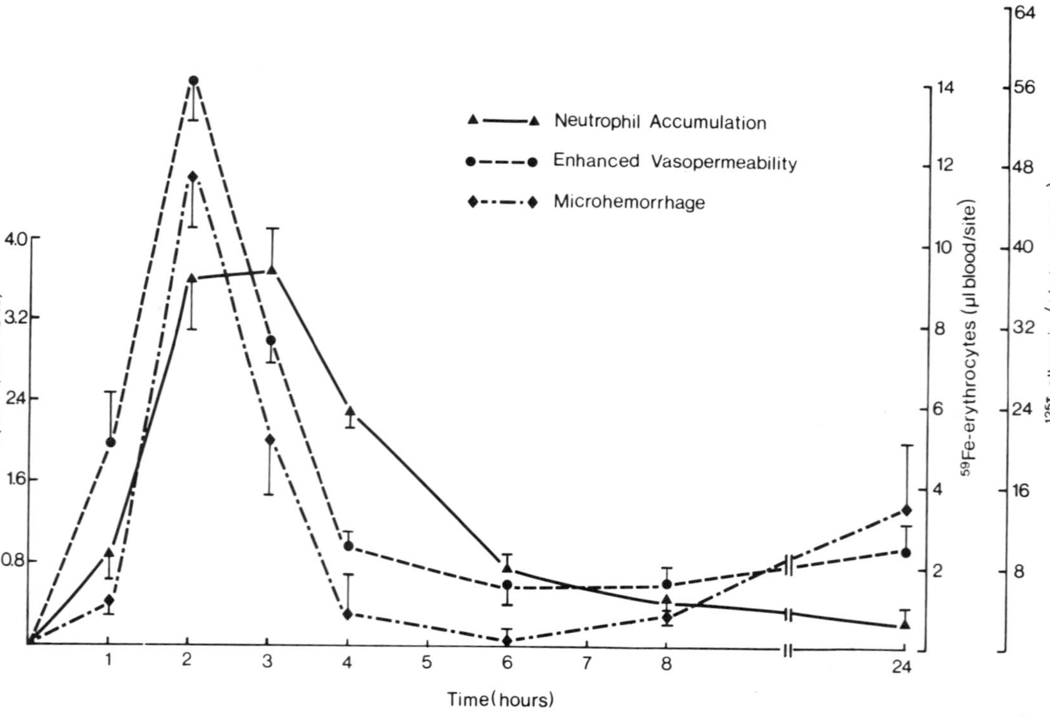

Figure 1. Quantitation and kinetics of inflammation induced by live *E. coli* (2×10^7 microorganisms/site) injected in triplicate at various times before death of the animal (time refers to age of the lesions). ^{59}Fe-labeled erythrocytes and ^{51}Cr-labeled neutrophils were injected intravenously 60 min before death and ^{125}I-labeled albumin was injected 15 min before death.

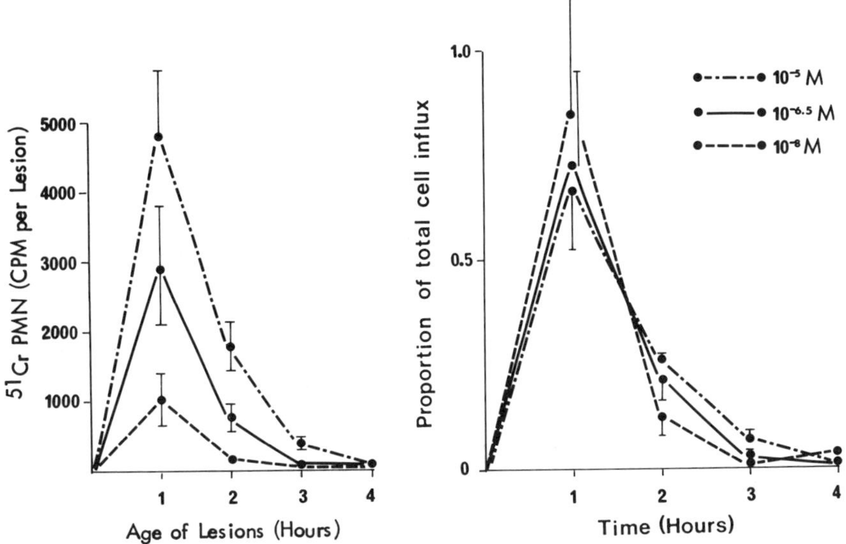

Figure 2. Kinetics of neutrophil emigration into sites injected with three concentrations of fMLP (*left*), with the results expressed as the proportion of the total neutrophil influx occurring each hour (*right*).

3. Mediators of Neutrophil Emigration

Chemotaxins as mediators of neutrophil emigration have been investigated thoroughly in the past 25 years, since the introduction of a reproducible *in vitro* quantitative assay for chemotaxis by Boyden (1962). Until then, innumerable mediators that had been proposed to induce neutrophil emigration were tested and in retrospect, many of the procedures seem to have induced activation of the complement system in fresh serum, which was present in most of the reaction mixtures tested (reviewed by Movat, 1985). In the course of these *in vitro* studies, a number of investigators injected the chemotaxins into experimental animals and observed neutrophil accumulation histologically, beginning with the observations of Jensen *et al.* (1969) soon after the discovery of C5a anaphylatoxin. There is little doubt that chemotaxins are important mediators of neutrophil emigration in many inflammatory reactions, particularly those in which the complement system is activated, for example, the formation of immune complexes.

As illustrated in Fig. 3, considerably less endotoxin can elicit a minimal neutrophil emigration than the stable form of C5a and other chemotaxins. This is one reason that the neutrophil emigration elicited by endotoxin could

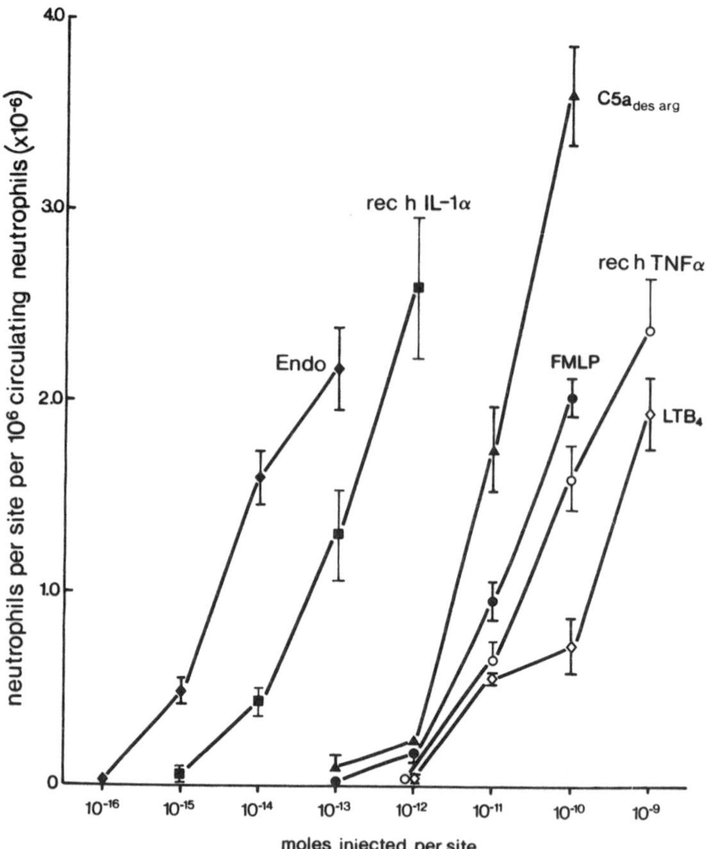

Figure 3. Neutrophil emigration into intradermal sites induced by a number of inflammatory agents. These dose–response experiments illustrate the potency of the agents tested.

not be mediated by serum-derived chemotaxins ($C5a_{des\ Arg}$) or cell-derived chemotaxins (LTB_4, PAF). If complement activation plays a role in endotoxin-induced neutrophil accumulation, it can only be a minor one. Whereas immune-complex-induced neutrophil accumulation was markedly inhibited in rabbits rendered hypocomplementemic with cobra venom factor (Crawford *et al.* 1985), decomplementation had no effect on the accumulation induced by killed *E. coli* (Kopaniak and Movat, 1983). This implies a mediating effect of complement that occurs only in the immune-complex-induced inflammatory reaction. On the other hand, *E. coli*-induced neutrophil infiltration was markedly diminished when the microorganisms were treated with polymyxin B or antiendotoxin antibodies, both known to inhibit endotoxin

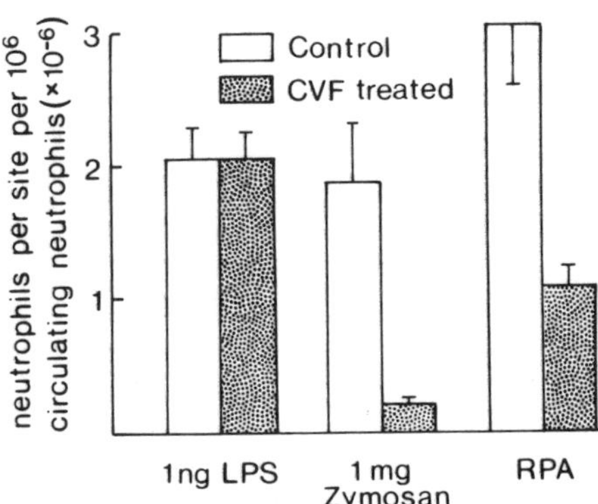

Figure 4. Neutrophil emigration into normal rabbits and into animals depleted of complement with cobra venom factor (CVF). While the neutrophil emigration induced by endotoxin (LPS) was not affected by the treatment, in CVF-treated rabbits the neutrophil response was diminished when zymosan was injected or a reverse passive Arthus reaction (RPA) was elicited in the skin.

(Issekutz *et al.*, 1982). Figure 4 illustrates the neutrophil influx into dermal sites in untreated rabbits and in animals depleted of complement with cobra venom factor. Neutrophil accumulation induced by zymosan or by the elicitation of a reversed passive Arthus reaction was considerably decreased compared to control in rabbits rendered hypocomplementemic with cobra venom factor. However, the decrease, and hence unavailability, of complement (primarily C5; see Crawford *et al.*, 1985) had no effect on the inflammatory reaction induced with endotoxin.

Next, we turned our attention more directly to potential intrinsic mediators released or synthesized and released by endotoxin. Crucial to these studies was an observation made earlier on desensitization or tachyphylaxis (Colditz and Movat, 1984a). Pertinent to these studies was the above-mentioned observation that all the inflammatory agents we had studied induced a transient emigration of neutrophils. When dermal sites were reinjected with the same chemotaxin and the neutrophil accumulation was quantitated with [^{51}Cr]neutrophils after the second injection, their number decreased compared to sites injected for the first time. This tachyphylaxis was chemotaxin-specific. In our first experiments, inflammatory lesions were initiated 8 h before sacrifice (fMLP or PAF) and restimulated 6 h later, and the accumulation of the neutrophils was quantitated during the last hour (the seventh to eighth hour of the experiment). Skin sites initiated with fMLP and restimulated witih fMLP and those initiated with PAF and restimulated with PAf exhibited tachyphylaxis or diminished neutrophil accumulations. In the same animals, sites injected first with PAF and restimulated with fMLP, or *vice versa*, showed normal responses (Fig. 5). Likewise, there was no cross-tachyphylaxis between any of the chemotaxins and endotoxins.

Experiments based on the hypothesis that neutrophil emigration induced by endotoxin is mediated by IL-1 seemed promising (Cybulsky *et al.*, 1985,

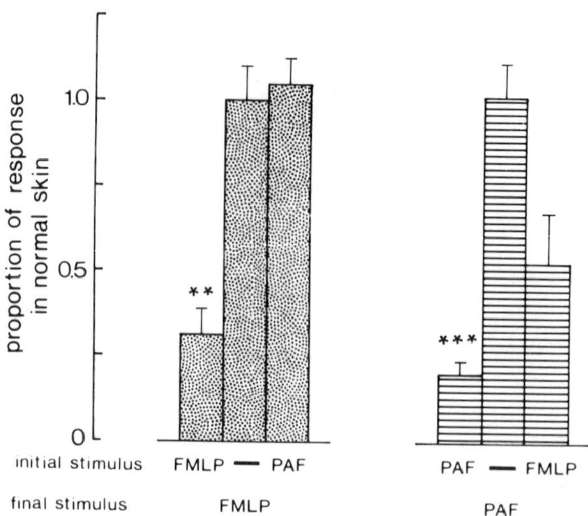

Figure 5. Tachyphylaxis experiment. Emigration of [^{51}Cr]neutrophils in dermal inflammatory sites after a second injection (final stimulus) of fMLP (10^{-6} M) or PAF (10^{-5} M). The first injections (initial stimuli) were given 8 h before sacrifice; the final ones 2 h before sacrifice, and neutrophil accumulation with radiolabeled cells was quantitated during the final hour of the experiment (**$p<0.01$; ***$p<0.001$).

1986a). They were based on the knowledge gradually acquired over the years that fever, granulocytosis, and other components of the acute-phase reaction elicited by endotoxin are mediated by IL-1, and that endotoxin is a potent stimulus for IL-1 synthesis by monocytes/macrophages and other cells, including endothelial cells (Dinarello, 1984, 1986). Parallel *in vitro* studies indicated that cultured endothelial cells stimulated with IL-1 or with TNF-α became adhesive for neutrophils (Bevilacqua *et al.*, 1985; Dunn and Fleming, 1985; Gamble *et al.*, 1985; Schleimer and Rutledge, 1986; Pohlman *et al.*, 1986). In addition to this endothelial-dependent (Bevilacqua *et al.*, 1987) adhesion of neutrophils, induced by both IL-1 and TNF-α, the latter cytokine was also capable of inducing a neutrophil-dependent adhesion (Gamble *et al.*, 1985; Harlan *et al.*, 1987).

Our first findings with human-monocyte-derived IL-1 were confirmed and extended with human recombinant IL-1 (Cybulsky *et al.*, 1987). As illustrated in Fig. 3, IL-1α was almost equipotent with endotoxin and, as described by Cybulsky *et al.* (1987), IL-1β was about one log-fold less potent than the α-variant. We also examined the kinetics of neutrophil emigration using a 30-min pulse of [^{51}Cr]neutrophils and made two observations. First, the rate of emigration with endotoxin lagged 30 min behind that induced with IL-1, implying synthesis of IL-1 (and probably TNF-α) when endotoxin is deposited into the dermis, before the cytokine(s) can act on the vascular endothelium. Second, the chemotaxin ZAP ($C5a_{des\ Arg}$) peaks at the same time as IL-1, but it initiates neutrophil emigraiton before IL-1. This is discussed and illustrated below in conjunction with more recent findings on protein-synthesis-dependent and -independent mechanisms of neutrophil emigration.

Tachyphylaxis experiments also supported the contention that IL-1 was the principal mediator of endotoxin-induced inflammation. A tachyphylaxis was already demonstrated against itself with the natural IL-1, but a cross-tachyphylaxis was also demonstrated with endotoxin (Cybulsky *et al.*, 1985, 1986*a*). When sites were first injected with endotoxin and then with recombinant IL-1, good suppression of the neutrophil influx occurred (Fig. 6). The converse (i.e., first IL-1 and then endotoxin injection) also induced cross-tachyphylaxis, but this was not always as pronounced. Pertinent to the observations on tachyphylaxis with respect to neutrophil emigration was the observation that, upon intradermal reinjection of endotoxin, both a systemic tachyphylaxis and a locally mediated tachyphylaxis were observed with respect to febrile responses in rabbits (Cybulsky *et al.*, 1987).

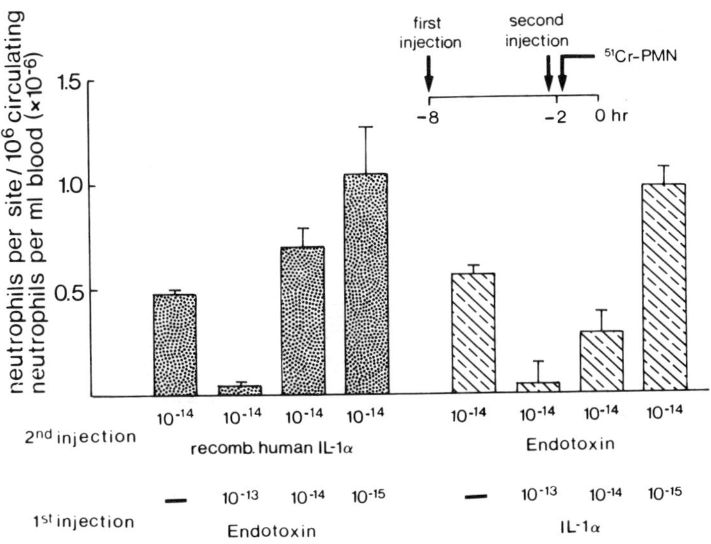

Figure 6. Tachyphylaxis experiments with rec h IL-1α and endotoxin. The timing of the injections is indicated (first injection at −8 h, second injejction at −2 h, and neutrophils labeled with ^{51}Cr circulated for 2 h prior to sacrifice). The doses (moles per site) are indicated for each injection, and the means and SEM of triplicate injections are plotted. Tachyphylaxis was not seen in sites prestimulated with a low (10^{-15} moles) dose.

Another inflammation-related phenomenon, the local Shwartzman reaction, was also investigated (Movat and Burrowes, 1985). It was observed that with relatively crude preparations of IL-1 (presumably containing TNF-α) a Shwartzman-like reaction could be "prepared." However, when highly purified recombinant IL-1 was injected intradermally as a preparative dose, a thrombohemorrhagic Shwartzman-like reaction could only be elicited when TNF-α was injected simultaneously (Movat *et al.*, 1987a). It had already been observed morphologically in these studies that before the intravenous challenge the inflammatory reaction at the local (preparative) injection site was very intense at sites injected with both IL-1 and TNF-α, compared to sites injected with only one of the two cytokines. These experiments were extended, using histology and quantitation with ^{51}Cr-labeled neutrophils, and a synergism between IL-1 and TNF-α was established with respect to both neutorphil accumulation and the Shwartzman-like reaction (Cybulsky *et al.*, 1987; Movat *et al.*, 1987b). When quantitated during peak emigration (2 h after the intradermal injections), the combined cytokines elicited a response that was 67% above the additive response of the individually administered IL-1 and TNF-α (Fig. 7). No such synergism was observed by combining

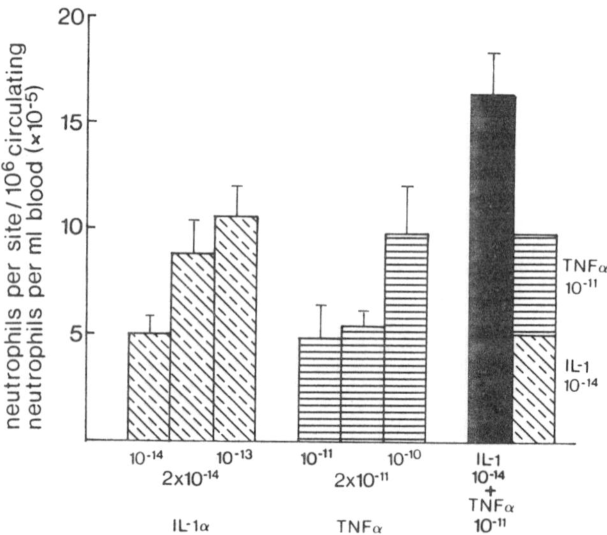

Figure 7. Neutrophil emigration elicited with IL-1, TNF-α, and a mixture of the two cytokines, quantitated with ^{51}Cr-labeled neutrophils in rabbit skin. The radiolabeled cells circulated for 2 h after the intradermal injections and entered the inflammatory lesions during that time period. The dose response for each of three concentrations of Il-1α and TNF-α is illustrated. Minimal emigration was induced with 10^{-14} moles of Il-1α and with 10^{-11} moles of TNF-α. When the two cytokines were injected together at those minimal concentrations, the neutrophil emigration was 67% greater than the additive response induced by each individual cytokine, implying a synergism between Il-1 and TNF-α.

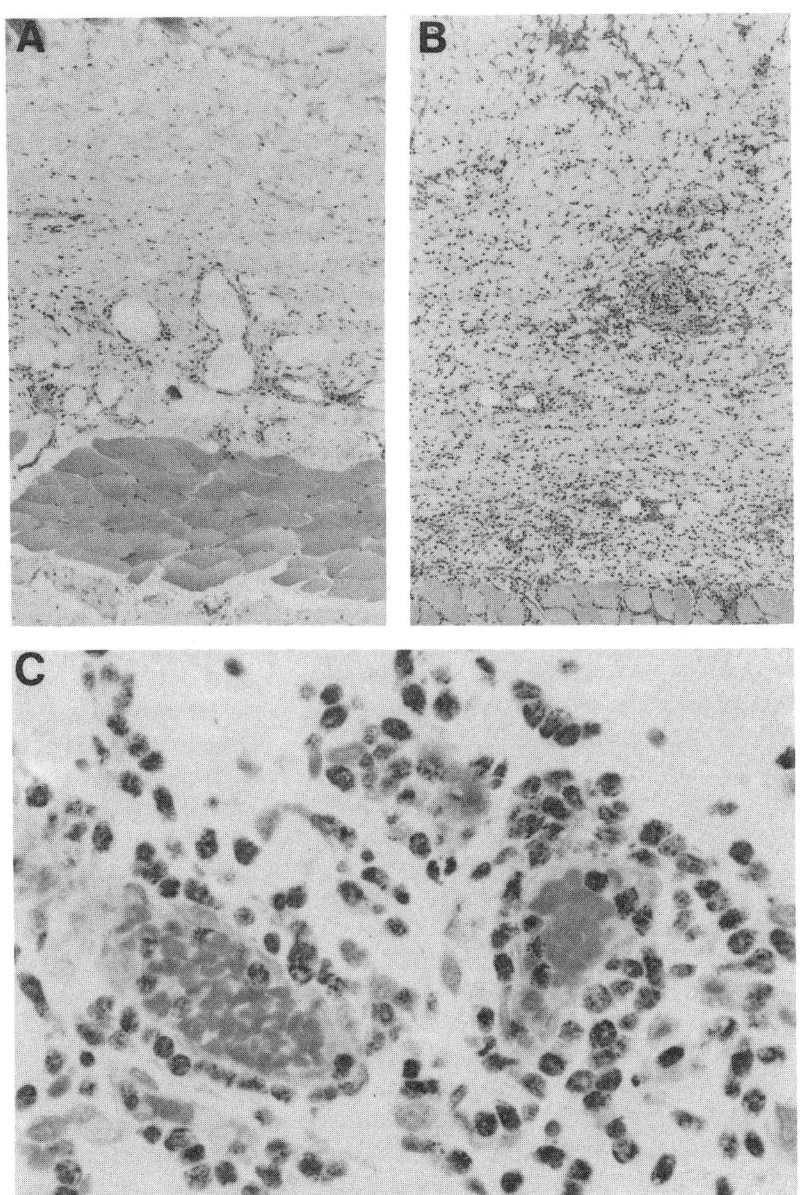

Figure 8. Histopathology of dermal sites injected 20 h before sacrifice with 0.5 μg of IL-1 (A), 0.5 μg of IL-1 and 1.0 μg of TNF-α (B), and a high magnification of a site injected with IL-1 and TNF-α to illustrate that the infiltrating cells are neutrophils (C). Sites injected with 1.0 μg of TNF-α showed fewer leukocytes in the lesion than sites injected with 0.5 μg of IL-1, and sites injected with endotoxin had about the same number of leukocytes as sites injected with IL-1 plus TNF-α. Magnifications: A and B—60×; C—480×.

IL-1 or TNF-α with fMLP. However, because leukocyte accumulation occurred over a longer time period at the sites prepared for a local Shwartzman reaction and because monocyte emigration continues beyond that of neutrophils (Issekutz *et al.*, 1981), the differences between sites injected with both IL-1 and TNF-α and sites injected with the individual cytokines were more impressive (Fig. 8). Because monocytes are an important source of IL-1 and TNF-α (Dinarello, 1986), as well as procoagulant activity, assessment of their accumulation at sites prepared with endotoxin may be important with respect to the subsequent development of a local Shwartzman reaction at such sites. Issekutz and Megyeri (1987) reported that endotoxin induced *in vivo*, in the pleural cavity, a "PMN infiltration-inducing activity," only some of which had the properties of IL-1 and most of which eluted during gel filtration with higher molecular weight fractions. More recently, however, Issekutz and Wankowicz (1987) reported a synergistic action between IL-1 and TNF-α with respect to neutrophil emigration, as described also in our studies.

There is a difference in the kinetics of neutrophil emigration into sites injected with a chemotaxin versus IL-1 or endotoxin (Fig. 9), as discussed briefly above in conjunction with neutrophil kinetics. With chemotaxins such as fMLP or $C5a_{des\ Arg}$, emigration is detectable within 30 min of injection; but with IL-1 or endotoxin, emigration does not start until 30–60 min after injection (Colditz, 1987; Cybulsky *et al.*, 1987). Figure 9 also shows a lag of about 30 min between IL-1 and endotoxin. The adhesion of neutrophils to vascular endothelium represents the first step in the emigration of these cells through the vessel wall. *In vitro* studies suggest that stimulation of endothelial cells by IL-1, TNF-α, or endotoxin leads to the expression of a new membrane protein on the endothelial cells, which renders these cells adhesive to neutrophils; this process can be blocked by inhibitors of protein synthesis (Bevilacqua *et al.*, 1985; Dunn and Fleming, 1985; Pohlman *et al.*, 1986; Schleimer and Rutledge, 1986).

Other mechanisms, also based on *in vitro* observations, imply an exclusively neutrophil-dependent adhesion to endothelium, in which chemotaxins or TNF-α act on the neutrophils to render them adhesive; and this process seems to be independent of protein synthesis (Pohlman *et al.*, 1986; Harlan *et al.*, 1987). This is difficult to visualize under *in vivo* conditions, in which a chemotaxin originating in the tissues could diffuse into the vessel wall or even the endothelium but could not reach the leukocytes in sufficient concentrations because it would be washed away by the flowing blood. However, recent *in vivo* studies suggest two mechanisms of neutrophil emigration, one protein synthesis dependent and one independent (McComb *et al.*, 1987). The potential role of protein synthesis was investigated by injecting the inflammatory agents intradermally, with or without an antibiotic (actinomycin D, cycloheximide, or puromycin), and quantitating neutrophil emigration

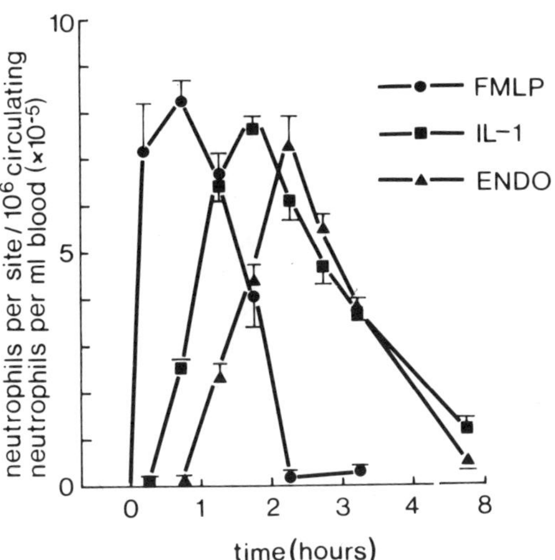

Figure 9. Kinetics of neutrophil emigration into intradermal sites injected with rec h IL-1α, 10^{-13} moles/site; endotoxin 10^{-13} moles/site; and fMLP 10^{-10} moles/site. Different sites were injected sequentially, and ^{51}Cr-labeled neutrophils were injected intravenously 30 min prior to sacrifice. The means and SEM of triplicate injections are plotted.

with a 30-min pulse of ^{51}Cr-labeled neutrophils. The antibiotics could inhibit the neutrophil emigration induced with IL-1 or endotoxin in a time- and dose-dependent manner, with greater than 90% inhibition at the time of peak emigration. In contrast, chemotaxin ($C5a_{des\ Arg}$, fMLP, LTB_4)-induced emigration of neutrophils could not be inhibited by the antibiotics (Fig. 10). These observations suggest two mechanisms of neutrophil emigration: one with a rapid onset and induced by chemotaxins was not protein synthesis dependent, and one with a slower onset and induced by IL-1 or endotoxin was dependent on protein synthesis.

4. Conclusions

When quantitated with ^{51}Cr-labeled cells, neutrophil emigration is a dynamic process, which under experimental conditions is transient, irrespective of the inflammatory stimulus used to elicit the response (*E. coli*, immune precipitates, $C5a_{des\ Arg}$, fMLP, PAF, LTB_4, endotoxin, IL-1). Of the well-defined inflammatory stimuli, endotoxin was the most potent in eliciting neutrophil emigration. Endotoxin mediates its effect primarily through generation of IL-1, based on the following observations: (1) IL-1 is the only substance that is almost equipotent as endotoxin in eliciting neutrophil emigration: (2) the onset and peak of endotoxin-induced neutrophil emigration lags approximately 30 min behind IL-1; (3) IL-1 shows cross-tachyphylaxis with endotoxin; and (4) unlike the neutrophil emigration elicited by chemotaxins, that induced by IL-1 or endotoxin is protein synthesis de-

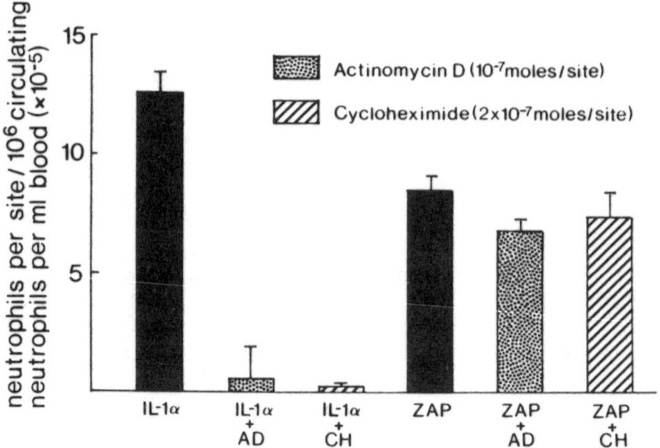

Figure 10. Neutrophil emigration induced by IL-1α (10^{-13} moles) or 33% zymosan activated plasma (ZAP; corresponding to 10^{-11} moles of $C5a_{des\ Arg}$). When the inflammatory agents injected were mixed with actinomycin D (10^{-7} moles) or cycloheximide (2×10^{-7} moles), neutrophil emigration (30-min pulse before sacrifice) induced by IL-1 was inhibited, but not that induced by ZAP.

pendent. In keeping with *in vitro* observations that endotoxin induces the synthesis and release of both IL-1 and TNF-α and that this may occur *in vivo*, IL-1 and TNF-α were found to act synergistically. This was observed with both the emigration of neutrophils and the induction of a local Shwartzman-like reaction.

ACKNOWLEDGMENTS. The authors wish to thank Dr. H. Michael Shepard (Genentech) for the generous supply of human recombinant TNF-α and Dr. Peter T. Lomedico (Hoffman–La Roche) for human recombinant IL-1α. We also thank Ms. Marica Michael and Mrs. Otti Freitag for secretarial and technical help. This research was supported by the Medical Research Council of Canada (MT-1251). Myron I. Cybulsky is a Postdoctoral Fellow of the Medical Research Council of Canada. M. K. William Chan is a graduate student supported by the Heart and Stroke Foundation of Ontario. Henry Z. Movat is a Career Investigator of the Medical Research Council of Canada.

References

Bevilacqua, M. P., Pober, J. S., Wheeler, M. E., Cotran, R. S., and Gimbrone, M. A., Jr., 1985, Interleukin 1 acts on cultured human vascular endothelium to increase the adhesion of polymorphonuclear leukocytes, monocytes and related leukocyte cell lines, *J. Clin. Invest.* **76:**2003–2011.

Bevilacqua, M. P., Wheeler, M. E., Pober, J. S., Fiers, W., Mendrick, D. L., Cotran, R. S., and Gimbrone, M. A., Jr., 1987, Endothelial-dependent mechanisms of leukocyte adhesion: Regulation by interleukin-1 and tumor necrosis factor, in: *Leukocyte Emigration and Its Sequelae* (H. Z. Movat, ed.), pp. 79–93, Karger, Basel.

Boyden, S., 1962, The chemotactic effect of mixtures of antibody and antigen on polymorphonuclear leucocytes, *J. Exp. Med.* **115:**453–466.

Colditz, I. G., 1985, Margination and emigration of leucocytes, *Surv. Synth. Pathol. Res.* **4:**44–68.

Colditz, I. G., 1987, Early accumulation of neutrophils in acute inflammatory lesions, in: *Leukocyte Emigration and Its Sequelae* (H. Z. Movat, ed.), pp. 14–23, Karger, Basal.

Colditz, I. G., and Movat, H. Z., 1984a, Desensitization of acute inflammatory lesions to chemotaxins and endotoxin, *J. Immunol.* **133:**2163–2168.

Colditz, I. G., and Movat, H. Z., 1984b, Kinetics of neutrophil accumulation in acute inflammatory lesions induced by chemotaxins and chemotaxinigens, *J. Immunol.* **133:**2169–2173.

Crawford, J. P., Movat, H. Z., Minta, J. O., and Opas, M., 1985, Acute inflammation induced by immune complexes in the microcirculation, *Exp. Mol. Pathol.* **42:**175–193.

Cybulsky, M. I., and Movat, H. Z., 1983, Application of ^{51}Cr-labeled PMN leukocytes in quantitating PMN kinetics in systemic and local inflammatory mediated processes. Effects of endotoxin, complement and interleukin 1, *Surv. Synth. Pathol. Res.* **1:**208–228.

Cybulsky, M. I., Colditz, I. G., and Movat, H. Z., 1985, Interleukin 1 activity in the local recruitment of PMNs: Its potential role in endotoxin-induced acute inflammation, *Fed. Proc.* **44:**1260 (Abstract).

Cybulsky, M. I., Colditz, I. G., and Movat, H. Z., 1986a, The role of interleukin-1 in neutrophil leukocyte emigration induced by endotoxin, *Am. J. Pathol.* **124:**367–372.

Cybulsky, M. I., Cybulsky, I. J., and Movat, H. Z., 1986b, Neutropenic responses to intradermal injections of *Escherichia coli*. Effects on the kinetics of polymorphonuclear leukocyte emigration, *Am. J. Pathol.* **124:**1–9.

Cybulsky, M. I., McComb, D. J., Dinarello, C. A., and Movat, H. Z., 1987, Mediation by interleukin-1 of neutrophil leukocyte emigration induced by endotoxin, in: *Leukocyte Emigration and Its Sequelae* (H. Z. Movat, ed.), pp. 38–50, Karger, Basel.

Dinarello, C., A., 1984, Interleukin 1, *Rev. Infect. Dis.* **6:**51–95.

Dinarello, C. A., 1986, Interleukin-1: Amino acid sequences, multiple biological activities and comparison with tumor necorsis factor (cachectin), *The Year in Immunology* **2:**68–89.

Dunn, C. J., and Fleming, W. E., 1985, The role of interleukin 1 in the inflammatory response with particular reference to endothelial cell–leukocyte adhesion, in: *The Physiologic, Metabolic, and Immunologic Actions of Interleukin 1* (M. J., Kluger, J. J., Oppenheim, and M. C. Powanda, eds.), pp. 45–416, Liss, New York.

Gamble, J. R., Harlan, J. H., Klebanoff, S. J., and Vadas, M. A., 1985, Stimulation of the adherence of neutrophils to umbilical vein endothelium by human recombinant tumor necrosis factor, *Proc. Natl. Acad. Sci. U.S.A.* **82:**8667–8671

Harlan, J. M., Schwartz, R. R., Wallis, W. J., and Pohlman, T. H., 1987, The role of neutrophil membrane proteins in neutrophil emigration, in: *Leukocyte Emigration and Its Sequelae* (H. Z. Movat, ed.), pp. 94–104, Karger, Basal.

Issekutz, A. C., and Bhimji, S., 1982, Role of endotoxin in the leukocyte infiltration accompanying *Escherichia cole* inflammation, *Infect. Immun.* **36:**558–566.

Issekutz, A. C., and Megyeri, P., 1987, Induction of leukocyte infiltration by endotoxin, in: *Leukocyte Emigration and Its Sequelae* (H. Z. Movat, ed.), pp. 24–37, Karger, Basel.

Issekutz, A. C., and Movat, H. Z., 1980, The *in vivo* quantitation and kinetics of rabbit neutrophil leukocyte accumulation in the skin, in response to chemotactic agents and *Escherichia coli, Lab. Invest.* **42:**310–317.

Issekutz, A. C., and Wankowicz, Z., 1987, Synergy between tumor necrosis factor (TNFα) and interleukin 1 (IL-1) in the induction of polymorphonuclear leukocyte emigration, *Fed. Proc.* **46:**737 (Abstract).

Issekutz, A. C., Movat, K. W., and Movat, H. Z., 1980, Enhancement of vascular permeability and haemorrhage-inducing activity of rabbit $C5a_{des\ Arg}$: Probable role of polymorphonuclear leucocyte lysosomes, *Clin. Exp. Immunol.***41:**512–520.

Issekutz, T. B., Issekutz, A. C., and Movat, H. Z., 1981, The *in vivo* quantitation and kinetics of monocyte emigration into inflammatory tissue, *Am. J. Pathol.* **103:**47–55.

Issekutz, A. C., Bhimji, S., and Bortolussi, R., 1982, Effect of immune serum or polymyxin B on *Escherichia coli*-induced inflammation and vascular injury, *Infect. Immun.* **36:**548–557.

Jensen, J. A., Snyderman, R., and Mergenhagen, S. E., 1969, Chemotactic activity, a property of guinea pig C′5-anaphylatoxin, in: *Cellular and Humoral Mechanisms of Anaphylaxis and Allergy* (H. Z. Movat, ed.), pp. 256–278, Karger, Basel.

Kopaniak, M. M., and Movat, H. Z., 1983, Kinetics of acute inflammation induced by *Escherichia coli* in rabbits. II. The effect of hyperimmunization, complement depletion, and depletion of leukocytes, *Am. J. Pathol.* **110:** 13–29.

Kopaniak, M. M., Issekutz, A. C., and Movat, H. Z., 1980, Kinetics of acute inflammation induced by *E. coli* in rabbits. Quantitation of blood flow, enhanced vascular permeability, hemorrhage, and leukocyte accumulation, *Am. J. Pathol.* **98:**485–498.

McComb, D. J., Cybulsky, M. I., and Movat, H. Z., 1987, PMN emigration: Protein synthesis dependent and independent mechanisms, *Fed. Proc.* **46:**1390 (Abstract).

Movat, H. Z., 1985, *The Inflammatory Reaction* pp. 1–365, Elsevier/North-Holland, Amsterdam.

Movat, H. Z., and Burrowes, C. E., 1985, Local Shwartzman reaction: Endotoxin-mediated inflammatory and thrombo-hemorrhagic lesions, in: *Handbook of Endotoxin, Volume 3, Cellular Biology of Endotoxin* (L. J. Berry, ed.), pp. 260–302, Elsevier/ North-Holland, Amsterdam.

Movat, H. Z., Burrowes, C. E., Cybulsky, M. I., and Dinarello, C. A., 1987a, Role of complement, interleukin-1 and tumor necrosis factor in a local Shwartzman-like reaction, in: *Leukocyte Emigration and Its Sequelae* (H. Z. Movat, ed.), pp.69–78, Karger, Basel.

Movat, H. Z., Burrowes, C. E., Cybulsky, M. I., and Dinarello, C. A., 1987b, Acute inflammation and Shwartzman-like reaction induced by interleukin-1 and tumor necrosis factor. Synergistic action of the cytokines in the induction of inflammation and microvascular injury, Am. J. Pathol. **129:** No. 3.

Pohlman, T. H., Stannes, K. A., Beatty, P. G., Ochs, H. D., and Harlan, J. M., 1986, An endothelial cell surface factor(s) induced *in vitro* by lipopolysaccharide, interleukin 1 and tumor necrosis factor-α increases neutrophil adherence by a CD_w18-dependent mechanism, *J. Immunol.* **136:**4548–4553.

Schleimer, R. P., and Rutledge, B. K., 1986, Cultured human vascular endothelial cells acquire adhesiveness for neutrophils after stimulation with interleukin 1, endotoxin and tumor-promoting phorbol diesters, *J. Immunol.* **136:**649–654.

4

The Role of Endothelium in Chronic Inflammation

MORRIS ZIFF and DRUIE CAVENDER

1. Introduction

The chronic inflammatory reaction that occurs in the sublining layer of the rheumatoid synovial membrane has the features of a cellular immune response. The response is of sufficient intensity and chronicity to result in the synthesis of large amounts of antibody. The amounts of antibody and the proportions of the immunoglobulin classes of antibody synthesized are similar to those produced by normal lymphoid tissue (Smiley *et al.*, 1968). The pattern of distribution of the mononuclear cells in rheumatoid synovial tissue into lymphocyte-rich areas and into transitional areas containing plasma cells and macrophages also has similarities to that in lymphoid tissue, in which lymphocyte-rich areas compose the paracortical region of the node, and antibody-producing cells with macrophages are present in the follicular and medullary regions.

It is likely that just as the maturation of lymphoid tissue requires persistent exposure to antigenic stimulation, so the development of lymphocyte-rich areas and transitional areas in the synovial membrane is a result of stimulation by an as yet unidentified antigen. Following such stimulation, a sequence of phenomena ensues that appears to be characteristic of chronic inflammation in general. In this sequence, a group of cytokines, derived from the inflammatory cells of the mononuclear cell infiltrate and the endothelial cells (ECs) of the postcapillary venules (PCVs), regulate the pattern of distribution of the infiltrating mononuclear cells. This regulation appears to be

MORRIS ZIFF and DRUIE CAVENDER • The University of Texas Health Science Center, Dallas, Texas 75235-9030.

highly dependent on the function of the ECs of the PCVs that sustain the infiltrate.

The rheumatoid synovial membrane is characterized by a lining layer that consists of macrophage-like (type A) and fibroblast-like (type B) cells. Beneath the lining layer, the cellular compostiton of the mononuclear cell infiltrate varies in different regions. There are areas in which small lymphocytes are densely infiltrated, which we have called lymphocyte-rich areas (Kobayashi and Ziff, 1973; Ishikawa and Ziff, (1976). At their margins, the lymphocyte-rich areas show a transition to areas that are rich in macrophages and often in plasma cells. We have referred to these as transitional areas. The density of plasma cells in certain areas is often high enough to merit their description as plasma-cell-rich areas. In addition, there are uninfiltrated interstitial areas.

2. *Lymphocyte-Rich and Transitional Areas of Synovial Membrane and Lymph Node*

The lymphocyte-rich areas contain the OKT4 helper cell, as do the paracortical areas of lymph node, as their predominant cell (Janossy *et al.*, 1981; Klareskog *et al.*, 1982; Meijer *et al.*, 1982; Kurosaka and Ziff, 1983). As in lymphoid tissue, T and B lymphocytes enter the lymphocyte-rich areas from PCVs with tall endothelium. The high density of lymphocytes surrounding tall endothelial PCVs is presumably a result of the heightened efficiency of the endothelium of such venules in facilitating the transendothelial migration of lymphocytes. Measurements of the relation between the tallness of the ECs of the PCVs and the composition of the surrounding perivascular infiltrates have shown that there is a highly significant correlation between the percentage of lymphocytes in the perivascular area and the tallness of the ECs (Iguchi and Ziff, 1986). Conversely, the greater the percentage of macrophages, plasma cells, and fibroblasts observed in the transitional areas, the flatter the endothelium.

2.1. *Steps in Lymphocyte Emigration*

Three consecutive steps are involved in the movement of a circulating lymphocyte toward a perivascular inflammatory focus: (1) binding of the lymphocyte to the EC, (2) movement of the lymphocyte through the endothelium and its underlying basement membrane to the perivascular space, and (3) migration of the lymphocyte to its appropriate microenvironment, a process known as ecotaxis. We have studied each of these steps in our laboratory.

2.2. *T Cell to Endothelial Cell Binding*

The binding of lymphocytes to ECs has been examined using two techniques. The first method, by Stamper and Woodruff (1976), uses lymphocytes that are permitted to adhere to the ECs of PCVs in frozen sections of lymphoid tissue; the number of lymphocytes remaining fixed to the ECs following washing of the sections is quantitated. The second method, currently used in our laboratory, employs lymphocytes labeled with chromium-51 (T cells were used in most experiments), which are incubated with monolayers of human umbilical vein endothelial cells (HUVECs). The nonadherent lymphocytes are removed by washing, and the percentage of bound T cells is determined by measuring the radioactivity of the lymphocytes adherent to the monolayers.

When the Stamper–Woodruff technique was used to suspend peripheral blood mononuclear cells (PBMCs) over frozen sections of rheumatoid synovial membrane, it was observed (Oppenheimer-Marks and Ziff, 1986) that lymphocytes adhered to the ECs of PCVs of lymphocyte-rich areas in a manner identical to that observed in the paracortical areas of normal lymphoid tissue, thus indicating a similarity in structure and function of the PCVs of the paracortical areas of lymph node and the lymphocyte-rich areas of rheumatoid synovium.

2.3. *Effect of Cytokines on T Cell to Endothelial Cell Binding*

Based on the assumption that the function of the endothelium of PCVs in areas of chronic inflammation might be regulated by cytokines produced in adjacent chronic inflammatory foci, monolayers of ECs were incubated in microtiter wells with supernatants of PHA-stimulated PBMC cultures or with supernatants of mixed lymphocyte cultures. Following incubation for 24 h, the supernatants were removed and suspensions of ^{51}Cr-labeled peripheral blood T lymphocytes were added to the cultures. Following removal of the nonadherent cells by washing, the percentage of adherent cells was measured by ^{51}Cr counting. It was observed that both the PHA supernatants and the mixed lymphocyte reaction supernatants stimulated the ECs to increase their capacity to bind T cells in a highly significant manner (Yu *et al.*, 1985). The observation that stimulatory activity was lost from the supernatants following heating at 56°C for 30 min, acidification at pH 2, or treatment with anti-IFN-γ serum suggested that IFN-γ was among the cytokines that contributed to the stimulation of the increased binding capacity of the ECs. Incubation of the monolayers with affinity-purified IFN-γ (Interferon Sciences, New

Brunswick, NJ) did in fact increase binding in a dose-dependent fashion. A significant increase in binding was noted with 10 U/mL of IFN-γ. The effect was apparent after 4 h; maximal increase in binding was noted after 24 h of incubation.

In view of the stimulatory activity of IFN-γ, a T cell lymphokine, it was also decided to examine the effects of macrophage-derived cytokines. In these experiments, monolayers of ECs or suspensions of T cells were incubated with the cytokine and washed prior to use in the binding assay. Increased binding occurred in a concentration-dependent fashion following incubation of ECs with interleukin-1 (IL-1) (Cavender *et al.*, 1986) and with tumor necrosis factor α (TNF-α) (Cavender *et al.*, 1987). Enhanced binding was noted with as little as 0.01 ng/mL of recombinant IL-1 α or β or recombinant tumor necrosis factor. Preincubation with T cells did not increase the percentage of binding.

2.4. *Effect of Lipopolysaccharide on T Cell to Endothelial Cell Binding*

Preincubation of EC monolayers with lipopolysaccharide (LPS) from a variety of bacterial sources also increased T cell–EC binding in a dose-dependent fashion. A significant increase in binding was observed with as little as 0.1 ng/mL of LPS. The effect was near maximal at a concentration of 0.1 μg/mL. The onset of the effect was very rapid and was observed after 2–3 min of preincubation with the ECs.

2.5. *Chemotaxis in the Emigration of Lymphocytes from the Postcapillary Venule*

It is likely that chemotaxis of lymphocytes plays a role in their movement through the PCV endothelium and underlying basement membrane. For this reason, an effort was made to identify a chemotactic agent for lymphocytes in synovial effusions from patients with classic rheumatoid arthritis. Using a modified Boyden chamber, it was observed that such effusions did exert a chemotactic effect on both B and T lymphocytes (Miossec *et al.*, 1986a). Fractionation of the synovial fluid on AcA 54 Ultrogel disclosed three peaks of chemotactic activity, with molecular weights of 60, 16, and 5 kD. When the fractions were assayed for IL-1 activity, there was a correspondence between chemotactic activity and IL-1 activity in the thymocyte proliferation assay for IL-1. On chromatofocusing, the 16-kD fraction showed IL-1 activity and chemotactic activity in fractions with the characterstic pI values of IL-1. This demonstration of IL-1 as a chemotactic agent in rheumatoid synovial fluid was consistent with our previous observation that ultrapure IL-1 acted as a chemotactic agent for lymphocytes (Miossec *et al.*, 1984).

2.6. Production of IL-1 by Human Endothelial Cells

Because ECs express a number of functions of the monocyte/macrophage lineage, EC supernatants were examined for the presence of IL-1. When EC cultures were carried out in the absence of serum, IL-1 activity was demonstrated in the supernatants, and upon stimulation with LPS the IL-1 activity was greatly increased (Misossec *et al.*, 1986b). This demonstration of the secretion of IL-1 by the ECs suggests that the ECs may exert a chemotactic effect on lymphocytes circulating in the PCVs through the release of IL-1, thus attracting these cells to the lining endothelium for binding.

2.7. Mechanisms of T Cell–Endothelial Cell Interaction

We have studied the T cell–EC receptor interaction involved in lymphocyte binding to EC. A triad of glycoprotein molecules, collectively known as the CwD-18 membrane antigen complex, which is important in various leukocyte adhesion processes, has recently been described (Springer, 1985). These molecules share a common beta subunit but have individual alpha subunits. Monoclonal antibodies (Moab) to one of the three molecules, the lymphocyte-function-associated molecule (LFA-1), inhibit a variety of T cell functions. We have recently shown that both the anti-β-chain monoclonal antibody, 60.3 (obtained from Dr. P. Beatty), and the anti-LFA-1 alpha chain Moab, TS 1/22 (obtained from Dr. T. Springer), markedly inhibit T cell binding to unstimulated endothelial cell monolayers *in vitro* (Haskard *et al.*, 1986). The action of the inhibiting Moab was observed when incubated with the T cell and not the EC, indicating that the LFA-1 antigen neutralized was present on the T cell. This is consistent with the known presence of the LFA-1 antigen on all leukocytes and its apparent absence on ECs. However, when the effect of LFA-1 Moab on the binding of T cells to IL-1 or LPS-stimulated ECs was investigated, the inhibition of binding that occurred appeared to be attributable only to inhibition of the baseline T cell–EC binding. We concluded from this that the augmented binding of T cells to ECs, which is induced by IL-1, TNF, IFN-γ, or LPS, involves a ligand–receptor interaction that is independent of LFA-1.

3. Discussion

The observations described emphasize the important role of cytokines in the emigration of lymphocytes from the blood into the perivascular space. The ability of ECs to bind lymphocytes prior to their emigration from PCVs is regulated by a number of mediators. The lymphokine IFN-γ and the monokines IL-1 and TNF act on the EC to increase its capacity to bind lymphocytes. Since binding of the lymphocyte is the initial step in its emigration, an increase in binding leads to an increase in the number of lymphocytes that emigrate.

Chemotaxis also plays a role in the mobilization of mononuclear cell infiltrates. A number of chemotactic agents for lymphocytes have been described, and these have had variable properties (Ward *et al.*, 1977; Cruikshank and Center, 1982; El-Naggar *et al.*, 1982; Van Epps *et al.*, 1983a,b; Kornfeld *et al.*, 1985; Potter and Van Epps, 1986). In the present research, IL-1 was demonstrated to be a chemotactic agent for both T cells and B cells (Miossec *et al.*, 1984). This finding suggests that IL-1, generated by activated macrophages in the synovial membrane, could attract lymphocytes along an IL-1 gradient toward the inflammatory infiltrate and in this way amplify the size of the infiltrate. In recent experiments, we and others (Stern *et al.*, 1985; Miossec *et al.*, 1986b) have shown that the EC itself, when appropriately stimulated, releases IL-1. The ability of the EC to generate IL-1 raises the possibility that lymphocytes, traveling through PCVs, may be attracted to the ECs by virtue of the chemotactic action of the locally released IL-1. As a result, the lymphocyte would have an increased probability of being bound to the EC. Following binding, it would pass through the endothelium to migrate to the inflammatory focus along an IL-1 gradient. Our recent observation that IFN-γ, preincubated with ECs, increases the amount of IL-1 subsequently secreted by these cells on stimulation by LPS (Miossec and Ziff, 1986) indicates that IFN-γ may play a role in enhancing the IL-1 secreting activity of the ECs *in vivo*.

The observations described indicate that the lymphokines and monokines generated by the chronic inflammatory infiltrate play an important part in the emigration of lymphocytes from the blood and in the regulation of their pattern of distribution in the perivascular space. Although the chronic inflammatory response may show variations in intensity and in the nature and distribution of its mononuclear cell components—reflecting, presumably, the properties of the antigen and the reactivity of the host—these features of the chronic inflammatory reaction appear to be regulated by cytokines produced by lymphocytes, macrophages, and ECs. These mediators may initiate chemotaxis of lymphocytes and monocytes toward the ECs and stimulate their binding to the endothelium, their emigration through the PCV wall, and their chemotaxis to the inflammatory focus.

References

Cavender, D., Haskard, D.O., Joseph, B., and Ziff, M., 1986, Interleukin-1 increases the binding of human B and T lymphocytes to EC monolayers, *J. Immunol.* **136**:203–207.

Cavender, D., Saegusa, Y., and Ziff, M., 1987, Stimulation of endothelial cell binding of lymphocytes by tumor necrosis factor, *J. Immunol.* **139**:1855–1860.

Cruikshank, W., and Center, D.M., 1982, Modulation of lymphocyte migration by human lymphokines. II. Purification of a lymphotactic factor, *J. Immunmol.* **128**:2569–2574.

El-Naggar, A., Van Epps, D.E., and Willims, R.C., Jr., 1982, A human lymphocyte chemotactic factor produced by the mixed lymphocyte reaction, *J. Lab. Clin. Med.* **100**:558–565.

Haskard, D., Cavender, D., Beatty, P., Springer, T., and Ziff, M., 1986, T lymphocyte adhesion to endothelial cells: Mechanisms demonstrated by anti-LFA-1 monoclonal antibodies, *J. Immunol.* **137**:2901–2906.

Iguchi, T., and Ziff, M., 1986, Electron microscopic study of rheumatoid synovial vasculature, *J. Clin. Invest.* **77**:355–361.

Ishikawa, H., and Ziff, M., 1976, Electron microscopic observations of immunoreactive cells in the rheumatoid synovial membrane, *Arthritis Rheum.* **19**:1–14.

Janossy, G., Panayi, G., Duke, O., Bofill, M., Poulter, L.W., and Goldstein, G., 1981, Rheumatoid arthritis: A disease of T-lymphocyte/macrophage immunoregulation, *Lancet* **ii**:839–842.

Klareskog, L., Forsum, U., Wigren, A., and Wigzell, H., 1982, Relationships between HLA-DR-expressing cells and T lymphocytes of different subsets in rheumatoid synovial tissue, *Scand. J. Immunol.* **15**:501–507.

Kobayashi, I., and Ziff, M., 1973, Electron microscopic studies of lymphoid cells in the rheumatoid synovial membrane, *Arthritis Rheum.* **16**:471–486.

Kornfeld, H., Berman, J.S., Beer, D.J., and Center, D.M. 1985, Induction of human T lymphocyte motility by interleukin 2, *J. Immunmol.* **134**:3887–3890.

Kurosaka, M., and Ziff, M., 1983, Immunoelectron microscopic study of the distribution of T cell subsets in rheumatoid synovium, *J. Exp. Med.* **158**:1191–1210.

Meijer, C. J. L. M., de Graaff-Reitsma, C. B., Lafeber, G. J. M., and Catz, A., 1982, *In situ* localization of lymphocyte subsets in synovial membranes of patients with rheumatoid arthritis with monoclonal antibodies, *J. Rheumatol.* **9**:359–365.

Miossec, P., and Ziff, M., 1986, Immune interferon enhances the production of interleukin 1 by human endothelial cells stimulated with lipopolysaccharide, *J. Immunol.* **137**:2848–2852.

Miossec, P., Yu, C.-L., and Ziff, M., 1984, Lymphocyte chemotactic activity of human interleukin 1, *J. Immunol.* **133**:2007–2011.

Miossec, P., Dinarello, C. A., and Ziff, M., 1986a, Interleukin 1 lymphocyte chemotactic activity in rheumatoid synovial fluid, *Arthritis Rheum.* **29**:461–470.

Miossec, P., Cavender, D. E., and Ziff, M., 1986b, Production of interleukin 1 by human endothelial cells, *J. Immunol.* **136**:2486–2491.

Oppenheimer-Marks, N., and Ziff, M., 1986, Binding of normal human mononuclear cells to blood vessels in rheumatoid synovial membrane, *Arthritis Rheum.* **29:**789–792.

Potter, J. W., and Van Epps, D. E., 1986, Human T-lymphocyte chemotactic activity: Nature and production in response to antigen, *Cell. Immunol.* **97:**59–66.

Smiley, J. D., Sachs, C., and Ziff, M., 1968, In vitro synthesis of immunoglobulin by rheumatoid synovial membrane, *J. Clin. Invest.* **47:**624–632.

Springer, T. A., 1985, The LFA-1, Mac-1 glycoprotein family and its deficiency in an inherited disease, *Fed. Proc.* **44:**2660–2663.

Stamper, H. B., and Woodruff, J. J., 1976, Lymphocyte homing into lymph nodes: In vitro demonstration of the selective affinity of recirculating lymphocytes for high-endothelial venules, *J. Exp. Med.* **144:**828–833.

Stern, D. M., Bank, I., Naworth, P. P., Cassimeris, J., Kisiel, W., Fenton, J. W. II, Dinarello, C. A., Chess, L., and Jaffee, E. A., 1985, Self-regulation of procoagulant events on the EC surface, *J. Exp. Med* **162:**1223–1235.

Van Epps, D. E., Potter, J. W., and Durant, D. A., 1983a, Production of a human T lymphocyte chemotactic factor by T cell subpopulations, *J. Immunol.* **130:**2727–2731.

Van Epps, D. E., Durant, D. A., and Potter, J. W., 1983b, Migration of human helper/inducer T cells in response to supernatants from Con A-stimulated suppressor/cytotoxic T cells, *J. Immunol.* **131:**697–700.

Ward, P. A., Unanue, E. R., Goralnick, S. J., and Schreiner, G. F., 1977, Chemotaxis of rat lymphocytes, *J. Immunol.* **119:**416–421.

Yu, C.-L., Haskard, D. O., Cavender, D., Johnson, A. R., and Ziff, M., 1985, Human gamma interferon increases the binding of T lymphocytes to endothelial cells, *Clin. Exp. Immunol.* **62:**544–560.

II

PEPTIDE MEDIATORS OF INFLAMMATION

5

Peptide Mediators of Inflammation

An Overview

STEPHEN M. KRANE

Some of us who have been attempting to understand the mechanisms of inflammation have focused on the nature of the interactions among component cells in inflammatory lesions. The deleterious consequences of chronic inflammatory diseases result from the effects of these interactions on resident cells of the involved tissues (Krane *et al.*, 1982). In most of these tissues, mesenchymal cells such as fibroblasts are the target cells, and monocytes and lymphocytes are the effector cells. It has become apparent, however, that the distinction between *effector* and *target* is blurred, since the so-called target cells can also profoundly influence functions of the effector cells. The interactions among these cells are mediated through direct cell contact or the release of soluble ligands that act by binding through specific cellular receptors.

The existence of these ligands was initially postulated on the basis of a biological response usually assayed on cells in culture; it was reasonable to assume that a single polypeptide product was responsible for the observed effect. Over the past several years, however, the purification, cloning, and sequencing of many of these polypeptides have provided definitive evidence that they exist and that they are distinct from each other (Oppenheim *et al.*, 1986; Old, 1987). Instead of a single product, multiple polypeptides have been identified that may serve similar functions. For example, the monocyte/macrophage has now been shown to produce three very different polypeptides with little or no amino acid sequence homology: interleukin-1 (IL-1α and IL-1β) and tumor necrosis factor (TNF-α), which have many similar

STEPHEN M. KRANE • Department of Medicine, Harvard Medical School and the Medical Services (Arthritis Unit), Massachusetts General Hospital, Boston, Massachusetts 02114.

effects on mesenchymal target cells (Beutler and Cerami, 1986; Oppenheim *et al.*, 1986). Whereas IL-1α and IL-1β probably act through the same receptor, TNF-α acts through a different receptor. In a recent review, Old (1987) emphasized that TNF-α, along with lymphotoxin (TNF-β), IL-1, and other polypeptide mediators, forms part of a network of interactive signals that orchestrate inflammatory and immunological events regulating the production and function of hematopoietic stem cells and their descendants, for example.

This network concept also applies to the replication and function of cells such as synovial fibroblasts and articular chondrocytes in the synovial lesion of rheumatoid arthritis. In studies from our laboratory, we found that cocultivation with monocyte/macrophages stimulated the synthesis and release of collagenase and prostaglandins (predominantly PGE_2 by rheumatoid synovial fibroblasts) (Dayer *et al.*, 1977a). Medium conditioned by the monocytes was also capable of producing such stimulation. It was subsequently demonstrated that these biological activities could be attributed to IL-1 (Mizel *et al.*, 1981). All the different effects on cells attributed to IL-1 can be reproduced using recombinant preparations, indicating that IL-1α and IL-1β are at least two of the components responsible for the actions of the soluble products released by mononuclear cells (Dinarello *et al.*, 1986). These mediators induce the synthesis of collagenase, not simply the release of preformed protein retained within the cells, as shown by the increased incorporation of [^{35}S] methionine into collagenase protein immunocomplexed with specific antibodies (McCroskery *et al.*, 1985). It has recently been found that recombinant IL-1 also increases the celluar levels of collagenase mRNA in human chondrocytes measured by hybridization with a nick-translated ^{32}P-labeled human cDNA probe (Stephenson *et al.*, 1987). It is likely that IL-1 acts to increase transcription of the procollagenase gene. TNF-α has also been shown to induce collagenase and PGE_2 production by synovial fibroblasts (Dayer *et al.*, 1985). In preliminary studies, we found that either TNF-α or TNF-β increases the synthesis of collagenase protein. It is therefore likely that these two polypeptide mediators also act by increasing transcription of the procollagenase gene as well as the gene for other metalloproteinases produced by these mesenchymal cells.

Products released by mononuclear cells have many profound and diverse effects on the target mesenchymal cells. These effects include changes in shape as well as replication and synthesis of matrix proteins. Preparations of monocyte-conditioned medium or cocultivation with monocytes also sensitize the target cells to exogenous PGE_2 under circumstances in which endogenous PGE_2 synthesis is inhibited by indomethacin (Dayer *et al.*, 1979; S. R. Goldring *et al.*, 1984). The most highly purified preparations of IL-1, as well as recombinant preparations, have now been shown to have these effects on cell replication (Schmidt *et al.*, 1982; Gowen *et al.*, 1985; Dinarello *et al.*, 1986; Rupp *et al.*, 1986). Furthermore, recombinant preparations of IL-1α or IL-1β, as well as several other polypeptide ligands, reproduce the effects of mononuclear-cell-conditioned medium in augmenting PGE_2-induced increases in cellular cyclic AMP content over and above those resulting from incubation with indomethacin alone (S. R. Goldring *et al.*, 1986, 1987a) IL-1 and TNF-α act in synergistic fashion in augmenting the synthesis of collagenase and PGE_2. This same synergism is observed in augmenting cellular responses to PGE_2 as measured by changes in the cellular content of cyclic AMP.

Many of the interactions of inflammatory and mesenchymal cells with their extracellular matrices have profound effects on their function, including modulation of attachment, spreading, differentiation, and replication. These responses may be exerted through specific receptors for these matrix proteins on the surface of cells. The most-studied receptors include those for fibronectin and other glycoproteins, which contain similar cellular recognition sites (Pytela *et al.*, 1985; Yamada *et al.*, 1985; Ruoslahti and Pierschbacher, 1986; Pytela *et al.*, 1986). The extracellular matrix may also function in a very different way; for example, by binding and therefore compartmentalizing growth factors—demonstrated recently for granulocyte macrophage colony-stimulating factor (GM-CSF)—the matrix may have an important function in the marrow microenvironment (Gordon *et al.*, 1987). On the other hand, inflammatory cells acting through their soluble polypeptide products may also profoundly modulate the levels of synthesis of extracellular matrix components. For example, crude mononuclear-cell-conditioned medium that contains IL-1 increases the synthesis of type I and III collagens and fibronectin by synovial fibroblastic cells, particularly if the ambient levels of PGE_2 (which suppress collagen synthesis) are lowered by incubation with indomethacin (Krane *et al.*, 1985). We can now show that collagen synthesis in cultured synovial or articular chondrocytes is *increased* by IL-1 (either IL-1α or IL-1β), but only if the cyclooxygenase is blocked and ambient levels of PGE_2 are low (M. B. Goldring and Krane, 1986). When recombinant IL-1 is incubated with these cells in the absence of indomethacin, the synthesis of type I and III collagens is *decreased*. Furthermore, the levels of type I and III procollagen mRNAs in the cells parallel those of the rates of collagen synthesis. The effects of IL-1 on type II collagen synthesis, which can be observed in cultured costal chondrocytes, appear to be much more complex. In cultures of costal chondrocytes for example, IL-1 suppresses type II collagen synthesis as well as levels of procollagen α1 (II) mRNA (M. B. Goldring *et al.*, 1987b). In contrast to the effects of IL-1 on type I and III collagens in other cells, however, the addition of indomethacin does not overcome this inhibition, but even potentiates it.

Thus, there is abundant evidence that the function of mesenchymal cells in the environment of inflammatory cells may be altered markedly as a result of the action of various cytokines produced by these inflammatory cells. In several circumstances, actions on the target cells are also modulated by the levels of cellular products, such as PGE_2, whose synthesis and release also result from the actions of the ligands on these cells. In matrix proteins, high levels of PGE_2 inhibit the synsthesis of type I and III collagens and fibronectin; the stimulatory effects of these ligands can only be observed when the synthesis of PGE_2 is blocked. Thus, the net effects of ligands such as IL-1 result not only from the specific signal transduction that follows binding through its receptors but also from autocrine effects that result from simultaneous stimulation of the synthesis of eicosanoids such as PGE_2, which then act on different receptors and induce different signals. Further complexities result from the actions of IL-1 and other polypeptides on modulating re-

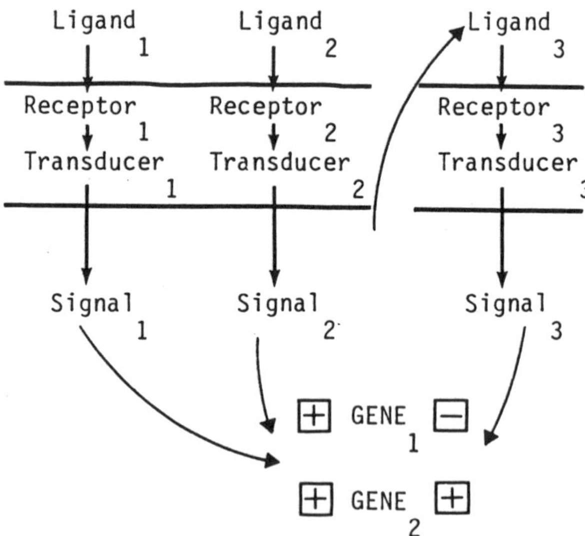

Figure 1. Model for effects of different ligands on cellular functions. In this scheme, two different ligands, acting through different receptors in the cell surface membrane and transducers, produce different signals that activate two different genes in a similar manner. $Ligand_1$ could be IL-1β and $ligand_2$ could be TNF-α, each acting through its unique receptor. In an alternative model, the transducer and signal could be identical. This shows that $ligand_3$ is made by the same cell and acts back on itself through a different receptor, transducer, and signal. $Ligand_3$ could be PGE_2. The signals produced by the interactions of IL-1β or TNF-α with their receptors could, in addition, affect the receptor and/or signal transduction for $ligand_3$ (in this case PGE_2). For example, IL-1β or TNF-α induces heterologous sensitization of PGE_2 responses in synovial fibroblasts in the presence of indomethacin.

sponses to the very ligands (PGE_2) whose synthesis they also induce. This sensitization effect of IL-1 on PGE_2 responses may be an important factor in determining how nonsteroidal anti-inflammatory agents act on cells. For example, if such drugs inhibit PGE_2 synthesis, yet at the same time markedly enhance the cellular responses to PGE_2, then the outcome may be exactly the opposite of that predicted from the blocking of synthesis alone.

It is still not clear how the changes induced by IL-1, TNF, and other inflammatory cytokines are regulated. As discussed previously, these mediators do appear to act by binding to specific high-affinity receptors present on the surface of different target cells (Fig 1). The interaction of the ligands with their receptors elicits signals that must ultimately act at the level of the promoters for the genes that are activated by these ligands. These genes include other mediators such as interferon-β, proteinases such as collagenase, gelatinase, and stromelysin, and components of the extracellular matrix such as collagens and fibronectin. How the intracellular signals are transduced remains to be demonstrated, but it is likely that they ultimately act through the production of *trans*-acting proteins that bind to specific sequences in the promoters of the respective genes. There are many other complexities introduced, for example, the receptor-mediated synthesis of eicosanoids such as PGE_2 that act through their own receptors to initiate yet another series of transduction events. These signals may operate in the same direction, as demonstrated in the regulation and expression of the proenkephalin gene, in which phorbol esters and agents that increase the cellular content of cyclic AMP together enhance the expression of this gene (Comb *et al.*, 1986). In other instances, these signals may operate in an opposing direction as illustrated by the effects of IL-1 and PGE_2 on the synthesis of type I and III collagens. A few of the complexities to be considered in the interactions of ligands and their target cells in inflammatory lesions are listed in Table I.

The following chapters will provide a detailed view of several of the most important mediators that regulate the function and proliferation of inflammatory and noninflammatory cells. These chapters are the work of individuals who were pioneers in the discovery and characterization of these

Table I. Some Generalizations Concerning the Interaction of Ligands and Target Cells in Inflammation[a]

1. Any given cellular response is subject to multiple controls.
2. Different ligands may produce similar effects in the same target cells.
3. Different target cells may respond differently to the same ligand.
4. Different ligands may interact with the same or different receptors.
5. Stimulation of the same pathway of signal transduction may result in different cellular responses, depending on interactions with other molecules and the type of cell involved.

[a]Adapted in part from Barnes (1986).

mediators. It is important to place these polypeptides in the perspective of the way they act in a complex network filled with amplification and inhibition loops.

ACKNOWLEDGMENTS. Original work presented here was supported by United States Public Health Service Grants AM-03564, AM-07258, and AM-03490. I thank Michele Angelo for preparation of the manuscript.

References

Barnes, D. M., 1986, How cells respond to signals, *Science* **234:**286–288.

Beutler, B., and Cerami, A., 1986, Cachectin and tumour necrosis factor as two sides of the same biological coin, *Nature* **320:**584–588.

Comb, M., Birnberg, N. C., Seasholtz, A., Herbert, E., and Goodman, H. M., 1986, A cyclic AMP-and phorbol ester-inducible DNA element, *Nature* **232:**353–356.

Dayer, J.-M., Robinson, D. R., and Krane, S. M., 1977a, Prostaglandin production by rheumatoid synovial cells. Stimulation by a factor from human mononuclear cells, *J. Exp. Med.* **145:**1399–1404.

Dayer, J.-M., Russell, R. G. G., and Krane, S. M., 1977b, Collagenase production by rheumatoid synovial cells: Stimulation by a human lymphocyte factor, *Science* **195:**181–183.

Dayer, J.-M., Goldring, S. R., Robinson, D. R., and Krane, S. M., 1979, Effects of human mononuclear cell factor on cultured rheumatoid synovial cells. Interactions of prostaglandin E_2 and cyclic adenosine 3′, 5′-monophosphate, *Biochim, Biophys, Acta* **586:**87–105.

Dayer, J.-M., Beutler, B., and Cerami, A., 1985, Cachectin/tumor necrosis factor stimulates collagenase and prostaglandin E_2 production by human synovial cells and dermal fibroblasts, *J. Exp. Med.* **162:**2163–2168.

Dinarello, C. A., Cannon, J. G., Mier, J. W., Bernheim, H. A., LoPreste, G., Lynn, D. L., Love, R. N., Webb, A. C., Auron, P. E., Reuben, R. C., Rich, A., Wolff, S. M., and Putney, S. D., 1986, Multiple biological activities of human recombinant interleukin 1, *J. Clin. Invest.* **77:**1734–1739.

Goldring, M. B., and Krane, S. M., 1986, Modulation of collagen synthesis in human chondrocyte cultures by interleukin 1, *J. Bone Mineral Res.* **1** (Suppl. 1):56 (Abstract).

Goldring, S. R., Dayer, J.-M., and Krane, S. M., 1984, Rheumatoid synovial cell hormone responses modulated by cell–cell interactions, *Inflammation* **8:**107–121.

Goldring, S. R., Roelke, M. S., Petrison, K. K., and Krane, S. M., 1986, Interleukin 1 (IL 1) mediates effects of soluble monocyte–macrophage products on responses to prostaglandin E_2 (PGE_2); A potential mechanism for regulating cellular activity at sites of inflammation, *J. Bone Mineral Res.* **1** (Suppl. 1):55 (Abstract).

Goldring, S. R., Roelke, M. S., Petrison, K. K., Evins, A. E., and Krane, S. M., 1987a, Mechanisms by which monocyte–macrophage products regulate responses of connective tissue (bone and synovial) cells to hormones, *J. Bone Mineral Res.* **2**(Suppl.1):239 (Abstract).

Goldring, M. B., Birkhead, J. R., Sandell, L. J., and Krane, S. M., 1987b, Differential effects of recombinant interleukin 1 and phorbol ester on collagen synthesis and procollagen mRNA levels in cultured human chondrocytes, *Arthritis Rheum.* **30:**(Suppl.):129 (Abstract).

Gordon, M. Y., Riley, G. P., Watt, S. M., and Greaves, M. F., 1987, Compartmentalization of a haematopoietic growth factor (GM-CSF) by glycosaminoglycans in the bone marrow microenvironment, *Nature* **326:**403–405.

Gowen, M., Wood, D. D., and Russell, R. G. G., 1985, Stimulation of the proliferation of human bone cells in vitro by human monocyte products with interleukin-1 activity, *J. Clin. Invest.* **75:**1223–1229.

Krane, S. M., Goldring, S. R., and Dayer, J.-M., 1982, Interactions among lymphocytes, monocytes, and other synovial cells in the rheumatoid synovium, *Lymphokines* **7:**75–136.

Krane, S. M., Dayer, J.-M., Simon, L. S., and Byrne, M. S., 1985, Mononuclear cell-conditioned medium containing mononuclear cell factor (MCF), homologous with interleukin 1, stimulates collagen and fibronectin synthesis by adherent rheumatoid synovial cells: Effects of prostaglandin E_2 and indomethacin, *Collagen Relat. Res.* **5:**99–117.

McCroskery, P.A., Arai, S., Amento, E. P., and Krane, S. M., 1985, Stimulation of procollagenase synthesis in human rheumatoid synovial fibroblasts by mononuclear cell factor/interleukin 1, *FEBS Lett.* **191:**7–12.

Mizel, S. B., Dayer, J.-M., Krane, S. M., and Mergenhagen, S. E., 1981, Stimulation of rheumatoid synovial cell collagenase and prostaglandin production by partially purified lymphocyte-activating factor (interleukin 1), *Proc. Natl. Acad. Sci. U.S.A.* **78:**2474–2477.

Old, L. J., 1987, Polypeptide mediator network, *Nature* **326:**330–331.

Oppenheim, J. J., Kovacs, E. J., Matsushima, K., and Durum, S. K., 1986, There is more than one interleukin 1, *Immunol. Today* **7:**45–56.

Pytela, R., Pierschbacher, M. D., and Ruoslahti, E., 1985, A 125/115-kDa cell surface receptor specific for vitronectin interacts with the arginine–glycine–aspartic acid adhesion sequence derived from fibronectin, *Proc. Natl. Acad. Sci. U.S.A.* **82:**5766–5770.

Pytela, R., Pierschbacher, M. D., Ginsberg, M. H., Plow, E. F., and Ruoslahti, E., 1986, Platelet membrane glycoprotein IIb/IIIa: Member of a family of Arg-Gly-Asp-specific adhesion receptors, *Science* **231:** 1559–1562.

Ruoslahti, E., and Pierschbacher, M. D., 1986, Arg-Gly-Asp: A versatile cell recognition signal, *Cell* **44:**517–518.

Rupp, E. A., Cameron, P. M., Ranawat, C. S., Schmidt, J. A., and Bayne, E. K., 1986, Specific bioactivities of monocyte-derived interleukin 1α and interleukin 1β are similar to each other on cultured murine thymocytes and on cultured human connective tissue cells, *J. Clin. Invest.* **78:**836–839.

Schmidt, J. A., Mizel, S. B., Cohen, D., and Green, I., 1982, Interleukin 1, a potential regulator of fibroblast proliferation, *J. Immunol.* **128:**2177–2182.

Stephenson, M. L., Goldring, M. B., Birkhead, J. R., Krane, S. M., Rahmsdorf, H. J., and Angel, P., 1987, Stimulation of procollagenase synthesis parallels increases in cellular procollagenase mRNA in human articular chondrocytes exposed to recombinant interleukin 1β or phorbol ester, *Biochem. Biophys. Res. Commun.* **144:**583–590.

Yamada, K. M., Akiyama, S. K., Hasegawa, T., Hasegawa, E., Humphries, M. J., Kennedy, D. W., Nagata, K., Urushihara, H., Olden, K., and Chen, W.-T., 1985, Recent advances in research on fibronectin and other cell attachment proteins, *J. Cell. Biochem.* **28:**79–97.

6

Interleukin-1

Biology and Molecular Biology

STEVEN B. MIZEL

1. Introduction

Interleukin-1 (IL-1) was originally described as a macrophage-derived low molecular weight (13–17 kd) polypeptide that acts as a comitogen with phytohemagglutinin (PHA) or concanavilin A (Con A) in cultures of murine thymocytes (Gery *et al.*, 1972). Subsequent studies revealed that IL-1 was stimulatory for a broad range of T-cell-dependent *in vitro* immune responses including antibody synthesis (Koopman *et al.*, 1978; Staruch and Wood, 1983), cytotoxic T cell activation (Farrar *et al.*, 1980), and mitogen- and antigen-induced T cell proliferation (Gery *et al.*, 1972; Mizel and Ben-Zvi, 1980; Mizel, 1982; Dinarello, 1984; Oppenheim *et al.*, 1986). More recent studies indicate that IL-1 may also play an essential role in thymocyte maturation (DeLuca and Mizel, 1986). B cells, as well as T cells, have been characterized as IL-1 responsive. For example, IL-1 can induce the synthesis of kappa light chains and surface immunoglobulin expression in pre-B cells (Giri *et al.*, 1984) and function as a comitogen in B cell proliferation (Howard *et al.*, 1983). Within the last few years, it has become evident that the biologic activity of IL-1 is not restricted to lymphocytes. Indeed, this mediator can modify the growth and secretory activity of a large number of cell types, including fibroblasts, hepatocytes, endothelial cells, chondrocytes, natural killer cells, synovial cells, and osteoclasts (Dinarello, 1984; Oppenheim *et al.*, 1986), which share a common involvement in immune or inflammatory responses. The overlapping responsiveness to IL-1 may pro-

STEVEN B. MIZEL • Department of Microbiology and Immunology, Wake Forest University Medical Center, Winston-Salem, North Carolina 27103.

vide one mechanism for the coordinated control of the diverse cell types that are involved in an immune or inflammatory response.

Our understanding of IL-1 biology has been accompanied by dramatic advances in IL-1 biochemistry and molecular biology. Murine IL-1 was first purified in 1981, using standard chromatographic procedures (Mizel and Mizel, 1981). Anti-IL-1 antibodies were subsequently generated and were used to purify murine IL-1, with a significant increase in overall yield of active protein (Mizel *et al.*, 1983). Using anti-IL-1 antibodies, Giri *et al.* (1984) demonstrated that murine IL-1 is initially synthesized as a 33,000 molecular weight precursor that is enzymaticaly cleaved to the lower molecular weight forms detected in the culture medium of stimulated cells. More recently, human IL-1 has been purified from the conditioned medium of cultures of stimulated peripheral blood mononuclear cells (Kronheim *et al.*, 1985; Van Damme *et al.*, 1985; Cameron *et al.*, 1986) as well as from a monocytic leukemia cell line (Knudsen *et al.*, 1986). However, a major breakthrough in IL-1 research was achieved with the cloning, sequencing, and expression of cDNAs for murine (Lomedico *et al.*, 1984) and human (Auron *et al.*, 1984; March *et al.*, 1985; Gubler *et al.*, 1986) IL-1. These studies revealed the existence of at least two forms of human IL-1, termed IL-1α (pI 5) and IL-1β (pI 7). The two forms exhibit a relatively low degree of overall homology at the protein level (26%), with a somewhat higher percentage of homology at the nucleotide level (45%). The complete nucleotide sequence of the human IL-1α gene has been reported (Furutani *et al.*, 1986). In a recent study, Gray and co-workers reported on the isolation of a cDNA for a murine form of IL-1β (Gray *et al.*, 1986).

Although IL-1 was initially described as a product of monocytes and macrophages, several studies have demonstrated that IL-1 or IL-1-like molecules are produced by a variety of cell types including Epstein–Barr virus-transformed B cells, normal B cells, dendritic cells, endothelial cells, mesangial cells, neutrophils (Dinarello *et al., 1984;* Oppenheim *et al.*, 1986), and even an IL-1-dependent T cell line (Oppenheim *et al.*, 1986; S. B. Mizel, unpublished observations). However, little if anything is known about the secretion and extracellular processing of IL-1 or IL-1-like molecules from these cell types. More generally, the physiological relevance of IL-1 release from these alternative cell sources is not presently clear. Does IL-1 play a general role in the basic communication pathways that exist between diverse types of immune and inflammatory cells? Additional studies are required to answer this question.

In contrast to many other aspects of IL-1 research, studies on the biochemical and molecular events that follow from the binding of IL-1 to its receptor are rather limited at this time. Stanton *et al.* (1986) have reported that IL-1 can stimulate an influx of Na^+ into 70Z/3 cells, an IL-1-responsive pre-B cell line, via stimulation of the amiloride-sensitive Na^+/H^+ antiport. Farrar and Hume (1985) recently reported that IL-1 can enhance the synthesis of 5- and 15-monohydroxyeicosatetraenoic acid (HETE), products of the lipoxygenase pathway, in a subclone of EL-4 thymoma cells. On the basis of their findings with pharmacological inhibitors, Farrar and Hume (1985) have advanced the view that 5-HETE and 15-HETE may play an important role in IL-1-mediated T cell activation. Dinarello and colleagues (1983) have also suggested that products of the lipoxygenase pathway may be involved in IL-1-stimulated thymocyte proliferation. However, our studies with the same set of inhibitors have revealed a direct correlation between the concentrations required for inhibition of IL-1-action and cellular toxicity (S. White and S. B. Mizel, unpublished observations). The results of Farrar and Hume (1985) and Dinarello *et al.* (1983) must thus be interpreted with some caution. Nonetheless, it is clear from several studies that IL-1 can stimulate lipid metabolism, for example, PGE_2 and prostacyclin synthesis (Mizel *et al.*, 1981; Postlethwaite *et al.*, 1983; Rossi *et al.*, 1985; Dukovich *et al.*, 1986). Additional studies are required to determine the precise mechanism(s) by which IL-1 participates in the activation of its target cells.

2. *IL-1 and T Cell Activation*

Is IL-1 essential for T cell activation, or does it simply serve to enhance responses initiated by the physical association of antigen-presenting cells and T cells? Since the initial description in 1973 of IL-1 as a comitogen for thymocyte proliferation, a large number of studies have focused on various aspects of IL-1 and T cell activation. However, in almost every case the studies have examined the effect of *exogenous* IL-1 and not *endogenous* IL-1. For example, in our studies on antigen-induced proliferation (Mizel and Ben-Zvi, 1980), we evaluated the effect of IL-1 on adherent accessory-cell-depleted cultures of lymph node T cells prepared from mice immunized with ovalbumin. The proliferation of these cells in response to ovalbumin was markedly reduced relative to unfractionated cells (Table I). However, the addition of IL-1 and antigen resulted in a complete restoration of antigen-specific T cell proliferation. If the T-cell-enriched population was "cleansed" of residual macrophages using a highly specific antimacrophage antibody preparation, IL-1 and antigen were no longer capable of restoring a proliferative response—clearly demonstrating a need for accessory cell antigen presentation. Similar experiments have been done with PHA or Con-A-induced splenic T cell proliferation: IL-1 has a restorative effect in macro-

Table I. Dependence of IL-1-Mediated Activation of Lymph Node Lymphocytes on Residual Macrophages

Lymph node lymphocytes	PC_x [b]	Additions	Δcpm[^{3}H]Tdr incorporation
Untreated	—	None	1,929 ± 602
		OVA[c]	3,903 ± 803
		IL-1[d]	2,420 ± 881
		OVA + IL-1	15,733 ± 2258
AMS treated[e]	—	None	587 ± 81
		OVA	1,833 ± 766
		IL-1	2,960 ± 1964
		OVA + IL-1	3,310 ± 1628
AMS treated	10^4	None	977 ± 631
		OVA	1,167 ± 469
		OVA + IL-1	15,082 ± 1227

[a] Adapted with permission from Mizel and Ben-Zvi (1981).
[b] Irradiated peritioneal macrophages from unprimed syngeneic mice.
[c] 50 μg/mL OVA.
[d] 25 units/mL IL-1
[e] Macrophage-depleted lymph node lymphocytes (1×10^7) were incubated for 30 min at 37°C with a 1 : 16 dilution of antimacrophage antibodies (AMS) and then for an additional 60 min with a 1 : 10 dilution of guinea pig complement. The cells were washed three times and resuspended to 2×10^6 viable cells/mL. Cells (4×10^5) were plated in each well and incubated for 72 h.

phage-depleted cultures. But this type of experiment does not answer a fundamental question: Is endogenously produced IL-1 essential to the induction of a response in unfractionated cultures of spleen cells or lymph node lymphocytes?

To evaluate this question, we examined the ability of anti-IL-1 IgG antibodies (Mizel *et al.*, 1983) to block antigen- and mitogen-induced proliferation in cultures of unfractionated spleen cells or lymph node lymphocytes. Our results clearly indicated that lectin-induced spleen cell proliferation is completely unaffected by concentrations of anti-IL-1 antibodies that would completely block the activity of exogenously supplied IL-1 (Table II). Identical results were obtained with proliferative responses induced with soluble protein antigens or in mixed lymphocyte reactions. Thus, one is left to consider two possibilities: IL-1 is not essential for T cell activation but rather plays an "adjuvant" role (and can thus function in a restorative capacity in the absence of sufficient numbers of antigen-presenting accessory cells), or IL-1 reaches the T cell by a mechanism that prevents antibody engagement. We are currently exploring experimental approaches to distinguish between these possible explanations.

In contrast to the situation with peripheral T cell activation, we have obtained evidence, in conjunction with Dominick DeLuca (Department of Biochemistry, Medical University of South Carolina), demonstrating that

Table II. Lack of Effect of Anti-IL-1 Antibodies on Mitogen-Induced Splenic T Cell Proliferation

Additions[a]	Δcpm[^{3}HTdR incorporation
PHA	113,314 ± 3720
PHA + anti-IL-1	116,144 ± 3864
Con A	139,254 ± 689
Con A + Anti-IL-1	134,955 ± 855

[a] C3H/HeJ spleen cells (2×10^5) were incubated with either 1.5 μg/mL Con A in the presence or absence of 30 μg/mL anti-IL-1 IgG for 72 h. During the last 6 h, the cells were pulsed with [^{3}H]TdR.

IL-1 may play an essential maturational role in the fetal thymus. In a prior study, DeLuca (1985) established a fetal thymus organ culture system using fetal thymus lobes from 14-day gestation mice and demonstrated that the development of functional T cells in these cultures was self-restricted. Furthermore, DeLuca showed that monoclonal anti-Ia antibodies inhibited the development of functional T cells as well as the expression of Ia on nonlymphoid cells recovered from fetal thymus organ cultures. In our collaborative study with DeLuca (DeLuca and Mizel, 1986), we found that the addition of purified recombinant IL-1 to antigen-Ia-treated cultures reversed the inhibition of T cell growth induced by anti-Ia (Table III). Il-1 also inducd the reexpression of Ia on the surfaces of nonlymphoid cells that could be recovered from the cultures. Furthermore, this action of IL-1 could be blocked by antibodies to gamma interferon (D. DeLuca and S. B. Mizel, unpublished observations), indicating that the lymphokine cascade (IL-1, Il-2, ǵamma interferon) may be operative in thymic maturation. Most importantly, however, we found that anti-IL-1 antibodies could inhibit T cell development in fetal thymus organ cultures and that the addition of IL-1 (in

Table III. Restoration of Thymocyte Development in Organ Cultures Treated with Anti-I-a by the Addition of Exogenous IL-1[a]

Experiment	Treatment	Cells ($\times 10^4$)	Control (%)
1	None	29.8	100
	Anti-I-a^k [b]	11.9	40.0
	Anti-I-a^k + IL-1[c]	27.0	90.6
2	None	29.8	100
	Anti-I-a^k [b]	7.9	26.5
	Anti-I-a^k + IL-1[c]	32.0	107.4

[a] Adapted with permission from DeLuca and Mizel (1986).
[b] Anti-I-a^k (0.08 μg/lobe), plus a amount of medium equal to the volume of IL-1 added, was added to CBA fetal thymus lobes on day 8 of culture.
[c] IL-1 (36 U) in 9 μL of RPMI 1640 was added to the cultures on day 8 for a final concentration of 12 U/mL.

Table IV. Inhibition of Thymocyte Development and MLC Reactivity by Anti-IL-1[a]

Experiment	Treatment	Cells (×10⁴)	MLC activity (cpm ± SEM)		
			Syngeneic	Allogeneic	Ratio
1	Control IgG[b]	48.9	1131 ± 28	3022 ± 71[c]	2.7
	Anti-IL-1[b]	28.8	1041 ± 366	259 ± 88	0.2
2	Goat IgG[b]	10.2	1041 ± 153	3424 ± 255[d]	3.3
	Anti-IL-1[b]	5.9	2443 ± 103	2048 ± 487	0.8

[a] Adapted with permission from DeLuca or Mizel (1986).
[b] Anti-IL IgG or control IgG (12 μ/mL).
[c] Cells derived from CBA fetal thymus organ cultures responding to mitomycin-C-treated BALB/c adult spleen stimulators.
[d] Cells derived from BALB/c fetal thymus organ cultures responding to mitomycin-C-treated CBA adult spleen stimulators.

excess of antibody) could circumvent the effect of anti-IL-1 antibodies (Table IV). These findings indicate that endogenously produced IL-1 may play an essential role in the growth and development of functional T cells in the fetal thymus.

Several studies have demonstrated that IL-1 can serve as a stimulant for the induction of IL-2 synthesis. For example, Gillis and Mizel (1981) found that the murine T cell line LBRM 33 1A5 is not capable of IL-2 production when incubated only with PHA. However, in the presence of PHA and IL-1, IL-2 production was markedly enhanced. In view of this observation and the well-characterized mitogenic action of IL-2 on T cells, it was generally assumed that the role of IL-1 in T cell proliferation was limited to its ability to induce IL-2 production. In view of the results of several recent studies on IL-1 and T cell activation, this hypothesis appears to be too limited.

In a collaborative study with Daniela Männel, Tibor Diamantstein, and Werner Falk (Männel *et al.*, 1985), we found that IL-1 and IL-2 exhibit a synergistic effect on thymocyte proliferation (Fig. 1). Indeed, the action of these two mediators occurs in the absence of PHA or Con A. A key feature of the IL-1–IL-2 synergism is the induction of IL-2 receptors on the responding cells. Fractionation studies revealed that the target cells reside in the peanut agglutinin-negative medullary fraction (Table V). These results indicate that IL-1 may do more than simply serve as one signal for the induction of IL-2 synthesis. Although our results indicate that IL-1 and IL-2 can induce the appearance of IL-2 receptors on thymocytes, there is some question as to whether IL-1 plays a role in the induction of these receptors on peripheral T cells. It is quite possible that some T cell populations may respond to IL-1 as one contributing signal in the induction of IL-2 receptors. For example in a collaborative study involving several laboratories (Kaye *et al.*, 1984) we found that D10.G4.1, a cloned T cell line with specificity for conalbumin,

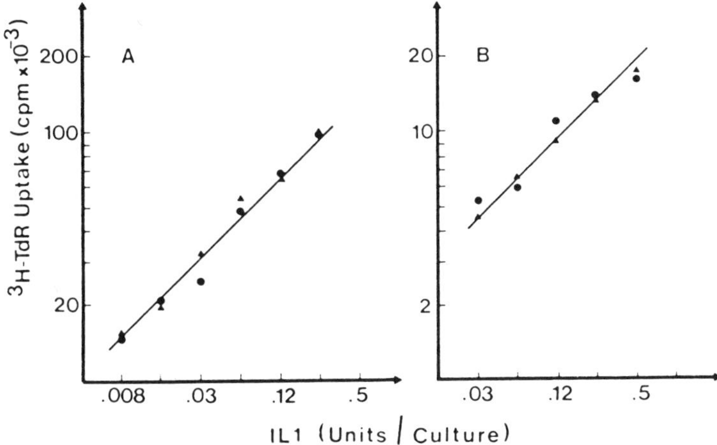

Figure 1. Induction of proliferative responses of (A) 3×10^5 and (B) 1×10^5 C3H/HeJ thymocytes with twofold serial dilution of purified IL-1. In (A), 10 U of partially purified murine IL-2 (●) or 5 U of recombinant human IL-2 (▲), and in (B), 5 U (●) or 20 U (▲) of recombinant human IL-2 were present in the cultures. Reprinted with permission from Männel *et al.* (1985).

exhibited the ability to generate IL-2 receptors in response to IL-1 and an anti-T cell receptor monoclonal antibody (3D3) (Fig. 2). These cells also exhibit a strong proliferative response in the presence of IL-1 and the 3D3 antibody or IL-1 and Con A. The conclusion that IL-1 enhancement of T cell activation may involve several actions is consistent with the observation that the enhancement of antibody synthesis by IL-1 may also involve a spectrum of actions, including a direct stimulatory action on B cell maturation (Giri *et al.*, 1984) and B cell growth (Howard *et al.*, 1983), as well as participating—either in an obligatory manner or as an adjuvant—in the induction and amplification of antigen-specific helper T cell populations.

Table V. IL-1-Induced Proliferation of Thymic Subpopulations[a]

Sample[b]	Unsorted	PNA$^-$	PNA$^+$
IL-2	7.7±0.7	16.6±1.3	1.2±0.1
IL-1	0.3±0.0	0.9±0.2	0.8±0.1
IL-2+IL-1	127.1±1.9	197.8±2.1	2.8±0.8
IL-2+Con A	85.8±2.8	141.8±7.0	2.3±0.1

[a] Adapted with permission from Männel *et al.*, (1985). Results are given as cpm ($\times 10^{-3}$) ± SD [^{3}H]TdR uptake.

[b] Cultures of 10^5 unseparated or PNA$^-$ or PNA$^+$ C3H/HeJ thymocytes were supplement with recombinant human IL-2 (20 U/mL) and/or purified murine IL-1 (0.5 U/mL) and/or Con A (1 μg/mL).

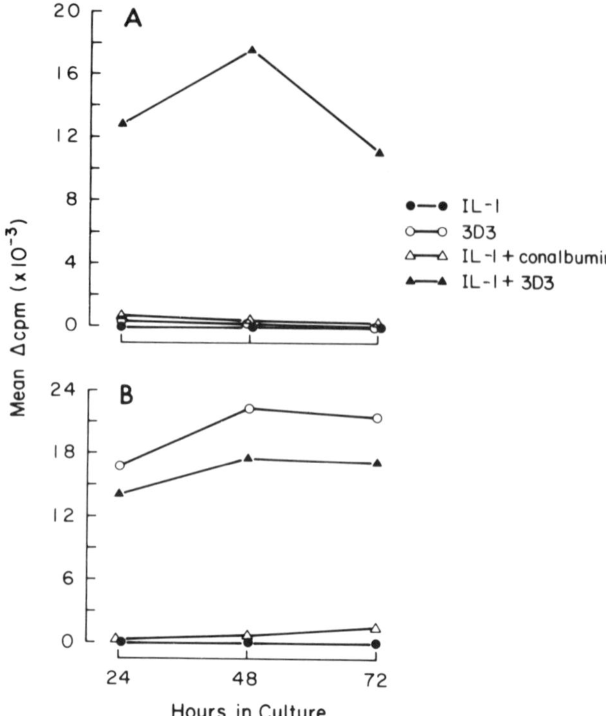

Figure 2. Kinetics of 3D3-induced IL-2 production and proliferation. (A) D10.G4.1 proliferation was assayed in 0.2-mL cultures containing 2×10^4 cloned T cells and IL-1 (●), 10 ng/mL 3D3 (○), IL-1 plus 200 μg/mL conalbumin (△), or IL-1 plus 10 ng/mL 3D3 (▲). Cells were pulsed with [^{3}H]TdR for the final 4 h of a 24-, 48-, 72-h culture period. (B) IL-2 production by D10.G4.1 was measured by a 24-h HT-2 proliferation assay. Reprinted with permission from Kaye *et al*. (1984).

From the previous discussion, it is evident that studies on the effect of IL-1 on T cell activation involve experiments in which IL-1 is examined as one of two required stimulatory signals, for example, in conjunction with PHA or Con A, or antigen, or IL-2. Can IL-1 have any effect on peripheral T cells in the absence of these other stimulatory molecules? Our earlier studies on the effect of IL-1 on the generation of human peripheral blood stable E-rosette-forming T cells (Ben-Zvi *et al*., 1981) support the notion that IL-1 can modulate at least one stage of T cell activation in the absence of antigen, mitogen, or exogenous IL-2.

The great majority of human T cells form rosettes with sheep red blood cells (E-rosettes) at 4°C. These rosettes readily dissociate when the temperature is raised to 37°C. However, incubation of cells in the presence of Con A or PHA or a number of other T-cell-activating substances results in the generation of large numbers of T cells expressing the capacity to form E-rosettes that are stable at 37°C. These findings have led to the proposal that the appearance of stable E-rosette receptor sites may be a reflection of T cell activation. We therefore decided to explore the effect of IL-1 on the appearance of this activation-associated receptor. We incubated human peripheral blood T cells with or without IL-1 for varying periods and determined the degree of stable E-rosette-forming cells. As shown in Table VI, IL-1 has a very pronounced effect on the appearance of this receptor; the frequency of these cells is in the same range (~60%) found in patients with rheumatoid arthritis, acute lymphatic leukemia, or chronic active hepatitis. Thus, our results indicate that IL-1 by itself can have an effect on the activation of peripheral T cells. However, we know very little about the relation of IL-1-induced activation to the overall requirements for moving resting T cells to IL-2 and IL-2 receptor synthesis and expression.

3. *IL-1 Receptor Studies*

Dower *et al.* (1985, 1986) provided the first evidence for the existence of a high-affinity receptor for IL-1. Their observations have been confirmed in several other recent studies (Kilian *et al.*, 1986; Lowenthal and MacDonald, 1986; Matushima *et al.*, 1986). Our studies (Mizel *et al.*, 1987) have focused on the events that follow the binding of IL-1 to its surface receptor. Using recombinant murine IL-1 in our binding studies, we have demonstrated that both EL-4 cells (Fig. 3) and Swiss 3T3 fibroblasts possess a single class of trypsin-sensitive, high-affinity receptors for IL-1. When cells are incubated with IL-1, the IL-1 receptor undergoes extensive ligand-

Table VI. Effect of IL-1 on the Generation of Stable E-Rosette-Forming Human T Cells[a]

Time of incubation[b] (h)	Stable E-rosette-forming cells[c] (%)	
	−IL-1	+IL-1
0	0	0
6	0	10
24	2±2	26±1
48	6±2	55±1
72	5±1	66±2

[a] Adapted with permission from Ben-Zvi *et. al.* (1981).
[b] 10^5 mononuclear cells were incubated with or without IL-1.
[c] The sample were tested for 37°C E-rosette-forming activity with sRBC. All cultures were incubated in triplicate. The results presented are the mean ± SE of two separate experiments.

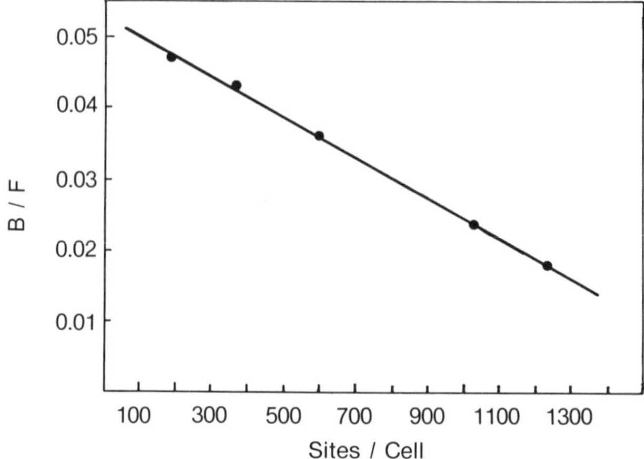

Figure 3. [^{125}I]IL-1 binding to EL-4 cells. (6×10^5/tube) were incubated with varying concentrations of labeled recombinant murine IL-1 in the presence or absence of an excess of unlabeled recombinant murine IL-1. Bound IL-1 was separated from free IL-1 by centrifugation through a silicone oil mixture.

induced down regulation. IL-1 itself is internalized at 37°C, with 50% of the surface-bound IL-1 internalized in 60–120 min. IL-1 does not undergo degradation for at least 6 h after internalization. Using electron microscopy and autoradiography, we observed several important features of the internalization process. When cells that bound [^{125}I]IL-1 at 4°C are shifted to 37°C, IL-1 moves from the cell membrane to the cytoplasm, where it is found in proximity to nuclei or within lysosomes. As the IL-1 moves out from the lysosome, it progressively accumulates in nuclei. Six hours after shifting cells to 37°C, 30–35% of the internalized [^{125} I]IL-1 is associated with the cell nucleus. When nuclei are isolated from such cells, they do indeed contain radiolabeled IL-1. The accumulation of relatively high levels of IL-1 in the nucleus raises the interesting possibility that IL-1 may interact in a specific manner not only with cell surface receptors but also with potentially important nuclear receptors. However, considerable work is required before any definitive conclusions can be drawn about the possible action of IL-1 in the nuclei of IL-1-responsive cells.

Unlike the IL-2 receptor, the IL-1 receptor appears to be constitutively produced by IL-1-responsive cells. In our studies with EL-4 cells, we tested the ability of gamma-interferon (IFN-γ, phorbol myristate acetate, the calcium ionophore A23187, and Con A to induce an increase in the expression of IL-1 receptors (Table VII). Our results were negative in each case. However, in a related series of experiments, we explored the ability of the

Table VII. Lack of Inducibility of the EL-4 – IL-1 Receptor

Stimulant	Specific [^{125}I]IL-1 binding (cpm)
None	1217
Con A (5 μg/mL)[a]	1053
Gamma interferon (100 units/mL)[a]	903
PMA (1×10^{-8} M)[b]	1296
A23187 (0.67 μg/mL)[b]	1187
PMA + A23187[b]	1154

[a] Overnight incubation at 37°C prior to analysis of IL-1 receptor levels.
[b] Four hours at 37°C prior to analysis of IL-1 receptor levels.

5-methyl cytosine analogue, azacytidine, to modify IL-1 receptor expression on EL-4 cells (M. Chedid and S. B. Mizel, unpublished observations). Azacytidine is incorporated into DNA but cannot be methylated (see review by Doerfler, 1983). This compound specifically inhibits methylation of cytosine residues in DNA and RNA and can cause the activation of previously quiescent genes or enhance the transcriptional activity associated with genes that are only weakly active. When EL-4 cells were incubated with 3×10^{-6} M azacytidine for 3 days and the level of IL-1 was analyzed in binding experiments (Fig. 4), we detected a significant increase in the expression of IL-1 receptors. The overall increase ranged from twofold to fourfold when considered in terms of the total number of cells in each assay. However, when we examined the expression of IL-1 receptors using fluorescein-labeled IL-1 and a FACS, we found that only 40% of the cells were viable after 3 days of azacytidine treatment, and the treated viable cells were clearly responsible for the increased expression of IL-1 receptors. Thus, the overall effect of azacytidine was to induce at least a tenfold increase in the expression of IL-1 receptors on cells surviving the exposure to this compound. When the IL-1 receptor gene is cloned, it will certainly be of interest to characterize the role of DNA methylation in the control of IL-1 receptor gene expression.

4. IL-1 Secretion and Extracellular Processing

In addition to our studies on the biology of IL-1 and its receptor, we have also focused on the fundamental problem of IL-1 synthesis and secretion. Our findings indicate that IL-1 may be one of a very small group of proteins that are released intact from viable cells and undergo processing from a precursor form in the extracellular environment (Suttles *et al.*, 1987).

The first step in the processing of precursors or many secreted proteins is the removal of an N-terminal signal peptide by a signal peptidase during

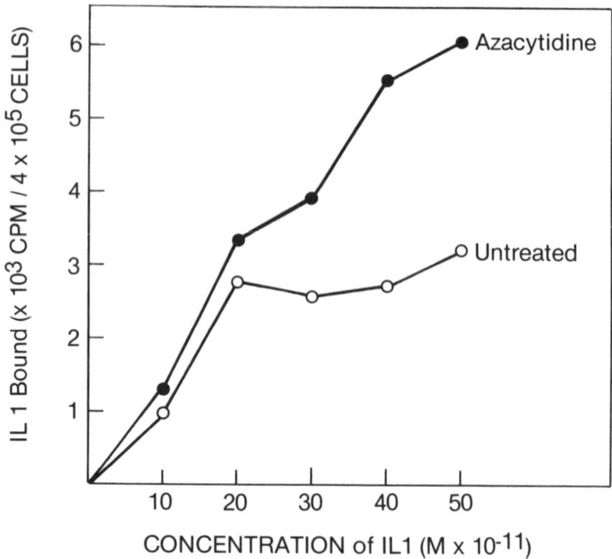

Figure 4. Effect of azacytidine on the expression of IL-1 receptors for EL-4 cells. EL-4 cells were incubated for 3 days with or without 3×10^{-6} M azacytidine, then washed and assayed for the level of IL-1 receptors.

translation and translocation into the lumen of the endoplasmic reticulum (Garoff, 1985). Cotranslational processing of this type can be studied *in vitro* by supplementing a rabbit reticulocyte lysate translation system with dog pancreas microsomal membranes. Although IL-1 has been thought of as a secretory hormone, an examination of amino acid sequences derived from the nucleotide sequences of the human and murine IL-1 precursor cDNAs (Auron *et al.*, 1984; Lomedico *et al.*, 1984; March *et al.*, 1985; Gray *et al.*, 1986) revealed that none of the IL-1 precursors exhibits the type of hydrophobic N-terminal signal peptide sequence that is characteristic of most secreted hormones and lymphokines. This observation prompted us to evaluate the issue of the presence or absence of a unique type of signal peptide in the IL-1 precursor. Poly-A$^+$ RNA was isolated from stimulated normal and cell line murine macrophages and translated in a rabbit reticulocyte lysate translation system in the presence or absence of dog pancreas microsomal membranes. There was no discernible difference in the molecular weight of the IL-1 synthesized in the presence of absence of microsomal membranes. In addition, there was no protection against trypsin degradation in the presence of the microsomal membranes. These results demonstrate that in addition to the absence of signal peptide cleavage there was no translocation of the IL-1 into the microsomes. In contrast, signal peptide cleavage and movement of another protein, prelactogen, into a trypsin-insensitive environment were detected in parallel experiments. The absence of any difference in the size of the IL-1 precursor obtained from biosynthetically labeled cells and the primary translation produce obtained *in vitro* in reticulocyte lysates also indicate that signal peptide processing is not taking place.

In addition to these experiments, we have also examined the fate of newly synthesized IL-1 precursor prior to processing and release from cells. In these studies, murine macrophages were stimulated to synthesize IL-1 (in the presence of [^{35}S]methionine) and subsequently fractionated into particulate and soluble fractions. We found that the majority of the newly synthesized IL-1 precursor remains in the soluble fraction of the cells for at least 4 h prior to associating with particulate structures (J. Giri and S. B. Mizel, unpublished observations). IL-1 release begins at this time. By 19 h, approximately 50% of the cell-associate IL-1 is associated with the particulate fraction of the cells. It thus appears that the IL-1 precursor does not follow the pathway of processing and secretion that is generally associated with secretory proteins (Garoff, 1985). Furthermore, it is quite likely that IL-1 may not be synthesized on membrane-bound polysomes. However, this issue has not yet been examined directly.

In very recent studies using a new method for stimulating the rapid release of IL-1 from macrophages, we have gained new insight into the pathway of IL-1 secretion and processing (Suttles *et al.*, 1987). In our earlier studies (Giri *et al.*, 1984), we found that the synthesis of the 33,000 molecular weight (MW) IL-1 precursor was essentially complete within 3–5 h after stimulating the cells. However, the appearance of low-molecular-weight IL-1 in the culture medium occurred over a far longer time course (12–24 h). In our initial analysis of the problem, we favored the view that the 33,000 MW IL-1 precursor might itself be released from cells and then processed to the 17,000–19,000 MW forms found in the culture medium (which is rich in secreted proteases such as elastase, collagenase, and plasminogen activator). We have now obtained evidence, using normal and cell line macrophages, that strongly supports this view.

We decided to examine the possible involvement of calcium in the release pathway for the IL-1 precursor. Our current studies analyze the form of IL-1 present in culture supernatants at very early times, that is, within the first minutes or hours after stimulating the cells with ionophore. By using culture supernatants that had been rapidly concentrated at low temperature, we increased our ability to detect very low levels of released ^{35}S-labeled IL-1. Normal and cell line macrophages were induced to synthesize the IL-1 precursor in the presence of [^{35}S]methionine, washed, and then incubated for 15 min in the presence or absence of A23187 or ionomycin—two well-defined calcium ionophores. The culture supernatants were collected and analyzed by immunoprecipitation–SDS gel electrophoresis for the release of processed IL-1. In the absence of either ionophore, stimulated cells released very low levels of IL-1 during the short incubation period. Nonetheless, we were able to detect released IL-1 and establish its predominant molecular weight as 33,000, not 17,000–19,000. With time, however, the 33,000 MW form was converted to the lower molecular weight forms. In contrast, incubation of cells in the presence of A23187 or ionomycin resulted in a dramatic increase in the appearance of the 33,000 and 17,000–19,000 MW forms of IL-1. The effect of the ionophores was blocked when cells were incubated in the presence of calcium chelating agents such as EDTA or EGTA. In contrast, the sodium ionophore, monensin, had no effect on IL-1 release. The action of calcium ionophores was quite rapid: within 1 min after addition to the cells, the 33,000 MW IL-1 was detected in the culture medium. During the next few minutes, the level of released 33,000 MW IL-1 continued to increase and the 17,000–19,000 MW forms began to appear. We have detected a number of additional extracellular intermediates between the 33,000 MW and 17,000–19,000 forms. These observations firmly establish that the 33,000 MW form of IL-1 is not processed intracellularly but is in fact released intact and subsequently converted to lower molecular weight forms by released or cell-surface-associated proteases. We are currently studying the nature of the extracellular cleavage events as well as the mechanism of calcium-ion-enhanced IL-1 secretion and extracellular processing.

In view of the possible existence of a membrane form of IL-1 (Kurt-Jones *et al.*, 1985a,b; Nagelherken and van Breda Vriesman, 1986), we evaluated the possibility that the released IL-1 was derived from a cell-surface-associated pool. Cells were first stimulated in the presence of labeled methionine to synthesize IL-1, then washed and treated with trypsin prior to addition of ionophore; the logic being that trypsin would remove any cell surface IL-1 and thus deplete the pool of material that was released in response to ionophore treatment. However, trypsin had no discernible effect on the level of released IL-1 following ionophore treatment. It thus appears quite likely that the released IL-1 is not derived from an external surface membrane pool. We also explored the possibility that the release of IL-1 was dependent on the integrity of microtubules and/or microfilaments. Cells were incubated with ionophore in the presence or absence of maximally disruptive concentrations of colchicine or cytochalasin B. The results of these experiments indicated that neither type of cytoskeletal structure was essential for ionophore-mediated IL-1 release.

Based on the experimental results described above, we favor the following view of IL-1 release and processing (Fig. 5):

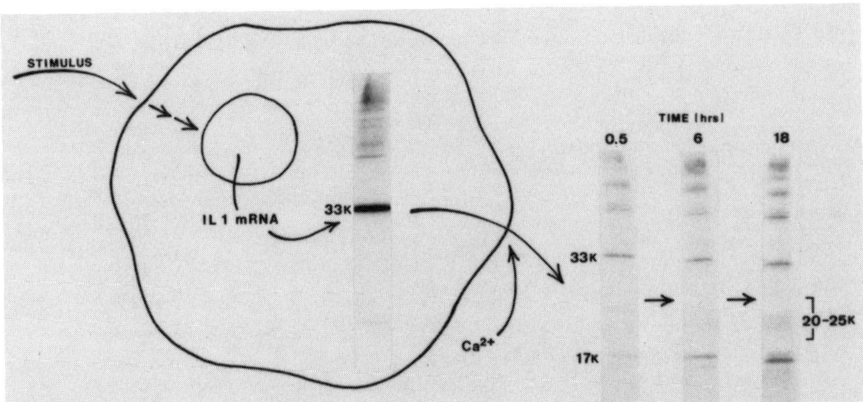

Figure 5. Pathway of secretion and extracellular processing of murine IL-1. Following stimulation, IL-1 mRNA is produced in murine macrophages and the 33,000 MW IL-1 precursor is synthesized. The complete precursor is then released from the cells, perhaps by a calcium-ion-dependent mechanism, and processed extracellularly through a series of intermediates ranging in molecular weight from 25,000 to 17,000.

1. Stimulation of cells results in a relatively rapid increase in the synthesis of IL-1 mRNA and the subsequent appearance of the 33,000 MW IL-1 precursor. The steady-state level of IL-1 mRNA reaches a peak within 3–5 h after stimulating the cells and then declines over the next 8–12 h. Our results indicate that the synthesis of the 33,000 MW form is almost complete within 5 h after stimulation of the cells.
2. Once the 33,000 MW precursor is released from the cells, it is extracellularly converted to 17,000–19,000 MW forms—a process that involves a number of intermediates.

5. Membrane IL-1

Several recent studies (Kurt-Jones *et al.*, 1985a, b; Nagelherken and van Breda Vriesman, 1986) have presented data in support of the existence of a membrane form of IL-1 (Table VIII). We have further examined this question using a murine macrophage cell line, $P388D_1$. Our results indicate that fixed, LPS-stimulated $P388D_1$ cells retain the ability to replace IL-1 in the murine thymocyte comitogenesis assay. Using these cells, we attempted to iodinate (using lactoperoxidase and galactose oxidase or iodogen) cell surface IL-1. Although a large number of surface proteins were iodinated, IL-1

Table VIII. Paraformaldehyde-Fixed Macrophages Have Cell-Associated IL-1[a]

	Δcmp[^{3}H]TdR incorporation[b]	
Listeria-pulsed, fixed macrophages	Thymocytes (PHA)[c]	*Listeria* T cells[d]
No addition	13,459 ± 2864	29,626 ± 1874
Control Ig, 25 μg/mL	11,027 ± 948	31,172 ± 2992
Anti-IL-1, 5 μg/mL	6,864 ± 874	12,236 ± 646
Anti-IL-1, 25 μg/mL	1,926 ± 211	1,283 ± 83
+ 2 units of IL-1/mL	5,576 ± 1041	9,392 ± 728
+ 4 Units of IL-1/mL	9,440 ± 508	21,644 ± 1483

[a] Adapted with permission from Kurt-Jones *et al.* (1985a).
[b] Data shown are as mean cpm ± SD of [^{3}H]TdR incorporation of triplicate cultures incubated for 48 h at 37°C, pulsed, and harvested 18–24 h later.
[c] A/J thymocytes (10^6 per well) with 5 μg/mL of PHA in 200 μL. Background [^{3}H]TdR incorporation was 900–1200 cpm with PHA.
[d] *Listeria* T cells (4×10^4 per well) in 200 μL. Background [^{3}H]TdR incorporation was 50–110 cpm.

was not among them. However, intracellular IL-1 could be easily iodinated. The lack of iodination is rather surprising if the putative surface IL-1 is accessible for biologic activity. It is also important to note that the biologically relevant portion of the IL-1 molecule has several available tyrosine residues.

In view of our negative results with surface iodination, we were also surprised by our observation that we could obtain IL-1 activity from mildly trypsinized, LPS-activated P388D$_1$ cells (the viability of the cells was unaffected by trypsin treatment). This activity was fully sensitive to anti-IL-1 antibodies. Obviously, the presence of a trypsin-sensitive form of IL-1 on cells would argue for the existence of a membrane form of IL-1. However, in subsequent experiments, we found that mild trypsin treatment actually serves as a stimulus for the rapid release of IL-1 from the intracellular environment. This obseration certainly clouds the interpretation of the initial experiments with trypsin. Furthermore, in a recent study on the immunocytochemical detection of IL-1 in human monocytes, Bayne *et al.* (1986) did not detect any surface immunofluorescence on stimulated monocytes when they used anti-IL-1 peptide antibodies. However, they did observe significant intracellular fluorescence with these same antibody preparations. Given these results, as well as the observations that IL-1 lacks a signal peptide, that it does not appear to undergo cotranslational movements into microsomes, and that it is found in the soluble fraction of cells immediately following its synthesis, we remain rather skeptical about th existence of a true membrane form of IL-1. It is quite possible that "membrane" IL-1 is in fact IL-1 in transit through the plasma membrane.

6. IL-1 and Inflammatory Cells

As evidenced by the results of numerous studies, IL-1 can enhance the functional activities and growth of a wide variety of inflammatory cells. Our studies have focused primarily on fibroblasts and the induction of proliferation and prostaglandin synthesis. Using purified recombinant IL-1, we have clearly demonstrated that this mediator can markedly enhance fibroblasts proliferation and prostaglandin production (Dukovich *et al.*, 1986). The effect of IL-1 on prostaglandin synthesis is concentration dependent and is first detected approximately 6 h after treating fibroblasts with IL-1 (S. B. Mizel, unpublished observations). Furthermore, active protein synthesis appears to be essential for IL-1 stimulation of prostaglandin synthesis.

In addition to fibroblasts, we have also demonstrated, in a collaborative study with Harvey Colten and colleagues, that IL-1 can induce a dose-dependent increase in serum amyloid A specific mRNA and an increase in serum amyloid A plasma protein concentration (Ramadori *et al.*, 1985). In primary hepatocyte cultures, IL-1 induced a dose- and time-dependent reversible increase in expression of the serum amyloid A and factor B genes and a decrease in albumin gene expression. Similar results have been obtained with human hepatoma cells (Perlmutter *et al.*, 1986). These observations indicate that a single mediator, IL-1, can induce increases and decreases in the expression of genes encoding liver-derived plasma proteins that are affected during the acute-phase response.

Several studies (Klempner *et al.*, 1979; Sauder *et al.*, 1984; Smith *et al.*, 1986) have suggested that IL-1 can modulate the functional activity of polymorphonuclear leukocytes (PMNs). In collaboration with Janelle A. Rhyne and Charles E. McCall, we have critically examined the possible action of IL-1 on PMNs using purified recombinant human IL-1. Our results clearly demonstrate that IL-1 does not have a direct effect on PMN: IL-1 did not stimulate degranulation, respiratory burst, chemotaxis, or glucose transport. In view of the lack of IL-1 action on PMNs, we were surprised to find that these cells express high-affinity IL-1 receptors (800 sites/cell $K_d = 0.74$ nM). Our results indicate that the action of IL-1 on acute neutrophilic inflammation must involve an intermediary cell population. The role of IL-1 receptors on PMNs is not known at this time. However, the possibility exists that these receptors may be involved at some earlier stage in PMN maturation.

7. Unanswered Questions

Progress in understanding the biology, biochemistry, and molecular biology of IL-1 has been remarkable during the last 5 years. The discovery of an IL-1 precursor, the cloning of multiple human and murine IL-1 genes, the characterization of the IL-1 receptor, and the elucidation of the broad range of IL-1 cell targets have not only provided us with a wealth of new data but have also moved this area of research from the confines of phenomenology to the broader realm of cellular and molecular endocrinology. But many questions remain to be answered.

The list of IL-1 cell targets is very large, extending from thymocytes to chondrocytes, B cells to natural killer cells, and synovial cells to endothelial cells. Yet we do not know which of these cell populations represent actual physiological or pathological targets of IL-1. The issue is clouded by the existence of several other macrophage-derived hormones that possess similar biological activities. In addition to IL-1, macrophages produce tumor necrosis factor, platelet-derived growth factor, and transforming growth factor β. All these factors possess proinflammatory activity. Are all these mediators active in a given inflammatory response, or are they differentially expressed at unique sites of inflammation? The answer to this question must await *in situ* experiments using molecular probes and specific antibodies.

The question of IL-1 and T cell activation also needs further examination. Do T cells have an absolute requirement for IL-1, or does this mediator function as an "enhancer" of T cell activation? How important is IL-1 in primary versus secondary T cell responses? Since most experiments with IL-1 involve secondary responses, it is likely that memory T cells may have a minimal requirment for IL-1. The recent report by Durum and co-workers (Oppenheim *et al.*, 1986) that several murine T cell lines produce IL-1 adds a fascinating twist to the general problem of IL-1 and T cell activation. Do normal T cells produce IL-1? If so, does the T-cell-derived IL-1 serve a role as an intracellular mediator of T cell activation, or must the IL-1 be secreted so as to permit an interaction with cell surface IL-1 receptors?

Advances in molecular biology have drastically accelerated progress in IL-1 research. We now have available to us milligram quantities of purified recombinant IL-1, monoclonal antibodies, molecular probes, and a vast array of cellular targets for *in vitro* study. The elucidation of the factors that control IL-1 gene expression, as well as the cloning and characterization of the IL-1 receptor, awaits us in the not too distant future. However, the ultimate understanding of IL-1 and its role in immunity and inflammation must await the results of future *in vivo* experiments.

ACKNOWLEDGMENTS. The work described here is the result of a number of collaborative efforts with Drs. Jill Suttles, Judith Giri, Mitchell Dukovich, Pat Kilian, Dominick DeLuca, Janelle Rhyne, Charles McCall, Amos Ben-Zvi, David Beller, Emil Unanue, Harvey Colten, Daniella Männel, and Marcio Chedid. Portions of these studies were supported by a grant from the National Science Foundation (DCB 8540479).

References

Auron, P. E., Webb, A. C., Rosenwasser, L. J., Mucci, S. F., Rich, A., Wolff, S. M., and Dinarello, C. A., 1984, Nucleotide sequence of human monocyte interleukin 1 precursor cDNA, *Proc. Natl. Acad. Sci. U.S.A.* **81:**7907–7911.

Bayne, E. K., Rupp, E. A., Limjvoco, G., Chin, J., and Schmidt, J. A., 1986, Immunocytochemical detection of interleukin-1 within stimulated human-monocytes, *J. Exp. Med.* **163:**1267–1280.

Ben-Zvi, A., Mizel, S. B., and Oppenheim, J. J., 1981, Generation of human peripheral blood stable E-rosette-forming T cells by interleukin 1, *Clin. Immunol.* **19:**330–337.

Cameron, P. M., Limjuco, G. A., Chin, J., Silberstein, L., and Schmidt, J. A., 1986, Purification to homogeneity and amino acid sequence analysis of two anionic species of human interleukin 1, *J. Exp. Med.* **164:**237–250.

DeLuca, D., 1985, Ia-positive non-lymphoid cells and T cell development in murine fetal thymocyte organ cultures: Monoclonal anti-Ia antibodies inhibit the development of T cells, *J. Immunol.* **136:**430–439.

DeLuca, D., and Mizel, S. B., 1986, I-A-positive nonlymphoid cells and T cell development in murine fetal thymus organ cultures: Interleukin 1 circumvents the block in T cell differentiation induced by monoclonal anti-I-A antibodies, *J. Immunol.* **137:**1435–1441.

Dinarello, C. A., 1984, Interleukin-1, *Rev. Infect. Dis.* **6:**51–95.

Dinarello, C. A., Marnoy, S. O., and Rosenwasser, J. L., 1983, Role of arachidonate metabolism in the immunoregulatory function of human leukocytic pyrogen/lymphocyte-activating factor/interleukin 1, *J. Immunol.* **130:**890–895.

Doerfler, W., 1983, DNA methylation and gene activity, *Ann. Rev. Biochem.* **52:**93–124.

Dower, S. K., Kronheim, S. R., March,C. J., Conlon, P. J., Hopp, T. P., Gillis, S., and Urdal, D. L., 1985, Detection and characterization of high affinity plasma membrane receptors for human interleukin 1, *J. Exp. Med.* **162:**501–515.

Dower, S. K., Call, S. M., Gillis, S., and Urdal, D. L., 1986, Similarity between the interleukin-1 receptors on a murine T-lymphoma cell-line and on a murine fibroblast, *Proc. Natl. Acad. Sci. U.S.A.* **83:**1060–1064.

Dukovich, M., Severin, J. M., White, S. J., Yamazaki, S., and Mizel, S. B., 1986, Stimulation of fibroblast proliferation and prostaglandin production by purified recombinant murine interleukin-1, *Clin. Immunol. Immunopathol.* **38:**381–389.

Farrar, W. L., and Hume, J. L., 1985, The role of arachidonic acid metabolism in the activities of interleukin 1 and 2, *J. Immunol.* **135:**1153–1159.

Farrar, W. L., Mizel, S. B., and Farrar, J. J., 1980, Participation of lymphocyte activation factor (interleukin 1) in the induction of cytotoxic T cell responses, *J. Immunol.* **124:**1371–1377.

Furutani, Y., Notake, M., Kukui, T., Ohue, M., Nomura, H., Yamada, M., and Nakamura, S., 1986, Complete nucleotide sequenc of the gene for human interleukin 1 alpha, *Nucleic Acids Res.* **14:**3167–3167.

Garoff, H., 1985, Using recombinant DNA techniques to study protein targeting in the eucaryotic cell, *Annu. Rev. Cell Biol.* **1:**403–445.

Gery, I., Gershon, R. K., and Waksman, B. H., 1972, Potentiation of the T-lymphocyte response to mitogens. I. The responding cell, *J. Exp. Med.* **136:**128–142.

Gillis, S., and Mizel, S. B., 1981, T-cell lymphoma model for the analysis of interleukin 1-mediated T-cell activation, *Proc. Natl. Acad. Sci. U.S.A.* **78:**1133–1137.

Giri, J. G., Kincade, P. W., and Mizel, S. B., 1984, Interleukin 1-mediated induction of κ-light chain synthesis and surface immunoglobulin expression of pre-B cells, *J. Immunol.* **132:**223–228.

Giri, J. G., Lomedico, P. L., and Mizel, S. B., 1985, Studies on the synthesis and secretion of interleukin 1. I. A 33,000 molecular weight precursor for interleukin 1, *J. Immunol.* **134:**343–349.

Gray, P. W., Glasiter, D., Chen, E., Goeddel, D. V., and Pennica, D., 1986, Two interleukin 1 genes in the mouse: Cloning and expression of the cDNA for murine interleukin 1β, *J. Immunol.* **137:**3644–3648.

Gubler, U., Chua, A. O., Stern, A. S., Hellman, C. P., Vitek, M. P., Dechiara, T. M., Benjamin, W. R., Collier, K. J., Dukovich, M., Familletti, P. C., Fiedler-Nagy, C., Jenson, J. Kaffla, K. Killian, P. L., Stremlo, D., Wittreich, B. H., Woele, D., Mizel, S. B., and Lomedico, P. T., 1986, Recombinant human interleukin 1α: Purification and biological characterization, *J. Immunol.* **136:**2492–2497.

Howard, M., Mizel, S. B., Lachman, L., Ansel, J., Johnson, B., and Paul W. E., 1983, Role of interleukin 1 in anti-immunoglobulin induced β cell proliferation, *J. Exp. Med.* **157:**1529–1543.

Kaye, J., Gillis, S., Mizel, S. B., Shevach, E. M., Malek, T. R., Dinarello, C. A., Lachman, L. L., and Janeway, C. A., Jr., 1984, Growth of a cloned helper T cell line induced by a monoclonal antibody specific for the antigen receptor: Interleukin 1 is required for the expression of receptors for interleukin 2, *J. Immunol.* **133:**1339–1345.

Kilian, P. L., Kaffka, K. L., Stern, A. S., Woehle, D., Benjamin, W. R., Dechiara, T. M., Gubler, U., Farrar, J. J., Mizel, S. B., and Lomedico, P. T., 1986. Interleukin 1α and interlukin 1β bind to the same receptor on T cells, *J. Immunol.* **136:**4509–4514.

Klempner, M. S., Dinarello, C. A., Henderson, W. K., and Gallin, J. I., 1979, Stimulation of neutrophil oxygen-dependent metabolism by human leukocytic pyrogen, *J. Clin. Invest.* **64:**996–1002.

Knudsen, P. J., Dinarello, C. A., and Strom, T. B., 1986, Purification and characterization of unique human interleukin 1 from the tumor cell line U937, *J. Immunol.* **136:**3311–3316.

Koopman, W. J., Farrar, J. J., and Fuller-Bonar, J., 1978, Evidence for the identification of lymphocyte activating factor as the adherent cell-derived mediator responsible for enhanced antibody synthesis by nude mouse spleen cells, *Cell. Immunol.* **35:**92–98.

Kronheim, S. R., March, C. J., Erb, S. K., Conlon, P. J., Mochizuki, D. Y., and Hopp, T. P., 1985, Human interleukin 1. Purification to homogeneity, *J. Exp. Med.* **161:**490–502.

Kurt-Jones, E. A., Beller, D. I., Mizel, S. B., and Unanue, E. R., 1985a, Identification of a membrane-associated interleukin 1 in macrophages, *Proc. Natl. Acad. Sci. U.S.A.* **82:**1204–1208.

Kurt-Jones, E. A., Keily, J. M., and Unanue, E. R., 1985b, Communication. Conditions required for expression of membrane IL 1 on B cells, *J. Immunol.* **135:**1548–1550.

Lomedico, P. L., Gubler, U., Hellmann, C. P., Dukovich, M., Giri, G., Pan, Y. E., Collier, K., Semionow, R., Chua, A. O., and Mizel, S. B., 1984, Cloning and expression of murine interleukin-1 cDNA in *Escherichia coli, Nature* **312:**458–462.

Lowenthal, J. W., and MacDonald, H. R., 1986, Binding and internalization of interleukin 1 by T cells: Direct evidence for high- and low-affinity classes of interleukin 1 receptor, *J. Exp. Med.* **164:**1060–1074.

Männel, D. N., Mizel, S. B., Diamantstein, T., and Falk, W., 1985, Induction of interleukin 2 responsiveness in thymocytes by synergistic action of interleukin 1 and interleukin 2, *J. Immunol.* **134:**3108–3110.

March, C. J., Mosley, B., Larsen, A., Cerretti, D. P., Braedt, G., Price, V., Gillis, S., Henney, C. S., Kronheim, S. R., Grabstein, K., Conlon, P. J., Hopp, T. P., and Cosmann, D., 1985, Cloning, sequence and expression of two distinct human interleukin-1 complementary DNAs, *Nature* **315:**641–646.

Matushima, K., Yodoi, J., Tagaya, Y., and Oppenheim, J. J., 1986, Down regulation of interleukin-1 (IL1) receptor expression by IL1 and fate of internalized ^{125}I-labeled IL1 β in a human large granular lymphocyte cell line, *J. Immunol.* **137:**3183-3188.

Mizel, S. B., 1982, Interleukin 1 and T-cell activation, *Immunol. Rev.* **63:**51–72.

Mizel, S. B., and Ben-Zvi, A., 1980, Studies on the role of lymphocyte activating factor (interleukin 1) in antigen-induced lymph node lymphocyte proliferation, *Cell. Immunol.* **54:**382–389.

Mizel, S. B., and Mizel, D., 1981, Purification to apparent homogeneity of murine interleukin 1, *J. Immunol.* **126:**834–837.

Mizel, S. B., Dayer, J. N., Krane, S. M., and Mergenhagen, S. E., 1981, Rheumatoid synovial cell collagenase and prostaglandin production by partially purified lymphocyte - activating factor (interleukin-1), *Proc. Natl. Acad. Sci. U.S.A.* **78:**2474–2477.

Mizel, S. B., Dukovich, M., and Rothstein, J., 1983, Preparation of goat antibodies against interleukin 1. Use of an immunoadsorbent to purify IL 1, *J. Immunol.* **131:**1834–1837.

Mizel, S. B., Kilian, P., Lewis, J. C., Paganelli, K. A., and Chizzonite, R. A., 1987, The interleukin 1 receptor. Dynamics of interleukin 1 binding and internalization in T cells and fibroblasts, *J. Immunol.* **138:**2906–2912.

Nagelherken, L. M., and van Breda Vriesman, P. J. C., 1986, Membrane-associated IL 1-like activity on rat dendritic cells. *J. Immunol.* **136:**2164–2170.

Oppenheim, J. J., Kovacs, E. J., Matsushima, K., and Durum, S. K., 1986, There is more than one interleukin-1, *Immunol. Today* **7:**45–56.

Perlmutter, D. H., Goldberger, G., Dinarello, C. A., Mizel, S. B., and Colten, H. R., 1986, Regulation of class III major histocompatibility complex gene products by interleukin-1, *Science* **232:**850–852.

Postlethwaite, A. E., Lachman, L. B., Mainardi, C. L., and Kang, A. H., 1983, Prostacyclin synthesis induced in vascular cells by interleukin-1, *J. Exp. Med.* **157:**801–806.

Ramadori, G., Sipe, J. D., Dinarello,C. A., Mizel, S. B., and Colten, H. R., 1985, Pretranslational modulation of acute phase hepatic protein synthesis by murine recombinant interleukin 1 (IL-1) and purified human IL-1, *J. Exp. Med.* **162:**930–942.

Rossi, V., Breviari, F., Ghezzi, P., Dejana, E., and Mantovan, A., 1985, Interleukin 1 stimulation of collagenase production by cultured fibroblasts, *Science* **229:**174–176.

Sauder, D. N., Mounessa, N. L., Katz, S., Dinarello, C. A., and Gallin, J. I., 1984, Chemotactic cytokines: The role of leukocytic pyrogen and epidermal cell thymocyte-activating factor in neutrophil chemotaxis, *J. Immunol.* **132:**828–832.

Smith, R. J., Bowman, B. J., and Speziale, S. C., 1986, Interleukin 1 stimulates granule exocytoxins from human neutrophils, *Int. J. Immunopharmacol.* **8:**33–46.

Stanton, T. H., Maynard, M., and Bomsztyk, K., 1986, Effect of interleukin-1 on intracellular concentration of sodium, calcium, and potassium in 702/3 cells, *J. Biol. Chem.* **261:**5699–5701.

Staruch, M. J., and Wood, D. D., 1983, The adjuvanticity of interleukin 1 in vivo, *J. Immunol.* **130:**2191–2194.

Suttles, J., Giri, J., and Mizel, S. B., 1987, Secretion of the murine interleukin 1 precursor. Evidence for a novel, processing independent secretory pathway in macrophages (submitted for publication).

Van Damme, J., DeLey, M., Opdenakker, G., Billiau, A., and DeSomer, P., 1985, Homogeneous interferon-inducing 22K-factor is related to endogenous pyrogen and interleukin 1, *Nature* **314:**266–268.

7

Structure–Function Relations for the Interleukin-2 Receptor

RICHARD J. ROBB

1. A Molecular Explanation for the Different Affinities of the IL-2 Receptor

The original study of IL-2 binding to activated lymphocytes (Robb *et al.*, 1981) described a high-affinity receptor–ligand interaction that fit nicely with the extremely low concentrations of IL-2 that were necessary to promote cellular proliferation. Subsequent investigations, however, indicated that the interaction of IL-2 with cells was not so simple. In particular, antibodies to the IL-2 receptor detected far more receptor molecules on the cell surface than high-affinity IL-2 binding sites (Leonard *et al.*, 1982). As IL-2 became more plentiful, the IL-2 binding assays were extended to higher concentrations of ligand, revealing a second class of low-affinity sites (Robb *et al.*, 1984). The latter sites appeared to account for most or all of the discrepancy between the number of receptors detected with antibodies and the original numerical estimates of high-affinity binding sites. The molecular basis for the two affinity classes of receptor, however, remained unclear. Possible explanations included: (1) the high- and low-affinity sites consisted of distinct IL-2 binding molecules; (2) the affinity difference occurred as a result of variable post-translational modifications of a single IL-2-binding molecule; (3) the difference arose from conformational changes in a single receptor protein, perhaps as a result of homodimer formation or association with a second receptor subunit; and (4) the difference was caused by the effect of a second receptor subunit, present in high-affinity binding sites, which partici-

RICHARD J. ROBB • Medical Products Department, Glenolden Laboratory, E. I. du Pont de Nemours & Company, Glenolden, Pennsylvania 19036.

pated directly in ligand binding. In the past year, several laboratories have obtained evidence that the last explanation is indeed the basis for the multiple affinities of the IL-2 receptor.

1.1. High- and Low-Affinity IL-2 Binding Sites: Role of the Tac (Alpha) Protein

A key contribution to the biochemical description of the IL-2 receptor was the demonstration by Leonard *et al.* (1982) that a monoclonal antibody (anti-Tac) raised against activated human T lymphocytes blocked IL-2 binding. The major molecule recognized by the antibody was a 55,000 MW glycoprotein, termed Tac. Further supporting the role of Tac as an IL-2 receptor was evidence that affinity supports coupled with either IL-2 or anti-Tac both captured the same 55,000 MW protein solubilized from receptor-positive cells (Robb and Greene, 1983). Shortly thereafter, Robb and colleagues (Robb *et al.*, 1984; Robb and Rusk, 1986) demonstrated that anti-Tac blocked IL-2 binding to both high- and low-affinity receptors. Therefore, the Tac molecule was implicated in both of these receptor subclasses.

1.2. The Two-Chain Hypothesis for the High-Affinity IL-2 Receptor

The idea that the difference between high- and low-affinity receptors was an intrinsic characteristic of the receptor binding sites was supported by examination of detergent-solubilized IL-2–receptor complexes. As shown in Fig. 1, when cells containing[^{125}I]IL-2 bound to high-affinity receptors were solubilized with detergent the IL-2 remained noncovalently associated with receptor molecules in a high molecular weight complex during size exclusion chromatography. When cells containing[^{125}I]IL-2 bound predominately to low-affinity receptors were solubilized in detergent, however, the IL-2 quickly dissociated and eluted in its normal position at 15,000 MW. Thus, the fundamental difference between high- and low-affinity IL-2–receptor complexes persisted after detergent solubilization of the ligand–receptor complex.

Assuming that the Tac protein was part of both high- and low-affinity receptors, attention focused on how the affinity of this molecule was modulated. When murine L cells were transfected with cDNA encoding the Tac protein, only a single class of low-affinity receptors was detected (Greene *et al.*, 1985; Hatakeyama *et al.*, 1985; Kondo *et al.*, 1986a). When the same DNA was used to transfect lymphocytes, however, both subclasses of receptors were expressed (Hatakeyama *et al.*, 1985; Kondo *et al.*, 1986a). Thus, something peculiar to the expression of Tac in a lymphocyte, as opposed to a fibroblast, enabled it to assume both high- and low-affinity binding site con-

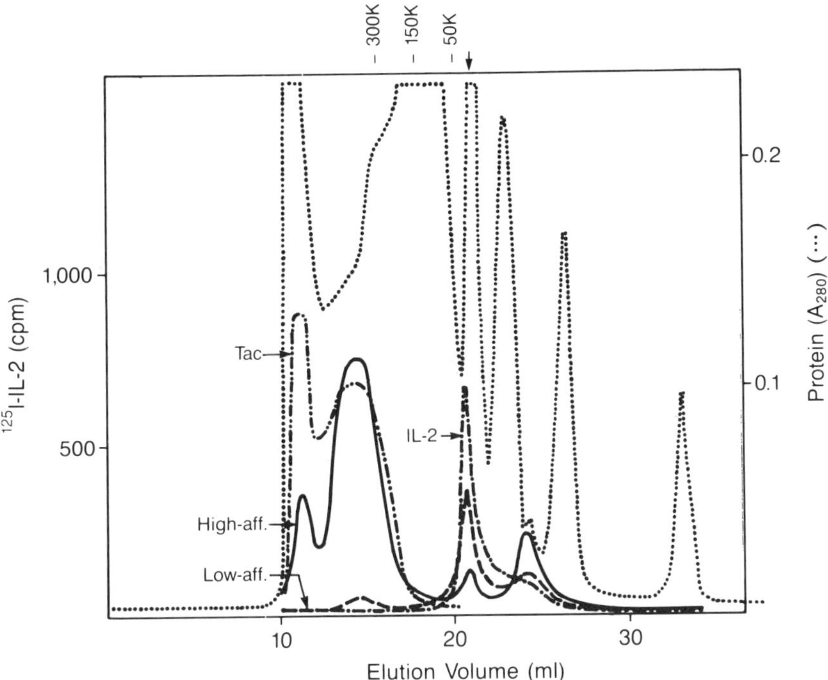

Figure 1. Superimposed results of size exclusion chromatography of IL-2 and noncovalent IL-2–receptor complexes. [^{125}I]IL-2 was added to HUT 102 cells under conditions favoring binding to either high-affinity or low-affinity receptor sites. After removing unbound factor, the cells were solubilized in detergent (1% octylglucoside). Insoluble material was removed using a Beckman airfuge and the supernatant was chromatographed on a Superose 6 sizing column (Pharmacia). Fractions from the column were examined for their optical density at 280 nm and for their level of [^{125}I]IL-2. Superimposed are the results for high-affinity (—) and low-affinity (– – –) IL-2 binding. Also shown is the elution profile of [^{125}I]IL-2 after it was mixed with detergent-extracted proteins from receptor-negative JURKAT cells (_._._._) and the elution profile of total Tac protein (_.._.._) extracted from the HUT 102 cells. The latter values were derived by blotting serial dilutions of each fraction on nitrocellulose, followed by quantitation with a combination of anti-Tac antibody and a horseradish peroxidase-coupled goat antimouse detection system (Bio-Rad).

figurations. The difference could have been due to variations in the posttranslational modifications of the Tac molecule or to the presence of associated receptor subunits. One approach used to resolve this uncertainty was to directly fuse cell membranes and measure the effect on receptor affinity (Robb, 1986). As diagrammed in Fig. 2, cell membranes containing the low-affinity murine equivalent of Tac protein (p55) were fused with cell membranes from human cells containing both high- and low-affinity receptors.

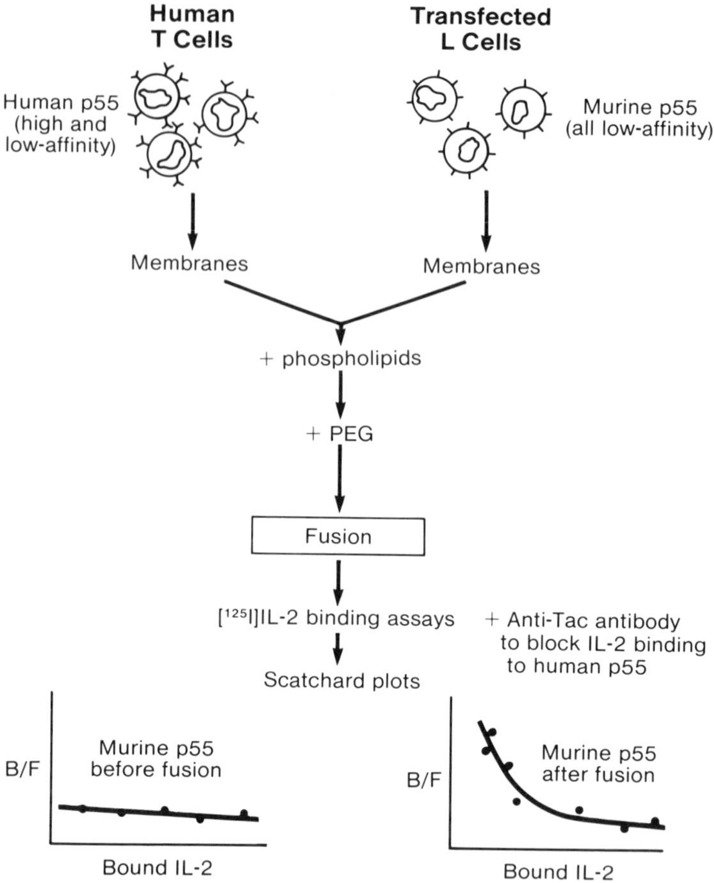

Figure 2. Experimental protocol to test the modulation of receptor affinity resulting from fusion of cell membranes. L cells were transfected with cDNA encoding the murine equivalent of the Tac p55 molecule (kindly provided by T. Malek and J. Miller). Such cells expressed large numbers of low-affinity IL-2 binding sites. A membrane preparation was then fused with membranes from human HUT 102 cells, which express both high- and low-affinity Tac (p55)-dependent binding sites. The fused membranes were tested in an $[^{125}I]$IL-2 binding assay in the presence of anti-Tac antibody. This antibody blocks binding of IL-2 to human Tac protein but not to the murine p55 equivalent. Therefore, any IL-2 binding that remained would represent the binding to murine p55-dependent receptor sites. If the murine receptor sites were unaffected by the fusion with human membranes, they would all remain low-affinity and the Scatchard plot of bound/free ligand versus bound ligand would be linear. On the other hand, if the murine p55 molecules combined with components in the human membranes that were responsible for modulating the affinity of p55 to a high-affinity state, then a mixture of high- and low-affinity sites would result, and the Scatchard plot would be curvilinear. In fact, the latter result was observed (Robb, 1986), demonstrating that high-affinity sites consist of the p55 molecule in combination with one or more other cell membrane components.

Following fusion, ligand binding to the Tac molecules derived from the human cell membranes was selectively blocked with species-specific anti-Tac antibody. If the human membranes provided an accessory molecule for conversion of the murine Tac equivalent to a high-affinity state, the resulting Scatchard binding curve was expected to be curvilinear (high- and low-affinity sites). If no modulation of the murine receptor molecules occurred, the curve would be linear. In fact, fusion with human cell membranes containing high-affinity receptors did indeed result in conversion of some of the murine receptor molecules to a high-affinity state (Robb, 1986). This evidence, together with the results from cDNA transfection experiments (Hatakeyama *et al.*, 1985; Kondo *et al.*, 1986a,b), resulted in the two-chain hypothesis for high-affinity IL-2 receptors. This hypothesis stated that high-affinity receptors consisted of Tac together with one or more accessory molecules that modulated the affinity of the binding site to a high-affinity state.

1.3. Characterization of a Novel IL-2 Binding Molecule (Beta)

Virtually all cellular responses to IL-2, including T and B cell proliferation and induction of lymphokine secretion, appeared to be dependent on the interaction of IL-2 with high-affinity receptors containing the Tac protein. Nevertheless, a number of observations were clearly inconsistent with the notion that only the Tac protein could bind IL-2. For example, Ortaldo *et al.* (1984) demonstrated that IL-2 increased the cytolytic activity of natural killer (NK) cells in a manner insensitive to inhibition by the anti-Tac antibody. Furthermore, our own studies indicated that large granular lymphocytes, a subpopulation of cells containing NK precursors, could bind small amounts of IL-2 in the presence of inhibitory levels of anti-Tac (R. J. Robb, unpublished observations). Similarly, Ralph *et al.* (1984) showed that high IL-2 levels caused a particular B cell line to secrete immunoglobulin despite the apparent absence of Tac on its surface. Attempts to characterize these non-Tac, IL-2-binding components, however, were complicated by their low number and the absence of appropriate antibodies. To overcome these obstacles, we took advantage of a cell line named YT, which had been isolated and characterized by Yodoi and colleagues (Yodoi *et al.*, 1985). These authors showed that YT cells could be induced by a number of agents to express relatively large numbers of high-affinity IL-2 receptors. Our own studies demonstrated that YT cells were also an ideal starting material for identifying and characterizing the non-Tac IL-2-binding component and studying its role in formation of high-affinity receptors.

Scatchard analysis of[^{125}I]IL-2 binding to unstimulated YT cells yielded a curvilinear plot indicative of two subclasses of receptor sites, one of a high affinity and one of a much lower affinity (Table I; Robb *et al.*, 1987). At first

Table I. Receptor Numbers and Affinities[a]

Cell type	Forskolin	Anti-Tac antibody	IL-2 Binding: High-affinity Sites/cell	High-affinity K_d (pM)	Intermediate-affinity Sites/cell	Intermediate-affinity K_d (pM)	Low-affinity Sites/cell	Low-affinity K_d (pM)	Anti-Tac binding (sites/cell)
YT	–	–	≤20	—	13,100	810	≤200	—	1,100
YT	–	+	1,160	17.4	14,200	820	≤200	—	≤20
YT	+ (24 h)	–	8,950	18.5	≤2,000	~800	≤1,000	11,000	10,200
YT	+ (24 h)	+	≤20	—	11,900	895	≤200	—	≤20
YT	+ (48 h)	–	9,200	19.3	≤2,000	~800	22,100	9,980	40,200

[a] Measurement of the binding [^{125}]IL-2 and [^{3}H]anti-Tac antibody to unstimulated and forskolin-activated YT cells in the absence and presence of anti-Tac antibody.

glance, this result was similar to that obtained with any activated T cell. When anti-Tac was included in the YT assays, however, only the high-affinity component of IL-2 binding was inhibited (Table I). The predominant lower-affinity binding component was unaffected. Moreover, the dissociation constant of the anti-Tac-resistant binding was 800 pM, distinctly different from the value (10–20 nM) usually found for low-affinity Tac receptors. Thus, YT cells expressed a high proportion of a non-Tac, intermediate-affinity structure, which for convenience we have termed beta (β). The Tac protein, as the original receptor subunit to be isolated and characterized, was termed alpha (α).

The unique nature of the non-Tac binding structure on YT cells was further substantiated by crosslinking studies. As shown in Fig. 3, chemical crosslinking of[^{125}I]IL-2 bound to the abundant low-affinity Tac receptors on the HUT 102 T cell line gave rise to a 70,000 molecular weight (MW) complex of Tac (55,000 MW) and IL-2 (15,000 MW). In contrast, crosslinking of IL-2 bound to receptors on unstimulated YT cells yielded a doublet on SDS–PAGE, which migrated at 83,000 and 90,000 MW. Immunoprecipitation with antibodies to Tac confirmed that the 70,000 MW complex contained the Tac protein, while the 83,90-kD doublet from unstimulated YT cells was not recognized by these reagents. Similar results have been reported independently by other groups for this and other cell lines (Sharon *et al.*, 1986; Tsudo *et al.*, 1986, Teshigawara *et al.*, 1987). Considering the size of IL-2 and of the 83,90-kD complexes derived from the β binding sites on YT cells, it would appear that the β protein(s) is 70,000–75,000 MW. In fact, Sharon *et al.* (1986) used a cleavable crosslinking agent to demonstrate that the IL-2–β complexes were derived from IL-2 and two proteins, one of 70,000 MW and one of 75,000 MW. Of considerable interest is whether these represent two distinct protein entities or a single protein that has a

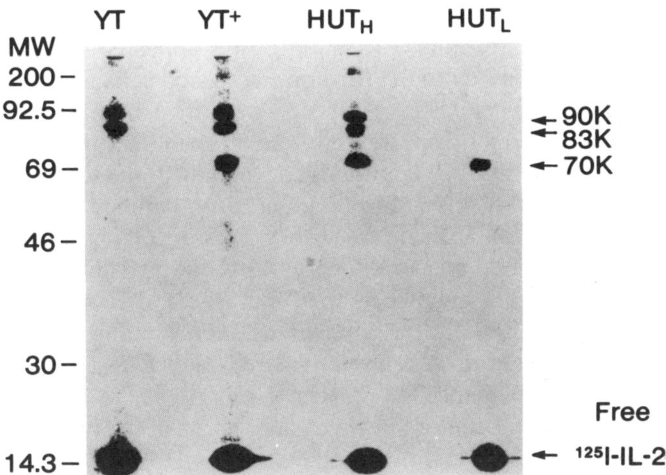

Figure 3. SDS–PAGE analysis of covalently crosslinked IL-2–receptor complexes derived from YT cells, forskolin-stimulated YT cells (YT$^+$), and HUT 102 cells, following IL-2 binding to high-affinity (HUT$_H$) and low-affinity (HUT$_L$) receptors. Reproduced from Robb *et al.* (1987).

variable carbohydrate content or other post-translational modification. If the former explanation is correct, each of the proteins might independently bind IL-2 with similar affinities. Alternatively, both proteins might be required at the same time to form a single β binding site. Digestion of the crosslinked IL-2–β complexes with a series of glycosidases reduced their size by 15–20% but did not totally eliminate the difference in size between the two bands in the doublet (Robb *et al.*, 1987). While the β proteins were thus heavily glycosylated, the source of their heterogeneous migration on SDS–PAGE remained unresolved.

1.4. Evidence that High-Affinity Binding Is the Result of an Alpha–IL-2–Beta Complex

Several lines of evidence suggest that the β receptor protein(s) provides the molecular basis that distinguishes high- and low-affinity IL-2 receptors. Analysis of unstimulated YT cells demonstrated that the number of high-affinity IL-2 binding sites was very similar to the number of Tac molecules detected with ^{3}H-labeled anti-Tac antibody (Table I; Robb *et al.*, 1987). When the YT cells were stimulated with forskolin for 24 h, the number of Tac molecules expressed on the surface increased dramatically. At the same time, there was a similar increase in the number of high-affinity binding sites and a reciprocal decrease in the number of intermediate-affinity β binding sites. Virtually no low-affinity Tac binding sites appeared. The addition of unlabeled anti-Tac antibody to the forskolin-activated cells reversed the situation, causing a disappearance of the newly formed high-affinity binding sites and a corresponding reappearance of the original intermediate-affinity β sites (Table I). These results suggested that the newly induced Tac proteins combined with the β proteins to form high-affinity binding sites, shifting the original β sites from an intermediate- to a high-affinity status. When the stimulation with forskolin was extended to 48 h, the Tac levels continued to rise. Consistent with the hypothesis that both Tac and β are necessary for high-affinity sites, however, the number of such binding sites reached a plateau at a value nearly equal to the original level of β binding sites.

Another line of evidence supporting the involvement of both α and β chains in high-affinity receptor sites was derived from the effects of various antibodies on high-, low-, and intermediate-affinity[^{125}I]IL-2 binding (Robb *et al.*, 1987). For these experiments, YT cells were used as a source of β binding sites, while the $HTLV_I$-transformed cell line MT-1 was used as a source of α binding sites. Although MT-1 cells express about 200,000 copies of Tac per cell, they apparently synthesize no β protein. The binding of low concentrations of IL-2 to the HUT cell line was used as a measure of high-affinity receptor sites. Not surprisingly, anti-Tac antibody blocked the binding of IL-2 to both low-affinity α sites on MT-1 cells and high-affinity sites (HUT_H) on HUT cells (Fig. 4). The antibody only marginally affected IL-2 binding to YT cells, owing to the preponderance of β sites on those cells. Thus, the high-affinity sites on the HUT cells clearly contained the Tac protein. Two antibodies to IL-2 were similarly used to examine the role of the β protein(s) in high-affinity sites. The monoclonal 1H11 antibody specifically recognizes residues 1–8 of glycosylated IL-2 (Robb *et al.*, 1983; Robb, 1985), while the $R135_{\alpha p8\text{-}27}$ antibody was selected on a peptide-affinity column to recognize a region of IL-2 consisting of residues 8–27 (Kuo and Robb, 1986). Both $R135_{\alpha p8\text{-}27}$ and F_{ab} fragments of 1H11 blocked IL-2 binding to β sites on YT (anti-Tac was included to eliminate the small contribution from Tac protein) but did not affect the binding of IL-2 to α sites on MT-1 cells. Although the 1H11 antibody was only partially effective in blocking high-affinity binding to HUT cells, the $R135_{\alpha p8\text{-}27}$ reagent totally inhibited such binding. Thus, the β protein(s) was also implicated in high-affinity receptor binding sites. A third anti-IL-2 reagent, 9B11, which binds an IL-2 segment containing residues 33–54 (Robb, 1985; Kuo and Robb, 1986), blocked all three types of binding. These results argue that high-affinity receptor sites are dependent on both the α and β receptor molecules. They also indicate that the α and β chains recognize IL-2 in fundamentally different ways. β–IL-2 interactions are sensitive to antibodies binding near

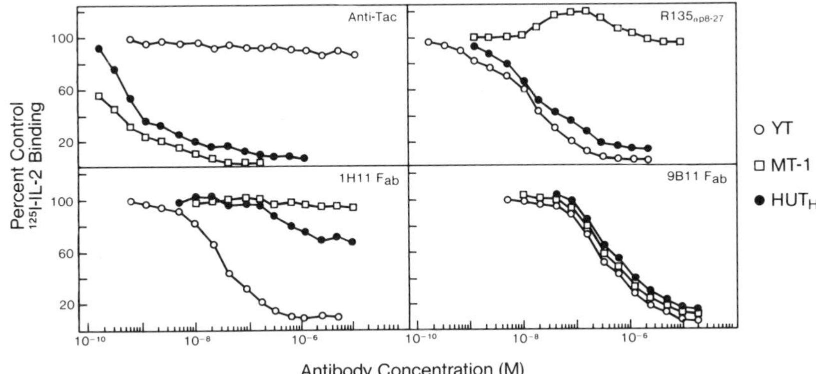

Figure 4. Effect of various antibodies on the binding of [^{125}I]IL-2 to YT cells (anti-Tac was added in the case of antibodies 1H11, 9B11, and R135 in order to convert all the receptors to intermediate-affinity β binding sites), to MT-1 cells (expressing only low-affinity Tac binding sites), and to HUT 102 cells under high-affinity binding conditions (HUT_H). 1H11 and 9B11 anti-IL-2 antibodies were tested as Fab fragments. $R135_{p8\text{-}27}$ is a rabbit polyclonal anti-IL-2 fractionated on an affinity support coupled with a peptide consisting of residues 8–27 of IL-2.

the N terminus of the molecule, while α–IL–2 interactions are not. Assuming that α and β chains provide two separate contact sites for the IL-2 molecule, the resulting affinity of a ternary α–IL-2–β complex would be increased enormously over the affinities of the individual receptor components.

A third form of experimental evidence supporting the notion of α and β involvement in high-affinity sites was derived from crosslinking studies. As mentioned previously, when[^{125}I]IL-2 was chemically crosslinked to low-affinity α sites, it gave rise to a 70,000 MW complex of Tac and IL-2, while when it was crosslinked to β sites, the 83,90-kD doublet resulted. When the IL-2 was selectively bound to high-affinity sites on HUT cells, all three sizes of crosslinked complexes were generated (Fig. 3). The same result was obtained for YT cells stimulated with forskolin to induce expression of the Tac protein and shift the β sites to a high-affinity status (Fig. 3). To demonstrate that the same proteins were involved in both HUT and YT cells, the IL-2–receptor complexes were cleaved with trypsin and the resulting fragments were examined on SDS–PAGE (Fig. 5). The α–IL-2 complexes from low-affinity HUT receptors gave rise to a characteristic 34,800 MW breakdown product. In contrast, the 83,90 kD β–IL-2 complexes gave rise to a 65,000 MW cleavage product. High-affinity IL-2–receptor complexes from HUT cells, which began with all three size bands, gave rise to both the 34,800 and 65,000 MW trypsin products. Thus, both α and β proteins took part in the high-affinity sites, and the β protein(s) on HUT cells appeared similar or identical to that on YT cells.

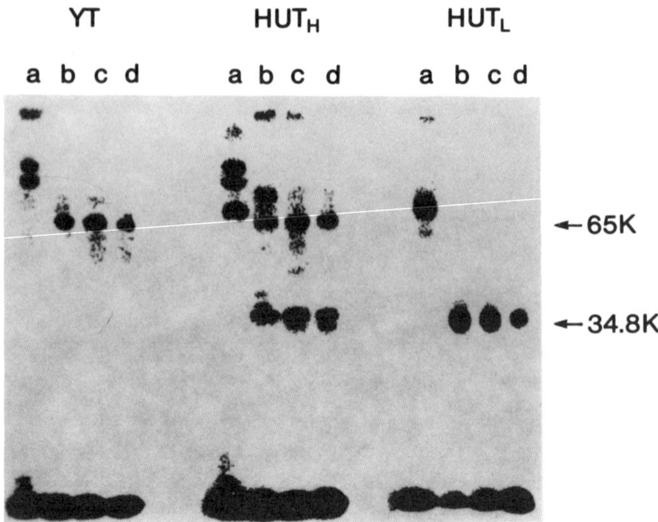

Figure 5. SDS–PAGE analysis of crosslinked IL-2–receptor complexes derived from YT cells and from HUT cells following high- and low-affinity IL-2 binding. The complexes were treated with increasing amounts of trypsin (lanes b,c,d) before electrophoresis. Tac–IL-2 complexes yielded a major breakdown product of 34,800 MW while β–IL-2 complexes yielded a major breakdown product at 65,000 MW. Reproduced from Robb *et al.* (1987).

2. *Multiple Forms of the IL-2 Receptor: Function and Cellular Distribution*

The identification of the two IL-2 binding structures, α and β, suggested that IL-2 receptors existed in at least three configurations (Fig. 6). Under this hypothesis, low-affinity receptors would consist of α or Tac protein. Intermediate-affinity receptors would consist of the β protein(s), and high-affinity sites would consist of a ternary complex of α, β, and IL-2. The close proximity of two ligand binding components in the latter sites would greatly decrease the rate of dissociation of IL-2 from the cell surface, giving rise to *high-affinity* binding. Having established a minimum of three different receptor configurations, the next logical step was to ask which forms of the receptor were active and on which cellular subpopulations they occurred. By drawing on our own data (Robb and Greene, 1987) and that of other laboratories (Dukovich *et al.*, 1987), it appears that the β receptor protein provides a key element necessary for a functional response. Thus, as detailed in Section 2.2, the β protein(s) can internalize bound IL-2 and is probably involved in initiating the signal transduction steps (i.e., activation of protein kinase C,

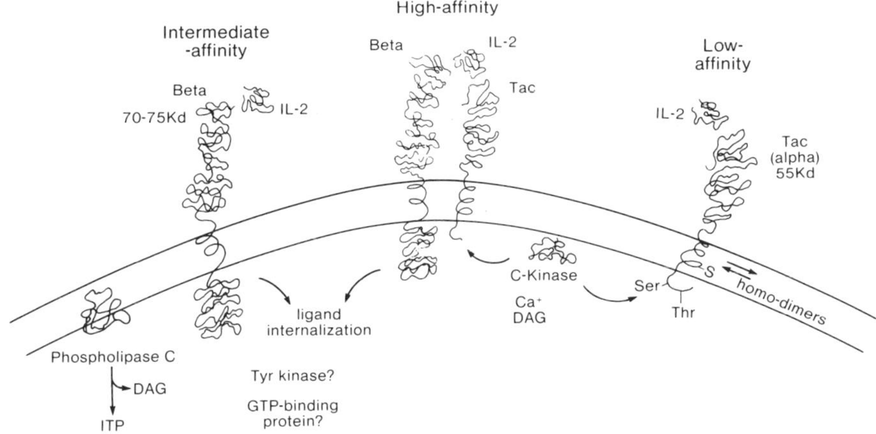

Figure 6. Diagram of three hypothetical forms of IL-2 receptor consisting of various combinations of the Tac (alpha) and β proteins. Also indicated are events sometimes associated with binding of IL-2 to the different receptor forms: Ligand internalization, activation of c-kinase by diacylglycerol (DAG) and phospholipase C (yielding DAG and inositol triphosphate, ITP), and phosphorylation of the Tac protein.

phospholipase C, GTP-binding, tyrosine kinase activity, and calcium mobilization), which may be necessary for the cellular response to IL-2 (see Fig. 6).

2.1. Type I Receptors: The Alpha (Tac) Chain

This type of receptor consists of the 55,000 MW Tac glycoprotein. Tac is found on the cell surface both as a monomeric chain and as a disulfide-linked homodimer (Kuo *et al.*, 1986). The functional significance of the latter structure, however, is still unclear. Tac protein binds IL-2 with a low affinity ($K_d \approx 10$–20 nM). It is very abundant (2×10^4 to 2×10^6 copies per cell) on antigen- or lectin-activated T cells and certain activated B cells and on all cell lines transformed with the $HTLV_I$ virus. Tac (or the murine equivalent) is also present on murine thymocytes at certain developmental stages, on certain activated monocytes, and on IL-2-treated NK cell and lymphokine-activated killer (LAK) cell populations. Tac is absent or present at low levels on *resting* lymphocytes, but its expression is upregulated by a wide variety of agents associated with cellular activation. IL-2 bound to the monomeric Tac protein is not internalized (Fujii *et al.*, 1986; Weissman *et al.*, 1986) and it remains unclear whether Tac alone can mediate a functional response at the high IL-2 concentrations necessary for its occupation. Based on its wide distribution (the above list is probably incomplete), however, it seems plausible that Tac will serve functions other than simply converting β binding sites to a high-affinity status.

2.2. *Type II Receptors: The Beta Chain*

This type of receptor consists of one or more glycoproteins of 70,000–75,000 MW β sites, have an intermediate affinity ($K_d \approx 0.8$–1.0 nM) for IL-2, and are found on a wide variety of cells before and after activation (Dukovich *et al.*, 1987). These include resting T cells, large granular lymphocytes, and a number of cell lines (Tsudo *et al.*, 1986; Teshigawara *et al.*, 1987; Dukovich *et al.*, 1987, Robb *et al.*, 1987). In general, however, the β sites are present at very low levels. This fact explains why they were originally overlooked during experiments employing cells with high levels of Tac protein (Robb and Greene, 1983; Robb *et al.*, 1984).

Several observations demonstrate that β protein(s) alone is capable of mediating IL-2-induced responses. For example, IL-2 bound to β sites was internalized at 37°C and targeted for lysosomal degradation with the same kinetics (half-time of 10–15 min for internalization; Robb and Greene, 1987) associated with high-affinity receptors (Robb *et al.*, 1981; Robb and Lin, 1984; Fujii *et al.*, 1986; Weissman *et al.*, 1986). In addition, the IL-2-induced increase of Tac expression on YT cells was shown to be unaffected by anti-Tac antibody, but it was inhibited dramatically by the 1H11 antibody, which selectively blocks binding to β sites (Table II). Similarly, antibodies that selectively blocked β site binding (i.e., 1H11 and $R135_{\alpha p8\text{-}27}$; see Section 1.4) inhibited the short-term activation of NK cell activity, while anti-Tac had a negligible effect (Table III; Ortaldo *et al.*, 1984). Finally, the 1H11 antibody, but not anti-Tac, was found to inhibit the increase of immunoglobulin secretion induced by IL-2 in the B cell line SKW6.4 (Ralph *et al.*,

Table II. Effect of 1H11 and Anti-Tac Antibodies on the IL-2-Induced Expression of Tac on TY Cells[a]

IL-2	1H11-1A5	Anti-Tac	Bound [^{125}I]7G7 Ab (cpm/10^6 cells)	Tac molecules per cell	Percentage of IL-2 control
–	–	–	280	560	–
+	–	–	6,200	12,400	100
+	+	–	1,550	3,100	25
+	–	+	6,500	13,000	105

[a] The expression of Tac antigen on YT cells treated with 200 pM IL-2 was measured using [^{125}I]7G7 antibody, which binds Tac at a site that does not overlap the binding sites of IL-2 or anti-Tac. Anti-Tac antibody or the anti-IL-2 antibody, 1H11, were included to determine their effect on Tac expression.

Table III. Effect of Anti-Tac and Anti-IL-2 Antibodies on the Cytolytic Activity of IL-2-Activated PBL[a]

Antibody	Lytic units per 10^7 cells	Percent inhibition
None	42.0	—
Control	42.0	0
Anti-Tac	41.0	0
1H11-1A5	13.4	68.1
R135αp8-27	0.8	98.1
9B11-1E5	8.0	81.0

[a] Peripheral blood lymphocytes were activated with 200 pm IL-2 for 5 h in the presence or absence of various antibody reagents. The cytotoxicity of the cells was then measured against an NK-sensitive target cell line in a standard ^{51} Cr release assay.

1984). Each of these responses thus appeared to be associated with non-Tac binding sites sharing certain characteristics with the β receptors. In the case of the NK precursors and the SKW6.4 cells, Dukovich *et al.* (1987) have shown that the IL-2 binding component responsible for the effects was the same size as the β protein(s) found on YT cells. Thus, the β receptor component, in addition to solving the mystery surrounding the structure of high-affinity binding sites, also provides the explanation for the observations (Ortaldo *et al.*, 1984; Ralph *et al.*, 1984) that originally supported the notion of a second, non-Tac IL-2 receptor.

2.3. Type III Receptors: The Alpha–IL-2–Beta Complex

In its simplest form, this type of receptor consists of a ternary complex of IL-2 together with the α and β receptor subunits. If each β binding site requires two different 70,000–75,000 MW proteins, however, it is possible that high-affinity sites are likewise composed of a more complex combination of subunits. Type III receptors bind IL-2 with a high affinity ($K_d \approx 5$–10 pM), which explains why they were the first form of the receptor detected (Robb *et al.*, 1981). Type III receptors, like the Tac protein, are associated with cellular activation, occurring on antigen- and lectin-treated T and B cells, IL-2-treated NK and LAK cells, and some $HTLV_I$-transformed cell lines. They are generally present in small numbers, being limited by the small amount of the β component. Responses of activated cells to low concentrations of IL-2, including proliferation, secretion of other lymphokines, and certain differentiation steps, have generally been attributed to IL-2 association with high-affinity receptors (Robb *et al.*, 1981; Ashwell *et al.*, 1986; Robb, 1984). Since these responses take hours or days to measure and since IL-2 binding to β receptors on resting cells generally results in the rapid expression of Tac protein and high-affinity α, β receptors, it is difficult to distinguish between responses that require the α, β structure and those for which β alone would be sufficient. Anti-Tac antibody totally blocks IL-2-dependent cellular proliferation at low IL-2 levels (Leonard *et al.*, 1982), but we have found that it is considerably less effective at higher IL-2 concentrations (100–600 pM) (R. J. Robb, unpublished observations). This result could simply indicate that inadequate amounts of antibody were used. Alternatively, at the higher IL-2 levels, where a significant proportion of intermediate β sites would be occupied, it is possible that the residual IL-2 binding to β alone might mediate proliferation. To resolve this question, conditions will have to be found in which proliferative responses of non-transformed, β-positive cells can be tested in the absence of expression of Tac antigen. If both the α and β chains are required for cellular proliferation, one would have to assume that signal transduction initiated by the β protein and the α, β complex differed in some fundamental way (i.e., perhaps a third receptor subunit associates with the α, β complex, but not with the β chain alone). If true, this situation would open up possibilities for modulating different responses to IL-2 based on selective manipulation of α–IL-2 or β–IL-2 interactions.

3. Ligand Binding and the Structure of the Tac Receptor Protein

The Tac protein consists of 251 amino acids and is encoded by a gene containing eight exon segments (Leonard *et al.*, 1985). Exon 1 encodes the leader sequence, exons 2–6 encode the extracellular region of the mature protein, exon 7 encodes the transmembrane region, and exon 8 encodes a small intracellular segment. The amino acid sequences encoded by exons 2 and 4 show a significant level of homology, suggesting that they represent protein domains derived from a common ancestor (Cosman *et al.*, 1984; Leonard *et al.*, 1985). Although the sequence of the Tac molecule is now known, much remains to be learned about the structural features of the molecule that contribute to its ability to bind IL-2 and to form high-affinity receptors with the β protein(s). To this end, we have initiated a number of investigations aimed at localizing the IL-2 binding site and examining the role of individual segments of the Tac molecule.

3.1. Localization of the IL-2 Binding Site

3.1.1. Antibody Inhibition Studies

Our first approach to localizing the IL-2 binding site on the Tac molecule was to prepare a series of rabbit polyclonal antibodies to synthetic fragments of the Tac protein (Kuo *et al.*, 1986). The synthetic peptides were synthesized from regions having a high average hydrophilicity. Following repeated immunization with peptide-conjugated KLH, sera were obtained that reacted in an ELISA assay with the respective peptide. Unfortunately, only a few of the sera reacted well with Tac protein on the surface of intact cells. These preparations, designated R234 and R239, recognized Tac segments consisting of residues 80–96 and 169–186, respectively (Kuo *et al.*, 1986). Each was tested for its ability to inhibit the binding of radiolabeled IL-2 to receptor-positive cells and the proliferative response of these cells to the growth factor, and each was found to be ineffective. Thus, these segments of the Tac molecule (amino acids 80–96 and 169–186) did not appear to overlap the ligand binding site.

3.1.2. Chemical Crosslinking Studies

As a second, more direct approach to localizing the IL-2 binding site on the Tac molecule, IL-2 was covalently crosslinked to the Tac receptor and the complex was fragmented to determine which Tac segment was attached to IL-2 (Kuo *et al.*, 1986). In this regard, we took advantage of an observation by Shackelford and Trowbridge (1986) that mild trypsin treatment cut the Tac molecule into two disulfide-linked pieces. Using our antibodies to synthetic Tac fragments, we determined that the larger tryptic piece (36,000 MW) was derived from the C terminus of the Tac molecule, while the smaller piece (26,000 MW) represented the N-terminal segment (Kuo *et al.*, 1986). Sequence analysis demonstrated that the principle cleavage site was between Arg_{83} and Lys_{84}.

Although the trypsin-generated pieces of Tac could be separated following reduction of disulfide bonds, this procedure also destroyed the ability of the pieces to bind IL-2. Therefore, IL-2 was first crosslinked to the receptor, followed by mild trypsin digestion of the covalent complex and SDS–PAGE analysis of the resulting fragments. Crosslinking of IL-2 to HUT cells at high IL-2 concentrations yielded mainly the 70,000 MW complex composed of IL-2 and the low-affinity Tac receptor, while crosslinking at low IL-2 concentrations yielded the 70,000 MW IL-2–Tac complex and the 80,000–90,000 MW bands corresponding to IL-2–β complexes (Kuo *et al.*, 1986). Under both sets of conditions, mild trypsin digestion yielded a major band at 34,800 MW on reduced SDS–PAGE. This trypsin-generated fragment must therefore have resulted from cleavage of the 70,000 MW IL-2–Tac complex. Treatment with an enzyme that removes Asn-linked carbohydrate demonstrated that the 34,800 MW complex contained two carbohydrate chains, each of 3000–4000 MW. This result indicated that the IL-2 was crosslinked to the trypsin-generated Tac segment composed of amino acids 1–83, since the only Asn-linked sugar in Tac occurs at amino acid positions 49 and 68 (Leonard *et al.*, 1984; Nikaido *et al.*, 1984).

The evidence indicated that IL-2 was preferentially crosslinked to a segment encoded by exons 2 (amino acids 1–64) and 3 (amino acids 65–101) of the Tac gene. Since $R234_{\alpha p80\text{-}96}$ did not affect IL-2 binding, we hypothesized that exon 2 provided at least some of the contact sites for IL-2 binding. The lack of crosslinking to the C-terminal tryptic piece, however, could merely have been due to an absence of appropriately spaced amino groups on the two molecules. Furthermore, since the amino acid segments encoded by exons 2 and 4 had significant sequence homology, it seemed reasonable that exon 4 might play some role in ligand contact.

3.2. Ligand Binding by Tac Protein Variants

Another approach used to localize the IL-2 binding site on the Tac molecule was the analysis of the functionality of mutant Tac molecules. Early studies of Tac protein expressed in mammalian cells transfected with Tac cDNA showed that such cells could bind IL-2. We sought to examine structure–function relations for Tac by modifying its cDNA and testing the expressed proteins.

Our initial modifications of Tac cDNA involved the removal of exons 4–8. Several forms of Tac mRNA are synthesized by activated human lymphocytes, with most of the variation due to different sites for polyadenylation (Leonard *et al.*, 1984). In addition, some of the mRNA is alternately spliced such that it is missing exon 4 (Leonard *et al.*, 1984). Early reports suggested that cells transfected with the corresponding exon 4-minus cDNA failed to express IL-2-binding molecules on their cell surface (Cosman *et al.*, 1984; Leonard *et al.*, 1984). To confirm and extend these reports, we transfected murine L cells with vectors containing full-length (pPK) and exon 4-minus (pCM) cDNA (Neeper *et al.*, 1987). Cells transfected with full-length cDNA expressed a Tac molecule on their surface that reacted with affinity supports coupled with IL-2 and the monoclonal antibodies anti-Tac and 7G7/B6 (see Fig. 7 for a summary of the results). Anti-Tac antibody binds Tac at a site overlapping the IL-2 binding site, while 7G7/B6 binds at a nonoverlapping site (Robb *et al.*, 1984; Rubin *et al.*, 1985a) Cells transfected with the exon 4-minus cDNA (pCM) made a protein of the expected size (35,000 MW) that was detected by its binding to antibodies made against synthetic Tac peptides. It did not, however, react with IL-2, anti-Tac, or 7G7/B6-coupled beads. The results thus demonstrated that structural elements encoded by exon 4 were essential to ligand binding despite the fact that IL-2 was not chemically crosslinked to Tac in this region.

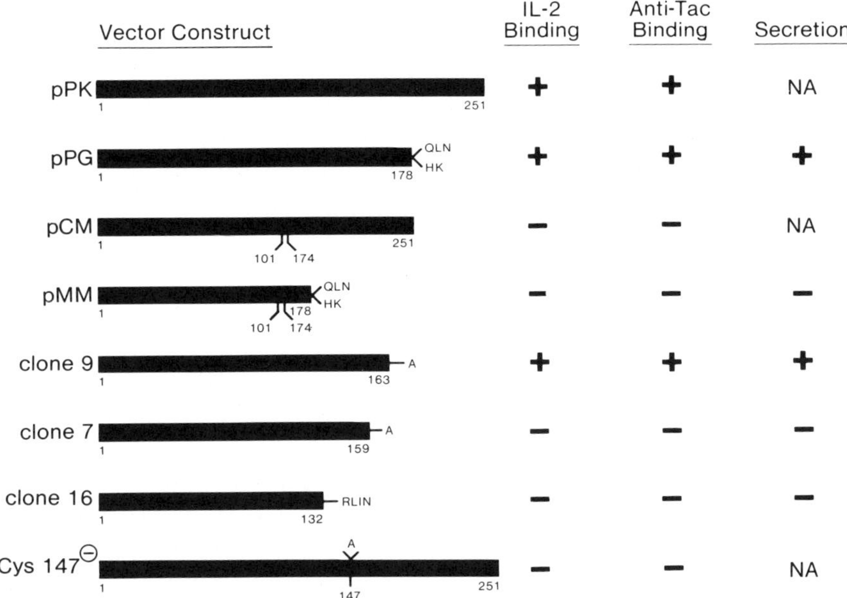

Figure 7. Summary of the results obtained with variant forms of the Tac protein expressed in murine L cells. Letters refer to the one-letter code for amino acids. The cells were tested for [^{125}I]IL-2 and [^{3}H]anti-Tac binding. In certain instances when the transmembrane segment (amino acids 220–238) of Tac was absent, the cells were also tested for their ability to secrete Tac protein (as detected using antipeptide antibodies R234 and R239). NA, not applicable.

To examine the contribution of exons 5–8, the full-length (pPK) and exon 4-minus (pCM) cDNA vectors were cut with the Mst-II restriction enzyme at a unique site near the beginning of exon 5, followed by insertion of a universal stop oligonucleotide (Neeper *et al.*, 1987). Cells transfected with the Mst-II-cut full-length cDNA (pPG) secreted a truncated Tac protein that bound to IL-2, anti-Tac, and 7G7/B6-coupled beads (Fig. 7). The cellular release of this form of Tac protein, which was missing the transmembrane and intracellular segments, confirmed the results of an earlier study (Traiger *et al.* 1986). In addition, its ability to bind IL-2 demonstrated that information encoded by exons 5–8 was not essential. Cells tranfected with the MST-II-cut version of exon 4-minus cDNA (pMM) produced a still smaller Tac protein that was unreactive with IL-2, anti-Tac, and 7G7/BG and that, in contrast to the product of the pPG vector, was not efficiently secreted.

Based on the above results, exon 4 clearly provided key information for ligand contact or for proper folding of the Tac protein. To determine what portion of exon 4 was essential, the pPG vector was digested with Bal 31 starting at a point within the universal stop oligonucleotide segment (L.-M. Kuo, M. P. Neeper, and R. J. Robb, unpublished observations). Following insertion of a new stop oligonucleotide segment, the resulting vectors were selected by size to encode a variety of cDNAs stopping at different points within exon 4. Transfection of L cells with these vectors demonstrated that a cDNA encoding Tac residues 1–163 resulted in the secretion of truncated, but active, Tac molecules, while one encoding residues 1–159 and one encoding residues 1–132 resulted in inactive protein molecules (Fig. 7). From these results, it appeared likely that Cys 163 was essential to the conformation of the Tac protein necessary for reaction with IL-2 and anti-Tac and for efficient secretion from the cell. As discussed in Section 3.3, Cys 163 is involved in a disulfide bond with Cys 131 and this bond is presumably essential to the folding of the molecule. The Cys 131–163 bond, however, is not the only essential element encoded by exon 4. The segments encoded by exons 2 and 4 are linked by a disulfide bond between Cys 3 and Cys 147 (also discussed in Section 3.3). The site-specific mutation of Cys 147 to an Ala residue also resulted in the complete loss of IL-2, anti-Tac, and 7G7/B6 reactivity (M. P. Neeper and R. J. Robb, unpublished observations) (Fig. 7). Thus, several elements of the exon 4-encoded segment are essential to proper protein folding or contact with the IL-2 ligand.

3.3. Disulfide Structure of the Tac Protein

The Tac receptor protein contains 12 cysteine residues, including one located in the transmembrane segment (Leonard *et al.*, 1984). Intramolecular disulfide structure is clearly essential to an active conformation of the molecular, since the interaction of Tac with IL-2, as well as its interaction with monoclonal antibodies anti-Tac and 7G7/B6, is substantially or totally destroyed by reducing agents. Intermolecular disulfide bonding, on the other hand, plays a role in the formation of Tac homodimers (Kuo *et al.*, 1986). Because of the importance of intramolecular disulfide bonds to the folding of the Tac protein, we sought to identify the cysteine residues involved.

A preliminary assignment of disulfide bonding was made based on the coelution of[^{35}S]Cys-containing fragments made from biosynthetically labeled Tac protein. The protein was derived from detergent-solubilized HUT 102 cells and was purified on an anti-Tac immunoaffinity column. It was cut by a combination of trypsin and chymotrypsin under conditions minimizing disulfide exchange and the resulting fragments were separated by reverse-phase HPLC (Fig. 8). Peptides in peaks containing the^{35}S tracer were sequenced to identify the fragments present. Peptides containing Cys 3 and Cys 147 consistently coeluted, as did fragments containing Cys 131 and 163 and

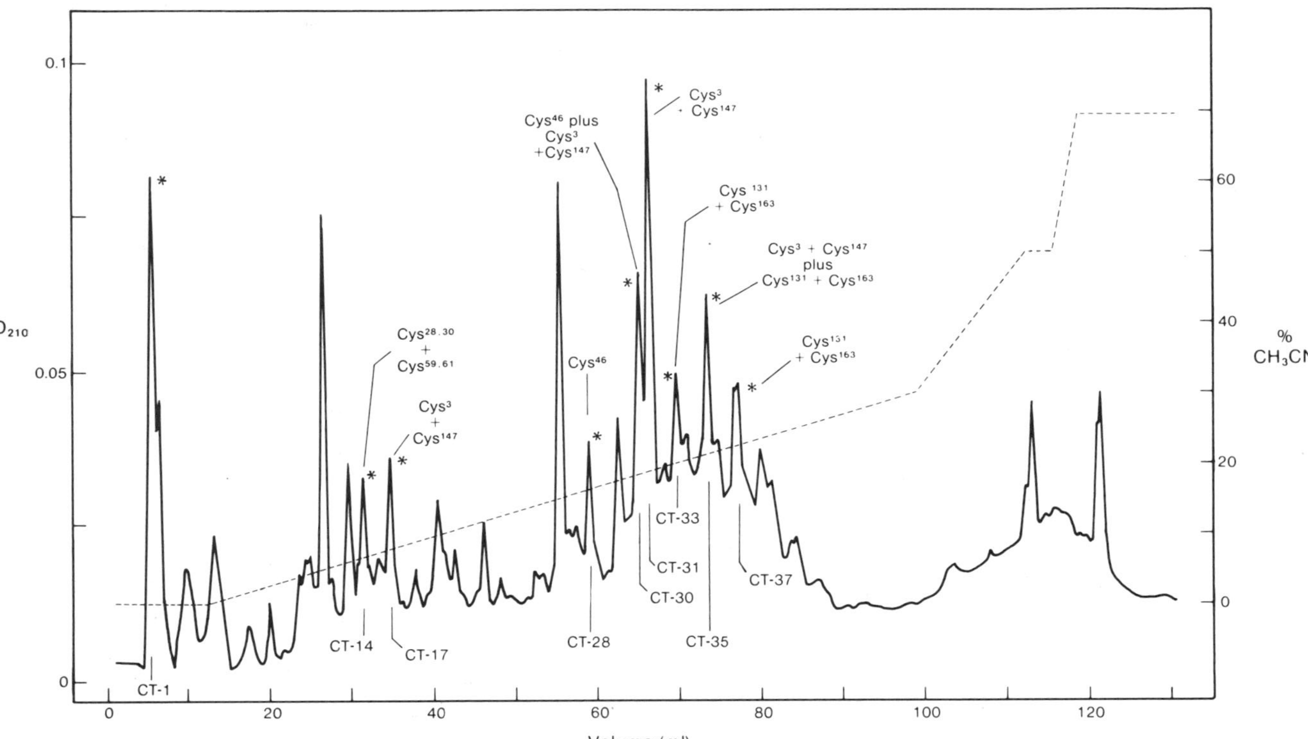

Figure 8. Reverse-phase chromatography of Tac peptides generated by a combination of chymotrypsin and trypsin cleavage. The peaks were identified by amino acid composition and sequence analyses. Trace amounts of Tac protein biosynthetically labeled with [^{35}S]Cys were included in the reaction to quickly identify peaks containing free or disulfide-bonded cysteine/cystine containing fragments.

fragments containing Cys 28,30 and Cys 59,61 (R. J. Robb and R. M. Kutny, unpublished observations). The fact that some of the disulfide-bonded peptide pairs appeared at several locations in the gradient was due in part to differences in the location of the enzymatic cleavages that generated the pieces. On this basis, it appeared that Cys 3 and 147 and Cys 131 and 163 formed disulfide bonds (Fig. 9). The designation of the Cys 3–147 bond was particularly noteworthy since it confirmed the trypsin cleavage data (Kuo *et al.*, 1986), which revealed that Tac fragments 1–83 and 84–251 were linked by a disulfide bond. The fragments containing Cys 28,30 and Cys 59,61 also appeared to be linked by one or two disulfide bonds. Based on the homology between exons 2 and 4, Cys 30 and 61 correspond to the disulfide-bonded Cys 131 and 163. We therefore hypothesize that Cys 30 and 61 are similarly linked (Fig. 9). We recovered only small quantities of the fragments containing Cys 46, 104, 192, and 225 and could not determine whether they were free or involved in disulfide bonds. Since Cys 225 is in the transmembrane segment, we suspect that it is not linked to another Cys residue in an intramolecular bond, although it might play a role in intermolecular disulfide-linked homodimers. Obviously, given the even number of cysteine residues, if Cys 225 is not involved in intramolecular bonding, then at least one other Cys (46, 104, or 192) is also free.

All 12 cysteine residues in the Tac protein are found in similar positions within the homologous murine protein (Miller *et al.*, 1985), further emphasizing their importance to the molecular conformation. Counterparts to cysteines 3, 30, 46, and 61 from exon 2 and cysteines 104, 131, 147, and 163 from exon 4 of the Tac protein are also found in the related sequence of the Ba fragment of human complement factor B (Leonard *et al.*, 1985). The significance of the configuration of exons 2 and 4 and the effect imparted by

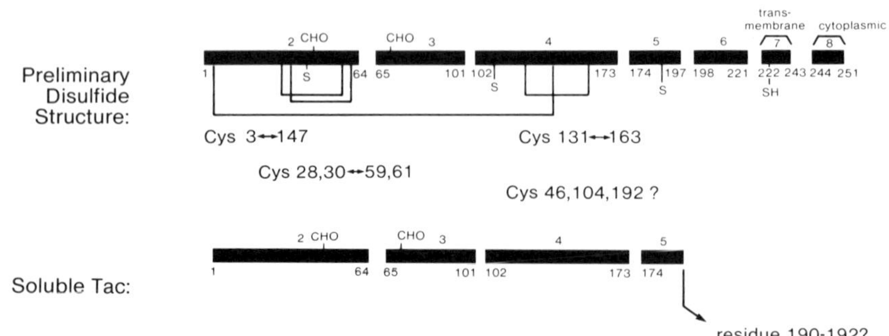

Figure 9. Preliminary results on the disulfide structure of Tac protein and on the sequence of soluble Tac released by HUT 102 cells.

the Cys 3–147 bond may therefore have implications for a number of important proteins.

3.4. Sequence Analysis of Soluble Tac Protein

In addition to its normal cell surface expression, Tac protein is also found in a soluble form in the *in vitro* culture supernatant of Tac-positive cells and in the sera of animals or human patients undergoing active immune responses or receiving IL-2 as part of an immunotherapy program (Rubin *et al.*, 1985b; Osawa *et al.*, 1986; Reske-Kunz *et al.*, 1987). The soluble form of the molecule was found to be 8,000–12,000 daltons smaller than the membrane-associated form and most of the difference appeared to be due to a loss of amino acids rather than carbohydrate (Rubin *et al.*, 1986). We have been interested in the phenomenon of soluble Tac from two standpoints. We were interested first in the affinity of the association between IL-2 and soluble Tac protein and second in the mechanism behind the cellular release of soluble Tac.

As a source of soluble Tac protein, we used the culture supernatant of HUT 102 cells. The protein was purified by a combination of anti-Tac immunoaffinity chromatography and reverse-phase HPLC (Robb and Kutny, 1987). Titration of soluble Tac in an [^{125}I]IL-2 cell binding assay demonstrated that it bound IL-2 with a K_d of approximately 10 nM, which was very similar to our estimate of the dissociation constant for low-affinity cellular Tac receptors. At high concentrations in patient sera, soluble Tac could potentially have an inhibitory effect on IL-2-dependent responses. For localized responses involving IL-2-producing and -responding cells that are in close proximity, however, the effect may be negligible. A final determination of the significance of soluble Tac should be forthcoming from *in vivo* analysis of animal model systems (i.e., mice treated with the soluble murine analogue).

With regard to the mechanism of release of soluble Tac, four different scenarios come readily to mind. First, the membrane and soluble forms of Tac could be encoded by separate genes. Second, soluble Tac could be produced as a result of alternate mRNA splicing. Third, soluble Tac could represent a proteolytic cleavage product of membrane-associated molecules, perhaps as part of their normal turnover. Finally, soluble Tac could be derived as a combination of mechanisms 2 and 3. Mechanism 1 was discarded on the basis that only a single structural gene exists for the Tac protein (Leonard *et al.*, 1985). To examine the contribution of the other mechanisms, we chose to analyze the structure of soluble Tac derived from HUT 102 cells. The purified soluble Tac protein reacted with rabbit polyclonal antibodies prepared against synthetic fragments from the molecule's extracellular segment, but not with an antibody to its intracellular segment (Robb and Kutny, 1987). Sequence analysis of the intact molecule further demonstrated that the protein began with the same N-terminal sequence as membrane-associated molecules. Thus, the release of soluble Tac was accompanied by the loss of C-terminal segments.

To complete the structural analysis of soluble Tac, the protein was reduced and alkylated and fragmented with trypsin. The resulting peptides were separated by reverse-phase HPLC and each peak was identified by amino acid composition and/or sequence analysis. Virtually all the fragments expected from within the first 192 amino acids were recovered, with no concrete evidence of unusual fragments that could have been derived by alternate mRNA splicing (Fig. 10). Carboxypeptidase Y digestion of soluble Tac released a mixture of amino acids consisting of Cys, Ser, Thr, and Glu, which matched the six residues found in positions 187–192. Thus, the evidence indicated that the predominant form of soluble Tac extended from

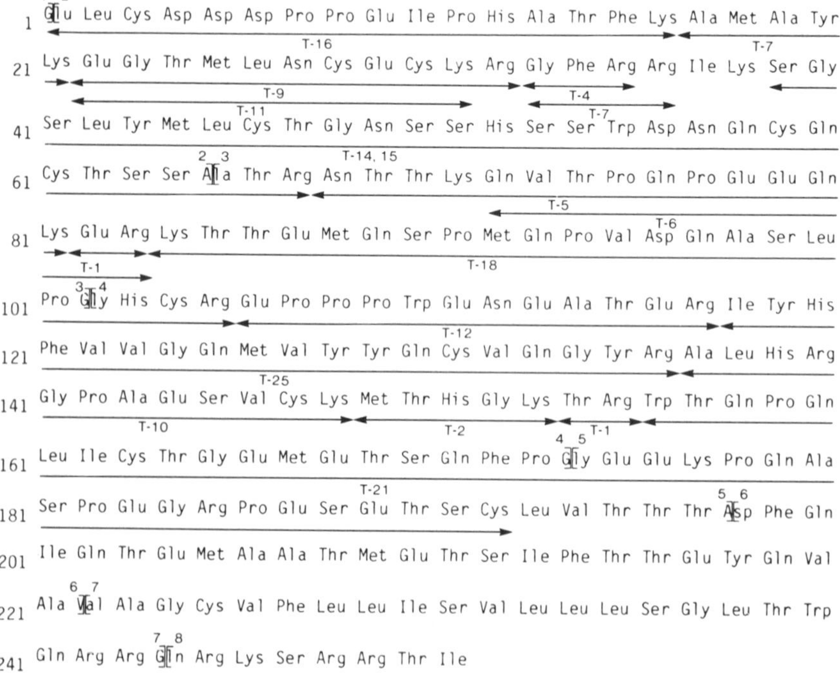

Figure 10. Sequence of the Tac protein showing exon junctions (brackets) and the positions of tryptic peptides isolated from soluble Tac protein released by HUT 102 cells. Reproduced from Robb and Kutny (1987).

position 1 to position 192 of the full-length molecule. The fact that the C-terminal tryptic fragment eluted in several positions on reverse-phase chromatography, however, suggested that there may have been heterogeneity in the C-terminal amino acid position. Alternatively, the heterogeneity may have been in the carbohydrate content of this fragment.

The sequence results from soluble Tac protein were consistent with proteolytic processing near the end of the segment encoded by exon 5. Since we have evidence that membrane-associated Tac molecules labeled by cell surface iodination were released in a soluble form, the data supported a proteolytic event occurring after cell surface expression. Nevertheless, while some soluble Tac may be so derived, we cannot comment on the contribution of soluble Tac made by alternate mRNA splicing or a combination of alternate splicing followed by proteolytic trimming. Moreover, the *in vitro* product of $HTLV_I$-transformed HUT cells may not reflect the structure of soluble Tac released by activated cells *in vivo*. Further studies of the processing of Tac mRNA and of the nature of the enzymes involved in proteolytic cleavage of the protein are clearly required before a definitive conclusion about its origins can be reached.

4. Concluding Remarks

In this chapter, I have presented evidence that indicates that IL-2 receptors exist in at least three forms, which differ in their composition, affinity for ligand, and ability to mediate cellular responses. The different forms of the receptor were composed of two glycoprotein subunits, alpha (Tac) and beta, each of which independently bound IL-2. The simultaneous interaction of IL-2 with both receptor subunits provided the elusive molecular basis for high-affinity IL-2 binding. Although both receptor subunits bound IL-2 when present alone, only the newly recognized β subunit appeared capable of independently mediating cellular responses. IL-2 binding to β receptor sites resulted in such events as ligand internalization, induction of Tac expression, short-term activation of NK cells, and immunoglobulin secretion by a particular B cell line.

In addition to discussing the subunit structure of the various forms of the IL-2 receptor, I described some of the requirements for the interaction of IL-2 with the Tac protein. Crosslinking studies demonstrated a close proximity between bound IL-2 and lysine residues found in Tac exons 2 and 3. Furthermore, analysis of ligand binding by artificial variants of the Tac protein demonstrated the importance of structural elements encoded by both exons 2 and 4, with special emphasis on the role of intramolecular disulfide bonding. Finally, sequence analysis indicated that the release of soluble Tac protein by activated lymphocytes was accompanied by some form of proteolytic processing.

Although our understanding of the structure–function relations in the IL-2–receptor system has improved considerably in the past 2 years, many areas remain to be explored. Of foremost importance is the need to clone and characterize the β receptor subunit. The β protein many possess a readily recognizable enzymatic activity (GTP binding, tyrosine kinase) that will account for the initial step in signal transduction following IL-2 binding. In addition, cloned β protein(s) should answer the question of whether β binding sites consist of a single protein or of two or more distinct molecules. The availability of β cDNA(s) will also make possible attempts to reconstitute high-affinity IL-2 binding sites in murine L cells expressing the Tac protein. This experiment will provide the final proof for the two-chain hypothesis of high-affinity IL-2 receptors. It will also provide a framework for determining if Tac–β complexes are preformed or occur only in the presence of IL-2 and if other unidentified receptor subunits interact with β or the Tac–IL-2–β complex in the signal transduction process. Finally, the availability of IL-2 and the two receptor subunits in highly purified forms will spur attempts to cocrystallize the ternary receptor–ligand complex. The resulting information could prove invaluable to designing mimics and antagonists of IL-2 that will be useful for the therapeutic manipulation of the immune system.

References

Ashwell, J. D., Robb, R. J., and Malek, T. R., 1986, Proliferation of T lymphocytes in response to interleukin 2 varies with their state of activation, *J. Immunol.* **137:**2572–2578.

Cosman, D., Cerretti, D. P., Larsen, A., Park, L., March, C., Dower, S., Gillis, S., and Urdal, D., 1984, Cloning, sequence and expression of human interleukin-2 receptor, *Nature* **312:**768–771.

Dukovich, M., Wano, Y., Le Thi Bich Thuy, Katz, P., Callen, B. R., Kerhl, J. H., and Greene, W. C., 1987, A second human interleukin 2 binding protein that may be a component of high-affinity IL-2 receptors, *Nature* **327:** 518–522.

Fujii, M., Sugamura, K., Sano, K., Nakai, M., Sugita, K., and Hinuma, Y., 1986, High-affinity receptor-mediated internalization and degradation of interleukin 2 in human T cells, *J. Exp. Med.* **163:**550–562.

Greene, W. C., Robb, R. J., Svetlik, P. B., Rusk, C. M., Depper, J. M., and Leonard, W. J., 1985, Stable expression of cDNA encoding the human interleukin 2 receptor in eucaryotic cells, *J. Exp. Med.* **162:**363–368.

Hatakeyama, M., Minamoto, S., Uchiyama, T., Hardy, R. R., Yamada, G., and Taniguchi, T., 1985, Reconstitution of functional receptor for human interleukin-2 in mouse cells, *Nature* **318:**467–470.

Kondo, S., Shimizu, A., Maeda, M., Tagaya, Y., Yodoi, J., and Honjo, T., 1986a, Expression of functional human interleukin 2 receptor in mouse T cells by cDNA transfection, *Nature* **320:**75–77.

Kondo, S., Shimizu, A., Saito, Y., Kinoshita, M., and Honjo, T., 1986b, Molecular basis for two different affinity states of the interleukin 2 receptor: Affinity conversion model, *Proc. Natl. Acad. Sci. U.S.A.* **83:**9026–9029.

Kuo, L.-M., and Robb, R. J., 1986, Structure–function relationships for the IL 2 receptor system. I. Localization of a receptor binding site on IL 2, *J. Immunol.* **137:**1538–1543.

Kuo, L.-M., Rusk, C. M., and Robb, R. J., 1986, Structure–function relationships for the IL 2–receptor system II. Localization of an IL 2 binding site in high- and low-affinity receptors, *J. Immunol.* **137:**1544–1551.

Leonard, W. J., Depper, J. M., Uchiyama, T., Smith, K. A., Waldmann, T. A., and Greene, W. C., 1982, A monoclonal antibody that appears to recognize the receptor for human T cell growth factor, *Nature* **300:**267–269.

Leonard, W. J., Depper, J. M., Crabtree, G. R., Rudikoff, S., Pumphrey, J., Robb, R. J., Krönke, M., Svetlik, P. B., Peffer, N. J., Waldmann, T. A., and Greene, W. C., 1984, Molecular cloning and expression of cDNAs for the human interleukin-2 receptor, *Nature* **311:**626–631.

Leonard, W. J., Depper, J. M., Kanehisa, M., Krönke, M., Peffer, N. J., Svetlik, P. B., Sullivan, M., and Greene, W. C., 1985, Structure of human interleukin-2 receptor gene, *Science* **230:**633–639.

Miller, J., Malek, T. R., Leonard, W. J., Greene, W. C., Shevach, E. M., and Germain, R. N., 1985, Nucleotide sequence and expression of a mouse interleukin 2 receptor cDNA, *J. Immunol.* **134:**4212–4217.

Neeper, M. P., Kuo, L.-M., Kiefer, M. C., and Robb, R. J., 1987, Structure–function relationships for the IL 2–receptor system. III. Tac protein missing amino acids 102–173 (exon 4) is unable to bind IL 2. Detection of spliced protein after L cell transfection, *J. Immunol.* **138:**3532–3538.

Nikaido, T., Shimizu, N., Ishida, N., Sabe, H., Teshigawara, K., Maeda, M., Uchiyama, T., Yodoi, J., and Honjo, T., 1984, Molecular cloning of cDNA encoding human interleukin-2 receptor, *Nature* **311:**631–635.

Ortaldo, J. R., Mason, A. T., Gerard, J. P., Henderson, L. E., Farrar, W., Hopkins, R. F., Herberman, R. B., and Rabin, H., 1984, Effects of natural and recombinant IL-2 on regulation of IFN_{γ} production and natural killer activity: Lack of involvement of the Tac antigen for these immunoregulatory effects, *J. Immunol.* **133:**779–783.

Osawa, H., Josimovic-Alasevic, O., and Diamantstein, T., 1986, Interleukin 2 receptors are released by cells *in vitro*. I. Detection of soluble IL 2 receptors in cell culture supernatants and in the serum of mice by an immunoradiometric assay. *Eur. J. Immunol.* **16:**467–469.

Ralph, P., Jeong, G., Welte, K., Mertelsmann, R., Rabin, H., Henderson, L. E., Souza, L. M., Boone, T. C., and Robb, R. J., 1984, Stimulation of immunoglobulin secretion in human B lymphocytes as a direct effect of high concentration of IL-2, *J. Immunol.* **133:**2442–2445.

Reske-Kunz, A. B., Osawa, H., Josimovic-Alasevic, O., Rude, E., and Diamantstein, T., 1987, Soluble interleukin 2 receptors are released by long-term-cultured insulin-specific T cells transiently after contact with antigen, *J. Immunol.* **138:**192–196.

Robb, R. J., 1984, Interleukin 2: The molecule and its function, *Immunol. Today* **5:**203–209.

Robb, R. J., 1985, Human interleukin 2, in: *Methods in Enzymology* (G. diSabato, J. J. Langone, and H. VanVunakis, eds.), Vol. 116, pp. 493–525, Academic Press, Orlando, FL.

Robb, R. J., 1986, Conversion of low-affinity interleukin 2 receptors to a high-affinity state following fusion of cell membranes, *Proc. Natl. Acad. Sci. U.S.A.* **83:**3992–3996.

Robb, R. J., and Greene, W. C., 1983, Direct demonstration of the identity of T cell growth factor binding protein and the Tac antigen, *J. Exp. Med.* **158:**1332–1337.

Robb, R. J., and Greene, W. C., 1987, Internalization of interleukin 2 is mediated by the beta chain of the high-affinity IL-2 receptor, *J. Exp. Med.* **165:**1201–1206.

Robb, R. J., and Kutny, R. M., 1987, Structure–function relationships for the IL 2–receptor system, IV. Analysis of the sequence and ligand-binding properties of soluble Tac protein *J. Immunol.* **139:**855–862.

Robb, R. J., and Lin, Y., 1984, T-cell growth factor: Purification, interaction with a cellular receptor, and *in vitro* synthesis, in: *Thymic Hormones and Lymphokines '83* (A. L. Goldstein, ed.), pp. 247–256, Plenum Press, New York.

Robb, R. J., and Rusk, C. M., 1986, High and low affinity receptors for interleukin 2: Implications of pronase, phorbol ester, and cell membrane studies upon the basis for differential ligand affinities, *J. Immunol.* **137:**142–149.

Robb, R. J., Munck, A., and Smith, K. A., 1981, T cell growth factor receptors: Quantitation, specificity, and biological relevance. *J. Exp. Med.* **154:**1455–1474.

Robb, R. J., Kuntny, R. M., Panico, M., Morris, H. R., DeGrado, W. F., and Chowdhry, V., 1983, Post-translational modification of human T-cell growth factor, *Biochem. Biophys. Res. Commun.* **116:**1049–1055.

Robb, R. J., Greene, W. C., and Rusk, C. M., 1984, Low and high affinity cellular receptors for interleukin 2: Implications for the level of Tac antigen, *J. Exp. Med.* **160:**1126–1146.

Robb, R. J., Rusk, C. M., Yodoi, J., and Greene, W. C., 1987, An interleukin 2 binding molecule distinct from the Tac protein: Analysis of its role in formation of high-affinity receptors, *Proc. Natl. Acad. Sci. U.S.A.* **84:**2002–2006.

Rubin, L. A., Kurman, C. C., Biddison, W. E., Goldman, N. D., and Nelson, D. L., 1985a, A monoclonal antibody, 7G7/B6, binds to an epitope on the human interleukin-2 (IL-2) receptor that is distinct from that recognized by IL-2 and anti-Tac, *Hybridoma* **4:**91–102.

Rubin, L. A., Kurman, C. C., Fritz, M. E., Biddison, W. E., Boutin, B., Yarchoan, R., and Nelson, D. L., 1985b, Soluble interleukin 2 receptors are released from activated human lymphoid cells *in vitro, J. Immunol.* **135:**3172–3177.

Rubin, L. A., Jay, G., and Nelson, D. L., 1986, The released interleukin 2 receptor binds interleukin 2 efficiently, *J. Immunol.* **137:**3841–3844.

Shackelford, D. A., and Trowbridge, I. S., 1986, Structural features of the human interleukin-2 receptor, in: *Advances in Gene Technology: Molecular Biology of the Endocrine System.* (D. Puett, F. Ahmed, S. Black, D. M. Lopez, H. H. Melner, W. A. Scott, and W. J. Whelan, eds.), pp. 390–391, Cambridge University Press, Cambridge, England.

Sharon, M., Klausner, R. D., Cullen, B. R., Chizzonite, R., and Leonard, W. J., 1986, Novel interleukin 2 receptor subunit detected by crosslinking under high-affinity conditions, *Science* **234:**859–864.

Teshigawara, K., Wang, H.-M., Kato, K., and Smith, K. A., 1987, Interleukin 2 high-affinity receptor expression requires two distinct binding proteins, *J. Exp. Med.* **165:**223–228.

Treiger, B. F., Leonard, W. J., Svetlik, P., Rubin, L. A., Nelson, D. L., and Greene, W. C., 1986, A secreted form of the human interleukin 2 receptor encoded by an "anchor minus" cDNA, *J. Immunol.* **136:**4099–4105.

Tsudo, M., Kozak, R. W., Goldman, C. K., and Waldmann, T. A., 1986, Demonstration of a new non-Tac peptide that binds interleukin 2: A potential participant in a multichain interleukin-2 receptor complex, *Proc. Natl. Acad. Sci. U.S.A.* **83:**9694–9698.

Weissman, A. M., Harford, J. B., Svetlik, P. B., Leonard, W. J., Depper, J. M., Waldmann, T. A., Greene, W. C., and Klausner, R. D., 1986, Only high-affinity receptors for interleukin 2 mediate internalization of ligand, *Proc. Natl. Acad. Sci. U.S.A.* **83:**1463–1468.

Yodoi, J., Teshigawara, K., Nikaido, T., Fukui, K., Noma, T., Honjo, T., Takigawa, M., Sasaki, M., Minato, N., Tsudo, M., Uchiyama, T., and Maeda, M., 1985, TCGF (IL-2)-receptor inducing factor(s). I. Regulation of IL-2 receptor on a natural killer-like cell line (YT cells), *J. Immunol.* **134:**1623–1630.

8

Cachectin (Tumor Necrosis Factor)

A Macrophage Protein that Induces a Catabolic State and Septic Shock in Infected Animals

KIRK R. MANOGUE and ANTHONY CERAMI

1. Introduction

Invasive stimuli, including parasitic, bacterial, or viral infection and neoplastic disease, precipitate catabolic changes in cellular metabolism and pathological alterations in the physiology of their mammalian hosts. These disruptions of normal homeostasis, left unresolved, can lead to the depletion of host energy stores advancing to wasting (cachexia), tissue damage, multiple organ system failure, shock, and death. Although such symptomatic patterns are all too familiar to clinicians, the underlying causes of this progressive decline have remained obscure. Because these patterns are frequently attributed to the metabolic demands or direct actions of the pathogen or its products, treatment has quite naturally focused on aggressive elimination of the invading agent and reactive support of failing host physiology. It now appears that invasive stimuli provide a trigger for the release of host-secreted cytokines, and the combined actions of these endogenous mediators elicit most or all of the biologic responses culminating in the pathophysiology of cachexia and shock. This new appreciation for the essential role of host-derived inflammatory mediators has provided a focus for new therapeutic approaches to a broad spectrum of invasive diseases.

The perceived role of bacterial endotoxin (lipopolysaccharide, LPS) in the pathogenesis of septic shock reflects these changing views. Experimental administration of LPS to animals can lead to septic shock, tissue damage to

KIRK R. MANOGUE and ANTHONY CERAMI • Laboratory of Medical Biochemistry, The Rockefeller University, New York, New York 10021.

most organ systems, and death. In fact, these characteristic inflammatory responses defined LPS as the bacterial principle conferring the toxicity of Gram-negative sepsis. Yet endotoxin is not directly toxic to most tissues, and a strain of mice (C3H/HeJ) is unaffected by endotoxin at doses that are lethal to a closely related strain, C3H/HeN. The transplantation of bone marrow from LPS-sensitive C3H/HeN donors to insensitive C3H/HeJ mice transfers LPS sensitivity to the previously resistant recipients and vice versa (Michalek *et al.*, 1980). These transplant results show that host factors of hematopoietic cell origin, rather than endotoxin itself, must account for LPS sensitivity. The macrophage was implicated as a source of such mediators (Vogel *et al.*, 1980; Ha *et al.*, 1983) and was eventually shown to be capable of producing, in response to LPS *in vitro*, a mediator that could kill endotoxin-resistant C3H/HeJ mice (Cerami *et al.*, 1985). Administration of this monokine, cachectin (tumor necrosis factor, TNF), which is now available as a recombinant peptide, resulted in pathophysiologic responses that very closely mimicked those attendant to experimental endotoxemia and bacteremia, including cachexia and septic shock. By specifically intervening in the host's cachectin response to endotoxin with cachectin-specific antibodies, many pathophysiologic responses are muted and death can be prevented (Beutler *et al.*, 1985d).

As cachectin's role in mediating specific components of inflammatory responses, both directly and in concert with other factors, continues to be defined, new targets should be revealed for therapeutic intervention in invasive disease.

2. The Parallel Histories of Cachectin and Tumor Necrosis Factor

The macrophage-derived cytokine *cachectin* was so named during its isolation by Cerami and colleagues in studies to determine the proximal causes of cachexia, a chronic wasting diathesis common to a variety of invasive disease states. Before dying from infections of the protozoal parasite *Trypanosoma brucei brucei*, rabbits commonly waste to less than half of their original body mass (Rouzer and Cerami, 1980). The metabolic demands of the light parasite burdens they harbored could not account for this severe weight loss. During the final few weeks of life, the infected rabbits became severely anorectic and exhibited a striking plasma lipemia (previously noted by Guy, 1975). The elevated plasma lipid levels were due to hypertriglyceridemia associated with very-low-density lipoprotein (VLDL). Rouzer and Cerami (1980) further determined that this seemingly paradoxical metabolic state—that animals were wasting to death while their blood carried elevated metabolic energy resources—had been caused by a profound systemic suppression of the enzyme lipoprotein lipase (triacylglyceroprotein acyl hydrolase EC 3.1.1.34, LPL), which is necessary for the clearance of triglycerides from the circulation.

A similar model of suppressed lipoprotein lipase activity resulting in a net serum lipemia had been noted in mice treated with endotoxin (bacterial lipopolysaccharide, LPS) (Sakaguchi and Sakaguchi, 1979)., Kawakami and Cerami (1981) exploited two genetically similar strains of mice—one sensitive to LPS (C3H/HeN) and the other resistant to LPS (C3H/HeJ)—to show that LPL suppression was caused by a transferable host-derived mediator(s) rather than the direct action of bacterial LPS. C3H/HeJ mice, which do not exhibit LPL suppression in response to endotoxin, do suffer suppression of adipose tissue LPL when injected with serum from C3H/HeN mice previously treated with endotoxin. The humoral factor conferring this effect was produced by peritoneal exudate cells from the endotoxin-sensitive but not the endotoxin-resistant strain, when challenged *in vitro* with LPS. The major factor conferring this effect was termed cachectin and was purified to homogeneity as a prominent secretory product of LPS-stimulated cells of the murine macrophage cell line RAW 264.7 (Beutler *et al.*, 1985b). Subsequent characterization of the 17-kD polypeptide revealed amino acid sequence homology and biological activity in common with human TNF (Beutler *et al.*, 1985a).

Both the long search to define TNF and the work to characterize its tumorolytic effects have been reviewed in detail elsewhere (Old, 1985; Paladino *et al.*, 1987) and will be summarized only briefly here. Dr. William Coley's observations at the turn of the century that bacterial infections or extracts could lead to the hemorrhagic necrosis of tumors led Shear and his colleagues to the isolation of the active bacterial agent—lipopolysaccharide—nearly 50 years later (Shear and Perrault, 1944). Within another 20 years they had demonstrated that serum from LPS-treated mice could be used to transfer hemorrhagic necrosis activity to another animal (O'Malley *et al*, 1962). In the early 1970s, complementary observations by Carswell *et al.* (1975) in the laboratory of Lloyd Old provided a definitive description of TNF as a serum-borne protein factor arising from endotoxin treatment of mice primed with the mycobacterium bacillus Calmette-Guerin (BCG), capable of inducing necrosis of tumors *in vivo* and exerting cytostatic or cytocidal effects on certain tumor cell lines *in vitro*. This work led to a standardized bioassay for TNF, depending on its *in vitro* cytolytic action on actinomycin-D-treated mouse fibroblast L-929 cells that allowed the eventual purification of TNF. Several lines of research showed that TNF was a cellular product of the monocyte/macrophage lineage and that similar bioactivities could be elicited from various species (Matthews, 1978; Mannel *et al.*, 1980; Matthews, 1981a,b; Zacharchuck *et al.*, 1983). TNF was first purified from the serum-free supernatants of a stimulated human promyelocytic HL-60 subclone by

Aggarwal *et al.* (1985c). Members of this group had already succeeded in purifying a related cytotoxic and tumorolytic factor, lymphotoxin (tumor necrosis factor β, LT) (Aggarwal *et al.*, 1984, 1985b), and they were able to demonstrate significant homology between the NH_2-terminal sequence of TNF and a fragment of LT.

Immediately thereafter, Beutler *et al.* (1985a) showed that the amino terminal sequence of purified murine cachectin was highly homologous to human TNF and that the two peptides shared a spectrum of bioactivities and immunological cross-reactivity. They suggested that these two monokines, one presumed to mediate cachexia or shock leading to death and the other thought to account for the necrosis of tumors without harming the host, were in fact one and the same protein. Independent genetic sequence analysis soon confirmed this hypothesis, uniting the pathological threat of cachectin and the therapeutic promise of TNF in a paradoxical biochemical identity.

3. Structure of the Cachectin/TNF Peptide, Message, and Gene

3.1. Several Mammalian Species Express Homologous Cachectins as Prohormones

The primary structure of cachectin/TNF peptides of three mammalian species—rabbit (Ito *et al.*, 1986a,b), mouse (Fransen *et al.*, 1985; Pennica *et al.*, 1985; Caput *et al.*, 1986), and human (Pennica *et al.*, 1984; Marmenout *et al.*, 1985; Nedwin *et al.*, 1985a; Shirai *et al.*, 1985; Wang *et al.*, 1985)—has now been deduced from cDNA clones, and the human sequence has been confirmed directly by protein sequencing (Aggarwal *et al.*, 1985c). In each species, cachectin is produced as a prohormone. Both the propeptide (which is cleaved at several sites to yield the active monokine) and the mature polypeptide are highly conserved.

Rabbit cachectin is produced as a 234 amino acid prohormone, and the mature cytokine consists of the C-terminal 154 residues, which do not provide a glycosylation site (Ito *et al.*, 1986b). The mature rabbit monokine is 82% homologous with human cachectin, with comparable nucleotide sequence homology in each of the four exons of the two species (Ito *et al.*, 1986a).

The murine cachectin translation product is a 235 amino acid polypeptide consisting of a 79 residue prosequence (3 amino acids longer than the human prosequence) followed by the mature hormone sequence of 156 amino acids (Fransen *et al.*, 1985). Natural murine TNF is glycosylated at a single N-linked site, an asparagine at position 7 of the mature murine TNF that precedes two serine residues (Asn-Ser-Ser). This arrangement conforms to the prototypic (Asp-X-Ser/Thr) eukaryotic glycosylation site triplet. This triplet site is not conserved in the human sequence (Fransen *et al.*, 1985); in fact, no such glycosylation sites occur in the human sequence and human cachectin is not glycosylated (Pennica *et al.*, 1985). The glycosylation of murine TNF includes sialic acid and galactosamine (Green *et al.*, 1976). The calculated molecular weight of murine TNF monomer is 17,200 daltons for the peptide portion alone (exclusive of glycosylation), and relative molecular weight estimates by SDS–PAGE under reducing conditions agree closely with this figure. Nonglycosylated murine TNF (recombinant from *E. coli*) (Pennica *et al.*, 1985) still exhibits potent cytotoxicity, but correct glycosylation may contribute to a species preference for the homologous cytokine.

There has been general agreement on the biochemical characterization of natural (nonrecombinant) cachectin/TNF (Ruff and Gifford, 1980; Beutler *et al.*, 1985b; Haranaka *et al.*, 1986). The monokine seems to associate into dimeric or higher oligomeric units in biological fluids or culture medium, and these associations apparently do not involve interchain disulfide bridging or any sort of covalent bond. The pI has been estimated at pH 3.9 by electrophoretic isoelectric focusing, and slightly higher estimates result from column chromatofocusing. The monokine has proved relatively stable over a pH range of 5.5–10 and unstable outside that range. Bioactivity is preserved after incubation for hours at 37–56°C but is rapidly lost at 100°C. Trypsin, protease, elastase, and α-chymotrypsin each destroy TNF activity, which is resistant to papain, neuraminidase, and pepsin.

Independent descriptions have closely agreed on the sequence of human cachectin cDNAs, predicting that a 233 amino acid prosequence will yield a 157 amino acid mature hormone after the 76 residue signal peptide has been removed. Relative to murine TNF, the human monokine has one additional amino acid, histidine, at position 73. This addition occurs between the two cysteine residues (positions 69 and 100 in the mature murine TNF, thus 69 and 101 in human) that apparently form an intrachain disulfide bond on the surface of the folded peptide. This disulfide bond has no measurable effect on the conformation of the molecule at room temperature and does not appear to be necessary for biological activity (Davis *et al.*, 1987). Cys^{69} and Cys^{101} appear to occupy the first and third positions (respectively) of β turns in the secondary structure of human TNF, which has been portrayed as a relatively α-helix-poor, β-sheet-rich peptide (Davis *et al.*, 1987). Recombinant human cachectin has recently been shown to associate into compactly arranged but noncovalently linked trimers in Tris-buffer with or without saline (Arakawa and Yphantis, 1987).

Independent comparisons of the rabbit, mouse, and human amino acid sequences repeatedly find that all are highly homologous (about 80% in both the propeptide and mature monokine). Similarly, hydrophobicity plots of rabbit and human cachectin (Ito *et al.*, 1986a) and mouse and human cachectin (Marmenout *et al.*, 1985) match in detail, both in the propeptide and mature monokine portions of the peptides. These interspecific similarities surely explain the general lack of species specificity for cachectin-mediated functions, particularly in animal-and cyto-toxicity assays. However, there is definitely a species preference in biologic potency (Smith *et al.*, 1986); that is, murine cachectin assayed against TNF-sensitive murine cell lines may be at least one order of magnitude more potent than heterologous (e.g., human) TNF. The reverse is also true, such that each hormone is more cytotoxic against cell lines from the homologous species. Furthermore, antisera raised against one species' cachectin commonly fail to cross-react as avidly with heterologous TNF.

Although produced and promptly secreted in abundance, cDNA cloning studies agree that cachectin bears an atypical secretory protein signal sequence. At present no biologic role for the propiece(s) is known, and the high degree of homology (about 80% for any comparison between mouse, human, and rabbit propeptides) in this portion of the precursor remains unexplained. The unusual length of TNF propeptides, the identical conservation of the first 14 amino acid residues between rabbit, murine, and human prosequences, and the presence of potential trypsin-sensitive cleavage doublets (Lys-Lys in human and rabbit at 18–19, and also Arg-Arg in human and mouse and Lys-Arg in rabbit at 27–28) suggest that a physiological role for the propeptide or specific fractions of it may yet be uncovered. Typical secretory protein signal sequences are much shorter (20–30 residues), contain one or more basic residues near the N terminus, and feature a highly hydrophobic region near the center. The TNF signal sequence is much longer than usual, contains a centrally located highly hydrophobic region (promouse Leu^{31} to Ile^{56}), and contains a variety of basic and acidic residues in the first 30 amino acids.

Cachectin bears marked homology (about 30%) to another cytokine, lymphotoxin (LT), which is a product not of the monocyte/macrophage lineage but of B-lymphoblastoid cells (reviewed in Aggarwal *et al.*, 1985c; Nedwin *et al.*, 1985a). LT is slightly larger than TNF (171 residues) and is associated with a much shorter propeptide sequence (34 amino acids). Neither human nor murine lymphotoxin possesses the intrachain disulfide bridge that seems characteristic of cachectin. Human LT has an asparagine-linked carbohydrate moiety and three methionine residues; neither feature occurs in human TNF. Nevertheless, the two cytokines are similar enough to compete for a common receptor and share biologic activities (Aggarwal *et al.*, 1985a).

3.2. *Cachectin Genes Exhibit Potential Regulatory Features*

The human TNF gene has been localized to the short arm of chromosome 6, near the LT locus (Nedwin *et al.*, 1985a). This proximity and the high degree of homology between these two sequences suggest that they are ancestrally related. In particular, the fourth exon of each gene (which encodes about 80% of the mature cytokines) may represent the product of an early tandem duplication event (Nedwin *et al.*, 1985a). The human TNF and LT genes lie between HLA-DR and HLA-A, linking them to the major histocompatibility complex (MHC) (Spies *et al.*, 1986). Murine cachectin and LT gene loci mirror the human arrangement. The two genes are tandemly arranged on chromosome 17 and are separated by only about 1100 base pairs (Nedospasov *et al.*, 1986). Like the human arrangement, murine TNF and LT genes are located within the MHC (H-2) and were found to map about 70 kD proximal to the *D* gene (Muller *et al.*, 1987). The consequences of the fact that these immunomodulatory inflammatory mediators are embedded within the MHC remain undetermined, but TNF does influence the expression of MHC surface antigens.

The intron/exon distributions of rabbit, murine, and human TNF genes are similar. The human cachectin gene contains three intervening sequences: the first interrupts the propeptide, the second occurs in the codon for the second residue of the mature cytokine, and the third occurs in the codon for residue 18 (Nedwin *et al.*, 1985a). Although not as conserved as the coding sequences, the 5′ and 3′ noncoding regions of rabbit and human also show significant homology, 69% and 64%, respectively, which implies conservation of sequences that may regulate TNF expression (Ito *et al.*, 1986a).

The spliced mRNA transcript of TNF contains a lengthy (33 nucleotide) 3′ untranslated sequence (a region not typically demonstrating high interspecific homologies) composed entirely of adenosine (A) and uridine (U) residues featuring repeated and overlapping copies of the consensus octamer UUAUUUAU (Caput *et al.*, 1986). This sequence, common to human and murine TNF messages, also appears in the mRNA molecules coding for other inflammatory mediators including lymphotoxin, IL-1, G/M-CSF, all subclasses of IFN, and several acute-phase proteins (Caput *et al.*, 1986). Post-transcriptional mechanisms recognizing this sequence are probably involved in the control of cachectin gene expression (Beutler *et al.*, 1986a; Caput *et al.*, 1986) and may help to shape the temporal pattern of cachectin release following invasive stimulation. Shaw and Kamen (1986) have shown that normally long-lived β-globin mRNA transcripts are markedly destabilized by insertion of the 3′ untranslated sequence from G/M-CSF, and the modified mRNA is strongly superinducible. Similar effects could be expected to predispose TNF mRNA to fit the characteristically transitory but vigorous burst of cachectin expression in response to LPS.

4. Regulation of Expression and Correlation with Disease States

In response to endotoxin challenge, *in vivo* or *in vitro*, macrophages promptly produce copious amounts of cachectin in a time-limited response.

Cachectin is one of the major LPS-inducible secretory proteins of the murine macrophage cell line RAW 264.7 and constitutes 1–2% of the total protein secreted by stimulated cells (Beutler *et al.*, 1985b). Although activated macrophages and monocytes appear to make the most significant cachectin response to invasive stimuli, natural killer cells, natural cytotoxic cells, and IL-3-dependent bone-marrow-derived mast cells appear to contribute to a lesser degree (Palladino *et al.*, 1987).

4.1. Bacterial, Viral, and Protozoal Stimuli Induce Cachectin

The induction of cachectin in macrophages can be amplified appreciably by "priming" with reticuloendothelial system activators such as *Cornybacterium parvum* or *Mycobacterium bovis* strain bacillus Calmette-Guerin (BCG) or zymosan 1–2 weeks before *in vivo* LPS challenge; or with thioglycolate broth or phorbol myristic acid several days before LPS challenge *in vitro*. It is now clear, however, that such priming is not necessary for secretion of TNF (Beutler *et al.*, 1985c; Sayers *et al.*, 1987). Gram-negative bacterial endotoxin (LPS) appears to be the most potent defined, naturally occurring stimulus for TNF production, and biosynthetic precursors of LPS elicit lesser TNF responses from bone marrow macrophages *in vitro*, roughly in proportion to their increasing resemblance to mature LPS, with the monosaccharide lipid X less effective than lipid A (Sayers *et al.*, 1987).

Hotez *et al.* (1984) have mimicked protozoal infection by exposing peritoneal exudate cells to lysates of *Trypanosoma brucei* or *Plasmodium berghei in vitro* and have shown that these biologic agents can also induce cachectin secretion. Wong and Goeddel (1986) showed that a variety of RNA or DNA viral particles or the virus-mimicking construct (poly)I–(poly)C (which is also a potent inducer of IFN-α and IFN-β) effectively trigger cultured peripheral blood leukocytes or the human promyelomonocytic leukemia cell line HL-60 to increase their synthesis of cachectin mRNA and produce TNF. These results suggest that cachectin will participate as an immunomodulator in the inflammatory host defense responses to many bacterial, parasitic, and viral diseases.

Beutler *et al.* (1985c) have characterized the scope of the cachectin response by rabbits to experimental administration of endotoxin. Cachectin was detectable in the circulation within minutes following intravenous exposure to LPS. Peak serum concentrations—which can be well in excess of the minimum required for biologic responses, including lethal injury—occurred 1–2 h later, and cachectin levels then rapidly fell. Rabbits in lethal endotoxic shock can attain peak circulating TNF concentrations in the range of several hundred nanomoles per liter, thus allowing a calculation that milligram amounts of cachectin were produced in response to LPS (Beutler *et al.*, 1985c).

4.2. Dexamethasone, IFN-γ, and the *lps^d* Mutation Regulate Cachectin

Beutler *et al.* (1986a) examined the secretion of cachectin by cultured peritoneal macrophages in response to *in vitro* induction with bacterial LPS. Peritoneal macrophages isolated from normal mice, whether primed with thioglycolate broth or not, expressed low but detectable amounts of cachectin mRNA, but immunoreactive cachectin protein could not be detected in either the medium or lysates of these noninduced cells (cryptic message). Stimulation *in vitro* with endotoxin increased the rate of cachectin mRNA transcription threefold (transcriptional activation), thus greatly increasing the amounts of cachectin mRNA and allowing the peritoneal macrophages to secrete large quantities of cachectin into the culture medium. Dexamethasone was shown to inhibit markedly (but not totally) both cryptic and LPS-activated cachectin gene transcription, and some transcriptional inhibition was apparent even when dexamethasone treatment followed LPS induction. Even more dramatic, however, was the complete absence of immunoreactive cachectin protein from the medium of LPS-induced macrophages that has been treated with dexamethasone prior to or at the same time as endotoxin. Dexamethasone treatment 2 h after LPS, however, had little effect and allowed the secretion of near-normal levels of the hormone. Thus, dexamethasone acts to inhibit cachectin production at both transcriptional and post-transcriptional levels, but once the post-transcriptional phase of cachectin biosynthesis is set in motion, dexamethasone is incapable of regulating the process. This *in vitro* glucocorticoid effect recalls the benefits of early clinical administration of glucocorticoids to septic patients. Early prophylactic treatment with glucocorticoids confers significant advantage in later sepsis, but administration of glucocorticoids after sepsis has become established does not improve, and may worsen, the deleterious pathobiology of sepsis. Beutler *et al.* (1986a) also examined the nature of endotoxin resistance in the C3H/HeJ mouse, which does not make cachectin in response to LPS. Relative to their endotoxin-sensitive C3H/HeN counterparts, macrophages from C3H/HeJ mice show a dual defect that acts to preclude expression of cachectin. A diminished quantity of TNF message is synthesized in response to low doses of LPS (transcriptional defect), but higher LPS doses do induce abundant TNF mRNA. Even when large quantities of TNF mRNA have been transcribed, C3H/HeJ macrophages neither secrete nor express immunoreactive cachectin peptide (post-transcriptional defect). These two lesions, combined with the inability to secrete other cytokines (IL-1, IFN, CSF) in response to endotoxin, comprise the *lps^d* mutation, which appears to be the result of a single mutational event on chromosome 4 in the C3H/HeJ mouse (reviewed in Beutler *et al.*, 1986a).

IFN-γ exerts a permissive influence on macrophages and monocytes, priming murine macrophages for greater secretion of cachectin in response to subsequent, especially suboptimal, LPS stimulation (Pace *et al.*, 1983a,b; Sayers *et al.*, 1987) and augmenting LPS-induced TNF secretion by human peripheral blood monocytes (Nedwin *et al.*, 1985b). IFN-γ alone does not induce cachectin (Beutler *et al.*, 1986b) but does increase the cachectin response to the relatively weak inducer, lipid X (Sayers *et al.*, 1987). This suggests that other priming agents, such as BCG, which reverse the lps^d mutation, may act through IFN-γ-mediated mechanisms (Vogel *et al.*, 1980; Sayers *et al.*, 1987). The permissive influence of IFN-γ enhances both the transcriptional and post-transcriptional phases of cachectin expression and can partially correct the lps^d phenotype, allowing normal expression of cachectin by C3H/HeJ macrophages in response to relatively low concentrations of LPS (Beutler *et al.*, 1986b). The parallels between the dual defects in cachectin expression, attributable to dexamethasone treatment or defined by the lps^d mutation, and the dual enhancements of TNF expression by INF-γ are striking and may point to a common target for therapeutic intervention in LPS-sensitive cells in order to limit or prevent life-threatening cachectin responses (Beutler *et al.*, 1986b).

4.3. Characterization and Biodistribution of the Systemic Cachectin Response

The *in vivo* elevation of circulating cachectin in response to bolus LPS is transient (Beutler *et al.*, 1985c). Freshly cultured macrophages and myelogenous cell lines (e.g., murine line RAW 264.7) also stop producing TNF despite continuing exposure to LPS *in vitro*, suggesting that the cessation of cachectin secretion appears to be more a function of regulation by the responding cells than a property of the stimulus (Beutler *et al.*, 1985b; Sayers *et al.*, 1987).

Several mechanisms that could contribute to the falling phase of the cachectin response have been described. The destabilization of cachectin mRNA mentioned above may help to temporally limit the period of cachectin synthesis. Also, the hormone is cleared rapidly from the blood to the tissues. Tracer amounts of intravenously administered radiolabeled cachectin were cleared with a half-life of about 6 min, most prominently by saturable mechanisms in the liver (31% of the total injected), gastrointestinal tract (9%), kidney (8%), and spleen (1%) (Beutler *et al.*, 1985c). Nonsaturable uptake by the lungs (2%) and particularly the skin (30%) also concentrated a large fraction of labeled cachectin from the blood, while only small amounts of label were eliminated in urine and none was excreted as intact cachectin. Within 20 min of injection, neither labeled cachectin nor degraded intermediate metabolites could be recovered from blood or extracted from tissues, indicating that tissue binding and rapid, rather complete metabolism may be the typical mode of clearing cachectin from the circulation (Beutler *et al.*, 1985c).

Sayers *et al.* (1987) have shown that pretreatment with LPS *in vitro* renders cachectin-producing cells refractory to subsequent LPS stimulation, with the second response greatly reduced or eliminated. Thus, decreases in circulating cachectin, despite continuing LPS stimulation, might also reflect the system-wide entry of TNF-producing macrophages into a refractory period. Nonetheless, continuing local responses by previously unstimulated or newly recruited macrophages might still mediate significant toxicity through action at paracrine range.

4.4. *Infusion of Cachectin Closely Mimics Endotoxemia*

The physiologic responses to infusion of high, but physiologically relevant, doses of heterologous (human) recombinant cachectin into animals have been examined in detail and closely compared to the derangements seen in response to endotoxin [Tracey *et al.*, 1986a (rats), 1987 (dogs)]. These experiments show that infusion of cachectin alone is sufficient to reproduce most or all of the pathobiological responses typical of lethal endotoxemia, including devastating vital organ dysfunction and death within minutes to hours. Rats infused with lethal doses (greater than 700 μg/kg) of recombinant humanTNF show abnormal symptoms including lethargy, piloerection without chills, bloody diarrhea, and tachypnea beginning within minutes of cachectin infusion. These outward signs are accompanied by severe metabolic acidosis and eventual death from respiratory arrest.

Postmortem examination revealed the overwhelming extent of inflammatory damage to vital organs. Gross inspection characterized the lungs as diffusely hyperemic, and histopathologic examination of focal hemorrhagic areas revealed occlusion of large arteries by polymorphonuclear leukocytic thrombi, with margination of polymorphonuclear cells through the walls of the pulmonary vessels accompanied by severe interstitial and peribronchiolar pneumonitis and marked thickening of the alveolar membranes.

Segmental ischemia and regional hemorrhage or necrosis of the bowel were obvious, with the cecum particularly involved. Microscopic inflammatory changes were apparent, even in less infarcted areas of the gut. Focal areas in which the intestinal epithelium had been shed were frequent, in addition to invasion of the submucosa and muscularis mucosa by polymorphonuclear leukocytes.

Overt distension and congestion of the kidneys were attributable to acute tubular necrosis involving both proximal and distal segments. Again, inflammatory cells were found within the tissue.

All these pathobiologic responses to infused cachectin (overt behavioral signs of distress, hypotension, transient hyperglycemia followed by a profound hypoglycemic state, hyperkalemia and metabolic acidosis, gross and microscopic postmortem pathology, vital organ failure, and lethal outcome) were prevented by pretreatment of the rats with a neutralizing mouse monoclonal antibody against recombinant human cachectin.

Tracey *et al.* (1987) further examined the dose-dependent hemodynamic and metabolic responses to infused recombinant human cachectin in dogs. Sublethal doses (10 μg/kg) produced transient hemodynamic and nonsignificant hormone responses. Lethal doses of rHuTNF precipitated progressive hypotension associated with decreased cardiac filling pressure and requiring aggressive fluid replacement in support of cardiac output. Despite this fluid resuscitation protocol and strong counterregulatory hormone responses (significantly elevated circulating epinephrine, norepinephrine, and cortisol), rHuTNF doses of 100 μg/kg were uniformly lethal, with terminal shock, respiratory failure, and death within 3 h. During the preterminal period, a significant elevation of blood pCO_2 was likewise unable to correct the development of profound metabolic acidosis. Postmortem examination revealed familiar and catastrophic patterns of cachectin poisoning, with gross congestion and focal hemorrhagic lesions of the lungs (thickened alveolar walls, peribronchiolar pneumonitis, interstitial inflammation, occlusive white blood cell thrombi in pulmonary vessels with margination of inflammatory leukocytes), kidneys (acute tubular necrosis and interstitial inflammation), and adrenals (hemorrhagic medullary necrosis with inflammation).

Cachectin has also been shown to be a potent endogenous pyrogen, eliciting a brisk febrile response first by direct action on the hypothalamus, where it increases PGE_2 synthesis in similar fashion to IL-1 (Dinarello *et al.*, 1986). At higher doses, a second fever peak occurred, which Dinarello *et al.* (1986) attributed to TNF-induced IL-1 secretion. This demonstration underscores the multiplicity of interacting responses that are induced by cachectin/TNF.

Many pathobiologic responses to invasive stimuli can be attributed to direct cellular effects of cachectin. Early passive immunization of mice with heterologous anticachectin antibodies can prevent death following subsequent injection with lethal doses of bacterial endotoxin (Beutler *et al.*, 1985d). This demonstration suggests that cachectin is a necessary mediator of the lethal host responses in experimental endotoxemia and that death might be prevented in cases of bacteremia or sepsis by inhibiting or neutralizing systemic exposure to cachectin. Yet undoubtedly harmful, even lethal, effects can also arise or be intensified by release of other host mediators in response to inflammatory stimuli or to the primary cachectin response. A list of such mediators should include, but may not be limited to, the interleukins, interferons, prostaglandins, leukotrienes, platelet activating factor, and the products of complement activation. Stevens *et al.* (1986) have demonstrated a contribution of the human complement component C5a to acute lethality from sepsis in primates. By passively immunizing against C5a, they were able to prevent death within 4 h, and to attenuate adult respiratory distress syndrome and some other systemic responses to bacteremia. Thus, cachectin, while sufficient to precipitate shock and death, may not be the only mediator conferring lethality in acute septic shock.

4.5. Cachectin in Disease States

Reports linking elevated circulating cachectin titers to specific human disease states are increasing with the development of more specific and sensitive assays for the monokine. Scuderi *et al.* (1986) used a sensitive cachectin ELISA to detect cachectin levels in patients with various parasitic diseases. Patients with kala azar (visceral leishmaniasis) or malaria had high circulating TNF levels, but significant elevations in basal TNF serum concentrations were not detected in patients suffering from other disease states (Scuderi *et al.*, 1986). Aderka *et al.* (1985) found that cultured blood mononuclear cells of cancer patients were more likely to produce cachectin spontaneously than cells of healthy controls and that the cells of cancer patients produced TNF in much higher amounts. Waage *et al.* (1987) used a very sensitive cachectin bioassay to measure blood levels of TNF in patients with meningococcal disease and found that detectable serum cachectin was highly correlated with a state of septic shock (17 of 18) and subsequent death (10 of 11); patients with serum cachectin levels in excess of 440 U/mL invariably died (Waage *et al.*, 1987).

5. Cachectin/TNF Receptor and Mechanism of Action

It is clear that most or all cell types possess specific high-affinity receptors for cachectin, yet the organ, tissue, and cellular consequences of binding the cytokine are diverse. Lung, kidney, and adrenal appear particularly sensitive to the destructive effects of high circulating cachectin levels; all show marked inflammatory responses, often culminating in frank necrotic disruption of tissue integrity. Yet other tissues do not seem to suffer overt damage when exposed to the same circulating cachectin levels.

Initial characterizations of cachectin binding suggested that there might be a single class of plasma-membrane-associated, high-affinity receptors common to different tissues and cell types (Beutler *et al.*, 1985b). As noted above, cachectin and lymphotoxin were shown to compete for the same receptor even though they share only limited homology (Aggarwal *et al.*, 1985a). Radiolabeled cachectin binds to cell surface receptors in a time-, temperature-, and concentration-dependent manner (Kull *et al.*, 1985). Cachectin binding is accompanied by rapid internalization (Baglioni *et al.*, 1985) and prompt degradation (Tsujimoto *et al.*, 1985).

Cells that respond to TNF have always been found to bind the hormone, but binding, internalization, and degradation can occur in cells from which no biologic response has been detected (Tsujimoto *et al.*, 1985; Lehmann and Droge, 1986). Such results suggest that TNF binding is necessary but not sufficient to account for cachectin-mediated cellular changes.

Several investigators have used crosslinking reagents to couple directly receptor (or receptor-associated) membrane proteins and radiolabeled cachectin ligand (Kull *et al.*, 1985; Israel *et al.*, 1986; Scheurich *et al.*, 1986; Smith *et al.*, 1986; Tsujimoto *et al.*, 1986b). These studies have generally agreed that more than one putative receptor protein is thereby labeled whether human or murine cells are tested. At least two protein–ligand complexes (at about 75 and 90 kD) were usually labeled, and as many as four have been detected (Creasey *et al.*, 1987). Corrected relative molecular weights to be assigned to the ligand-free receptor polypeptides have been difficult to calculate, because they depend on (as yet unproven) assumptions about the stoichiometry of cachectin binding and the multimeric association of cachectin monomer in culture medium and after receptor binding. Tsujimoto *et al.* (1986b) suggests that the human U-937 cell TNF receptor is a glycoprotein.

These investigations of binding kinetics of cachectin/TNF receptors for their ligand have agreed in general that several hundred to several thousand cell surface receptors are available, depending on the cell type studied. The affinity of these receptors is high, with 50% occupancy in the low or subpicomolar range, which is appropriate to mediate cachectin effects at physiological concentrations of the ligand. Relatively few of these receptors (about 5–10%) need to be occupied in order to exert the biological effect, again depending on the specific cell type and response being studied. The number of cell surface TNF receptors can be strongly upregulated by pretreatment with IFN-γ, which may underlie some of the synergies noted for the two cytokines (Aggarwal *et al.*, 1985a; Israel *et al.*, 1986), but IFN-γ also seems to exert synergistic effects distal to (Scheurich *et al.*, 1986) or separate from (Tsujimoto *et al.*, 1986a) TNF binding. In summary, biological changes in response to cachectin appear to be mediated by high-affinity cell surface receptors, but transduction and effector mechanisms subsequent to ligand binding remain obscure.

6. Biological Activities: Parallels between Disease States, in Vitro Effects, and in Vivo Models

The framework of evidence that cachectin mediates the deleterious consequences of invasive disease is being built on data from three complementary approaches. First, clinical profiles of human septic patients reveal characteristic and potentially life-threatening alterations in metabolism, physiology, and tissue morphology that resemble cachectin-induced changes. New, more sensitive assays are now allowing the magnitude and time course of cachectin responses to be accurately quantitated in critically ill patients. To a large degree, these metabolic changes are mimicked by the administration of endotoxin to healthy volunteers, and the cachectin responses of LPS-treated humans are already being characterized. Second, cellular *in vitro* models of tissue metabolic processes analogous to those in normal and disease states have been constructed in which the influence of recombinant cachectin and other experimentally administered reagents can be assessed independently and in combination. The availability of recombinant cachectin and anticachectin antibodies has been invaluable in isolating and defining the cachectin-specific effects in these model systems. Third, the effects of administering invasive stimuli, recombinant cachectin, and anticachectin antibodies in animal models is helping to define the role of cachectin as a critical early mediator of acute, subacute, and chronic pathological alterations of host homeostasis associated with sepsis or bacteremia and other invasive diseases.

6.1. Adipocytes

The effort to define cachectin as an inflammatory mediator began with characterization of the metabolic derangements underlying cachexia and included the description of a net catabolic state with accumulation of high levels of lipid in the blood. Torti *et al.* (1985) provided an *in vitro* model of cachexia in the TA1 adipocyte cell line. After reaching confluency in culture, changes in gene activation of TA1 cells lead to the expression of several adipocyte-specific mRNAs that are not expressed by the preadipocytes. Treatment with cachectin did not affect cell growth or viability but completely inhibited the differentiation-induced expression of the adipose-specific genes, while expression of other, "housekeeping" genes (such as β-actin) was unaffected. Furthermore, cachectin prevented the accumulation of lipid within differentiating TA1 preadipocytes and caused mature, lipid-laden adipocytes to lose their stored triglycerides and decrease adipose-specific mRNAs by more than 90%. Pekala *et al.* (1983) demonstrated that a mediator from LPS-induced exudate cells could markedly reduce the activity of two key enzymes essential for *de novo* fatty acid biosynthesis in differentiating 3T3-L1 preadipocytes, in part by specifically reducing synthesis of the enzymes. Patton *et al.* (1986) have shown that physiologically appropriate doses of recombinant cachectin reduce anabolic and increase catabolic processes in differentiated 3T3-L1 adipocytes. Specific effects included the reduced incorporation of acetate for lipid synthesis, the inhibition of intracellular and heparin-releasable lipoprotein lipase activity, thus decreasing triglyceride uptake, and increased lipolysis resulting in an increased release of fatty acids into the medium. The combined catabolic and antianabolic effects of cachectin on adipose cells illustrate the double-edged sword of cachectin effects. In the short term, cachectin should provide a helpful mobilization of energy reserves for the acute metabolic demands of inflammatory responses for host defense, but persisting cachectin secretion could also contribute to the cachexia of chronic disease states, with the eventual exhaustion of metabolic reserves and death of the host.

6.2. Skeletal Muscle and Muscle Cell Lines

Cachexia includes not only changes in adipose tissue metabolism but also wasting of lean body mass. Differentiated L6 myotubes in culture respond to LPS-stimulated macrophage supernatants with an immediate acceleration of glycogenolysis marked by a sharp increase in fructose 2,6-bisphosphate, glycogen depletion, and increased lactate production (Lee *et al.*, 1987). Hours later in the response, enhanced glucose uptake via protein-synthesis-dependent increased expression of hexose transporters in the L6 cell plasma membrane begins to rebuild glycogen supplies. Lee *et al.* (1987) showed that cachectin qualitatively accounted for these effects but that other inflammatory mediators, as yet poorly characterized, quantitatively share and synergize with cachectin the ability to effect these changes in glucose metabolism.

Muscle cells also show changes in membrane polarization that may underlie alterations in the physiology of skeletal muscle noted in septic states. Critical illness in humans is frequently accompanied by a marked shift of Na^+ to the third space with loss of cellular potassium, and a positive shift in net fluid balance with intravascular dehydration. In addition, the resting skeletal muscle membrane polarization of critically ill patients is often reduced (Cunningham *et al.*, 1971). Although no mechanism linking these changes in fluid and ion distribution to changes in the skeletal muscle membrane has been established, cachectin may contribute to these effects directly. Recombinant human cachectin induces a dose-dependent reduction of skeletal muscle membrane polarization *in vitro* and *in vivo*, and this response has been used to construct a sensitive cachectin bioassay (Tracey *et al.*, 1986b). Isolated rat muscle cell fibers responded to low doses of recombinant human cachectin (rHuTNF) by proportional decreases in membrane polarization, and similar reductions were recorded following incubation with the serum of critically ill patients (Tracey *et al.*, 1986b) These responses to patients' sera were specifically inhibited by anticachectin antibodies. Dogs treated with recombinant human cachectin showed reduced resting hindlimb muscle transmembrane potentials and increased lactate release after infusion of rHu-TNF and before other physiological changes cascaded to full-blown shock (Tracey *et al.*, 1987). Cachectin thus seems to participate in the regulation of skeletal muscle membrane polarization and to mediate changes in the mobilization and utilization of energy stores by muscle cells.

6.3. Endothelial Cells, Procoagulant Activity, and Leukocyte Adhesion

The tissue-specific effects of circulating cachectin may, in many cases, depend on tissue access across the vascular endothelium. The normal endothelial surface acts to minimize coagulation of circulating blood, and patency of the vascular endothelium restricts the penetration of fluid, blood-borne cells, and macromolecules into the tissues. Exposure to endotoxin following injury or in the course of bacterial sepsis is frequently accompanied by disseminated intravascular coagulation and inflammatory responses leading to increased capillary permeability, with leakage of macromolecules into the tissue space. Treatment of cultured endothelial cells with recombinant human cachectin leads to a rapid and transient shift in surface properties favoring procoagulant activity for the deposition of fibrin; these protein-synthesis-dependent changes involve both the induction of tissue factor procoagulant activity (not normally expressed on endothelial surfaces) and the suppression of the cofactor activity supporting protein-C-mediated antithrombic mechanisms that normally prevent clot formation within the vascular bed (Nawroth and Stern, 1986). Bevilacqua *et al.* (1986) showed that the induction of tissue-factor-like procoagulant activity is independent of, but additive to, similar induction by IL-1 (Bevilacqua *et al.*, 1984). Taylor *et al.* (1987) have provided evidence that adequate activated protein C anticoagulant activity can protect against coagulopathy and death in a primate model of lethal bacteremic shock, suggesting that changes in endothelial surface protein C activity may be among the most toxic inflammatory responses during bacteremic shock.

Of course, the endothelium plays additional (noncoagulostatic) roles in response to invasive stimuli, particularly as a focus for inflammatory leukocyte responses. Cachectin rapidly stimulates the *de novo* synthesis and expression of endothelial cell surface antigens that promote the adhesion of inflammatory leukocytes to cultured endothelial cells (Gamble *et al.*, 1985; Pohlman *et al.*, 1986), one of many bioactivities shared with IL-1 (Bevilacqua *et al.*, 1985) and LT (Pober *et al.*, 1987). These surface effects can apparently be amplified by the cachectin-stimulated release of IL-1 from endothelial cells (Libby *et al.*, 1986; Nawroth *et al.*, 1986). A recent report by Pober *et al.* (1987) has extended these observations and shown that comparable doses of TNF or LT induce similar antigens on cultured human endothelial cells. Surface antigen induction includes a rapid, transient increase in a surface antigen recognized by monoclonal antibody H4/18 (Pober *et al.*, 1986a) and a rapid, sustained increase in the lymphocyte adhesion structure ICAM-1 (Pober *et al.*, 1986b). Furthermore, the TNF- and LT-induced responses seem to be independently regulated from similar antigen expression induced by IL-1α and IL-1β. Endothelium–leukocyte adhesion is, of course, a reflection of surface properties of both cells involved. Neutrophil adhesiveness for cells of the luminal surface is independently increased by cachectin treatment through CDw18-dependent effects on the neutrophils (Pohlman *et al.*, 1986). Cachectin-mediated effects on endothelial cells also include enhanced surface expression of HLA-A,B antigens (Collins *et al.*, 1986), which could further modulate endothelial interactions with inflammatory cells. The enhancement of HLA-A,B expression occurs over a much longer time period (days versus hours) than the leukocyte adhesion antigens, and although LT shares this ability, the IL-1s do not (Pober *et al.*, 1987).

Cachectin directly and in combination with IFN-γ modulates the morphology of cultured endothelial cells, resulting in elongation and overlapping of cells in confluency, rearrangement of their actin filaments, and a loss of stainable fibronectin matrix (Stolpen *et al.*, 1986), and changes the appearance of the cultured endothelial monolayer from epithelioid to fibroblastoid (Pober *et al.*, 1987). In addition, the exposure of vascular endothelial cells to circulating cachectin should serve to recruit more leukocytes into the inflammatory response by endothelial production of granulocyte and macrophage colony-stimulating factor (Broudy *et al.*, 1986).

Such effects may well underlie some of the changes in potency, activation, and proliferative state of the vascular endothelium at loci of inflammatory stimulation. Cachectin-mediated endothelial changes would seem to contribute to the increased capillary permeability and leakage of sepsis, the localization and margination of leukocytes at sites of inflammation, and the tissue damage and hemorrhagic necrosis following systemic exposure to endotoxin or cachectin. The observation that TNF- or LT-dependent hemorrhagic necrosis of transplantable Meth A sarcoma tumors occurred only when the tumors were located so as to be vascularized (i.e., subcutaneous but not intraperitoneal) further emphasizes the role of endothelial–leukocytic activation on cachectin-mediated toxicity for normal and neoplastic tissues (Palladino *et al.*, 1987).

6.4. Bone, Cartilage, and Fibroblasts

The cytokines cachectin and IL-1 have each been shown to promote soft tissue necrosis and bone remodeling processes typical of abcess formation or inflammatory resorption of cartilage and bone, as in rheumatoid arthritis. Cultured human synovial cells respond to cachectin with increased production of collagenase and prostaglandin E_2 (Dayer *et al.*, 1985), a feature shared with IL-1 and quite possibly amplified by cachectin-mediated IL-1 production by fibroblasts (Le *et al.*, 1987). Cachectin is somewhat less potent than IL-1 in causing the limited proteolysis and subsequent release of proteoglycans from cartilage explants while simultaneously inhibiting the synthesis of new proteoglycan, with the two cytokines apparently exerting their effects through two distinct receptors (Saklatvala, 1986), Treatment of fetal bone with TNF and LT stimulates bone resorption with calcium release, increases the number of multinucleated osteoclasts, decreases the amount of mineralized bone matrix, and decreases bone collagen synthesis by osteoblasts in treated explants (Bertolini *et al.*, 1986). It will be interesting to determine the local roles of cachectin and IL-1 in normal and pathological regulation of bone homeostasis and joint remodeling.

Normal fibroblasts from differing tissue sources have revealed a spectrum of responses to cachectin. Dermal fibroblasts mirror the cachectin-mediated increases in production of collagenase and PGE_2 shown by synovial cells (Dayer *et al.*, 1985). Elias *et al.* (1987) have recently shown that prostaglandin production by cultured normal lung fibroblasts is stimulated both by TNF and more potently by IL-1 and that these cytokines synergize to increase further prostaglandin production that is almost entirely due to augmented PGE_2 production. At low concentrations, cachectin has been shown to augment the growth of normal diploid fibroblasts deriving from several tissues (Sugarman *et al.*, 1985; Vilcek *et al.*, 1986), and cachectin provides a protective antiviral activity for certain fibroblast lines, in which there is a powerful synergy with IFN-γ (Mestan *et al.*, 1986; Wong and Goeddel, 1986). Cachectin stimulates fibroblasts to synthesize IFN-β_2, and this protein in turn mediates antiviral effects and antagonizes cachectin-induced mitogenesis (Kohase *et al.*, 1986). Fibroblasts also respond to cachectin treatment with the elaboration of other cytokines including IL-1 (Le *et al.*, 1987) and G/M-CSF (Munker *et al.*, 1986), which could modify tissue responses to cachectin alone. Expression of HLA-A,B antigens on the surface of fibroblasts is increased by cachectin (Collins *et al.*, 1986), providing a mechanism to involve the immune system further in inflammatory responses to invasive stimuli.

6.5. *Liver and Hepatocytes*

The host response to inflammation features dramatic changes in the state of the liver as reflected by altered concentrations of certain liver-derived plasma proteins, the so-called acute-phase proteins. A direct physiologic role for cachectin remains to be established in hepatic aspects of the acute-phase response, but cachectin does elicit many of the component alterations in acute-phase protein concentrations in cultured hepatocytes (Perlmutter *et al.*, 1986). The profiles of positively, negatively, and noninfluenced acute-phase proteins are strikingly similar for both cachectin and IL-1, although each is thought to act through a distinct receptor (Perlmutter *et al.*, 1986).

6.6. *Leukocytes*

A great deal of attention has quite naturally been focused on the interaction of cachectin with immune system cells and other cytokines. In part, this simply reflects the recognition of cachectin as an inflammatory mediator and the desire to pinpoint both its independent and its synergistic roles in inflammatory responses. Simple descriptions have not been forthcoming, however, which further emphasizes the central, wide-reaching, and interactive role of cachectin in determining the scope and focus of leukocyte participation in inflammatory responses.

One hallmark of spesis and bacteremia is the margination and activation responses of neutrophils and macrophages, which in systemic responses may be reflected in net changes in the number of circulating leukocytes. Cachectin directly promotes degranulation, lysosomal enzyme release, and the generation of reactive oxygen intermediates by respiratory burst activity in isolated neutrophils (Klebanoff *et al.*, 1986; Tsujimoto *et al.*, 1986c). These effects are potentiated by providing substrates for phagocytosis, which cachectin also enhances (Shalaby *et al.*, 1985; Tsujimoto *et al.*, 1986c). Cachectin improves antigen-dependent cell cytotoxicity (Shalaby *et al.*, 1985; Klebanoff *et al.*, 1986), increases surfaces display of C3bi and attachment to endothelial cells (Gamble *et al.*, 1985), and mediates enhanced fungal growth inhibition (Djeu *et al.*, 1986) by neutrophils. Degliantoni *et al.* (1985) have attributed both natural-killer-cell-derived cytotoxic factor and colony-inhibiting factor activities to TNF.

Cachectin influences its own principal cells of origin, the monocytes and macrophages, providing striking examples of autocrine regulation at several levels to influence the number of cells responding to invasive stimuli and to shape the quality, quantity, and efficacy of the inflammatory response. Either human peripheral blood mononuclear cells (Takeda *et al.*, 1986) or lymphocytes (Leung and Chiao, 1985) can be stimulated to produce factors that induce the differentiation of myelogenous cell lines along the monocyte/macrophage pathway, and one of these differentiation-inducing factors has been identified as cachectin. Thus, macrophages responding to invasive stimuli by secreting cachectin may well recruit additional monocyte and macrophage cells through inducing their differentiation from precursor populations.

Exposure to cachectin elicits from macrophages a variety of responses typical of an ''activated'' state. Cachectin (and lymphotoxin, but not IL-1) primes macrophages for reactive oxygen intermediate responses to subsequent triggering stimuli (Shparber and Nathan, 1986); and cachectin alone and more potently in synergy with IFN-γ induces macrophages to display I-A surface antigen (Chang and Lee, 1986). Cachectin serves as an autocrine immunomodulator, increasing the production of IL-1 and PGE_2 by cultured macrophages (Bachwich *et al.*, 1986). Cachectin-mediated tumor cell cytotoxicity is enhanced by pretreatment of activated macrophages with cachectin or other cytokines (IFN-γ or IL-1), whose effects are also attributable to cachectin (Philip and Epstein, 1986). Local production of TNF by activated macrophages can also be expected to attract additional monocytes and polymorphonuclear leukocytes to the site of inflammation through inherent chemotactic activity for these cells (Ming *et al.*, 1987). These demonstrations reinforce the notion that cachectin produced locally could influence the scope of inflammatory responses at a distance. Cachectin provides the autocrine induction of an antimicrobial state, inhibiting the multiplication of *T. cruzi* within cultured macrophages (De Tito *et al.*, 1986), and cachectin-containing serum has been shown to kill malarial parasites *in vitro* (Taverne *et al.*, 1981; Haidaris *et al.*, 1983). The antiparasitic benefits of cachectin are not limited to its direct and autocrine actions; cachectin also improves the antiparasitic performance of eosinophils against schistosome larvae (Silberstein and David, 1986).

The diversity of these immunomodulatory responses suggests a central role for cachectin in both humoral and cell-mediated immune and inflammatory responses to invasive stimuli, but it does not promise that the scope of influence cachectin exerts will be easily defined or narrowly limited. Cachectin/TNF must now be considered a primary and pleiotropic inflammatory mediator with acute, subacute, and chronic effects on cellular, physiologic, and organismal processes. Dissociation and therapeutic control of these diverse mechanism promise to be a great challenge, but the potential for prevention and cure of a wide variety of disease states argues that we should continue the effort.

References

Aderka, D., Fisher, S., Levo, Y., Holtman, H., Hahn, T., and Wallach, D., 1985, Cachectin/tumor necrosis factor production by cancer patients, *Lancet* **i:**1190.

Aggarwal, B. B., Moffat, T., and Harkins, R. N., 1984, Human lymphotoxin, *J. Biol. Chem.* **259:**686–691

Aggarwal, B. B., Eessalu, T. E., and Hass, P. E., 1985a, Characterization of receptors for human tumour necrosis factor and their regulation by γ-interferon, *Nature* **318:**665–667.

Aggarwal, B. B., Henzel, W. J., Moffat, B., Kohr, W. J., and Harkins, R. N., 1985b, Primary structure of human lymphotoxin derived from 1788 lymphoblastoid cell line, *J. Biol. Chem.* **260:**2334–2344.

Aggarwal, B. B., Kohr, W. J., Hass, P. E., Moffat, B., Spencer, S. A., Henzel, W. J., Bringman, T. S., Nedwin, G. E., Goeddel, D. V., and Harkins, R. N., 1985c, Human tumor necrosis factor: Production, purification, and characterization, *J. Biol. Chem.* **260:**2345–2354.

Arakawa, T., and Yphantis, D. A., 1987, Molecular weight of recombinant human tumor necrosis factor-α. *J. Biol. Chem.* **262:**7484–7485.

Bachwich, P. R., Chensue, S. W., Larrick, J. W., and Kunkel, S. W., 1986, Tumor necrosis factor stimulates interleukin-1 and prostaglandin E_2 production in resting macrophages, *Biochem. Biophys. Res. Commun.* **136:**94–101.

Baglioni, C., McCandless, S., Tavernier, J., and Fiers, W., 1985, Binding of human tumor necrosis factor to high affinity receptors on HeLa and lymphoblastoid cells sensitive to growth inhibition, *J. Biol. Chem.* **260:**13395–13397.

Bertolini, D. R., Nedwin, G. E., Bringman, T. S., Smith, D. D., and Mundy, G. R., 1986, Stimulation of bone resorption and inhibition of bone formation in vitro by human tumour necrosis factors, *Nature* **319:**516–518.

Beutler, B., Greenwald, D., Hulmes, J. D., Chang, M., Pan, Y.-C. E., Mathison, J., Ulevitch, R., and Cerami, A., 1985a, Identity of tumour necrosis factor and the macrophage-secreted factor cachectin, *Nature* **316:**552–554.

Beutler, B., Mahoney, J., Le Trang, N., Pekala, P., and Cerami, A., 1985b, Purification of cachectin, a lipoprotein lipase-suppressing hormone secreted by endotoxin-induced RAW 264.7 cells, *J. Exp. Med.* **161:**984–995.

Beutler, B., Milksark, I. W., and Cerami, A., 1985c, Cachectin/tumor necrosis factor: Production, distribution, and metabolic fate *in vivo, J. Immunol.* **135:**3972–3977.

Beutler, B., Milsark, I. W., and Cerami, A., 1985d, Passive immunization against cachectin/ tumor necrosis factor protects mice from lethal effect of endotoxin, *Science* **229:**869–871.

Beutler, B., Krochin, N., Milsark, I. W.,, Luedke, C., and Cerami, A., 1986a, Control of cachectin (tumor necrosis factor) synthesis: Mechanisms of endotoxin resistance, *Science* **232:**977–980.

Beutler, B., Tkacenko, V., Milsark, I., Krochin, N., and Cerami, A., 1986b, Effect of γ interferon on cachectin expression by mononuclear phagocytes, *J. Exp. Med.* **164:**1791–1796.

Bevilacqua, M. P., Pober, J. S., Majeau, G. R., Cotran, R. S., and Gimbrone, M. A., Jr., 1984, Interleukin 1 (IL-1) induces biosynthesis and cell surface expression of procoagulant activity in human vascular endothelial cells, *J. Exp. Med.* **160:**618–623.

Bevilacqua, M. P., Pober, J. S., Wheeler, M. E., Cotran, R. S., and Gimbrone, M. A., Jr., 1985, Interleukin 1 acts on cultured human vascular endothelium to increase the adhesion of polymorphonuclear leukocytes, monocytes, and related leukocyte cell lines, *J. Clin. Invest.* **76:**2003–2011.

Bevilacqua, M. P., Pober, J. S., Majeau, G. R., Fiers, W., Cotran, R. S., and Gimbrone, M. A., Jr., 1986, Recombinant tumor necrosis factor induces procoagulant activity in cultured human vascular endothelium: Characterization and comparison with the actions of interleukin 1, *Proc. Natl. Acad. Sci. U.S.A.* **83:**4533–4537.

Broudy, V. C., Kaushansky, K., Segal, G. M., Harlan, J. M., and Adamson, J. W., 1986, Tumor necrosis factor type α stimulates human endothelial cells to produce granulocyte/ macrophage colony-stimulating factor, *Proc. Natl. Acad. Sci. U.S.A.* **83:**7467–7471.

Caput, D., Beutler, B., Hartog, K., Thayer, R., Brown-Shirmer, S., and Cerami, A., 1986, Identification of a common nucleotide sequence in the 3′-untranslated region of mRNA molecules specifying inflammatory mediators, *Proc. Natl. Acad. Sci. U.S.A.* **83:**1670–1674.

Carswell, E. A., Old, L. J., Kassel, R. L., Green, S., Fiore, N., and Williamson, B., 1975, An endotoxin-induced serum factor that causes necrosis of tumors, *Proc. Natl. Acad. Sci. U.S.A.* **72:**3666–3670.

Cerami, A., Ikeda, Y., Le Trang, N., Hotez, P. J., and Beutler, B., 1985, Weight loss associated with an endotoxin-induced mediator from peritoneal macrophages: The role of cachectin (tumor necrosis factor), *Immunol. Lett.* **11:**173–177.

Chang, R. J., and Lee S. H., 1986, Effects of interferon-γ and tumor necrosis factor-γ on the expression of an Ia antigen on a murine macrophage cell line, *J. Immunol.* **137:**2853–2856.

Collins, T., Lapierre, L. A., Fiers, W., Strominger, J. L., and Pober, J. S., 1986, Recombinant human tumor necrosis factor increases mRNA levels and surface expression of HLA-A,B antigens in vascular endothelial cells and dermal fibroblasts *in vitro, Proc. Natl. Acad. Sci. U.S.A.* **83:**446–450.

Creasey, A. A., Yamamoto, R., and Vitt, C. R., 1987, A high molecular weight component of the human tumor necrosis factor receptor is associated with toxicity, *Proc. Natl. Acad. Sci. U.S.A.* **84:**3293–3297.

Cunningham, S. N., Jr., Carter, N. W., Rector, F. C., Jr., and Seldin, D. W., 1971, Resting transmembrane potential in normal subjects and severely ill patients, *J. Clin. Invest.* **50:**49–50.

Davis, J. M., Narachi, M. A., Alton, N. K., and Arakawa, T., 1987, Structure of human tumor necrosis factor α derived from recombinant DNA, *Biochemistry* **26:**1322–1326.

Dayer, J. M. Beutler, B., and Cerami, A., 1985, Cachectin/tumor necrosis factor stimulates collagenase and prostaglandin E_2 production by human synovial cells and dermal fibroblasts, *J. Exp. Med.* **162:**2163–2168.

Degliantoni, G., Murphy, M., Kobayashi, M., Francis, M. K., Perussia, B., and Trinchieri, G., 1985, Natural killer (NK) cell-derived hematopoietic colony-inhibiting activity and NK cytotoxic factor. Relationship with tumor necrosis factor and synergism with immune interferon, *J. Exp. Med.* **162:**1512–1530.

De Titto, E. H., Catterall, J. R., and Remington, J. S., 1986, Activity of recombinant tumor necrosis factor on *Toxoplasma gondii* and *Trypanosoma cruzi, J. Immunol.* **137:**1342–1345.

Dinarello, C. A., Cannon, J. G., Wolff, S. M., Bernheim, H. A., Beutler, B., Cerami, A., Figari, I. S., Palladino, M. A., Jr., and O'Connor, J. V., 1986, Tumor necrosis factor (cachectin) is an endogenous pyrogen and induces production of interleukin-1, *J. Exp. Med.* **163:**1433–1450.

Djeu, J. Y., Blanchard, D. K., Halkias, D., and Friedman, H., 1986, Growth inhibition of *Candida albicans* by human polymorphonuclear neutrophils: Activation by interferon-γ and tumor necrosis factor, *J. Immunol.* **137:**2980–2984.

Elias, J. A., Gustilo, K., Baeder, W., and Freunlich, B., 1987, Synergistic stimulation of fibroblast prostaglandin release by recombinant interleukin 1 and tumor necrosis factor, *J. Immunol.* **138:**3812–3816.

Fransen, L., Muller, R., Marmenout, A., Tavernier, J., Van der Heyden, J., Kawashima, E., Chollet, A., Tizard, R., Van Heuverswyn, H., Van Vliet, A., Ruysschaert, M. R., and Fiers, W., 1985, Molecular cloning of mouse tumour necrosis factor cDNA and its eukaryotic expression, *Nucleic Acids Res.* **13:**4417–4429.

Gamble, J. R., Harlan, J. M., Klebanoff, S. J., and Vadas, M. A., 1985, Stimulation of the adherence of neutrophils to umbilical vein endothelium by human recombinant tumor necrosis factor, *Proc. Natl. Acad. Sci. U.S.A.* **82:**8667–8671.

Green, S., Dobrjansky, A., Carswell, E. A., Kassel, R. L., Old, L. J., Fiore, N., and Schwartz, M. K., 1976, Partial purification of a serum factor that causes necrosis of tumors, *Proc. Natl. Acad. Sci. U.S.A.* **73:**381–385.

Guy, M. W., 1975, Serum and tissue fluid lipids in rabbits experimentally infected with *Trypanosoma brucei, Trans. R. Soc. Trop. Med. Hyg.* **69:**429.

Ha, D. K. K., Gardner, I. D., and Lawton, J. W. M., 1983, Characterization of macrophage function in *Mycobacterium lepraemurium*-infected mice: Sensitivity of mice to endotoxin and release of mediators and lysosomal enzymes after endotoxin treatment, *Parasite Immunol.* **5:**513–526.

Haidaris, C. G., Haynes, J. D., Meltzer, M. S., and Allison,A. C., 1983, Serum containing tumor necrosis factor is cytotoxic for the human malaria parasite *Plasmodium falciparum, Infect. Immun.* **42:**385–393.

Haranaka, K., Carswell, E. A., Williamson, B. D., Prendergast, J. S., Satomi, N., and Old, L. J., 1986, Purification, characterization, and antitumor activity of nonrecombinant mouse tumor necrosis factor, *Proc. Natl. Acad. Sci. U.S.A.* **83:**3949–3953.

Hotez, P. J., Le Trang, N., Fairlamb, A. H., and Cerami, A., 1984, Lipoprotein lipase suppression in 3T3-L1 cells by a haematoprotozoan-induced mediator from peritoneal exudate cells, *Parasite Immunol.* **6:**203–209.

Israel, S., Hahn, T., Holtmann, H., and Wallach, D., 1986, Binding of human TNF-α to high-affinity cell surface receptors: Effect of IFN, *Immunol. Lett.* **12:**217–224.

Ito, H., Shirai, T., Yamamoto, S., Akira, M., Kawahara, S., Todd, C. W., and Wallace, R. B., 1986a, Molecular cloning of the gene encoding rabbit tumor necrosis factor, *DNA* **5:**157–165.

Ito, H., Yamamoto, S., Kuroda, S., Sakamoto, H., Kajihara, J., Kiyota, T., Hayashi, H., Kato, M., and Seko, M., 1986b, Molecular cloning and expression in *Escherichia coli* of the cDNA coding of rabbit tumor necrosis factor, *DNA* **5:**149–156.

Kawakami, M., and Cerami, A., 1981, Studies of endotoxin-induced decrease in lipoprotein lipase activity, *J. Exp. Med.* **154:**631–639.

Klebanoff, S. J., Vadas, M. A., Harlan, J. M., Sparks, L. H., Gamble, J. R., Agosti, J. M., and Waltersdorph, A. M., 1986, Stimulation of neutrophils by tumor necrosis factor, *J. Immunol.* **136:**4220–4225.

Kohase, M., Henriksen-DeStefano, D., May, L. T., Vilcek, J., and Sehgal, P. B., 1986, Induction of β_2-interferon by tumor necrosis factor: A homeostatic mechanism in the control of cell proliferation, *Cell* **45:**659–666.

Kull, F. C., Jr., Jacobs, S., and Cuatrecasas, P., 1985, Cellular receptor for ^{125}I-labeled tumor necrosis factor: Specific binding, affinity labeling, and relationship to sensitivity, *Proc. Natl. Acad. Sci. U.S.A.* **82:**5756–5760.

Le, J., Weinstein, D., Gubler, U., and Vilcek, J., 1987, Induction of membrane-associated interleukin 1 by tumor necrosis factor in human fibroblasts, *J. Immunol.* **138:**2137–2142.

Lee, M. D., Zentella, A., Vine, W., Pekala, P. H., and Cerami, A., 1987, Effect of endotoxin-induced monokines on glucose metabolism in the muscle cell line L6, *Proc. Natl. Acad. Sci. U.S.A.* **84:**2590–2594.

Lehmann, V., and Droge, W., 1986, Demonstration of membrane receptors for human natural and recombinant ^{125}I-labeled tumor necrosis factor on HeLa cell clones and their role in tumor cell sensitivity, *Eur. J., Biochem.* **158:**1–5.

Leung, D. Y. M., and Chiao, J. W., 1985, Human leukemia cell maturation induced by a T-cell lymphokine isolated from medium conditioned by normal lymphocytes, *Proc. Natl. Acad. Sci. U.S.A.* **82:**1209–1213.

Libby, P., Ordovas, J. M., Auger, K. R., Robbins, A. H., Biriyi, L. K., and Dinarello, C. A., 1986, Endotoxin and tumor necrosis factor induce interleukin-1 gene expression in adult human vascular endothelial cells, *Am. J. Pathol.* **124:**179–185.

Mannel, D. N., Moore, R. N., and Mergenhagen, S. E., 1980, Macrophages as a source of tumoricidal activity (tumor-necrotizing factor), *Infect. Immun.* **30:**523–530.

Marmenout, A., Fransen, L., Tavernier, J., Van Der Heyden, J., Tijard, R., Kawashima, E., Shaw, A., Johnson, M.-J., Semon, D., Muller, R., Ruysschaert, M.-R., Van Vliet, A., and Fiers, W., 1985, Molecular cloning and expression of human tumor necrosis factor and comparison with mouse tumor necrosis factor, *Eur. J., Biochem.* **152:**512–519.

Matthews, N., 1978, Tumour-necrosis factor from the rabbit. II. Production by monocytes, *Br. J. Cancer* **38:**310–315.

Matthews, N., 1981a, Production of an anti-tumour cytotoxin by human monocytes, *Immunology* **44:**135–142.

Matthews, N., 1981b, Tumour-necrosis factor from the rabbit. V. Syntheses *in vitro* by mononuclear phagocytes from various tissues of normal and BCG-injected rabbits, *Br. J. Cancer* **44:**418–424.

Mestan, J., Digel, W., Mittnacht, S., Hillen, H., Blohm, D., Moller, A., Jacobsen, H., and Kirchner, H., 1986, Antiviral effects of recombinant tumor necrosis factor in vitro, *Nature* **323:**816–819.

Michalek, S. M., Moore, R. N., McGhee, J. R., Rosenstreich, D. L., and Mergenhagen, S. E., 1980, The primary role of lymphoreticular cells in the mediation of host responses to bacterial endotoxin, *J. Infect. Dis.* **141:**55–63.

Ming, W. J., Bersan, L., and Mantovani, A., 1987, Tumor necrosis factor is chemotactic for monocytes and polymorphonuclear leukocytes, *J. Immunol.* **138:**1469–1474.

Muller, U., Jongeneel, C. V., Nedospasov, S. A., Lindahl, K. F., and Steinmetz, M., 1987, Tumour necrosis factor and lymphotoxin genes map close to H-2D in the mouse major histocompatibility complex, *Nature* **325:**265–267.

Munker, R., Gasson, J., Ogawa, M., and Koeffler, H. P., 1986, Recombinant human TNF induces production of granulocyte–monocyte colony-stimulating factor, *Nature* **323:**79–82.

Nawroth, P. P., and Stern, D. M., 1986, Modulation of endothelial cell hemostatic properties by tumor necrosis factor, *J. Exp. Med.* **163:**740–745.

Nawroth, P. P., Bank, I., Handley, D., Cassimeris, J., Chess, L., and Stern, D., 1986, Tumor necrosis factor/cachectin interacts with endothelial cell receptors to induce release of interleukin 1, *J. Exp. Med.* **163:**1363–1375.

Nedwin, G. E., Naylor, S. L., Sakaguchi, A. Y., Smith, D., Jarrett-Nedwin, J., Pennica, D., Goeddel, D. V., and Gray, P. W., 1985a, Human lymphotoxin and tumor necrosis factor genes: Structure, homology and chromosomal localization, *Nucleic Acids Res.* **13:**6361–6367.

Nedwin, G. E., Svedersky, L. P., Bringman, T. S., Palladino, M. A., Jr., and Goeddel, D, V., 1985b, Effect of interleukin 2, interferon-γ and mitogens on the production of tumor necrosis factors α and β. *J. Immunol.* **135:**2492–2497.

Nedospasov, S. A., Hirt, B., Shakhov, A. N., Dobrynin, V. N., Kawashima, E., Accolla, R. S., and Jongeneel, C. V., 1986, The genes for tumor necrosis factor (TNF-alpha) and lymphotoxin (TNF-beta) are tandemly arranged on chromosome 17 of the mouse, *Nucleic Acids Res.* **14:**7713–7725.

Old, L. J., 1985, Tumor necrosis factor (TNF), *Science* **230:**630–632.

O'Malley, W. E., Achinstein, B., and Shear, M. J., 1962, Action of bacterial polysaccharide on tumors. II. Damage of sarcoma 37 by serum of mice treated with *Serratia marcescens* polysaccharide, and induced tolerance, *J. Natl. Cancer Inst.* **29:**1169–1175.

Pace, J. L., Russell, S. W., Schreiber, R. D., Altman, A., and Katz, D. H., 1983a, Macrophage activation: Priming activity from a T-cell hybridoma is attributable to interferon-γ, *Proc. Natl. Acad. Sci. U.S.A.* **80:**3782–3876.

Pace, J. L., Russell, S. W., Torres, B. A., Johnson, H. M., and Gray, P. W., 1983b, Recombinant mouse γ interferon induces the priming step in macrophage activation for tumor cell killing, *J. Immunol.* **130:**2011–2013.

Palladino, M. A., Jr., Shalaby, M. R., Kramer, S. M., Ferraiolo, B. L., Baughman, R. A., Deleo, A. B., Crase, D., Marafino, B., Aggarwal, B. B., Figari, I. S., Liggitt, D., and Patton, J. S., 1987, Characterization of the antitumor activities of human tumor necrosis factor-α and the comparison with other cytokines: Induction of tumor-specific immunity, *J. Immunol.* **138:**4023–4032.

Patton, J. S., Shepard, H. M., Wilking, H., Lewis, G., Aggarwal, B. B., Eessalu, T. E., Gavin, L. A., and Grunfeld, G., 1986, Interferons and tumor necrosis factors have similar catabolic effects on 3T3 L1 cells, *Proc. Natl. Acad. Sci. U.S.A.* **83:**8313–8317.

Pekala, P. H., Kawakami, M., Angus, C. W., Lane, M. D., and Cerami, A., 1983, Selective inhibition of synthesis of enzymes for de novo fatty acids biosynthesis by an endotoxin-induced mediator from exudate cells, *Proc. Natl. Acad. Sci. U.S.A.* **80:**2743–2747.

Pennica, D., Nedwin, G. E., Hayflick, J. S., Seeburg, P. H., Derynck, R., Palladino, M. A., Kohr, W. J., Aggarwal, B. B., and Goeddel, D. V., 1984, Human tumour necrosis factor: Precursor structure, expression and homology to lymphotoxin, *Nature* **312:**724–729.

Pennica, D., Hayflick, J. S., Bringman, T. S., Palladino, M. A., and Goeddel, D. V., 1985, Cloning and expression in *Escherichia coli* of the cDNA for murine tumor necrosis factor, *Proc. Natl. Acad. Sci. U.S.A.* **82:**6060–6064.

Perlmutter, D. H., Dinarello, C. A., Punsa, P. I., and Colten, H. R., 1986, Cachectin/tumor necrosis factor regulates hepatic acute-phase gene expression, *J. Clin. Invest.* **78:**1349–1354.

Philip, R., and Epstein, L. B., 1986, Tumour necrosis factor as immunomodulator and mediator of monocyte cytoxicity induced by itself, γ-interferon and interleukin-1, *Nature* **323:**86–89.

Pober, J. S., Bevilacqua, M. P., Mendrick, D. L., Lapierre, L. A., Fiers, W., and Gimbrone, M. A., Jr., 1986a, Two distinct monokines, interleukin 1 and tumor necrosis factor, each independently induce biosynthesis and transient expression of the same antigen on the surface of cultured human vascular endothelial cells, *J. Immunol.* **136:**1680–1687.

Pober, J. S., Gimbrone, M. A., Jr., Lapierre, L. A., Mendrick, D. L., Fiers, W., Rothlein, R., and Springer, T. A., 1986b, Overlapping patterns of activation of human endothelial cells by interleukin 1, tumor necrosis factor, and immune interferon, *J. Immunol.* **137:**1893–1896.

Pober, J. S., Lapierre, L. A., Stophen, A. H., Brock, T. A., Springer, T. A., Fiers, W., Bevilacqua, M. P., Mendrick, D. L., and Gimbrone, M. A., Jr., 1987, Activation of cultured human endothelial cells by recombinant lymphotoxin: Comparison with tumor necrosis factor and interleukin 1 species, *J. Immunol.* **138:**3319–3324.

Pohlman, T. H., Stanness, K. A., Beatty, P. G., Ochs, H. D., and Harlan, J. M., 1986, An endothelial cell surface factor(s) induced *in vitro* by lipopolysaccharide, interleukin 1, and tumor necrosis factor-α increases neutrophil adherence by a CDw18-dependent mechanism, *J. Immunol.* **136:**4548–4553.

Rouzer, C., and Cerami, A., 1980, Hypertriglyceridemia associated with *Trypanosoma brucei brucei* infection in rabbits: Role of defective triglyceride removal, *Mol. Biochem. Parasitol.7* **2:**31–38.

Ruff, M. R., and Gifford, G. E., 1980, Purification and physico-chemical characterization of rabbit tumor necrosis factor, *J. Immunol.* **125:**1671–1677.

Sakaguchi, O., and Sakaguchi, S., 1979, Alterations in lipid metabolism in mice injected with endotoxin, *Microbiol. Immunol.* **23:**71–74.

Saklatvala, J., 1986, Tumour necrosis factor-α stimulates resorption and inhibits synthesis of proteoglycan in cartilage, *Nature* **322:**547–549.

Sayers, T. J., Macher, I., Chung, J., and Kugler, E., 1987, The production of tumor necrosis factor by mouse bone marrow-derived macrophages in response to bacterial lipopolysaccharide and chemically synthesized monosaccharide precursor, *J. Immunol.* **136:**2935–2940.

Scheurich, P., Ucer, U., Kronke, M., and Pfizenmaier, K., 1986, Quantification and characterization of high-affinity membrane receptors for tumor necrosis factor on human leukemic cell lines, *Int. J. Cancer* **38:**127–133.

Scuderi, P., Lam, K. S., Ryan, K. J., Petersen, E., Sterling, K. E., Finley, P. R., Ray, G. C., Slymen, D. J., and Salmon, S. E., 1986, Raised serum levels of tumour necrosis factor in parasitic infections, *Lancet* **ii:**1364–1365.

Shalaby, M. R., Aggarwal, B. B., Rinderknecht, E., Svedersky, L. P., Finkle, B. S., and Palladino, M. A., Jr., 1985, Activation of human polymorphonuclear neutrophil functions by interferon-γ and tumor necrosis factors, *J. Immunol.* **135:**2069–2073.

Shaw, G., and Kamen, R., 1986, A. conserved AU sequence from the 3′ untranslated region of GM-CSF mRNA mediates selective mRNA degradation, *Cell* **46:**659–667.

Shear, M. J., and Perrault, A., 1944, Chemical treatment of tumors, IX. Reactions of mice with primary subcutaneous tumors to injection of a hemorrhage-producing bacterial polysaccharide, *J. Natl. Cancer Inst.* **4:**461–475.

Shirai, T., Yamaguchi, H., Ito, H., Todd, C. W., and Wallace, R. B., 1985, Cloning and expression in *Escherichia coli* of the gene for human tumour necrosis factor, *Nature* **313:**803–806.

Shparber, M., and Nathan, C., 1986, Autocrine activation of macrophages by recombinant tumor necrosis factor (rTNFα) but not recombinant interleukin-1(rIL-1), *Blood* **68:**86a.

Silberstein, D. S., and David, J. R., 1986, Tumor necrosis factor enhances eosinophil toxicity to *Schistosoma mansoni* larvae, *Proc. Natl. Acad. Sci. U.S.A.* **83:**1055–1059.

Smith, R. A., Kirstein, M., Fiers, W., and Baglioni, C., 1986, Species specificity of human and murine tumor necrosis factor, *J. Biol. Chem.* **261:**14871–14874.

Spies, T., Morton, C. C., Nedospasov, S. A., Fiers, W., Pious, D., and Strominger, J. L., 1986, Genes for the tumor necrosis factors α and β are linked to the human major histocompatibility complex, *Proc. Natl. Acad. Sci. U.S.A.* **83:**8699–8702.

Stevens, J. H., O'Hanley, P., Shapiro, J. M., Mihm, F. G., Satoh, P. S., Collins, J. A., Raffin, T. A., 1986, Effects of anti-C5a antibodies on the adult respiratory distress syndrome in septic primates.*J. Clin. Invest.* **77:**1812–1816.

Stolpen, A. H., Guinan, E. C., Fiers, W., and Pober, J. S., 1986, Recombinant tumor necrosis factor and immune interferon act singly and in combination to reorganize human vascular endothelial cell monolayers, *Am. J. Pathol.* **123:**16–23.

Sugarman, B. J., Aggarwal, B. B., Hass, P. E., Figar, I. S., Palladino, M. A., Jr., and Shepard, H. M., 1985, Recombinant human tumor necrosis factor-α: Effects on proliferation of normal and transformed cells in vitro, *Science* **230:**943–945.

Takeda, K., Iwamoto, S., Sugimoto, H., Takuma, T., Kawatani, N., Noda, M., Masaki, A., Morise, H., Arimura, H., and Konno, K., 1986, Identity of differentiation-inducing factor and tumour necrosis factor, *Nature* **323:**338–340.

Taverne, J., Dockrell, H. M., and Playfair, J. H. L., 1981, Endotoxin-induced serum factor kills malarial parasites *in vitro, Infect. Immun.* **33:**83–89.

Taylor, F. B., Jr., Chang, A., Esmon, C. T., D'Angelo, A., Vigano-D'Angelo, S., and Blick. K. E., 1987, Protein C prevents coagulopathic and lethal effects of *Escherichia coli* infusion in the baboon, *J. Clin. Invest.* **79:**918–925.

Torti, F. M., Dieckmann, B., Beutler, B., Cerami, A., and Ringold, G. M., 1985, A macrophage factor inhibits adipocyte gene expression: An *in vitro* model of cachexia, *Science* **229:**867–869.

Tracey, K. J., Beutler, B., Lowry, S. F., Merryweather, J., Wolpe, S., Milsark, I. W., Harir, R. J., Fahey III, T. J., Zentella, A., Albert, J. D., Shires, G. T., and Cerami, A., 1986a,

Shock and tissue injury induced by recombinant human cachectin, *Science* **234:**470–474.
Tracey, K. J., Lowry, S. F., Beutler, B., Cerami, A., Albert, J. D., and Shires, G. T., 1986b, Cachectin/tumor necrosis factor mediates changes of skeletal muscle plasma membrane potential, *J. Exp. Med.* **164:**1368–1373.

Tracey, K. J., Lowry, S. F., Fahey III, T. J., Albert, J. D., Fong, Y., Hesse, D., Beutler, B., Manogue, K. R., Calvano, S., Wei, H., Cerami, A., and Shires, G. T., 1987, Cachectin/tumor necrosis factor induces lethal shock and stress hormone responses in the dog, *Surg. Gynecol. Obstet.* **164:**415–422.

Tsujimoto, M., Yip, Y. K., and Vilcek, J., 1985, Tumor necrosis factor: Specific binding and internalization in sensitive and resistant cells, *Proc. Natl. Acad. Sci. U.S.A.* **82:**7626–7630.

Tsujimoto, M., Feinman, R., and Vilcek, J., 1986a, Differential effects of type 1 IFN and IFN-γ on the binding of tumor necrosis factor to receptors in two human cell lines, *J. Immunol.* **137:**2272–2276.

Tsujimoto, M., Feinman, R., Kohase, M., and Vilcek, J., 1986b, Characterization and affinity crosslinking of receptors for tumor necrosis factor on human cells, *Arch, Biochem. Biophys.* **249:**563–568.

Tsujimoto, M., Yokota, S., Vilcek, J., and Weissman, G., 1986c, Tumor necrosis factor provokes superoxide anion generation from neutrophils, *Biochem. Biophys. Res. Commun.* **137:**1094–1100.

Vilcek, J., Palombella, V. J., Henriksen-DeStefano, D., Swenson, C., Feinman, R., Hirai, M., and Tsujimoto, M., 1986, Fibroblast growth enhancing activity of tumor necrosis factor and its relationship to other polypeptide growth factors, *J. Exp. Med.* **163:**632–643.

Vogel, S. N., Moore, R. N., Sipe, J. D., and Rosenstreich, D. L., 1980, BCG-induced enhancement of endotoxin sensitivity in C3H/HeJ mice, *J. Immunol.* **124:**2004–2009.

Waage, A., Halstensen, A., and Espevik, T., 1987, Association between tumor necrosis factor in serum and fatal outcome in patients with meningococcal disease, *Lancet* **i:**335–357.

Wang, A. M., Creasey, A. A., Ladner, M. B., Lin, L. S., Strickler, J., Van Arsdell, J. N., Yamamoto, R., and Mark, D. F., 1985, Molecular cloning of the complementary DNA for human tumor necrosis factor, *Science* **228:**149–154.

Wong, G. H. W., and Goeddel, D. V., 1986, Tumour necrosis factors α and β inhibit virus replication and synergize with interferons, *Nature* **323:**819–822.

Zacharchuk, C. M., Drysdale, B. E., Mayer, M. M., and Shin, H. S., 1983, Macrophage-mediated cytotoxicity: Role of a soluble macrophage cytotoxic factor similar to lymphotoxin and tumor necrosis factor, *Proc. Natl. Acad. Sci. U.S.A.* **80:**6341–6345.

III

BIOSYNTHESIS AND RELEASE OF LIPID MEDIATORS OF INFLAMMATION

9

Arachidonic Acid Metabolism in Tissue Injury

STEPHEN M. SPAETHE and PHILIP NEEDLEMAN

1. Introduction

The significance of enhanced arachidonic acid metabolism in a variety of inflammatory disorders has been the focus of intense investigation for over 20 years. Several types of tissue injury, including hydronephrosis (Morrison *et al.*, 1977; Nishikawa *et al.*, 1977), glomerulonephritis (Lianos *et al.*, 1983), renal vein constriction (Zipser *et al.*, 1980), myocardial infarction (Evers *et al.*, 1985), ulcerative colitis (Zipser *et al.*, 1985), rheumatoid arthritis (McGuire *et al.*, 1982), and pulmonary fibrosis (Clark *et al.*, 1983), are characterized by marked quantitative, as well as qualitative, changes in arachidonic acid metabolism. In each case, the exaggerated eicosanoid production is associated with the invasion of inflammatory cells into the site of injury. A major interest in this laboratory has been understanding the temporal relations between inflammatory cell influx and enhanced arachidonate metabolism. Additionally, attempts have been made to correlate the biochemical events with the pathophysiological response observed following the onset of tissue injury.

2. Exaggerated Eicosanoid Production in Hydronephrosis

2.1. Arachidonic Acid Metabolism in Normal and Hydronephrotic Kidney

The capacity of renal tissue to produce prostaglandins (PG) has long been known. Prostaglandin-like substances were identified in the renal

STEPHEN M. SPAETHE and PHILIP NEEDLEMAN • Department of Pharmacology, Washington University School of Medicine, St. Louis, Missouri 63110.

venous effluent from dog kidney, (1) during nerve stimulation (Dunham and Zimmerman, 1970), (2) during renal ischemia (McGiff *et al.*, 1970a; Jaffe *et al.*, 1972), (3) upon infusion of norepinephrine (Dunham and Zimmerman, 1970) or bradykinin (McGiff *et al.*, 1972), or (4) upon infusion of angiotensin II (McGiff *et al.*, 1970b), which is blocked by pretreatment with indomethacin (Aiken and Vane, 1971). Subsequent experiments in the isolated perfused rabbit kidney have shown similar findings. Intra-arterial administration of catecholamines or renal nerve stimulation resulted in the release of prostaglandins via alpha adrenergic receptor activation, while PG release in response to angiotensin II was independent of alpha receptor activation but rather dependent on stimulation of specific angiotensin receptors (Needleman *et al.*, 1973, 1974). The primary site of prostaglandin biosynthesis in these "normal" kidney preparations was demonstrated to be the renal medulla in both the dog (Crowshaw *et al.*, 1970) and the rabbit (Crowshaw, 1971).

The primary renal prostaglandin released in response to the various stimuli is PGE_2, a vasodilator (Daniels *et al.*, 1967) and natriuretic (Johnson *et al.*, 1967). This prostanoid is thought to be important in the regulation of intrarenal blood flow (McGriff *et al.*, 1974) and maintenance of glomerular filtration, especially in the presence of enhanced vasoconstrictor activity (Dibona, 1986). Elevation of ureteral pressure in the cat inhibits vasoconstriction elicited by either neuronal stimulation or exogenous catecholamines (Schramm and Carlson, 1975). Inhibition of prostaglandin synthesis diminished the inhibitory effects of ureteral pressure elevations on renal resistance during nerve stimulation, indicating a compensatory production of the endogenously produced vasodepressor PGE_2. Vaughn *et al.* (1971) have demonstrated that following unilateral ureteral obstruction (a model of hydronephrosis) in the dog, there was a transient increase in renal blood flow. Pretreatment with indomethacin prior to ureter ligation resulted in a fall in total renal blood flow, and the characteristic ipsilateral renal vasodilation did not occur (Allen *et al.*, 1978). Furthermore, the ureteral pressure remained significantly lower in the indomethacin pretreated animals, suggesting that prostaglandin-dependent vasodilation occurs in response to ureteral occlusion.

In the rabbit, complete unilateral ureteral obstruction resulted in increased basal PGE_2 production. Additionally, the responsiveness of PGE_2 biosynthesis to stimulatory agents such as bradykinin (BK), angiotensin II, and norepinephrine is enhanced in the perfused hydronephrotic kidney (HNK) when compared with the contralateral (CLK) or a normal perfused kidney (Nishikawa *et al.*, 1977). Moreover, since indomethacin treatment increased basal perfusion pressure, basal resistance in this experimental surgical model of hydronephrosis in the rabbit is dependent at least in part on endogenous vasodepressor prostaglandin production.

While the production of vasodepressor prostaglandins may explain the fall in resistance seen in the initial phases of ureter obstruction, it does not explain the progressive fall in renal blood flow and glomerular filtration rate observed in dogs following 18 h of ureteral ligation (Moody *et al.*, 1975). Moody *et al.* (1977) showed that intrarenal infusion of an angiotensin II receptor blocker failed to ablate the decreased ipsilateral renal blood flow and increased renal resistance following ureteral obstruction. Morrison *et al.* (1977) showed that the rabbit HNK (but not the CLK) has the capacity to produce thromboxane A_2 in response to agonist (BK) stimulation. Associated with this increase in TxA_2 production is a marked increase in perfusion pressure that can be blocked by pretreatment with a selective thromboxane synthetase inhibitor, OKY1581 (Miyamoto *et al.*, 1980; Kawasaki and Needleman, 1982). Moreover, the major site of thromboxane A_2 synthase activity is localized in the cortex of the HNK (Morrison *et al.*, 1978), the site of preglomerular vasoconstriction. Intra-arterial infusion of imidazole, a selective thromboxane synthetase inhibitor (Needleman *et al.*, 1977), above the renal arteries of rats following release of the unilateral ureteral obstruction resulted in increased GFR and renal blood flow.

2.2. Accumulation of Inflammatory Cells in Hydronephrosis

Histologically, the exaggerated arachidonic metabolism seen in hydronephrosis is associated with a profound proliferation of fibroblast-like interstitial cells and the invasion of inflammatory cells, most notably the influx of mononuclear cells into the cortex and outer medulla (Nagle *et al.*, 1973, 1976; Okegawa *et al.*, 1983a). Using indium-111-labeled leukocytes and platelets, we examined the inflammatory cellular changes that occur in hydronephrosis (Mathias *et al.*, 1987). Following unilateral ureter ligation in rabbit, there was an immediate (within 10 min) accumulation (2.5-fold) of platelets in the HNK compared to the control CLK, which was maximal within 30 min and returned toward control levels by 2 h. If platelets were administered at 6 h. after ureter ligation, there was no preferential accumulation. Upon administration of III ^{111}In-labeled polymorphonuclear leukocytes (PMNs) at time of obstruction, radioactivity accumulated in the HNK by 12 h. and was maximal at 24 h. before returning to basal levels (ratio of HNK to CLK of 1) by 72 h. Monocyte infiltration was first detectable after 24 h. of ureter obstruction and was maintained at maximal levels throughout the time course of study (>5 days. Fig. 1). These studies demonstrate a temporal sequence of events in inflammatory cell influx following unilateral ureter obstruction, which is the orderly attachment of platelets, PMNs, and finally monocytes. The rapid attachment of platelets was surprising, since ureter obstruction occurs at the level of the bladder and neither renal vasculature nor

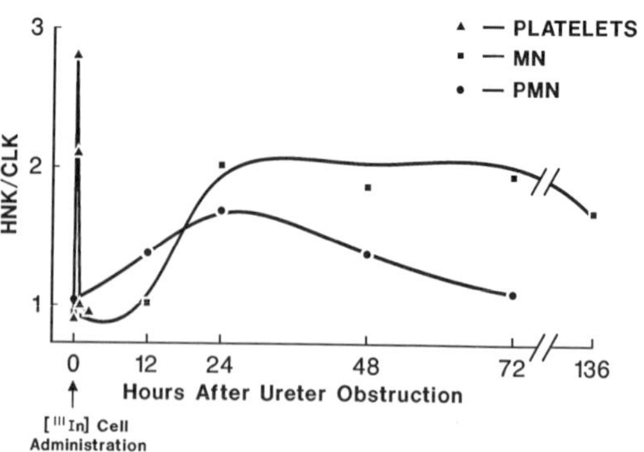

Figure 1. Ratio of radioactivity present in the rabbit HNK relative to CLK from imaging data on a computer. The gamma imaging occurred at times indicated after ureteral ligation. Administration of ^{111}In labeled platelets, monocytes, or PMNS was at time 0. The data are normalized relative to the CLK (=1.0).

renal capsule was exposed, suggesting that the increase in cortical interstitial pressure that occurs might also result in renal vascular damage.

We hypothesize that the initial platelet adhesion would be accompanied by the release of the vasopressor TxA_2, as well as the potent PMN chemotactant, platelet-derived growth factor. Indeed, the initial leukocyte appearing at the site of tissue injury is the PMN that possesses the synthetic capacity for LTB_4 production. The result of this potent monocyte chemotactic agent could be the recruitment of circulating blood monocytes into the site of injury for scavenger and repair functions. Additionally, we hypothesize that the invading blood monocytes are intimately involved in the exaggerated eicosanoid production in the HNK.

2.3. *Role of Tissue Macrophages*

Evidence to support the intimate relation of monocyte invasion to enhanced arachidonate metabolism is copious. Ureteral obstruction in the cat does not result in exaggerated agonist-stimulated eicosanoid production, and inflammatory cell influx is absent (Reingold *et al.*, 1981). Similarly, in rabbits made leukopenic prior to ureter obstruction (Lefkowith *et al.*, 1984), agonist-stimulated arachidonate metabolism was not enhanced compared to the CLK, and cortical microsomal thromboxane synthase activity was not induced. These results also implicate the macrophages as the cellular origin of TxA_2 production in the HNK. Furthermore, following a 3- or 10-day release of the ureter obstruction in normal rabbits, enhanced bradykinin-stimulated prostanoid production is markedly diminished, associated histologically by the loss of tissue macrophages.

The hypothesis that the tissue macrophage is important in regulating the changes in the arachidonate metabolic profile was further tested by the use of endotoxin (lipopolysaccharide, LPS), a macrophage agonist for eicosanoid production (Feuerstein *et al.*, 1981; Halushka *et al.*, 1981), which resuslts in these cells becoming cytolytic (Taffet and Russell, 1981). Administration of LPS to the *ex vivo* perfused rabbit kidney resulted in an immediate and sustained release of TxA_2, and PGE_2 from the HNK but not the CLK (or the postobstructed rabbit HNK or the cat HNK), consistent with the absence of the target cell (Okegawa *et al.*, 1983b). One hour of LPS pretreatment *in vivo* prior to removal of the HNK for *ex vivo* perfusion markedly enhanced (four- to ten-fold) subsequent BK-stimulated PGE_2 and TxB_2 production, while the CLK was unaffected by this maneuver (Lefkowith *et al.*, 1984).

2.4. Monokines

Macrophages have been shown to modulate actively fibroblast proliferation and PGE_2 production (Korn *et al.*, 1980). Monocytes, when stimulated, release a variety of products including arachidonic acid metabolites (Humes *et al.*, 1977), components of complement (Einstein *et al.*, 1976), and monokines (Gery and Waksman, 1972). The supernatant of adherent human peripheral mononuclear cells contains a factor(s) that results in a marked stimulation of PGE_2 production by dermal fibroblasts (Korn *et al.*, 1980), gingival fibroblasts (DiSouza *et al.*, 1981), synovial cells (Dayer *et al.*, 1977), and lung fibroblasts (Clark *et al.*, 1983).

Studies in this laboratory have examined the effect of the presence of monocytes and monocyte-conditioned media on the arachidonate metabolism of other cells types. Cell cultures from explants of the rabbit HNK cortex contained fibroblasts and macrophages, while the CLK contained only fibroblasts. Stimulation of these cells cultures with endotoxin resulted in release of PGE_2 from the HNK cultures only and a marked ability to release PGE_2 in response to bradykinin by the HNK compared to the CLK was also observed. Following passage of the HNK culture explants, macrophages were lost from the colony, as well as the responsiveness of the remaining fibroblasts to LPS stimulation. Conditioned media from adherent rabbit peripheral blood monocytes termed mononuclear cell factor (MNCF) were able to stimulate PGE_2 production by both HNK and CLK cell cultures (Jonas *et al.*, 1984). Further evaluation of this mononuclear cell factor showed that incubation with fibroblasts resulted in a marked increase in both basal- (100-fold) and agonist-stimulated (8–12-fold) PGE_2 production by the fibroblast (Jonas and Needleman, 1984). This effect was blocked by inhibition of protein synthesis or pretreatment with ibuprofen and was determined kinetically to be due to an increase in total microsomal cyclooxygenase activity (i.e. increased V_{max}, no change in K_m). Similar effects of MNCF on prostaglandin production have been reported for smooth muscle cells and endothelial cells (Albrightson *et al.*, 1985a). Attempts to characterize MNCF showed that the dose- and time-dependent effects of this factor on the cell types studied were mimicked by treatment with purified recombinant human interleukin-1 (IL-1). When the MNCF was subjected to gel filtration chromatography, it coeluted with recombinant IL-1 (M_r 12,000–18,000 daltons). The purified monokine contained fibroblast and endothelial cell prostaglandin stimulating activity and IL-1 activity (as determined by D 10 lymphocyte proliferation) (Albrightson *et al.*, 1985b). These data suggest that factors released by the monocyte have profound effects on fibroblasts and other cell types, including the induction of cyclooxygenase. Additionally, at least one of the mediators of this phenomenon is IL-1. We hypothesize that similar interaction occurs in the HNK between the cortical tissue macrophage and the fibroblast-like interstitial cells. Endotoxin pretreatment *in vivo* activates the tissue macrophages in the obstructed kidney, possibly leading to the release of IL-1. This monokine could stimulate adjacent interstitial cells, resulting in the induction of enzymes involved in arachidonic acid metabolism.

Macrophages contain not only cyclooxygenase but also 5-lipoxygenase and have been shown to produce leukotrienes *in vitro* (Borgeat and Samuelsson, 1979; Fels *et al.*, 1982; Scott *et al.*, 1982; Goldyne *et al.*, 1984; Hsueh *et al.*, 1985). Stimulation of human monocytes with the chemotactic *N*-formylmethionyl-leucyl-phenylalanine (fMLP) results in the synthesis of the leukotrienes LTB_4 and LTC_4 (Williams *et al.*, 1986). Additionally, cortical explant cultures from rabbit hydronephrotic kidney, which contain tissue macrophages, in addition to fibroblast-like interstitial cells, produce leukotrienes in response to the calcium ionophore A23187 (Dannon *et al.*, 1986). Leukotrienes may indeed represent an important eicosanoid product of the HNK. The potential pathophysiologic importance of leukotriene synthesis in renal inflammation is suggested by their ability to contract smooth muscle and induce vascular leakage, and by their chemotactic activity for inflammatory cells.

2.5. *Production of Leukotrienes in Hydronephrosis*

To evaluate the potential of the rabbit HNK to synthesize and release leukotrienes, LPS was administered *in vivo* 1 hr prior to removal of the kidney for *ex vivo* perfusion. In addition to enhanced BK- and fMLP-stimulated release of the cyclooxygenase products, prostaglandin E_2, thromboxane A_2, and prostacyclin, the hydronephrotic kidney, but not the CLK, releases prodigious amounts of leukotrienes in response to this inflammatory cell agonist fMLP (Albrightson *et al.*, 1987). Bradykinin does not directly stimulate macrophages, monocytes, or cultured resident peritoneal macrophages to synthesize eicosanoids. We believe a major cellular source of leukotriene production to be the invading tissue macrophage. This is consistent with the observation that the CLK and the postobstructed HNK, which lack macrophages, did not release significant amounts of leukotrienes in response to fMLP (Albrightson *et al.*, 1987; Spaethe *et al.*, 1987).

Associated with the administration of fMLP is an increase in perfusion pressure by the HNK. This increased resistance in the perfused HNK was blocked not only by the peptidoleukotriene receptor antagonist FPL55712 but also by the thromboxane synthase inhibitor OKY1581 (Albrightson *et al.*, 1987). Exogenous administration of LTC_4 also resulted in a vasopressor response in the HNK but not the CLK, which was a result of the production of TxA_2, since OKY1581 completely abolished the pressure increase. This newly acquired ability to synthesize and release leukotrienes by the injured tissue may have an important impact on renal function via TxA_2 synthesis. Renal vasoconstriction could reduce renal blood flow, resulting in tissue ischemia and exacerbated tissue damage. Likewise, the unmasking or leukotriene production in the HNK could enhance vascular leakage and stimulate inflammatory cell chemotaxis, resulting in further tissue damage.

3. Essential Fatty Acid Deficiency and Arachidonic Acid Metabolism in Inflammation

3.1. Biochemical Effects

Mammals have an absolute requirement for dietary (n-6) fatty acids (Holman, 1969). Following dietary deprivation of these fatty acids, a state of essential fatty acid (EFA) deficiency is induced that is characterized by growth retardation (Holman, 1969), a scaly dermatitis with increased transepidermal water loss (Holman, 1969), and depressed inflammatory responsiveness (Bonta and Parnham, 1982). Biochemically, this state of EFA deficiency is characterized by a decrease in n-6 and n-3 fatty acid and an accumulation of n-9 fatty acids, notably 20:3 (n-9) or Mead acid, which is not a constituent of normal tissue (Fulco and Mead, 1959). As a result of a general decrease in tissue levels of arachidonate, EFA deficiency is a useful tool for examining the role of arachidonic acid and its metabolites in various physiologic and pathologic states.

A notable effect of EFA deficiency is the ability of this state to abrogate the inflammatory nephritis seen in a murine model of systemic lupus erythematosus. Dietary depletion of arachidonic acid precursor pools in these lupus mice successfully suppressed the histological evidence of inflammation in the glomerulus and the accompanied proteinuria, and it prolonged survival (Hurd *et al.*, 1981; Prickett *et al.*, 1981). Likewise, selective depletion of leukocytes with antimacrophage antiserum prevents the glomerular injury seen in a variety of experimental glomerulonephritides (Holdsworth *et al.*, 1981). A number of other investigations have also implicated eicosanoids in the pathogenesis of glomerulonephritis (Kaizu *et al.*, 1985; Lianos *et al.*, 1985; Patrono *et al.*, 1985; Kelley *et al.*, 1986).

Since the macrophage is an important cellular constituent of this inflammatory lesion, our laboratory has examined the effect of EFA deficiency on the eicosanoid metabolism and function of these cells. The resident peritoneal macrophages from exhibited mice deprived of dietary EFA exhibited depletion of phospholipid arachidonate, and an accumulation of 20:3 (n-9); phosphatidylinositol was preferentially affected (Lefkowith *et al.*, 1987). Stimulation of these cells with zymosan resulted in the production of markedly less LTC_4 and LTB_4 than macrophages from control animals. Synthesis of LTC_3 from 20:3 (n-9) was also observed; however, no LTB_3 was detected and total peptidoleukotriene (LTC_3 plus LTC_4) was reduced. Therefore, a major constituent of the anti-inflammatory activity associated with EFA deficiency appears to be due to decreased leukotriene production, resulting in decreased chemotactic activity via reduced LTB_4 production, and may also be a consequence of decreased LTC_4-mediated vasoconstrictor activity and reduced plasma exudation (Dahlen *et al.*, 1981), as well as diminished attachment of macrophages to glomeruli associated with decreased lipoxygenase activity (Baud *et al.*, 1985).

3.2. Functional Effects

Macrophages isolated from EFA-deficient mice demonstrated a marked impairment in receptor-mediated pinocytosis, while phagocytosis was unaffected (Lefkowith *et al.*, 1987). The deficiency in the ability of EFA-deficient macrophages to pinocytose may also contribute to the anti-inflammatory effect. Indeed, receptor-mediated endocytosis has been implicated in antigen processing (Allen and Unanue, 1984) and in the internalization of immune complexes via the FC receptor (Mellman *et al.*, 1984).

Both resident glomerular macrophage population and prostanoid production are thought to be significant factors in the pathogenesis of glomerulonephritis. Examination of the effect of EFA deficiency on macrophages isolated from either Lewis or Sprague–Dawley rats revealed a marked reduction in both cortical and resident glomerular macrophage number and a decrease in the glomerular Ia^+ cell population (Lefkowith and Schreiner, 1987). The circulating monocyte number remained unchanged.

Stimulation of glomerular macrophages with angiotensin II resulted in an approximate threefold increase in PGE_2 and TxB_2 release by cells from normal rats. EFA deficiency reduced both basal- and agonist-induced eicosanoid production (Lefkowith and Schreiner, 1987). Dietary (n-6), but not (n-3), fatty acid supplementation reversed both the reduction in glomerular macrophage number and the decreased arachidonic acid metabolism associated with EFA deficiency. Collectively, these data suggest that not only modulation of glomerular macrophage number but also reduction in macrophage functionality (i.e., eicosanoid production and pinocytosis) may contribute to sparing effects associated with dietary fatty acid restriction in lupus glomerulonephritis. Reduction in the synthesis of the chemotactic leukotrienes, LTB_4, could result in attenuation of leukocyte influx and subsequent monocyte activation associated with the pathogenesis of experimental glomerulonephritis (Schreiner *et al.*, 1978; Holdsworth *et al.*, 1981).

3.3. Effect of EFA Deficiency on Hydronephrosis

As previously discussed, the influx of circulating monocytes into the kidney following ureter ligation appears to be intimately associated with the exaggerated eicosanoid production observed in hydronephrosis. To examine this hypothesis, we have employed the essential fatty acid deficiency (EFAD) as an experimental manipulation in evaluating the role of inflammatory cells in the onset and maintenance of the exaggerated eicosanoid production in hydronephrosis. Weanling rabbits were deprived of dietary essential fatty acids for 3 months prior to unilateral ureter obstruction and removal of the kidneys for *ex vivo* perfusion. This dietary manipulation resulted in a reduction of renal cortical arachidonate that was preferential for the phosphatidylinositol phospholipid (Spaethe *et al.*, 1987). As previously discussed, unilateral ureteral obstruction results in the prodigious production of PGE_2 and thromboxane A_2 in response to bradykinin in the HNK from LPS-pretreated rabbits (Lefkowith *et al.*, 1984), as well as the unmasking of leukotriene synthesis in response to the chemotactic peptide fMLP (Albrightson *et al.*, 1987). Bradykinin- and fMLP-stimulated arachidonic acid metabolism was dramatically suppressed (Figs. 2 and 3) in the EFAD HNK (Spaethe *et al.*, 1987). Similar reduction in agonist-stimulated eicosanoid release was observed in isolated perfused EFA-deficient mouse heart and kidney compared to control organs (Lefkowith *et al.*, 1985).

In agreement with the anti-inflammatory properties associated with EFAD, there was a reduced influx of monocytes into the HNK renal cortex following ureteral obstruction in the EFA-deficient rabbit. There was an accumulation of 20:3 (n-9) Mead acid in the phospholipid pool of the renal cortex of these animals. We hypothesize as to the sparing effect of EFAD in hydronephrosis; the initial stimulus following ureter obstruction may be the release of platelet-derived growth factor to recruit PMNs. This stimulus for PMNs could result in the release of fatty acids from the phospholipid pools, including 20:3 (n-9), which is metabolized to LTA_3 by 5-lipoxygenase.

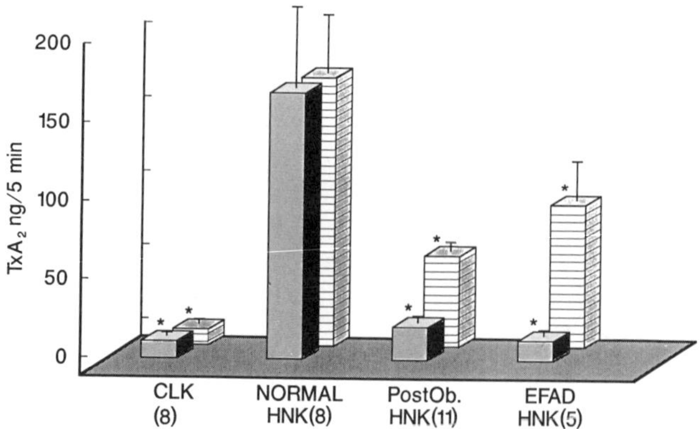

Figure 2. Agonist (BK or fMLP)-induced release of TxA_2 from isolated perfused rabbit kidney. Animals were pretreated with LPS (8 mg/kg i.v.) 1 h prior to removal of kidney for *ex vivo* perfusion. Fifty milliliters (5 min) of renal venous effluent were collected for quantitation of agonist-mediated release of TxA_2 by RIA. The results are presented as mean $\pm$SEM, and the value in parentheses indicates the number of animals studied indicates different from normal HNK, $p<0.05$).

LTA_3 covalently binds LTA_4 hydrase, resulting in inhibition of this enzyme (Evans *et al.*, 1985). As a result of decreased formation of the monocyte chemotactant LTB_4, the ability of the HNK to perpetuate the inflammatory response might be lost.

4. Arachidonic Acid Metabolism in Myocardial Infarction

Myocardial infarction, like hydronephrosis, is an inflammatory condition characterized by exaggerated arachidonic acid metabolism—both cyclooxygenase (McCluskey *et al.*, 1982) and lipoxygenase (Mullane *et al.*, 1984; McCluskey *et al.*, 1985) in the canine models examined. Associated with this exaggerated eicosanoid production is the sequential invasion of blood-borne inflammatory cells, predominantly PMNs and macrophages (Mallory *et al.*, 1979). Leukocyte depletion prior to coronary occlusion results in a reduction in infarct size (Romson *et al.*, 1983). These observations suggest a predominant role of circulating leukocytes and their metabolic products in the modulation of cardiac injury. In the infarcted tissue there is liberation of arachidonate from phospholipids, which results in cyclooxygenase-dependent PGE and TxA_2 production (Mullane *et al.*, 1984; McCluskey *et al.*, 1985). The significance of lipoxygenase products is suggested by the obser-

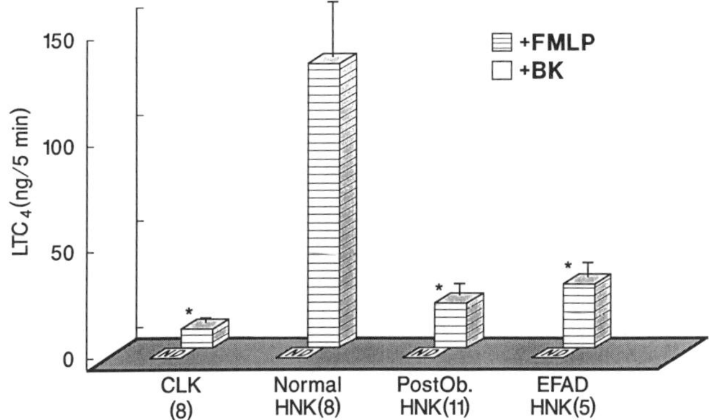

Figure 3. Agonist (BK or fMLP)-induced release of LTC_4 from isolated perfused rabbit kidney. Animals were pretreated with LPS (8 mg/kg i.v.) 1 h prior to removal of kidney for *ex vivo* perfusion. Fifty milliliters (5 min) of renal venous effluent were collected for quantitation of agonist-mediated release of LTC_4 by IRA. The results are presented as mean ± SEM, and the value in parentheses indicates the number of animals studied (*indicates different from normal HNK, $p<0.05$).

vation that BW-755C, a combined cyclooxygenase–lipoxygenase inhibitor, reduces canine infarct size and incidence of ventricular dysrhythmias (Mullane and Moncada, 1982; Jolly and Lucchesi, 1983; Mullane *et al.*, 1984). In contrast, cyclooxygenase inhibition with aspirin (Bonow *et al.*, 1981), naproxen (Bolli *et al.*, 1981), and indomethacin (Jugdutt *et al.*, 1979) lack myocardial sparing effects.

4.1. Eicosanoid Production in Rabbit Myocardial Ischemia

We have developed a rabbit model of *in vivo* left ventricular myocardial infarction, followed by *ex vivo* perfusion to identify and quantitate agonist-stimulated eicosanoid production. Stimulation of this isolated perfused heart (removed 1 day or 4 days after circumflex infarction and reperfusion) with the inflammatory cell agonist fMLP resulted in the release of leukotrienes B_4, C_4, and D_4, and prodigious production of TxA_2, PGI_2, and PGE_2. Associated with the release of these arachidonate metabolites was a profound coronary vasoconstriction that was blocked by FPL 55712 (Evers *et al.*, 1985). The release of the leukotriene C_4 from the right atria of the damaged tissue was significantly greater than from control hearts, and the accompanying vasoconstriction was absent in the control preparation. Following administration of bradykinin to the infarcted heart, detectable leukotrienes were not produced. However, there was a tenfold increase in TxA_2, PGI_2, and PGE_2 released compared to control tissue (Evers *et al.*, 1985). These data further implicate a vasoactive lipoxygenase product in mediating pressor activity in the infarcted tissue.

4.2. Biochemical, Morphological, and Anatomical Considerations

The exaggerated eicosanoid production in the infarcted rabbit heart is associated with the invasion of inflammatory cells into the infarcted tissue (Evers *et al.*, 1985). As in hydronephrosis, this influx in leukocytes is temporally associated with the onset of enhanced arachidonate metabolism. The obligatory relation between inflammatory cell influx and exaggerated eicosanoid synthesis is further strengthened by the observation that, 30 days after infarction, inflammatory cells have disappeared from the infarcted tissue; this is associated with a marked reduction in prostanoid synthesis (Evers *et al.*, 1987).

Further investigation of the exaggerated arachidonate metabolism in the rabbit myocardial infarct revealed, surprisingly, that the site of cardiac arachidonate metabolizing enzymes, of agonist-stimulated arachidonate metabolism, and of the fMLP-induced coronary vasoconstriction was concentrated in the right atrium, a site distant from the zone of injury (Evers *et al.*, 1987). Furthermore, enzymatic studies demonstrated that the enhanced conversion of arachidonic acid by the right atria of infarcted hearts was not due to enzyme induction; there was no increase in V_{max} for PGE_2 or TxB_2 production compared to control tissue (and K_m was unchanged). This suggests that the enhanced atrial prostanoid production occurring following the reversible coronary ligation results from increased arachidonate availability, resulting either from increased phospholipase activity and/or increased agonist receptor recognition in the coronary arteries of the infarcted tissue.

The enhanced eicosanoid production associated with myocardial infarction, as well as other models of tissue injury, has previously been attributed to the influx of inflammatory cells that infiltrate the injured tissue. Consistent with this hypothesis, there was increased agonist-stimulated TxA_2 formation in the ventricles of infarcted hearts. Surprisingly, however, most of the augmented agonist (BK and fMLP)-stimulated eicosanoid production was confined to right atrial tissue, an area devoid of inflammatory infiltrate. The mechanism by which damaged ventricular tissue influences atrial arachidonate metabolism remains unclear.

The potential for the production of the inflammatory cell chemotactic leukotriene LTB_4 does exist in cardiac tissue. Myocardial infarction may serve as the initial stimulus for LTB_4 formation for recruitment of inflammatory cells into the damaged tissue. Additionally, production of the vasopressor leukotrienes C_4 and D_4 may further increase myocardial damage by exacerbating the ongoing ischemia. Therefore, inhibition of the synthesis of these leukotrienes may be a useful therapeutic intervention in limiting myocardial infarct size.

References

Aiken, J. R., and Vane, J. R., 1971, Blockade of angiotensin-induced prostaglandin release from dog kidney by indomethacin, *Pharmacologis* **13:**293.

Albrightson, C. R., Baenziger, N. L., and Needleman, P., 1985a, A mononuclear cell factor is a potent agonist of prostaglandin biosynthesis in cultured human fibroblasts, smooth muscle and endothelial cells, in: *Advances in Prostaglandin, Thromboxane and Leukotriene Research* (O. Hayaishi and S. Yamamoto, eds.), Vol. 15, pp 213–216, Raven Press, New York.

Albrightson, C. R., Baenziger, N. L., and Needleman, P., 1985b, Exaggerated human vascular cell prostaglandin biosynthesis mediated by monocytes: Role of monokines and interleukin 1, *J. Immunol.* **135:**1872–1877.

Albrightson, C. R., Evers, A. S., Griffin, A. C., and Needleman, P., 1987, The effect of endogenously produced leukotrienes and thromboxane on renal vascular resistance in rabbit hydronephrosis, *Cir. Res.* **G1:**514–522.

Allen, J. T., Vaughn, E. D., and Gillenwater, J. H., 1978, The effect of indomethacin on renal blood flow and ureteral pressure in unilateral ureteral obstruction in awake dogs, *Invest. Urol.* **15:**324–327.

Allen, P. M., and Unanue, E. R., 1984, Differential requirements for antigen processing by macrophages for lysozyme-specific T cell hybridomas, *J. Immunol.* **132:**1077–1079.

Baud, L., Sraer, J., Delarue, F., Bens, M., Balavoine, F., Schlondorff, D., Ardaillou, R., and Sraer, J. D., 1985, Lipoxygenase products mediate the attachment of rat macrophages to glomeruli *in vitro, Kidney Int.* **27:**855–863.

Bolli, R., Goldstein, R. E., Davenport, N., and Epstein, S. E., 1981, Influence of sulfinpyrazone and naproxen on infarct size in the dog, *Am. J. Cardiol.* **47:**841–847.

Bonow, R. O., Lipson, L. C., Sheehan, F. H., Cappuro, N. L., Isner, J. M., Roberts, W. C., Goldstein, R. E., and Epstein, S. E., 1981, Lack of effect of aspirin on myocardial infarct size in the dog, *Am. J. Cardiol.* **47:**258–264.

Bonta, I. L., and Parnham, M. J., 1982, Prostaglandin, essential fatty acids and cell–tissue interactions in immun-inflammation, *Prog. Lipid Res.* **20:**617–623.

Borgeat, P., and Samuelsson, P., 1979, Transformation of arachidonic acid by rabbit polymorphonuclear leukocytes. Formation of a novel dihydroxy-eicosatetraenoic acid, *J. Biol. Chem.* **254:**2643–2646.

Clark, J. G., Kostal, K. M., and Marino, B. A., 1983, Bleomycin-induced pulmonary fibrosis in hamsters, *J. Clin. Invest.* **72:**2082–2091.

Crowshaw, K., 1971, Prostaglandin biosynthesis forms endogenous precursors in the rabbit kidney, *Nature New Biol.* **231:**240–241.

Crowshaw, K., McGiff, J. C., Strand, J. C., Lonigro, A. J., and Terragna, N. A., 1970, Prostaglandins in dog renal medulla, *J. Pharm. Pharmacol.* **22:**302–303.

Dahlen, S. E., Bjork, J., Hedquist, P., Arfors, K. E., Hammarstrom, S., Lindgren, J. A., and Sammuelsson, B., 1981, Leukotrienes promote plasma leakage and leukocyte adhesion in postcapillary venules: *In vivo* effects with relevance to the acute inflammatory response, *Proc. Natl. Acad. Sci. U.S.A.* **78:**3887–3891.

Daniels, E. G., Hinman, J. W., Leach, B. E., and Muirhead, E. E., 1967, Identification of prostaglandin E_2 as the principal vasodepressor lipid of rabbit renal medulla, *Nature* **215:**1298–1299.

Danon, A., Zenser, T. V., Thomasson, D. L., Palmier, M. O., and Davis, B. B., 1986, Eicosanoid synthesis by rabbit hydronephrotic cortical interstitial cells in culture, *J. Pharmacol. Exp. Ther.* **238:**95–99.

Dayer, J.-M., Robinson, D. R., and Krane, S. M., 1977, Prostaglandin production by rheumatoid synovial cells. Stimulation by a factor from human mononuclear cells, *J. Exp. Med.* **145:**1399–1404.

Dibona, G. F., 1986, Prostaglandins and non-steroidal anti-inflammatory drugs; effects on renal hemodynamics, *Am. J. Med* **80**(Suppl. 1A):12–21.

DiSouza, S. M., Englis, D. J., Clark, A., and Russell, R. G., 1981, Stimulation of production of prostaglandin E in gingival cells exposed to products of human blood mononuclear cells, *Biochem, J.* **198:**391–396.

Dunham, E. W., and Zimmerman, B. G., 1970, Release of prostaglandin-like material from dog kidney during nerve stimulation, *Am. J. Physiol.* **219:**1279–1285.

Einstein, L. P., Schneeberger, E. E., and Colten, H. R., 1976, Synthesis of the second component of complement by long-term primary culture of human monocytes, *J. Exp. Med.* **143:**114–126.

Evans, J. F., Nathaniel, D. J., Zamboni, R. J., and Ford-Hutchinson, A. W., 1985, Leukotriene A_3: A poor substrate but a potent inhibitor of rat and human neutrophil leukotriene A_4 hydrolase, *J. Biol. Chem* **260:**10966–10970.

Evers, A. S., Murphree, S., Saffitz, J. E., Jakschik, B. A., and Needleman, P., 1985, Effects of endogenously produced leukotrienes, thromboxane, and prostaglandins on coronary vascular resistance in rabbit myocardial infarction, *J. Clin. Invest.* **75:**992–999.

Evers, A. S., Dunkel, C. G., Saffitz, J. E., and Needleman, P., 1987, Exaggerated atrial arachidonate metabolism in rabbit left ventricular myocardial infarction, *J. Clin. Invest* **79:**155–162.

Fels, A. O. S., Pawlowski, N. A., Cramer, E. B., King, T. K. C., Cohn, Z. A., and Scott, W. A., 1982, Human alveolar macrophages produce leukotriene B_4, *Proc. Natl. Acad. Sci U.S.A.* **79:**7866–7870.

Feuerstein, N., Foegh, M., and Ramwell, P. W., 1981, Recently reported stimulation of TxB_2 and 6-keto$PGF_{1\alpha}$ synthesis by rat peritoneal macrophages incubated with *E. coli* 055:BS lipopolysaccharide, *Br. J. Pharmacol.* **72:**389–391.

Fulco, A. J., and Mead, J. F., 1959, Metabolism of essential fatty acids, *J. Biol. Chem.* **234:**1411–1416.

Gery, I., and Wasksman, B. H., 1972, Potentiation of the T-lymphocyte response to mitogens. II. The cellular source of potentiating mediator(s), *J. Exp. Med.* **1361:**143–155.

Goldyne, M. E., Burrish, G. F., Poubelle, P., and Borgeat, P., 1984, Arachiodonic acid metabolism among human mononuclear leukocytes. Lipoxygenase-related pathways, *J. Biol. Chem.* **259:**8815–8819.

Halushka, P. V., Cook, J. A., and Wise, W. C., 1981, Thromboxane A_2 and prostacyclin production by lipopolysaccharide-stimulated peritoneal macrophages, *J. Reticuloendothel. Soc.* **30:**445–450.

Holdsworth, S. R., Neale, T. J., and Wilson, C. B., 1981, Abrogation of macrophage-dependent injury in experimental glomerulonephritis, *J. Clin. Inves* **68:**689–698.

Holman, R. T., 1969, Essential fatty acid deficiency, *Prog. Chem. Fats Other Lipids* **9:**275–348.

Hsueh, W., Sun, F. F., and Henderson, S., 1985, The biosynthesis of leukotriene B_4, the predominant lipoxygenase products in rabbit alveolar macrophages, is enhanced during immune activation, *Biochim. Biophys. Acta* **835:**92–97.

Humes, S. L., Bonney, R. J., Pelus, L., Dahlgren, M. E., Sadowski, S. J., Kuehl, F. A., and

Davis P., 1977, Macrophages synthesize and release prostaglandins in response to inflammatory stimuli, *Nature* **269:**149–151.

Hurd, E. R., Johnston, J. M., Okita, J. R., MacDonald, P. C., Ziff, M., and Gilliam, J. N, 1981, Prevention of glomerulonephritis and prolonged survival in New Zealand Black/New Zealand White F_1 hybrid mice fed an essential fatty acid-deficient diet, *J. Clin. Invest.* **67:**476–485.

Jaffe, B. M., Parker, C. W., Marshall, G. R., and Needleman, P., 1972, The renal concentrations of prostaglandin E in acute and chronic renal ischemia, *Biochem. Biophys. Res. Commun.* **49:**799–805.

Johnson, H. H., Herzog, J. P., and Lauler, D. P., 1967, Effect of prostaglandin E_1 on renal hemodynamics, sodium and water excretion, *Am. J. Physiol.* **213:**939–946.

Jolly, S. R., and Lucchesi, B. R., 1983, Effect of BW755C in an occlusion–reperfusion model of ischemic myocardial injury, *Am. Heart J.* **106:**8–13.

Jonas, P. E., and Needleman, P., 1984, Mechanism of enhanced fibroblast arachidonic acid metabolism by mononuclear cell factor, *J. Clin. Invest.* **74:**2249–2253.

Jonas, P. E., Leahy, K. M., DeSchryver-Kecskemeti, K., and Needleman, P., 1984, Cellular interactions and exaggerated arachidonic acid metabolism in rabbit renal injury, *J. Leukocyte Biol.* **35:**55–64.

Jugdutt, B. I., Hutchins, G. M., Bulkley, B. J., Pitt, B., and Becker, L. C., 1979, Effect of indomethacin on collateral blood flow and infarct size in the conscious dog, *Circulation* **59:**734–743.

Kaizu, K., Marsh, D., Zipser, R., and Glassock, R. J., 1985, Role of prostaglandins and angiotensin II in experimental glomerulonephritis, *Kidney Int.* **28:**629–635.

Kawasaki, A., and Needleman, P., 1982, Contribution of thromboxane to renal resistance changes in the isolated perfused hydronephrotic rabbit kidney, *Circ. Res.* **50:**486–490.

Kelley, V. E., Snever, S., and Musinski, S., 1986, Increased renal thromboxane production in murine lupus nephritis, *J. Clin. Invest.* **77:**252–259.

Korn, J. H., Halushka, P. V., and LeRoy, E. C., 1980, Mononuclear cell modulation of connective tissue function: Suppression of fibroblast growth by stimulation of endogenous prostaglandin production, *J. Clin. Invest.* **65:**543–554.

Lefkowith, J. B., and Schreiner, G., 1987, Essential fatty acid deficiency depletes rat glomeruli of resident macrophages and inhibits angiotensin II-induced eicosanoid synthesis *J. Clin. Invest.* **80:**947–956.

Lefkowith, J. B., Okegawa, T., DeSchryver-Kecskemeti, K., and Needleman, P., 1984, Macrophage-dependent arachidonate metabolite in hydrophrosis, *Kidney Int.* **26:**10–17.

Lefkowith, J. B., Flippo, V., Sprecher, H., and Needleman, P., 1985, Paradoxical conservation of cardiac and renal arachidonate content in essential fatty acid deficiency, *J. Biol. Chem.* **260:**15736–15744.

Lefkowith, J. B., Jackschik, B. A., Stahl, P., and Needleman, P., 1987, Metabolic and functional alteration in macrophages induced by essential fatty acid deficiency, *J. Biol. Chem.* **262:**6668–6675.

Lianos, E. A., Giuseppe, G. A., and Dunn, M. J., 1983, Glomerular prostaglandin and thromboxane synthesis in rat nephrotoxic serum nephritis, *J. Clin. Invest.* **721:**439–1448.

Lianos, E. A., Rahman, M. A., and Dunn, M. J., 1985, Glomerular arachidonate lipoxygenation in rat nephrotoxic serum nephritis, *J. Clin. Invest.* **76:**1355–1359.

Mallory, G. P., White, P., and Sakedo-Salgar, T., 1979, The speed of healing of myocardial infarction. A study of the pathologic anatomy in seventy-two cases, *Am. Heart J.* **18:**647–671.

Mathias, C. J., Welch, M. J., Schwartz, D., Spaethe, S. M., and Needleman, P., 1987, In-111 labeled cells to differentiate thc sequential blood cell invasion of the injured rabbit kidney *in vivo J. Nuc. Med.,* (in press).

McCluskey, E. R., Corr, P. B., Lee B. I., Saffitz, J. E., and Needleman, P., 1982, The arachidonic acid metabolic capacity of canine myocardium is increased during healing of acute myocardial infarction, *Circ. Res.* **51:**743–750.

McCluskey, E. R., Murphee, S., Saffitz, J. E., Morrison, A. R., and Needleman, P., 1985, Temporal changes in 12-HETE formation in two models of canine myocardial infarction, *Prostaglandins* **29:**387–403.

McGiff, J. C., Crowshaw, K.,Terragno, N. A., Lonigro, A. J., Strand, J. C., Williamson, M. A., Lee, J. B., and Ng, K. D. F., 1970a, Prostaglandin-like substances appearing in canine renal venous blood during renal ischemia, *Circ. Res.* **27:**765–789.

McGiff, J. C., Crowshaw, K., Terragno, N. A., and Lonigro, A. J., 1970b, Release of prostaglandin-like substances into renal venous blood in response to angiotensin-II, *Circ. Res.* **26** (Suppl. I) :1121–1130.

McGiff, J. C., Terragno, N. A., Malik, K. U., and Lonigro, A. J., 1972, Release of prostaglandin E-like substance from canine kidney by bradykinin, *Circ. Res* **31:**36–43.

McGiff, J. C., Crowshaw, k., and Itskovitz, H. D., 1974, Prostaglandins and renal function, *Fed. Proc.* **33:**39–47.

McGuire, M. K., Meats, J. E., Ebsworth, N. M., Harvey, L., Murphy, G., Russell, G. G., and Reynolds, J. J., 1982, Properties of rheumatoid and normal synovial tissue in vitro and cells derived from them. Production of prostaglandins and collagenase in response to factors derived from cultured blook mononuclear cells and from synovium, *Rheumatol. Int.* **2:**113–120.

Mellman, I., Plutner, H., and Ukkonen, P., 1984, Internalization and rapid processing of macrophage F_c receptors tugged with monovalent antireceptor antibody: Possible role of a prelysosomal compartment, *J. Cell Biol.* **98:**1163–1169.

Miyamoto, T., Taniguchi, K., Tanonchi, T., and Hirata, F., 1980, Selective inhibition of thromboxane synthetase. Pyridine and its derivatives, in: *Advances in Prostaglandin Research,* (B. Samuelsson, P. Ramwell, and P. Paoletti, eds.), Vol.6, pp. 443–445, Raven Press, New York.

Moody, T. E., Vaughn, E. D., and Gillenwart, J. Y, 1975, Relationship between renal blood flow and ureteral pressure during eighteen hours of total unilateral occlusion. Implications for changing sites of renal resistance, *Invest. Urol.* **13:**246–251.

Moody, T. E., Vaughn, E. D., Jr., Wyker, A. T., and Gillenwater, J. Y., 1977, The role of intrarenal angiotensin II in the hemodynamic response to unilateral obstructive uropathy, *Invest. Urol.* **14:**390–397.

Morrison, A. R., Nishikawa, K., and Needleman, P., 1977, Unmasking of thromboxane A_2 synthesis by ureteral obstruction in the rabbit kidney, *Nature* **267:**259–260.

Morrison, A. R., Nishikawa, K., and Needleman, P., 1978, Thromboxane A_2 biosynthesis in the ureter obstructed isolated perfused kidney of the rabbit, *J. Pharmacol. Exp. Ther.* **205:**1–8.

Mullane, K. M., and Moncaa, S., 1982, The salvage of ischemic myocardium by BW-755C in anesthetized dogs, Prostaglandins **24:**255–266.

Mullane, K. M., Read, N., Salmon, J. A., and Moncada, S., 1984, Role of leukocytes in acute myocardial infarction in anesthetized dogs: Relationship to myocardial salvage by anti-inflammatory drugs, *J. Parmacol. Exp. Ther.* **228:**510–522.

Nagle, R. B., Bulger, R. E., Culter, R. E., Jervis, H. R., and Benditt, E. P., 1973, Unilateral obstructive nephropathy in the rabbit: I. Early morphologic, physiologic, and histochemical changes, *Lab Invest.* **28:**456–467.

Nagle, R. B., Johnson, M. E., and Jervis, H. R., 1976, Proliferation of renal interstitial cells following injury induced by ureteral obstruction, *Lab. Invest.* **35:**18–22.

Needleman, P., Kauffman, A. H., Douglas, J. R., Johnson, J. R., and Marshall, G. R., 1973, Specific stimulation and inhibition of renal prostaglandin release by angiotensin analogs, *Am. J. Physiol.* **224:**1415–1419.

Needleman, P., Douglas, J. R., Jakschik, B., Stoecklein, P. B., and Johnson, E. M., 1974.

Release of renal prostaglandins by catecholamines: Relationship to renal endocrine function, *J. Pharmacol. Exp. Ther.* **188:**453–460.

Needleman, P., Raz, A., Ferrendelli, J. A., and Minkes, M., 1977, Application of imidazole as a selective inhibitor of thromboxane synthetase in human platelets, *Proc. Natl. Acad. Sci. U.S.A.* **74:**1716–1720.

Nishikawa, K., Morrison, A. R., and Needleman, P., 1977, Exaggerated prostaglandin biosynthesis and its influence on renal resistance in the isolated hydronephrotic rabbit kidney, *J. Clin. Invest,* **59:**1143–1150.

Okegawa, T., Jonas, P. E., DeSchryver, K., Kawasaki, A., and Needleman, P., 1983a, Metabolic and cellular alterations underlying the exaggerated renal prostaglandin and thromboxane synthesis in ureter obstruction in rabbits. Inflammatory response involving fibroblasts and mononuclear cells. *J. Clin. Invest.* **71:**81–90.

Okegawa, T., DeSchryver-Kecskemeti, K., and Neddleman, P., 1983b, Endotoxin induces chronic prostaglandin and thromboxane synthesis from ureter-obstructed kidneys: Role of inflammatory cells, *J. Pharmacol. Exp. Ther.* **225:**213–218.

Patrono, C., Ciabattoni, G., Remuzzi, G., Gotti, E., Bombardieri, S., Di Munno, O., Tartarelli, G., Cinotti, G. A., Simonetti, B. M., and Pierucci, A., 1985, Functional significance of renal prostacyclin and thromboxane A_2 production in patients with systemic lupus erythematosus, *J. Clin. Invest.* **76:**1011–1018.

Prickett, J. D., Robinson, D. R., and Steinberg, A. D., 1981, Dietary enrichment with the polyunsaturated fatty acid eicosapentaenoic acid prevents proteinuria and prolongs survival in NZBxNZW/F_1 mice, *J. Clin. Invest.* **68:**556–559.

Reingold, D. F., Waters, S., Holmberg, S., and Needleman, P., 1981, Differential biosynthesis of prostaglandins by hydronephrotic rabbit and cat kidneys, *J. Pharmacol. Exp. Ther.* **216:**510–515.

Romson, J., Hook, B., Krunkel, S. L., Abrams, G. D., Schork, M. A., and Lucchesi, B. R., 1983, Reduction of the extent of ischemic myocardial injury by neutrophil depletion in the dog, *Circulation* **67:**1016–1023.

Schramm, L. P., and Carlson, D. E., 1975, Inhibition of renal vasoconstriction by elevated ureteral pressure, *Am. J. Physiol.* **228:**1126–1133.

Schreiner, G. F., Cotran, R. S., Pardo, V., and Unanue, E. R., 1978, A mononuclear cell component in experimental immunological glomerulonephritis, *J. Exp. Med.* **147:**369–384.

Scott, W. A., Pawlowski, N. A., Andreach, M., and Cohn, Z. A., 1982, Resting macrophages produce distinct metabolites from exogenous arachidonic acid, *J. Exp. Med.* **155:**535–547.

Spaethe, S. M., Freed, M. S., DeSchryver-Kecskemeti, K., Lefkowith, J. B., and Needleman, P., 1987, Essential fatty acid deficiency reduces the inflammatory cell invasion in rabbit hydronephrosis resulting in suppression of the exaggerated eicosanoid synthesis *J. Pharmacol. Exper. Thes.* (in press).

Taffet, S. M., and Russell, S. W., 1981, Macrophage-mediated tumor cell killing: Regulation of expression of cytolytic activity by prostaglandin E, *J. Immunol.* **126:**434–427.

Vaughn, E. D., Shenasky, J. H., II, and Gillenwater, J. H., 1971, Mechanism of acute hemodynamic response to ureteral occlusion, *Invest, Urol.* **9:**109–118.

Williams, J. D., Robin, J. L., Lewis, R. A., Lee, T. H. and Austin, K. F., 1986, Generation of leukotrienes by human monocytes pretreated with cytochalasin B and stimulated with formly-methionyl-leucyl-phenylalanine, *J. Immunol.* **136:**2169–2173.

Zipser, R., Meyers, S., and Needleman, P., 1980, Exaggerated prostaglandin and thromboxane synthesis in the rabbit with renal vein constriction, *Circ. Res.* **47:**231–237.

Zipser, R. D., Patterson, J. B., Kao, H. W., Hauser, C. J., and Locke, R., 1985, Hypersensitive prostaglandin and thromboxane response to hormones in rabbit colitis, *Am. J. Physiol.* **249:**G457–G463.

10

Lipoxygenase Metabolites

Chemistry and Biochemistry

JOSHUA ROKACH and BRIAN J. FITZSIMMONS

1. Introduction

The lipoxygenase-derived metabolites of arachidonic acid have been the subject of intense interest and research over the last decade, since the discovery of this pathway in mammals (Borgeat and Samuelsson, 1979). This important biochemical pathway is summarized in Fig. 1. The initial step in the lipoxygenase reaction is the conversion of a Z, Z1,4 diene to E, Z1,3 diene with the introduction of a molecule of oxygen to give a hydroperoxide (Fig. 2). In the case of arachidonic acid, this reaction can result in the incorporation of oxygen at C-5, -8, -9, -11, -12, or -15. In humans and other mammals, products of lipoxygenation at C-5, -11, -12, and -15 have been detected, with those that result from the 11-lipoxygenase reaction being known as products of the cyclooxygenase pathway. The products of the other three lipoxygenase enzymes retain the linear backbone of arachidonic acid and are referred to as the lipoxygenase products.

The initial product of the lipoxygenase enzyme is a *h*ydro*p*eroxy*e*icosa*t*etra*e*noic acid, or HPETE, and can suffer one of three fates: (1) reduction of the hydroperoxide to the hydroxy moiety to give HETE; (2) further lipoxygenation to give di- and trioxygenated metabolites; and (3) a stereospecific dehydration that extends the conjugation by a further olefinic unit and generates an epoxide. The above transformations are not mutually exclusive; therefore, sequential combinations can and do occur, lending to the complexity of this pathway.

JOSHUA ROKACH and BRIAN J. FITZSIMMONS • Merck Frosst Canada, Inc., Pointe-Claire-Dorval-Quebec, Canada H9R 4P8.

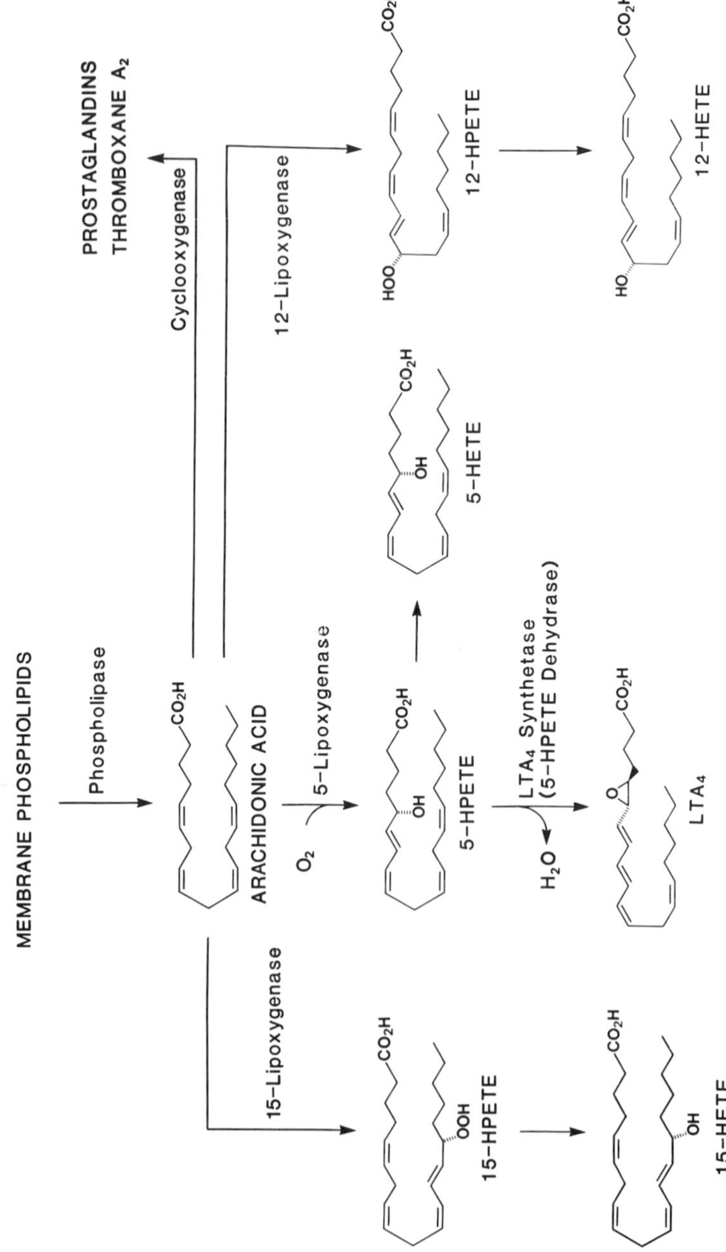

MEMBRANE PHOSPHOLIPIDS
Phospholipase
ARACHIDONIC ACID
Cyclooxygenase
PROSTAGLANDINS
THROMBOXANE A_2
12-Lipoxygenase
12-HPETE
12-HETE
5-Lipoxygenase
O_2
5-HPETE
5-HETE
LTA_4 Synthetase
(5-HPETE Dehydrase)
H_2O
LTA_4
15-Lipoxygenase
15-HPETE
15-HETE
CO_2H
HOO
HO
OH
OOH
O

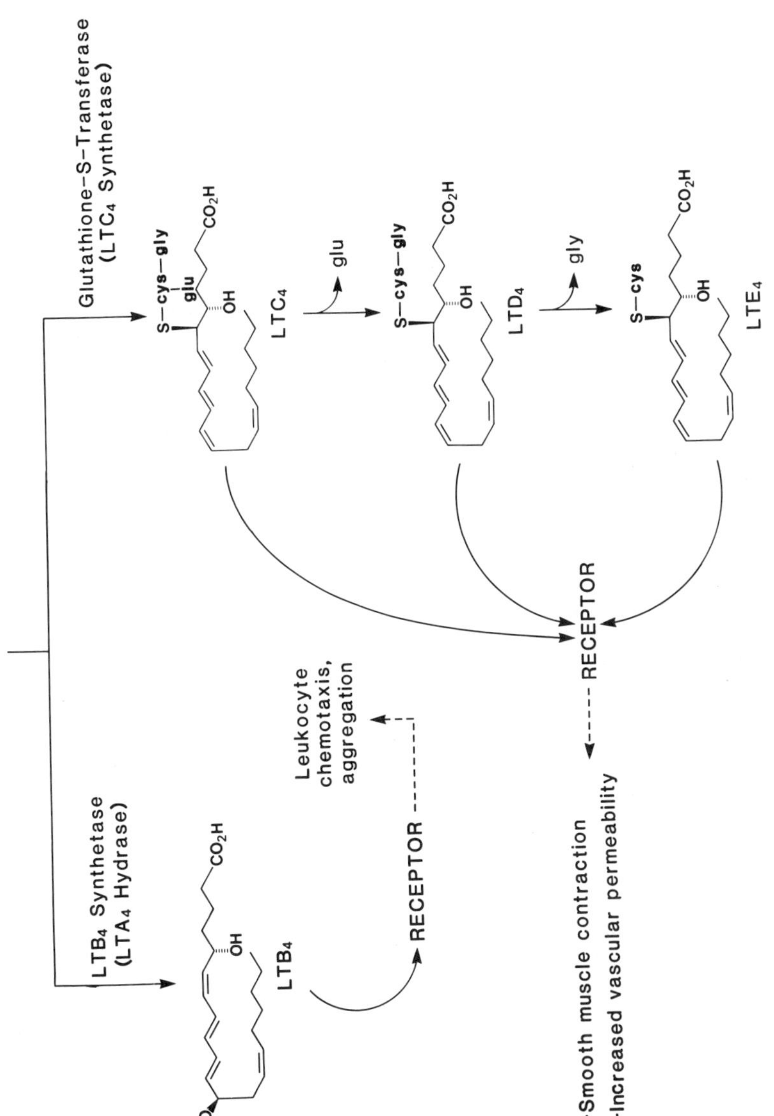

Figure 1. Lipoxygenase pathway of the arachidonic acid cascade.

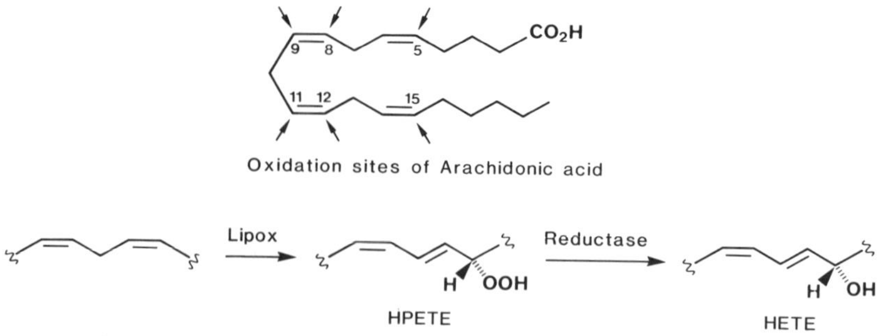

Oxidation sites of Arachidonic acid

Lipox
Reductase
HPETE
HETE

Figure 2. Lipoxygenation reaction.

The lipoxygenase products arising from the 5-lipoxygenase have received the most attention because of their potent and potentially important biological effects. Of these, products the biologically most potent molecules come from a common biosynthetic intermediate that is produced by the dehydration of 5-HPETE (Fig. 1). This molecule, known as leukotriene A_4 (LTA_4), is extremely unstable in an aqueous milieu and is rapidly transformed further. Direct opening of the epoxide at the more reactive allylic center (C-6) by glutathione gives the peptido-leukotriene C_4 (LTC_4), while conjugate hydrolysis with a stereospecific shifting of the triene gives the dihydroxy-leukotriene B_4 (LTB_4). Leukotriene C_4 is transformed into leukotriene D_4 and leukotriene E_4 by the sequential loss of the glutamate and glycine residues from the peptide portion of the molecule.

The peptido-leukotrienes (LTC_4, D_4, and E_4) were initially detected as potent spasmogens on smooth muscle (Kellaway and Trethewie, 1940; Brocklehurst, 1960). Because of this and other data, they have subsequently been implicated in numerous disease states, including allergic asthma (Ford-Hutchinson, 1985; Hedqvist *et al.*, 1985). On the other hand, leukotriene B_4 is a potent chemotactic agent (Ford-Hutchinson. 1983) that is possibly involved in inflammatory conditions such as psoriasis and inflammatory bowel disease (Klickstein *et al.*, 1980; Stenson and Lobos, 1982; Grabbe *et al.*, 1984. The lipoxygenase-derived metabolites of arachidonic acid in general and of leukotrienes specifically possess a wide diversity of biological properties in addition to those mentioned previously. However, the chronicling of these properties is beyond the scope and intent of this chapter. Suffice it to say that the myriad of physiological and biochemical events elicited by the various lipoxygenase products have made this pathway a prime target for pharmaceutical intervention.

2. General Overview

In spite of their interesting and potentially important biological properties and their comparable structural simplicity relative to their cyclooxygenase-derived cousins, these biomolecules were structural enigmas until recently. For example, the peptido-leukotrienes (LTC_4, D_4, and E_4) were first described collectively as the *s*low *r*eacting *s*ubstance of *a*naphylaxis, or SRS-A, almost 50 years ago (Kellaway and Trethewie, 1940); however, the structures of the components were not elucidated until less than 10 years ago (Lewis *et al.*, 1980; Morris *et al.*, 1980). A similar story applies to leukotriene B_4; it was known as various chemotactic factors until the recent elucidation of its structure (Borgeat and Samuelsson, 1979). There were several reasons for delay in the structural elucidation of these compounds. First, these products are present in biological systems in minute amounts, so that obtaining sufficient pure material for even the most simple structural analytical techniques was a maor undertaking. Second, because of the physical nature of these molecules, purification by classical techniques proved very difficult. Therefore, the structural elucidation of these molecules had to wait for the right combination of techniques and disciplines to come along. Two analytical techniques that have played the greatest role in the initial detection and determination of the basic structures of these molecules are high-pressure (or performance) liquid chromatography (HPLC), especially reverse-phase HPLC, and gas chromatography–mass spectroscopy (GC-MS). HPLC allowed the isolation and purification of the minute quantities present, while GC-MS permitted the basic connectivity of these molecules to be determined on nanogram quantities. However, these techniques only provided a crude structure. The refining of the structure required the involvement of an additional area of expertise—that of synthetic organic chemistry.

Once the basic structure of the compound was known, the role of the synthetic chemist became crucial. By preparing biosynthetically rational variations of the basic structure, the chemist initially provided structurally unambiguous, pure standards for comparison with the naturally derived material in terms of both physical and biological properties, thus allowing positive and complete identification of the compounds. Historically, the groundbreaking structural determinations in the lipoxygenase pathway were of the peptido-leukotrienes and leukotriene B_4. These determinations which were made possible by the availability of pure synthetic material and representative syntheses of LTD_4 (Rokach *et al.*, 1981) and LTB_4 (Guindon *et al.*, 1982), are shown in Figs. 3 and 4, respectively.

While the contribution of synthetic chemistry initially dealt with the identification of the lipoxygenase metabolites, it did not stop there. For example, in the course of the synthesis of the peptido-leukotrienes, leukotriene

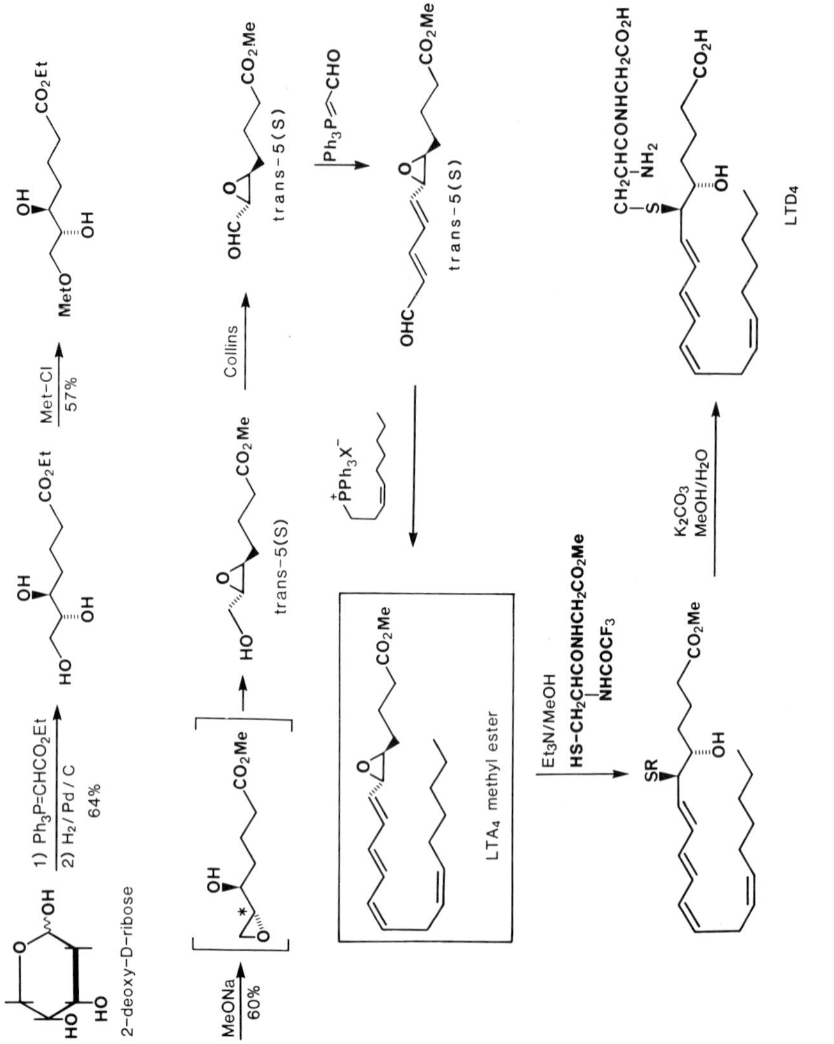

Figure 3. Synthesis of leukotriene A_4 and leukotriene D_4 from 2-deoxy-D-ribose.

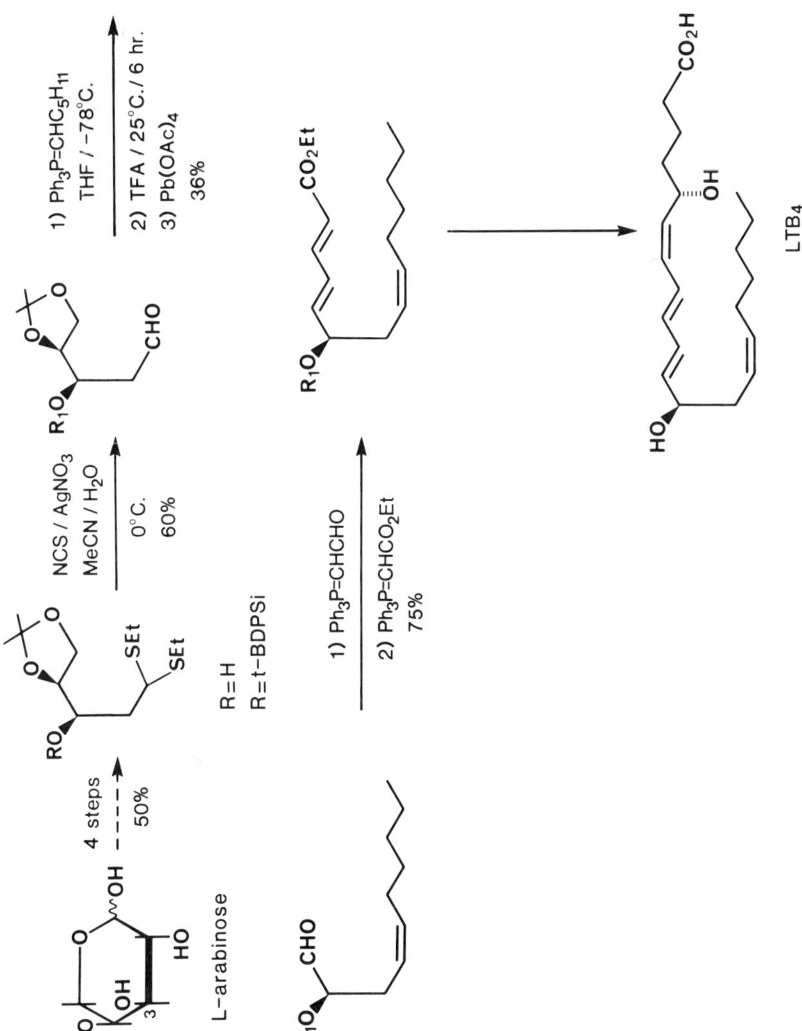

Figure 4. Synthesis of leukotriene B_4. (R = t − BDPSi.)

A_4 was prepared as a synthetic intermediate. This molecule had been proposed as an intermediate in the biosynthesis of LTB_4 and the peptido-leukotrienes, based on circumstantial evidence. However, owing to its exquisite sensitivity to nonenzymatic aqueous hydrolysis, it had never been isolated. Therefore. the availability of synthetic material allowed the obtaining of direct evidence for the intermediacy of LTA_4. Synthetic LTA_4 has also facilitated the isolation of the enzymes that convert LTA_4 into LTB_4 (Evans *et al.*, 1985) and LTC_4 (Mannervik *et al.*, 1984). The synthesis of the leukotrienes also allowed hitherto unheard of quantities of these compounds to be prepared, thus allowing comprehensive studies of the biological properties of these interesting molecules.

After the initial detection and synthesis of the peptido-leukotrienes LTA_4 and LTB_4, a long string of reports of other metabolites from the 5-, 12-, and 15-lipoxygenase enzymes have followed, the most recent example being a new class of arachidonic acid metabolites called the lipoxins (Serhan *et al.*, 1984a,b). The elucidation of the structures of these compounds and the biosynthetic pathways by which they are formed is an excellent demonstration of the role that synthetic chemistry plays in elucidating structure and biosynthesis.

Synthetic chemistry has performed many other roles in addition to identifying the principal components of lipoxygenase pathway. Because of the flexibility of the synthetic routes used, modifications have provided specifically labeled materials such as 14, 15-dideuterio LTB_4, whose synthesis is illustrated in Fig. 5. Substitution of tritium for deuterium in the semireduction of the acetylene yields radioisotope-labeled material. The deuterated material has been used as an internal standard for mass-spectroscopy-based assays of LTB_4 in biological samples, while the tritiated material has been used in radioimmunoassays, receptor purifications, and metabolism studies. Similarly, tritiated LTA_4, LTC_4, LTD_4, and LTE_4 have also been prepared and put to comparable uses.

In addition to providing the radioisotopically labeled materials for radioimmunoassay, synthetic compounds have made the assays possible at a much more basic level; without synthetic material, raising antibodies to these compounds would have been almost impossible. Synthetic chemistry has also provided analogues of the natural products, thus allowing study of the structural requirements for biological activity. This will be illustrated by the study of LTB_4 and its analogues.

When the principal products, structure, and biosynthesis of a pathway have been elucidated the quest is far from complete. The question then becomes, "What is the fate of the initial products, especially the biologically potent ones?" Synthetically produced materials have been instrumental in determining the structure of the metabolites and the specificity of the cata-

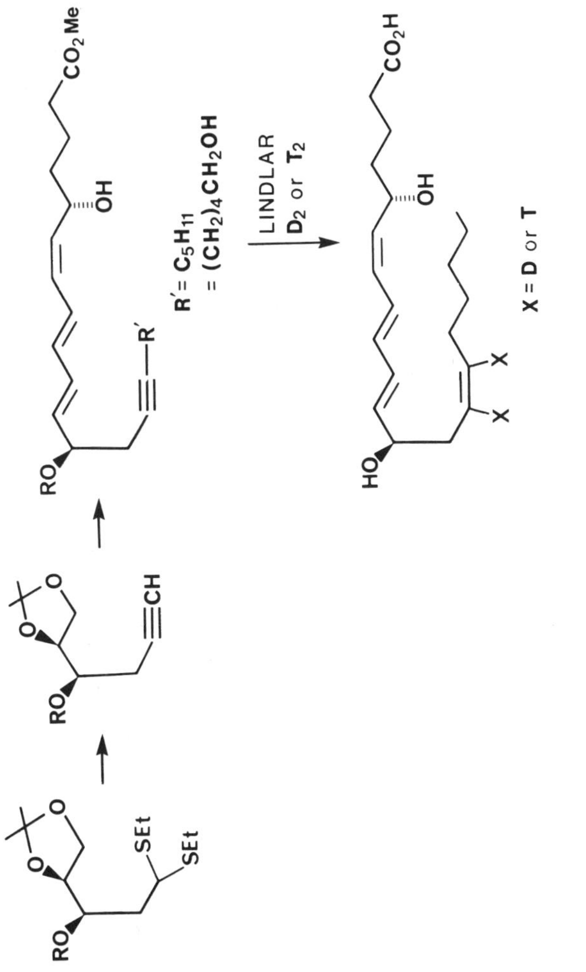

Figure 5. Synthesis of heavy-isotope-labeled leukotriene B_4. (X = D,T.)

bolizing enzymes by the availability of pure synthetic standards and their structural analogues. Since many, if not all, of these metabolites have greatly diminished biological activites, the detection of these metabolites has largely been due to the use of synthetically prepared radiolabeled materials. This is particularly true in examining the *in vivo* metabolism of the leukotrienes.

One can therefore say that synthetic chemistry has played, and continues to play, a major role in research directed toward the lipoxygenase pathway. In the following sections, several examples are described in detail to add the flavor of this area of endeavor to the general contributions discussed above.

3. Specific Examples

3.1 Determination of Structure and Biosynthesis: The Lipoxins

The isolation of a new class of arachidonic acid metabolites was recently reported by Serhan *et al.*, (1984a, b) and the names lipoxin A and lipoxin B were assigned to the two substructural groups. These novel trihydroxy tetraenes were also reported to possess numerous intriguing biological properties. The structures shown in Fig. 6 were proposed for these metabolites; however owing to the minute quantities isolated, the relative stereochemistry of the hydroxyl groups and the double-bond geometries of these compounds could not be determined.

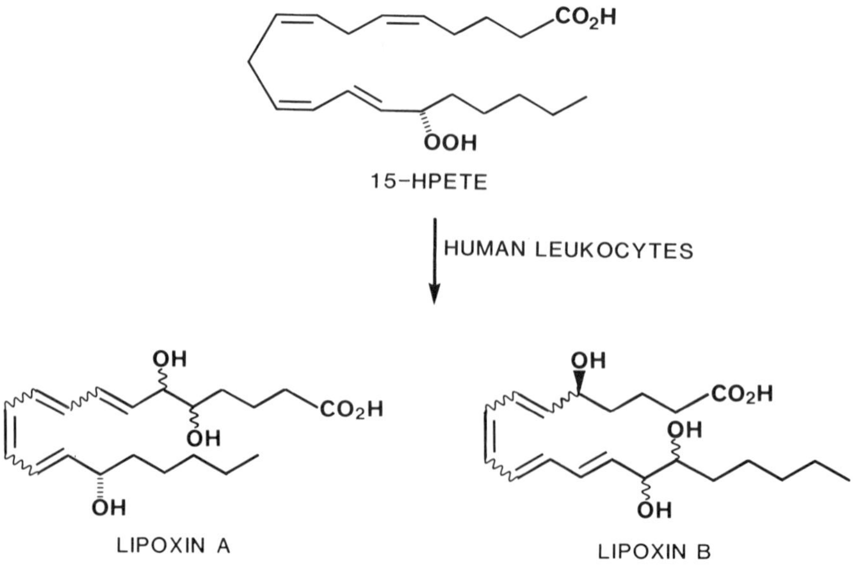

Figure 6. Initially proposed structures of lipoxin A and lipoxin B.

Because of the small quantities produced biologically, a synthetic program was initiated—first to prepare various biochemically rational isomers of lipoxin A and lipoxin B and to identify the natural products by comparison with these synthetic standards, and secondly, with the identification of the natural products in hand, to elucidate the biosynthetic route by which these compounds are formed.

Since the preparation of all possible combinations of double-bond geometry and relative hydroxyl group stereochemistry would involve the synthesis of 128 compounds, we examined possible biosynthetic routes to choose biorational isomers as synthetic targets (Adams *et al.*, 1985). The first postulate considered by which the lipoxins could be produced (Fig. 7) was that successive enzymatic oxidations at C-15 and C-5 of arachidonic acid would yield the known *5S,15S*-diHPETE. A third lipoxygenation was then envisaged. Based on the stereochemical outcome of a similar lipoxygenation, the stereochemistry of the new asymmetric center at C-6 in lipoxin A or at C-14 in lipoxin B was predicted to be of the *R* absolute configuration. The tetraene geometry would be as shown in Figure 7, 7-*E*, 9-*E*, 11-*Z*, 13-*E* for lipoxin A*2* and 6-*E*, 8-*Z*, 10-*E*, 12-*E* for lipoxin B*8*.

In the alternative biosynthetic route, 5*S*,15*S*-diHPETE would also be formed as a common intermediate. However, by analogy to the formation of leukotriene A_4 (LTA_4), this intermediate could undergo a stereospecific enzymatic dehydration to produce a 5*S*,6*S* epoxide with concomitant formation of a tetraene of 7-*E*, 9-*E*, 11-*Z*, 13-*E* geometry, or a 14*S*,15*S epoxide with a 6-E*, 8-*Z*, 10-*E*, 12-*E* tetraene. These epoxides could be expected to undergo further transformations to produce the lipoxins. An enzymatic hydrolysis of the epoxide *9* at C-6 would lead to lipoxin A, while nonenzymatic homoconjugate addition of water at C-14 would produce all *trans*-lipoxin Bs. Enzymatic formation of lipoxin A should leave the tetraene geometry intact, while nonenzymatic hydrolysis of the tetraene epoxide would form the all-*trans*-tetraene and a diastereomeric mixture of lipoxin Bs epimeric at C-14 and the two all-*trans*-lipoxin A's by analogy to the nonenzymatic hydrolysis of LTA_4. Similarly, the 14, 15-tetraene epoxide *10* could yield the two *cis*-lipoxin Bs *7* and *8* and the all-*trans*lipoxin As and Bs *3, 4* and *5, 6,* respectively. Therefore, these eight possible isomers were prepared by total asymmetric synthesis, as illustrated in Fig. 8. It is important to note that all the asymmetric centers were obtained from carbohydrate precursors, thus eliminating any possible ambiguities in the final structures.

The lipoxins isolated from the incubation of 15-HPETE with human leukocytes were found by RP-HPLC to be a mixture of several isomers (Adams *et al.*, 1985; Fitzsimmons and Rokach, 1985; Fitzsimmons *et al.*, 1985; Leblanc *et al.*, 1985; Serhan *et al.*, 1986a,b). By comparison to synthetic standards, these isomers were identified as the 5*S*, 14*R*, 15*S* all-*trans* and 5*S*, 14*S*, 15*S* all-*trans*-lipoxin B isomers *5* and *6*, the two all *trans*-

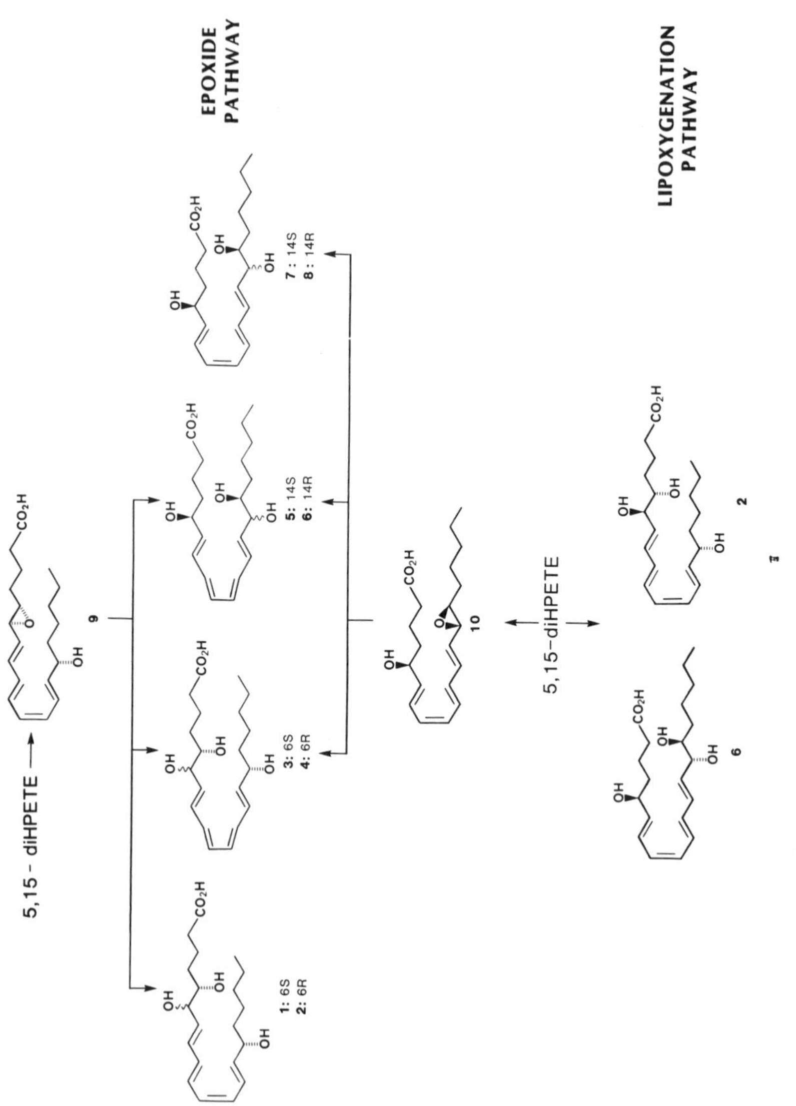

Figure 7. Two biosynthetic postulates for the biosynthesis of lipoxins.

lipoxin A isomers *3* and *4*, the 5*S*, 6*R*, 15*S* 11-*cis*-lipoxin A 2, and the 5*S*, 6*S*, 15*S* 11-*cis*-lipoxin A isomer *1*, which was the first of these compounds positively identified in these laboratories. It is worthwhile to note that while neither of the 8-*cis*-lipoxin B isomers *7* or *8* were detected in our initial experiments, 5*S* 14*R*, 15*S* 8-*cis*-lipoxin B *8* was detected in a subsequent incubation. The failure to detect this isomer in the previous incubations was not due to its instability as determined by subjecting the synthetic standard to the isolation conditions and recovering it intact. The original identification of the lipoxin isomers relied on two sequential HPLC steps with different solvents. However, we have recently developed an HPLC system that separates seven of the eight synthetic lipoxin isomers, thus greatly simplifying the analysis of lipoxin-containing mixtures. A sample trace is shown in Fig. 9. Having identified the leukocyte-derived lipoxins, we turned our attention to elucidating the biosynthesis of these compounds.

The first postulate, shown in Fig. 8, was that the lipoxins were formed as a result of three successive lipoxygenations, the first lipoxygenation being that which produced the 15-HPETE used in the incubation. This pathway should lead to the 11-*cis*-lipoxin A and 8-*cis*-lipoxin B isomers *2* and *8*, respectively. To determine whether three successive lipoxygenations of arachidonic acid were feasible and to test the prediction of the outcome of such an event, arachidonic acid was incubated with commercial soybean lipoxygenase. We were gratified to find that, upon incubation of arachidonic acid with soybean lipoxygenase, two tetraene triol metabolites were observed. Comparing these compounds with synthetic standards, we found 5*S*, 6*R*, 15*S* 11-*cis*-lipoxin A *2* and 5*S*, 14*R*, 15*S* 8-*cis*-lipoxin B *8*, as predicted.

Although 5*S*, 6*R*, 15*S*, 11-*cis*-lipoxin A is found in the leukocyte-derived lipoxins, 5*S*, 14*R*, 15*S* 8-*cis*-lipoxin B *8* was only found in one preparation of leukocyte-derived material and other isomers were always produced in comparable or greater amounts. This led us to consider the second biosynthetic pathway to account for the observed mixture of lipoxins.

The second biosynthetic postulate was that 15-HPETE could be acted on by the enzymatic machinery that converts arachidonic acid into leukotriene A_4 (LTA_4) to give the epoxides *9* and/or *10*. These epoxides could then undergo enzymatic or nonenzymatic hydrolysis to give the lipoxins. By this postulate the 11-*cis*-lipoxin A isomers *1* and *2* and the 8-*cis*-lipoxin B isomers *7* and *8* would be the result of enzymatic vicinal hydrolysis of the epoxides *9* and *10*, respectively, while the all-*trans*-lipoxin A and all-*trans*-lipoxin B isomers would be from a nonenzymatic hydrolysis of the tetraene epoxides. The formation of all-*trans*-lipoxin B isomers from tetraene epoxide *9* would be directly analogous to the production of the all-*trans*-LTB_4s by the ω-attack of water in the nonenzymatic hydrolysis of LTA_4. Conversely, enzymatic ω-addition of water to epoxide *9* could lead to an 8-*cis*-lipoxin B isomer.

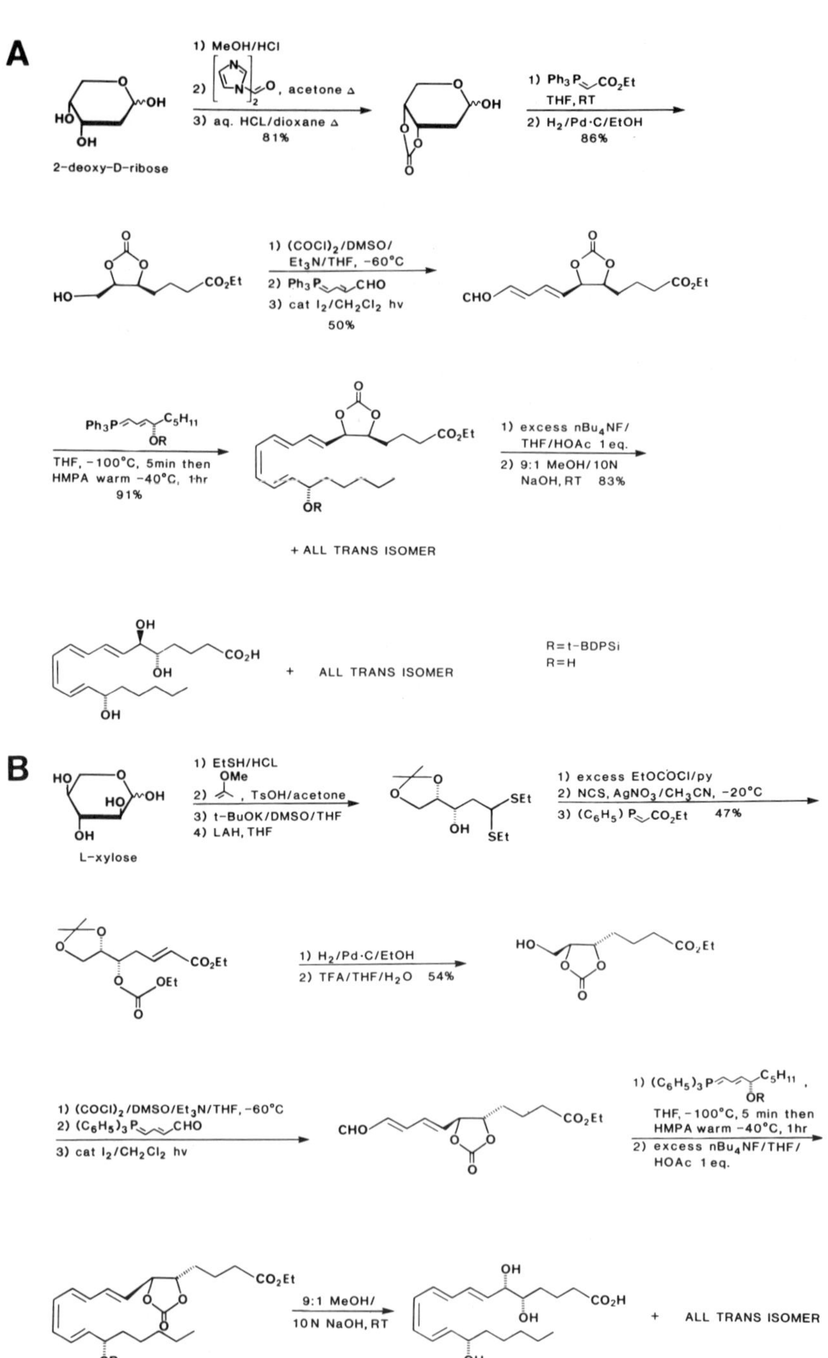

Figure 8. Synthesis of the eight biorational lipoxin isomers. (R = t − BDPSi, H.)

C

2-deoxy-D-ribose

1) t-BDPSiCl/Et_3N
2) Ph_3P=
3) H_2/Pd C

RO HO OH

76%

1) [imidazole]$_2$C=O
2) Bu_4NF

HO

75%

1) $(COCl)_2$, Et_3N, THF, DMSO
2) Ph_3P=CHO
3) I_2, hv

OHC

47%

BuLi

OR CO$_2$Et + OR CO$_2$Et

1 : 1
92%

1) nBu_4NF
2) K_2CO_3, MeOH-H_2O

OH S CO$_2$H OH R S OH + OH S CO$_2$H OH R S OH

D

ref. 168

SEt OH SEt

1) t-BDPSiCl/Et_3N
2) NCS/$AgNO_3$
3) Ph_3P=
4) H_2/Pd•C
5) nBu_4NF
6) Cl-CO_2Et

OEt

56%

TFA/THF-H_2O

HO

76%

1) $(COCl)_2$, Et_3N, THF, DMSO
2) Ph_3P=CHO
3) I_2, hv

OHC

47%

BuLi

OR CO$_2$Et + OR CO$_2$Et

1 : 1
92%

1) nBu_4NF
2) K_2CO_3, MeOH-H_2O

OH CO$_2$H OH OH + OH S CO$_2$H OH S S OH

Figure 8. (Continued)

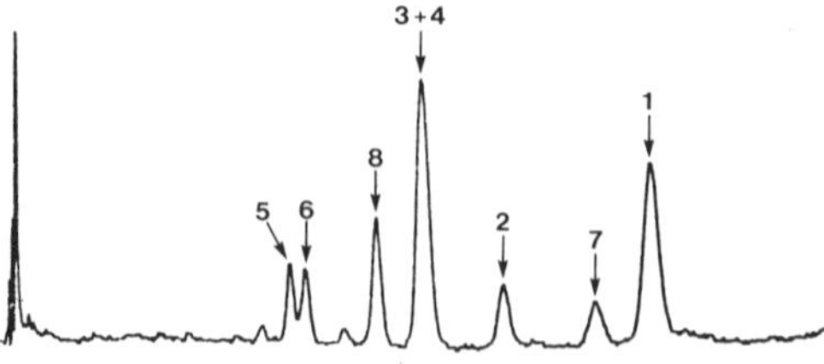

Figure 9. RP-HPLC of synthetic standards: 57.5:21.25:21.25:0.05, water/methanol/acetonitrile/acetic acid; Waters Novapak C_{18} column, flow = 1.0 mL/min, λ = 301nm. (Numbers refer to the structures in Fig. 7.)

To test this postulate, synthetically prepared tetraene 5,6 epoxide *9* was prepared (Adams *et al.*, 1984), as shown in Fig. 10, and incubated with native and denatured LTA_4 hydrolase (Evans *et al.*, 1985) of varying purity. In all cases, no products were formed with the active enzyme that were not also formed with the denatured enzyme, as demonstrated by the superimposable HPLC traces from these incubations.

Since there are other enzymes in human leukocytes that carry out epoxide hydrolysis, the tetraene epoxide *9* was incubated with leukocytes under the identical conditions used for the generation of the lipoxins from 15-HPETE, and the products were isolated and analyzed as for the lipoxins. The products isolated from this incubation were the same as those obtained when 15-HPETE was incubated with the leukocytes. Although this result provides strong evidence for the intermediacy of the tetraene epoxide in the biosynthesis of the lipoxins, we desired further substantiation. Since the incubation is quenched with methanol, we looked for the methanol opening products of the tetraene epoxide. Upon analysis of the less polar products of the 15-HPETE and tetraene epoxide incubations, we did indeed find compounds that appeared to be methanol–epoxide adducts (methoxy-lipoxins) by virtue of their less polar nature and ultraviolet spectra ($\lambda_{max} \approx$ 302nm). To verify that these compounds were methanol adducts, the tetraene epoxide *9* was treated with acidic methanol and the products were then analyzed by RP-HPLC (methanol/water/acetic acid 70:30:0.05 0.05% HOAc), ^{1}H-NMR, UV spectroscopy, and gas chromatography–mass spectroscopy. The products obtained were not only the same as those in both leukocyte incubations but they were also produced in the same relative amounts. Therefore, it was concluded that the tetraene epoxide *9* is produced from 15-HPETE by the leukocytes and that this tetraene epoxide gives rise to the majority of the lipoxins.

On the basis of the second biosynthetic postulate, and thus the intermediacy of the tetraene epoxide *9*, it was felt that the all-*trans*-lipoxin isomers were likely to be formed by nonenzymatic hydrolysis of the tetraene epoxide. Therefore, the tetraene epoxide *9* was subjected to the conditions of the leukocyte incubation but in the absence of cells. To our surprise, we found that these conditions produced not only the all-*trans*-isomers, but all the products from the incubation of 15-HPETE with leukocytes, with the exception of 8-*cis*-lipoxin B isomer *8*.

These data led us to conclude that the six lipoxin isomers found in all incubations are principally nonenzymatic hydrolysis products of the tetraene epoxide *9*. The 5*S*, 14*R*, 15*S*, 8-*cis*-lipoxin B *8*, found in a subsequent incubation could be formed by triple lipoxygenation or via the 14,15 tetraene epoxide *10*, as shown in Fig. 8.

In collaboration with Yamamoto and Ueda (Rokach *et al.*, 1987), we sought to clarify the other possibilities available for lipoxin biosynthesis and found that both of the tetraene epoxide and triple lipoxygenation pathways were possible in other biochemical systems. Incubation of 5,15-diHPETE with purified porcine 12-lipoxygenase yielded a mixture of products, with the predominant product being 5*S*, 14*R*, 15*S* 8-*cis*-lipoxin B *8* with lesser amounts of the all-*trans*-lipoxin As *3* and *4*, all-*trans*-lipoxin Bs *6* and *7*, and 5*S*, 14*S*, 15*S* 8-*cis*-lipoxin B *7* (Fig. 11, trace A). The exclusion of oxygen from the incubation mixture led to a diminution only of the major peak (trace B), which in combination with the analogous reaction of 15-HPETE with the same enzyme led us to conclude that the 5*S*, 14*R*, 15*S* 8-*cis*-lipoxin B *8* was being formed to a large extent by triple lipoxygenation. The minor products, by analogy to the 15-HPETE/12-lipoxygenase reaction, would then be due to

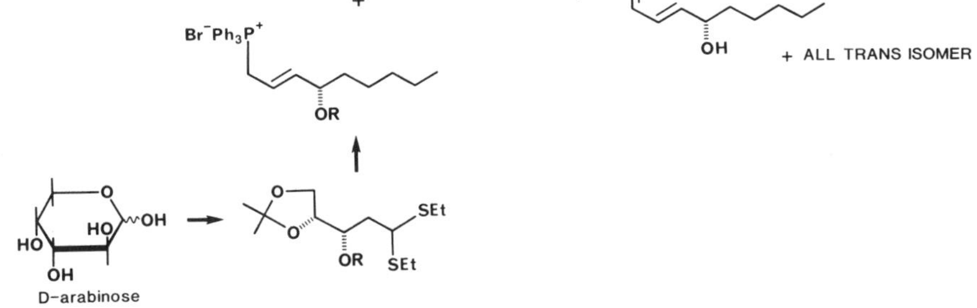

Figure 10. Synthesis of epoxide *9* proposed as a biosynthetic intermediate to the lipoxins (R = t − BDPSi, H).

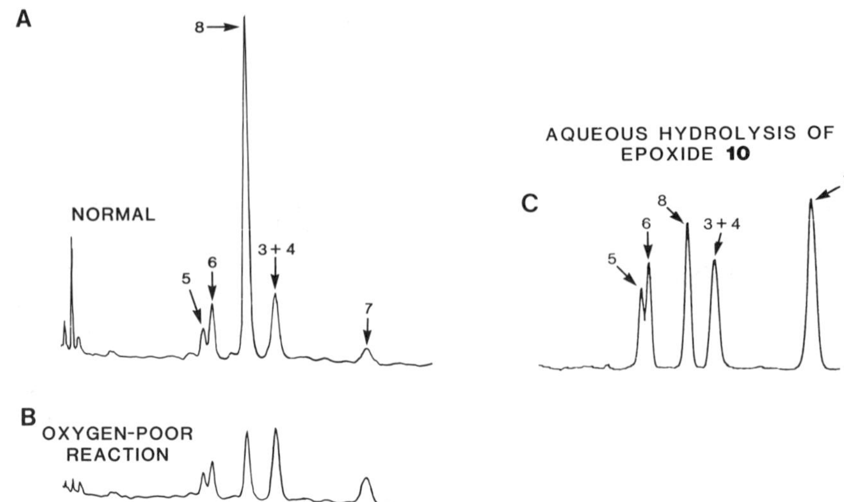

Figure 11. Traces A and B: lipoxins produced by the incubation of 5,15-d:HPETE with purified 12-lipoxygenase. Trace A: oxygen-rich incubation. Trace B: oxygen-poor incubation. Trace C: aqueous hydrolysis product profile from epoxide *10*. (Numbers refer to structures in Fig. 7.)

the formation and nonenzymatic hydrolysis of the 14,15 tetraene epoxide *10*. Therefore, the epoxide *10* was prepared by total synthesis (Fig. 12), and its nonenzymatic hydrolysis was studied. In this case the 8-*cis*-lipoxin B isomer *8* was nonpredominant and was produced in approximately equal amounts to 5*S*, 14*S*, 15*S* 8-*cis*-lipoxin B *7* (Fig. 11, trace C), thus substantiating the conclusion that the 14*R* isomer in the 12-lipoxygenase reaction was principally due to triple lipoxygenation. These data indicated that the 5*S*, 14*R*, 15*S* 8-*cis*-lipoxin B *8* in the leukocyte preparation could be formed by either of these pathways.

Alternatively, incubation of 15-HPETE or 5,15-diHPETE with purified porcine 5-lipoxygenase produces the six lipoxin isomers found in all human leukocyte preparations (Fig. 13). However, in this case, the predominant pathway is via the 5,6 tetraene epoxide, rather than triple lipoxygenation as shown by the product profile and by oxygen exclusion experiments. The 8-*cis*-lipoxin B isomers *7* and *8* do not appear to be produced by this route.

In conclusion, seven lipoxin isomers have now been identified from the incubation of 15-HPETE with human leukocytes. Of these isomers, the 11-*cis*-lipoxin A isomers *1* and *2* and the all-*trans*-lipoxin A and B isomers *3*, *4*, *7*, and *8* appear to be formed predominately by the nonenzymatic hydrolysis of the 5,6 tetraene epoxide *9*, produced by the action of the 5-lipoxygenase enzyme, while the origin of 5*S*, 14*R*, 15*S* 8-*cis*-lipoxin B *8* is

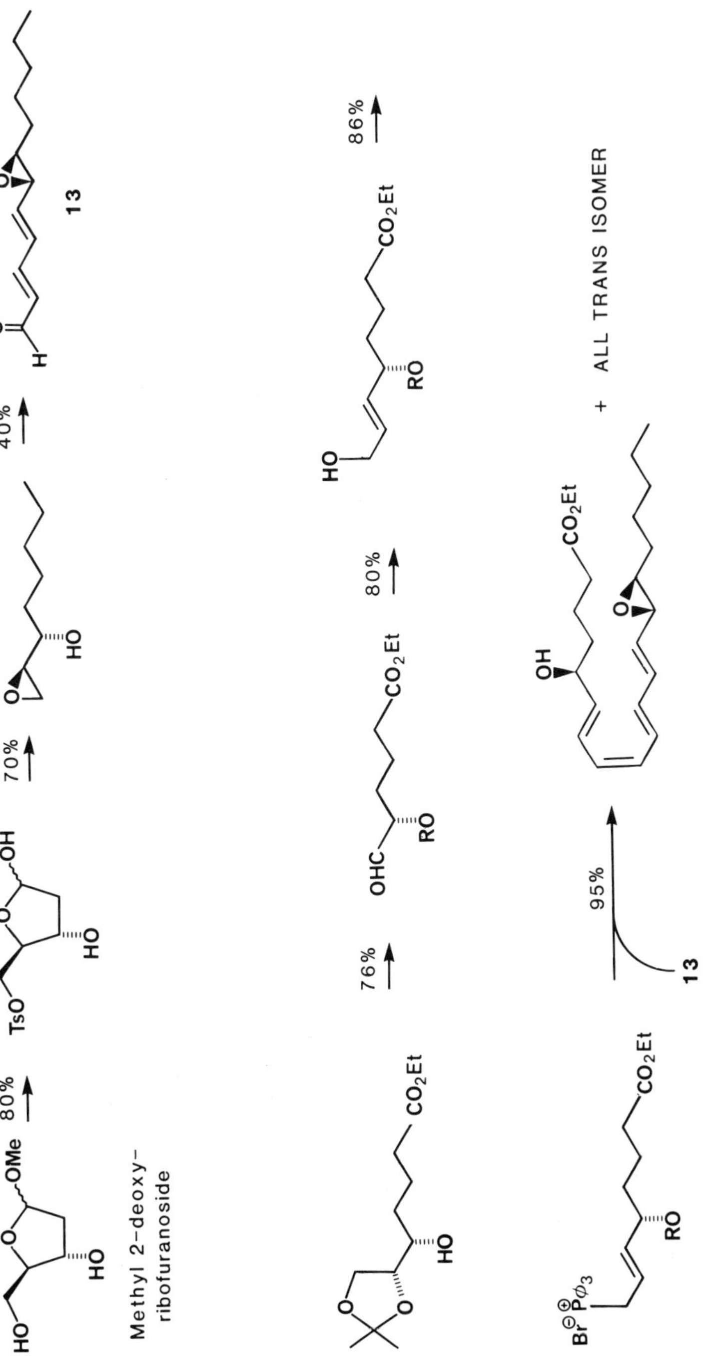

Figure 12. Synthesis of epoxide *10.*

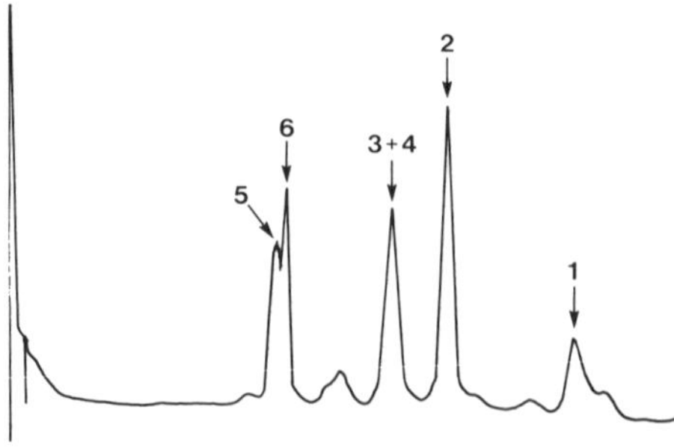

Figure 13. Lipoxins produced by the incubation of 15-HPETE with purified 5-lipoxygenase. (Numbers refer to the structures in Fig. 7.)

uncertain. However, the experiments with purified 12-lipoxygenase indicate that this isomer can arise from triple lipoxygenation or via the 14,15 tetraene epoxide *10*.

3.2. Analogues as Probes for Structural Specificity

Leukotriene B_4 (5(S),12(R)-dihydroxy 6(Z),8(E),10(E),14(Z)-eicosatetraenoic acid) is a potent inducer of the aggregation, chemokinesis, and chemotaxis of polymorphonuclear leukocytes (Ford-Hutchinson *et al.*, 1980; Bray *et al.*, 1981a,b). Specific receptors for leukotriene B_4 have been characterized on rat (Kreisle *et al.*, 1985) and human (Goldman and Goetzl, 1982; Kreisle and Parker, 1983; Goldman and Goetzl, 1984; Lin *et al.*, 1984) leukocytes and also in rat and human leukocyte membrane preparations (Charleson *et al.*, 1986; O'Flaherty *et al.*, 1986). Excellent correlation between specific high-affinity binding of leukotriene B_4 to leukocyte membrane receptor preparations and two leukocyte functions—namely, aggregation and chemokinesis—have been demonstrated (Charleson *et al.*, 1986). Various leukotriene B_4 analogues have previously been synthesized and their potencies to activate some neutrophil responses described (Charleson *et al.*, 1986). No studies have investigated the importance of hydroxyl functionalities with chemically pure isomers with the correct *cis–trans–trans* triene geometry of leukotriene B_4. Here we describe the synthesis of such leukotriene B_4 analogues and the importance of the two hydroxyl functionalities in the binding of leukotriene B_4 to the leukocyte leukotriene B_4 receptor.

Five leukotriene B_4 analogues were synthesized with modifications to the hydroxyl positions (Figs. 14 and 15). These leukotriene B_4 analogues were assayed for agonist action in a rat leukocyte aggregation assay and for binding to leukotriene B_4 receptors on membrane preparations from both rat and human leukocytes. The results are illustrated in Figs. 16 and 17 and in summary form with other previously reported and analogues in Table I. Figure 16 shows the rat leukocyte aggregation responses of the leukotriene B_4

analogues and leukotriene B_4 itself. Figure 17 shows the inhibition of [^{3}H]leukotriene B_4 binding to a human leukocyte membrane preparation by leukotriene B_4 and the deoxy and epimeric hydroxyl leukotriene B_4 analogues. As can be seen from Table I, very similar results were obtained for inhibition of [^{3}H]leukotriene B_4 binding to rat leukocyte membrane preparations.

Previous studies have shown that triene geometry is a very important determinant of both functional and binding efficacy (Lewis *et al.*, 1981; Hoffstein *et al.*, 1986), but although such studies have suggested that the stereochemistry and presence of the hydroxyl group are important, only the all-*trans*-leukotriene B_4 isomers (Lewis *et al.*, 1981), the double lipoxygenation 5*S*, 12*S* product LTB_x with *trans–cis–trans* triene geometry (Lewis *et al.*, 1981), or mixed epimers (Hoffstein *et al.*, 1986) have been studied. Here we have systematically investigated the synthetically pure isomers of leukotriene B_4 with the same triene geometry as leukotriene B_4 but with either

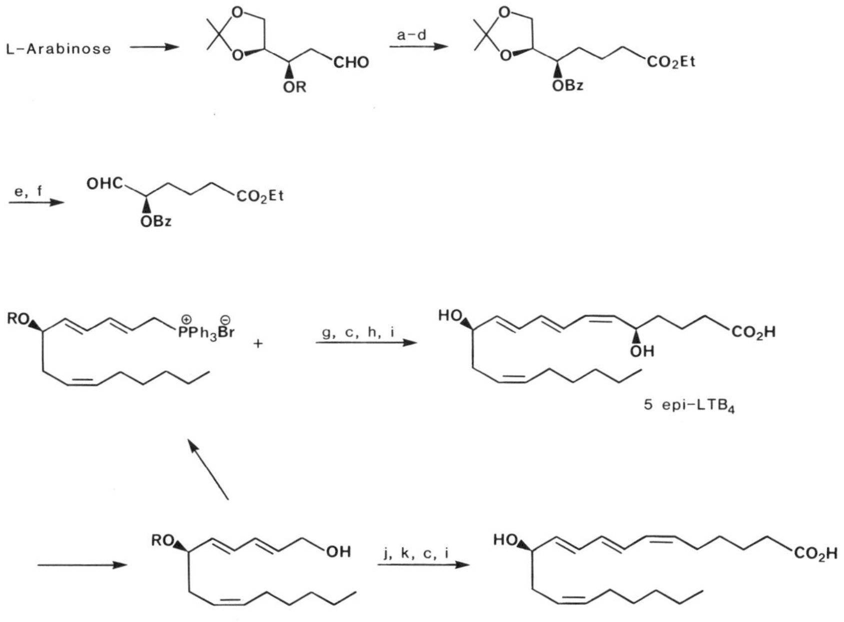

Figure 14. Syntheses of 5-deoxy- and 5-epi-LTB_4: (a) PhP$=CO_2$Et, CH_2Cl_2, room temperature. (b) H_2/Pd·C, 40 psi, A_cOET. (c)*n*Bu_4NF, THF; HPLC purification (6—7, 8—9). (d) BzCl, py, O°C to room temperature. (e) TFA, THF, H_2O. (f) $Pb(OAc)_4$, CH_2Cl_2, 4ml40°C. (g) *n*BuLi, THF, HMPA, −78°C—0°C. (h) K_2CO_3/MeOH, room temperature. (i) NaOH, MeOH, H_2O, 0°C. (j) PCC, CH_2Cl_2, room temperature. (k) $BrPPh_3$ $(CH_2)_5$ CO_2H, LiHMDS 2equiv, THF, HMPA, −78°C—0°C; $(MeO)_2$ SO_2/$MeHCO_3$, 0°C. (R$=$t$-$BuPh_2Si.)

Table I. Relative Potencies of LTB_4 And Its Analogues

Compound	Aggregation rat PMN IC_{50} (nM)	Binding rat PMN membranes IC_{50} (nM)	Binding human PMN membranes IC_{50} (nM)
LTB_4	1.0	1.0	0.5
5-deoxy-LTB_4	4.0	2.5	4.0
12-epi-LTB_4	25	8	11
5-epi-LTB_4	90	130	100
12-deoxy-LTB_4	400	120	140
5, 12-deoxy-LTB_4	8000	2200	870
6-*trans*-LTB_4	60	46	50
12*R*-HETE	600	50	20
12*R*-HETE	6000	1000	200
5*S*, 12*S*-LTB_x	—	550	—
6-*trans*-12-epi-LTB_4	—	3000	—

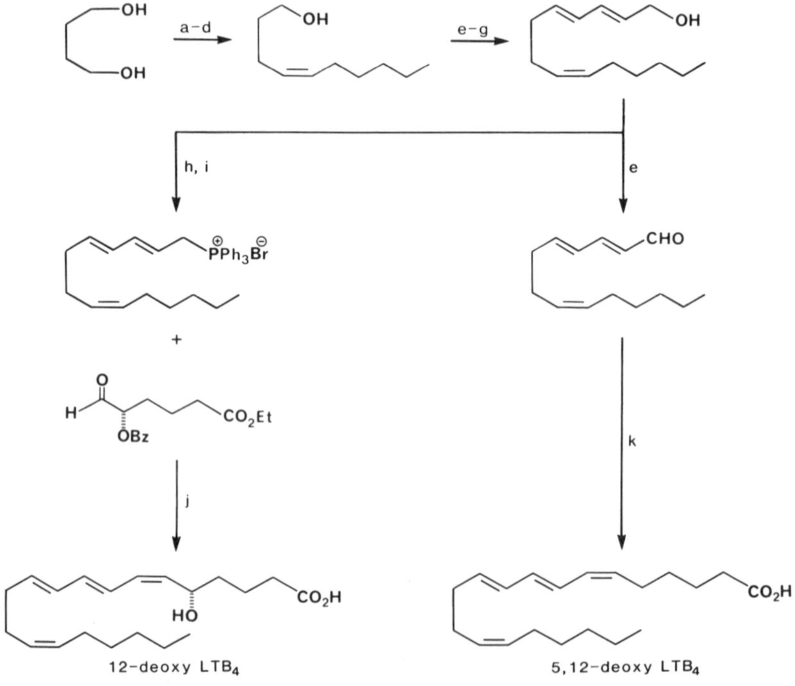

Figure 15. Synthesis of 12-deoxy- and 5, 12-dideoxy-LTB_4: (a) MeOTrCl, py, 0°C. (b) $(COCl)_2$, DMSO, Et_3N, THF, −60°C. (c) $BrPPh_3$ $(CH_2)_5$ CH_3, BuLi, THF, −78°C—0°C. (d) TFA, THF, H_2O. (e) PCC, CH_2Cl_2, room temperature. (f) $(EtO)_2$ PO—CH_2=CH—CO_2Et, NaH, toluene, (g) AlH_3· ⅓Et_2O, THF, 0°C. (h) CBr_4, DIPHOS, CH_2Cl_2, 0°C. (i) Ph_3P, CH_3CN, room temperature. (j) g (Fig. 14), HPLC purification, hand i (Fig. 14). (k) Figure 14, HPLC purification, i (Fig. 14).

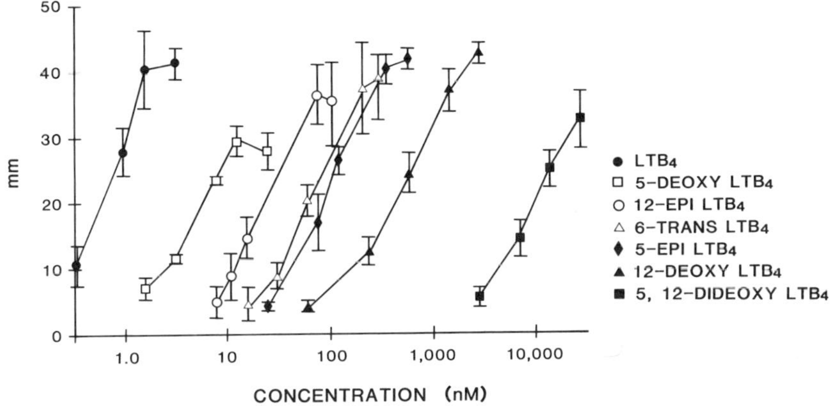

Figure 16. Aggregation of rat leukocytes by LTB_4 and its analogues.

different hydroxyl group stereochemistry or the complete absence of hydroxyl groups. We have unexpectedly discovered that the C-5 hydroxyl does not play a very important role in binding or activation at the leukotriene B_4 receptor. Clearly, however, the constraints of the receptor binding site do not favor the epimeric C-5 hydroxyl. The C-12 hydroxyl apparently contributes more to the binding of leukotriene B_4 to its receptor than does the C-5 hydroxyl, but the stereochemistry of the C-12 hydroxyl is more flexible in terms of binding site interaction.

Only one antagonist of the leukotriene B_4 receptor has been reported (Namiki *et al.*, 1986). Dimethylamide leukotriene B_4 (Showell *et al.*, 1982)

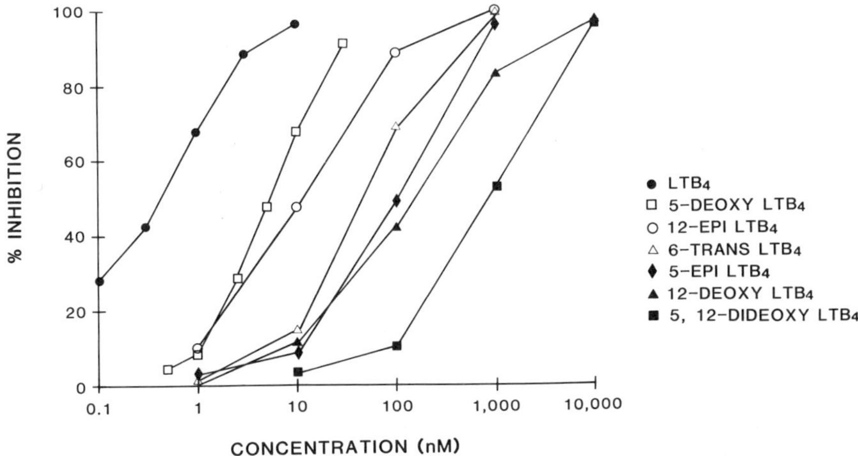

Figure 17. Competition for binding to human leukocyte LTB_4 receptors between $[^3H]LTB_4$ and its analogues.

and 5*S*,12*S*-diHETE (Feinmark *et al.*, 1981) and 12*R*-HETE are partial agonists with antagonist activities at subagonist doses. In order to rationally design leukotriene B_4 antagonists studies such as the one we describe here, it is critically important to have a clear definition of the interaction of the leukotriene B_4 receptor with leukotriene B_4.

3.3. Determination of in Vivo Metabolism of the Peptido-Leukotrienes

The peptido-leukotrienes LTC_4, LTD_4, and LTE_4 are sequential metabolites of arachidonic acid produced via the 5-lipoxygenase pathway (Samuelsson, 1983) (Fig. 18). They have been shown individually to have potent biological actions and are thought to be involved in a number of pathophysiological disorders *in vivo* (Ford-Hutchinson and Letts, 1986). Despite their potent actions *in vivo*, little is known about the mechanisms of their detoxification and subsequent routes of elimination.

The lack of knowledge in this area is in no way indicative of a lack of effort, but rather of the complexity of the question. However, considerable progress has recently been made in this area of research. Part of the problem encountered early on was that the mode of elimination of the leukotrienes was not known; however, it has recently been determined that a major pathway for excretion of the peptido-leukotrienes and their metabolites is via bile (Orning *et al.*, 1986). This was found initially in the rat and subsequently in primates. The initial biliary metabolite of the peptido-leukotrienes identified is *N*-acylated LTE_4 (*N*-acetyl-LTE_4) (Hagmann *et al.*, 1986). This compound has been identified in the bile and feces of the rat. Using synthetically prepared *N*-acetyl-LTE_4, this metabolite has subsequently been shown to have weak biological activity when compared to its biological precursors, which is consistent with *N*-acetylation as a mechanism of deactivation (Foster *et al.*, 1986).

However, *N*-acetyl-LTE_4 is unlikely to be the end point of the metabolic pathway of the peptido-leukotrienes. Therefore, further possible metabolic fates for the peptido-leukotrienes were considered. Two structural modifying pathways were thought possible: (1) the oxidation of the sulfur atom converts the sulfide link of the parent to a sulfoxide, with this in turn being oxidized further to the sulfone (Fig. 19); and (2) the oxidation of the C-20 carbon from a methyl to a hydroxymethyl and the oxidation of this to a carboxy moiety. To this end, the sulfoxide and sulfone analogues of the peptido-leukotrienes were prepared from the parent molecules and the 20-hydroxy and 20-carboxy materials were prepared by total synthesis (illustrated in Fig. 20). With these possible metabolites in hand, we set out to investigate the metabolism of the peptido-leukotrienes in the rat.

The intravenous administration of [^{3}H]leukotrienes to the anesthetized rat resulted in a time-related biliary excretion of leukotriene metabolites as

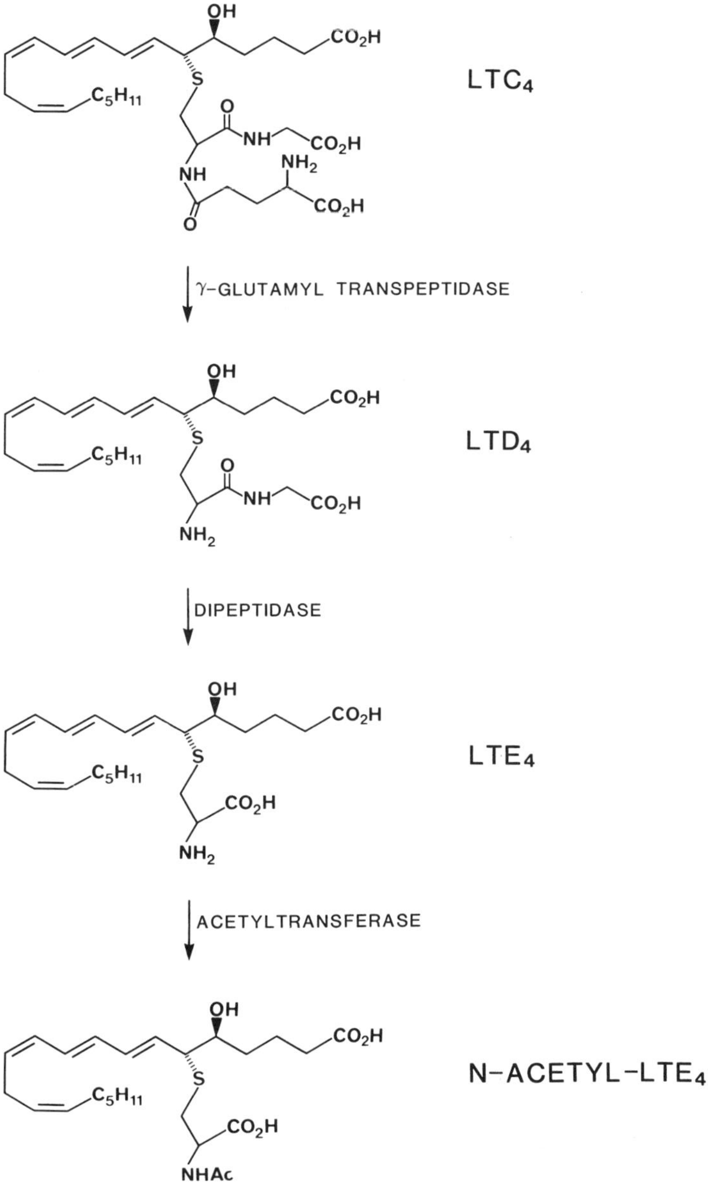
OH
CO_2H
S
C_5H_{11}
O
NH
CO_2H
NH_2
NH
CO_2H
O
LTC4
γ-GLUTAMYL TRANSPEPTIDASE
LTD4
DIPEPTIDASE
LTE4
ACETYLTRANSFERASE
NHAc
N-ACETYL-LTE4

Figure 18. Metabolism of the peptido-leukotrienes by modification of the peptide portion.

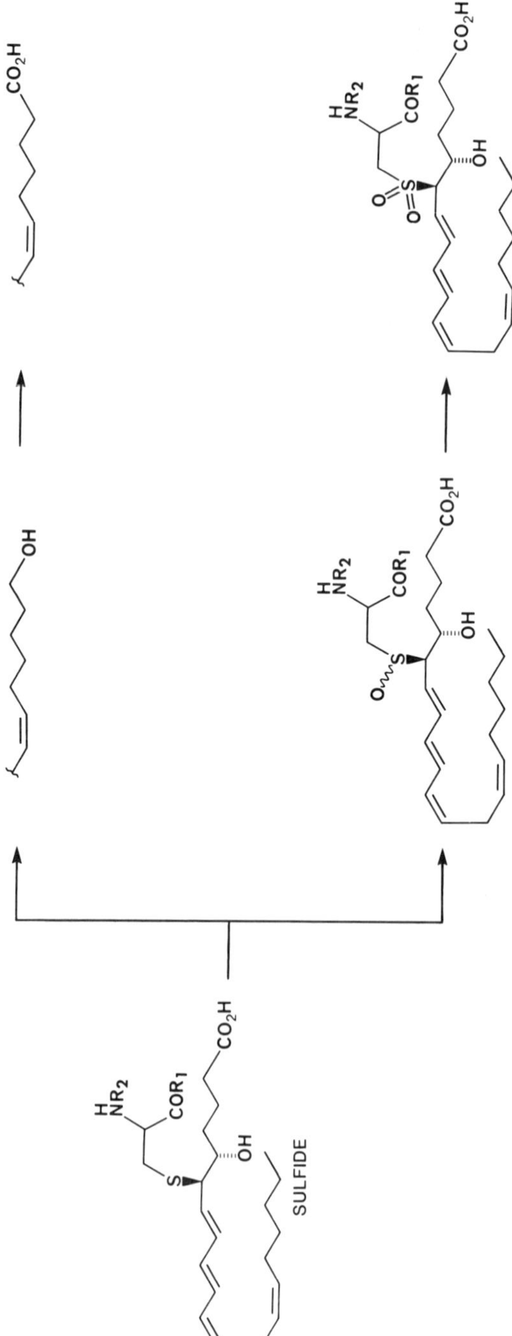

Figure 19. Possible oxidative metabolic pathways of the peptido-leukotrienes.

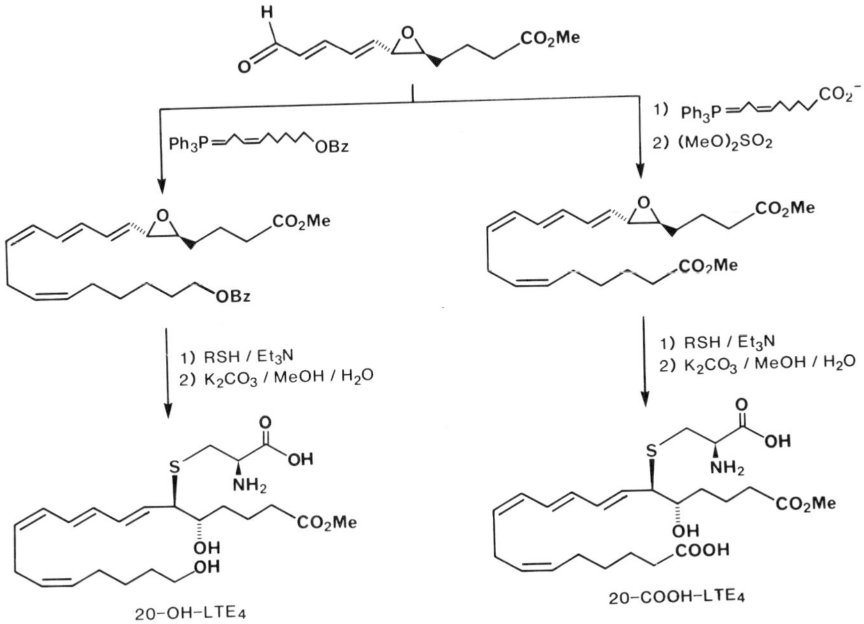

Figure 20. Synthesis of 20OH— and 20 COOH—LTE_4.

illustrated in Figs. 21–24. It can be seen that the majority of the radioactivity was eliminated within 30 min for each of the leukotrienes. [^{3}H]LTC_4 (2.6×10^{-11} moles, $n=6$) administration gave a $69 \pm 4.1\%$ recovery of radioactivity over 60 min. [^{3}H]LTD_4 (2.5×10^{-11} moles, $n=4$) gave a $61.8 \pm 7.5\%$ recovery. [^{3}H]LTE_4 (2.5×10^{-11} moles, $n=4$) and [^{3}H]*N*-acetyl-LTE_4 (2.1×10^{-11} moles, $n=5$) also showed significant biliary elimination, giving $52.2 \pm 1.5\%$ and $36.9 \pm 4.6\%$ recovery, respectively.

Examination of the hepatic lipoxygenase metabolites at two time points after administration ($t_{1/2}$ and $t_{3/4}$) shows significant metabolism of the leukotrienes (results illustrated in Figs. 21–24). All samples were purified by RP-HPLC and identified by chromatography with authentic standards in two solvent systems. Figure 21 shows the results obtained for [^{3}H]LTC_4. The metabolic profiles at $t_{1/2}$ (11 min) showed a small amount of LTC_4 and *N*-acetyl-LTE_4 present, with the major metabolite being LTD_4. No LTE_4 was detected. The profile at $t_{3/4}$ (21 min) shows further degradation of these metabolites, with minimal amounts of LTC_4 present and the majority of the identified radioactivity corresponding with *N*-acetyl-LTE_4 and LTD_4. In this fraction there is a small peak (at 25 min) that corresponds to the retention time of LTE_4. The peak at 8 min is an artifact that has been identified with

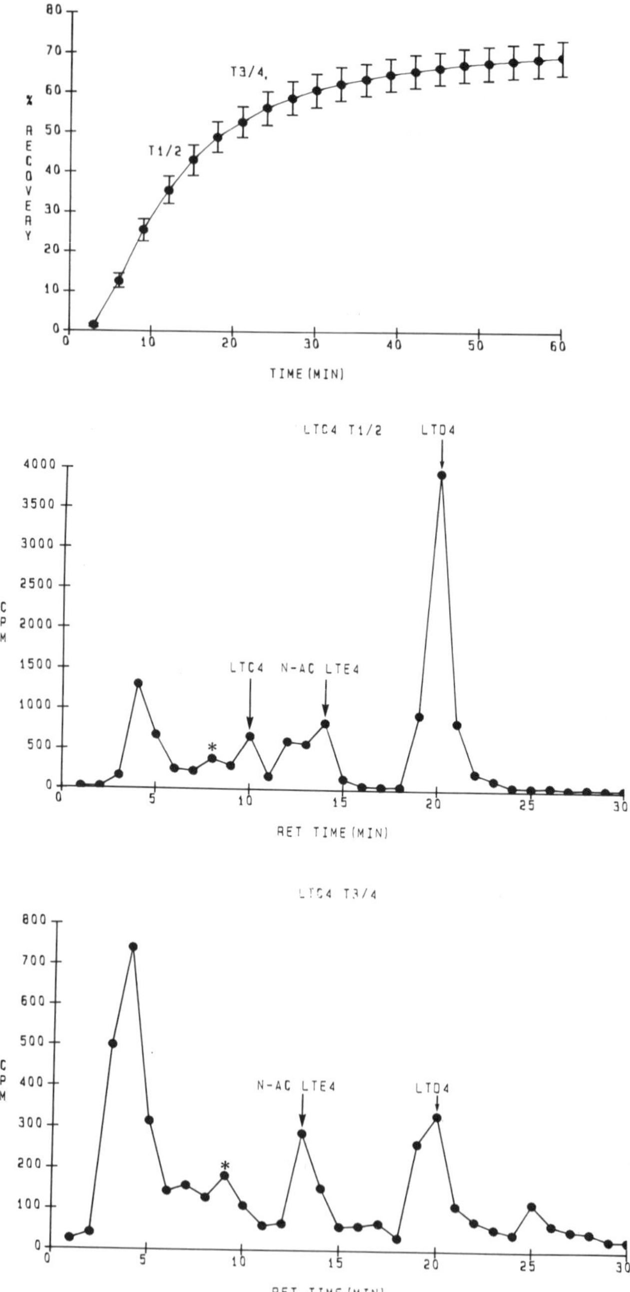

Figure 21. Metabolic profile of leukotriene-derived material in the bile of the anesthetized rat after LTC_4 (i.v.).

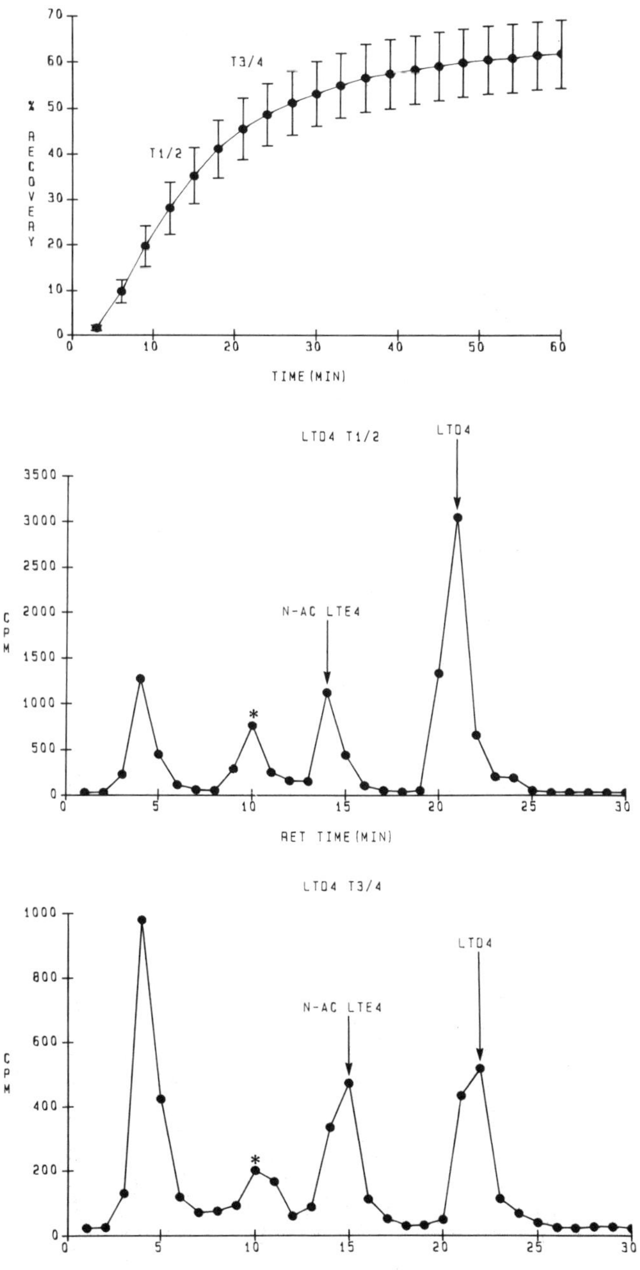

Figure 22. Metabolic profile of leukotriene-derived material in the bile of the anesthetized rat after LTD_4 (i.v.).

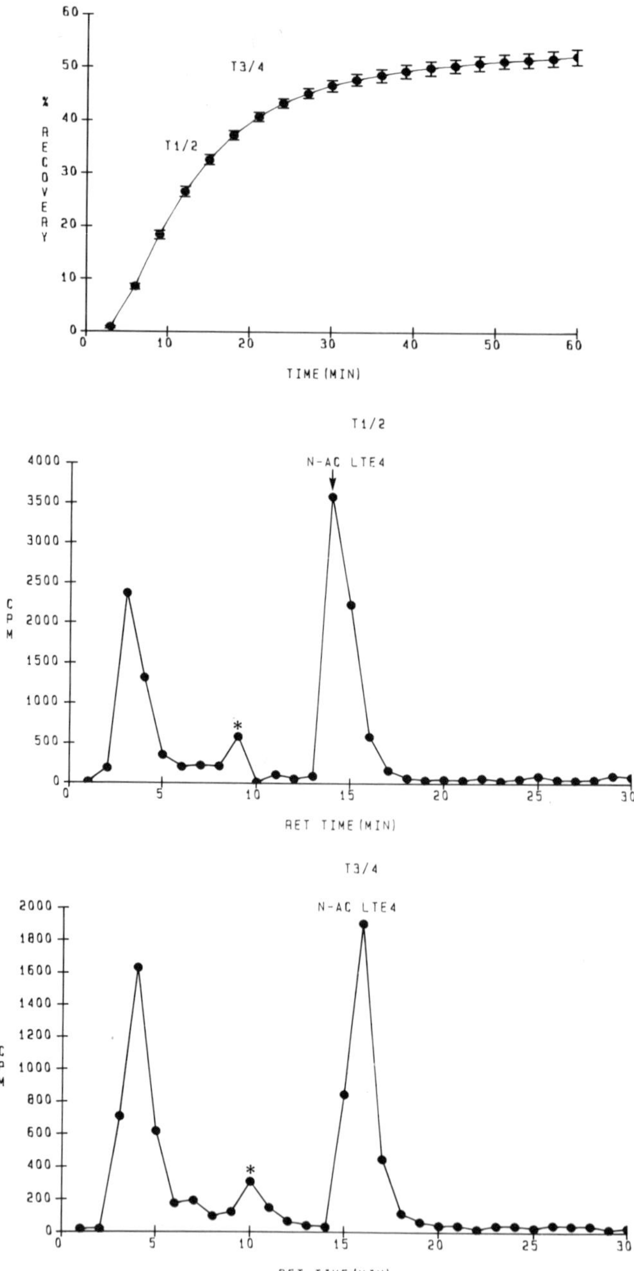

Figure 23. Metabolic profile of leukotriene-derived material in the bile of the anesthetized rat after LTE_4 (i.v.).

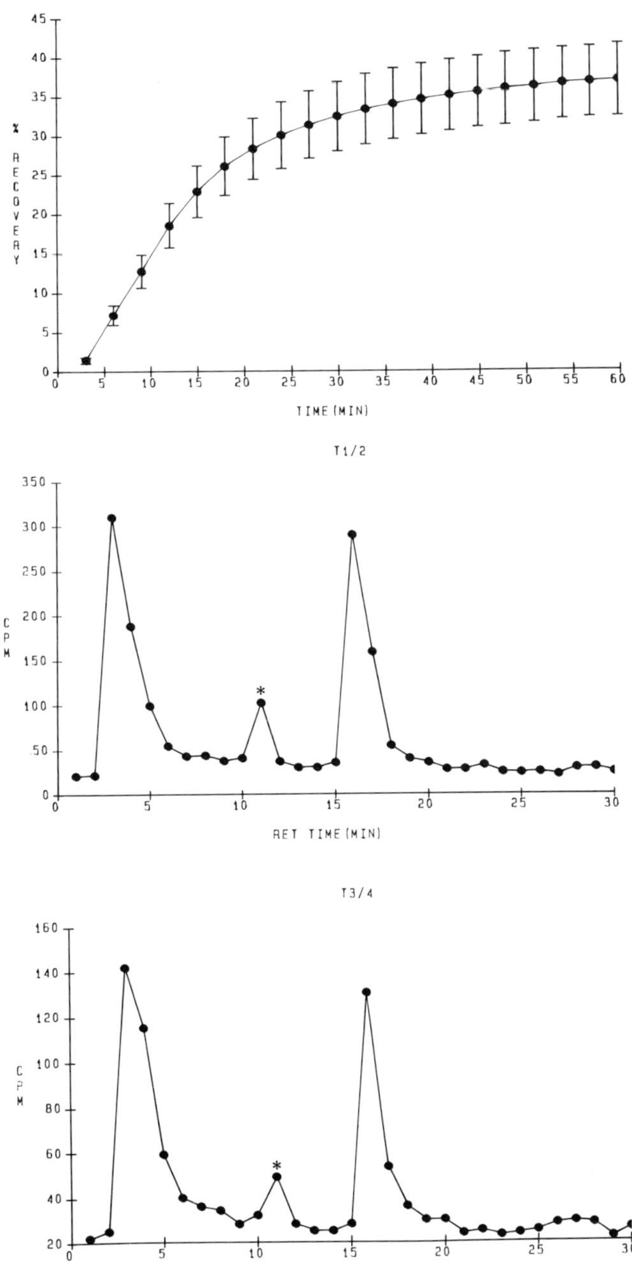

Figure 24. Metabolic profile of leukotriene-derived material in the bile of the anesthetized rat after NAc LTE_4 S (i.v.).

the extraction procedure. Finally, there is radioactivity (15.4% at $t_{1/2}$ and 25.3% at $t_{3/4}$) at the solvent front.

The intravenous administration of [^{3}H]LTD_4 (Fig. 22) shows a pattern similar to the results obtained with LTC_4. LTD_4 was the main metabolite at $t_{1/2}$ (11 min), with *N*-acetyl-LTE_4 present in smaller but significant amounts. At $t_{3/4}$ (21 min), LTD_4 was further degraded, with *N*-acetyl-LTE_4 present in equal amounts. Again, there was no significant amount of LTE_4 detected. The radioactivity at the solvent front is significant (15.6% of observed radioactivity at $t_{1/2}$ and 26.6% at $t_{3/4}$) and appears to be time related.

Further work with [^{3}H]LTE_4 (Fig. 23) showed that the only major metabolite identified was *N*-acetyl-LTE_4 ($t_{1/2}$ = 12 min; $t_{3/4}$ = 21 min). In addition, the intravenous administration of [^{3}H]*N*-acetyl-LTE_4 (Fig. 24) showed a pattern almost identical to that of LTE_4 ($t_{1/2}$ = 11 min; $t_{3/4}$ = 24 min). Again, substantial amounts of radioactivity were detected at the solvent front in both cases.

These results show that in the anesthetized rat the intravenous administration of a bolus dose of the leukotrienes results in significant biliary excretion of the particular leukotriene and/or its metabolic product(s). It was observed that LTD_4 was a major metabolite at $t_{1/2}$ after either LTC_4 or LTD_4 administration, while LTC_4 was only a minor product at $t_{1/2}$ (after LTC_4 administration). This implies that breakdown of LTC_4 to LTD_4 is quite fast, while LTD_4 metabolism is much slower.

Examination of the metabolism of LTD_4, or LTC_4, showed LTE_4 to be a minor metabolite. In addition, the intravenous administration of LTE_4 showed no evidence of biliary LTE_4, with much of the radioactivity being recovered as *N*-acetyl-LTE_4, suggesting that *N*-acetylation of LTE_4 is quite efficient in this species. In light of this fast conversion, it is interesting that LTE_4 and *N*-acetyl-LTE_4 showed significantly different biliary recoveries after intravenous administration ($p<0.05$, unpaired Students *t*-test). This would further suggest that for optimal elimination the acetylation process should not occur until after the metabolite is well advanced into the hepatic system. It is generally considered that hepatic *N*-acetylation takes place in the cytosol of the liver.

Of note in these experiments is the consistent observation of radioactivity coeluting near the solvent front of each of the HPLC systems. These radioactivities, which comprised up to 30% of radioactivity, appeared to be time dependent. By modifiying the HPLC conditions, this peak was resolved into several components (Fig. 25), one of which has been identified as 20 CO_2HNAc LTE_4. This ω-oxidation of NAc LTE_4 *in vivo* is in accordance with the recently published data by Orning (1987), in which the transformation of NAc LTE_4 to 20 OHNAc LTE_4 by rat liver microsomes is reported.

In conclusion, a major route of elimination of each of the peptide leukotrienes from the systemic circulation of the rat is via hepatic detoxification. In

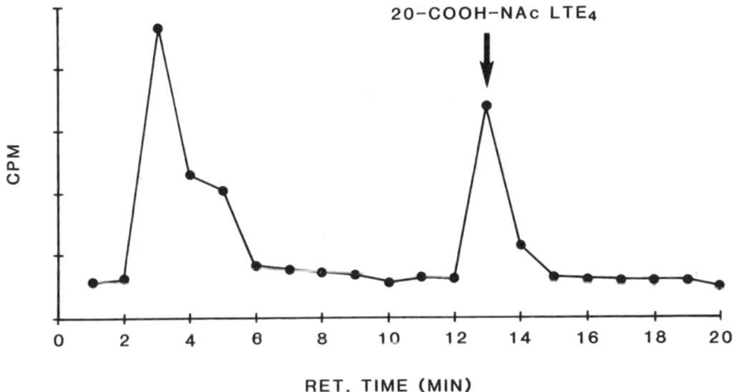

Figure 25. $20CO_2HNAc$ leukotriene E_4. 45% $MeOH/H_2O$; waters novapak C18; flow equals 0.5 ml/min.

this species, the conversion of LTC_4 to LTD_4 is quite fast, while LTD_4 metabolism is much slower. LTE_4 is efficiently converted to the biologically less active *N*-acetyl-LTE_4 and this material is further metabolized by oxidative pathways.

4. Conclusions

The lipoxygenase-derived metabolites of arachidonic acid have become targets of intense research because of their myriad of potent biological properties. However, it is unlikely that the progress in this field would have occurred if not for the contributions of synthetic chemists. The role of the chemist has been pivotal in the structural and biosynthetic elucidation of these potent molecules, the cataloging of their biological properties, and the determination of their catabolism, among many other contributions. Synthetic chemistry will continue to play a role in the above-mentioned areas in the future, which for this area of research is still wide open. The true role of lipoxygenase products in various disease states is presently unknown. However, the efforts of many scientists from many disciplines will aid in this discovery, and in the process will continue to expand scientific endeavor.

References

Adams, J., Fitzsimmons, B. J., and Rokach, J., 1984, Synthesis of lipoxins: Total synthesis of conjugated trihydroxy eicosatetraenoic acids, *Tetrahedron Lett.* **25:**4713–4716.

Adams, J., Fitzsimmons, B. J., Girard, Y., Leblanc, Y., Evans, J. F., and Rokach J., 1985,

Enantiospecific and stereospecific synthesis of lipoxin A. Stereochemical assignment of the natural lipoxin A and its possible biosynthesis, *J. Am. Chem. Soc.* **107:**464–469.

Borgeat, P., and Samuelsson, B., 1979, Arachidonic acid metabolism in polymorphonuclear leukocytes: Effects of ionophore A23187, *Proc. Natl. Acad. Sci. U.S.A.* **76:**2148–2152.

Bray, M. A., Cunningham, F. M., Ford-Hutchinson, A. W., and Smith, M. J. H., 1981a, Leukotriene B_4: A mediator of vascular permeability, *Br. J. Pharmacol.* **72:**483–486.

Bray, M. A., Ford-Hutchinson, A. W., and Smith, M. J. H., 1981b, Leukotriene B_4: An inflammatory mediator in vivo, *Prostaglandins* **22:**213–222.

Brocklehurst, W. E., 1960, The release of histamine and formation of a slow-reacting substance (SRS-A) during anaphylactic shock, *J. Physiol. (Lond.)* **151:**416–435.

Charleson, S., Evans, J. F., Zamboni, R. J., Leblanc., Y., Fitzsimmons, B. J., Léveillé, C., Dupuis, P., and Ford-Hutchinson, A. W., 1986, Leukotriene B_3, leukotriene B_4 and leukotriene B_5; binding to leukotriene B_4 receptors on rat and human leucocyte membranes, *Prostaglandins* **32:**503–516.

Evans, J. F., Dupuis, P., and Ford-Hutchinson, A. W., 1985, Purification and characterisation of leukotriene A_4 hydrolase from rat neutrophils, *Biochim. Biophys. Acta* **840:**43–50.

Feinmark, S. J., Lindgren, J. A., Claesson, H. E., Malmsten, C., and Samuelsson, B., 1981, Stimulation of human leukocyte degranulation by leukotriene B_4 and its omega-oxidized metabolites, *FEBS Lett.* **136:**141–144.

Fitzsimmons, B. J., and Rokach, J., 1985, On the biosynthesis and the structure of lipoxin B, *Tetrahedron Lett.* **26:**3939–3942.

Fitzsimmons, B. J., Adams, J., Evans, J. F., Leblanc, Y., and Rokach, J., 1985, The lipoxins. Stereochemical identification and determination of their biosynthesis, *J. Biol. Chem.* **260:**13008–13012.

Ford-Hutchinson, A. W., 1983, The role of leukotriene B_4 as a mediator of leukocyte function, *Agents Actions [Suppl.]* **12:**154–166.

Ford-Hutchinson, A. W., 1985, Leukotrienes as mediators of human disease, *Adv. Prostaglandin Thromboxane Leukotriene Res.* **15:**353–355.

Ford-Hutchinson, A. W., and Letts, L. G., 1986, Biological actions of leukotrienes, *Hypertension Suppl. II* **8:**1144–1149.

Ford-Hutchinson, A. W., Bray, M. A., Doig, M. V., Shipley, M. E., and Smith, M. J. H., 1980, Leukotriene B, a potent chemokinetic and aggregating substance released from polymorphonuclear leukocytes, *Nature* **286:**264–265.

Foster, A., Fitzsimmons, B., and Letts, L. G., 1986, The synthesis of *N*-acetyl-leukotriene E_4 and its effects on cardiovascular and respiratory function of the anesthetized pig, *Prostaglandins* **31**(6):1077–1086.

Goldman, D. W., and Goetzl, E. J., 1982, Specific binding of leukotriene B_4 to receptors on human polymorphonuclear leukocytes, *J. Immunol.* **129:**1600–1604.

Goldman, D. W., and Goetzl, E. J., 1984, Heterogeneity of human polymorphonuclear leukocyte receptors for leukotriene B_4. Identification of a subset of high affinity receptors that transduce the chemotactic response. *J. Exp. Med.* **159:**1027–1041.

Grabbe, J., Czarnetzki, B. M., Rosenbach, T., and Mardin, M., 1984, Identification of chemotactic lipoxygenase products of arachidonate metabolism in psoriatic skin, *J. Invest. Dermatol.* **82:**477–479.

Guindon, Y., Zamboni, R., Lau, C. K., and Rokach. J. 1982, Stereospecific synthesis of leukotriene B_4 (LTB_4), *Tetrahedron Lett.* **23:**739–742.

Hagmann, W., Denzlinger, C., Rapp, S., Weckbecker, G., and Keppler, D., 1986, Identification of the major endogenous leukotriene metabolite in the bile of rats as *N*-acetyl leukotriene E_4, *Prostaglandins* **31**(2):239–251.

Hedqvist, P., Dahlén, S. E., and Palmertz, U., 1985, Leukotrienes as mediators of airway anaphylaxis, *Adv. Prostaglandin Thromboxane Leukotriene Res.* **15:**345–348.

Hoffstein, S. T., Manzi, R. M., Razgaitis, K. A., Bender, P. E., and Gleason, J., 1986, Structural requirements for chemotactic activity of leukotriene B_4 (LTB_4), *Prostaglandins* **31:**205–215.

Kellaway, C. H., and Trethewie, E. R., 1940, The liberation of a slow-reacting smooth-muscle-stimulating substance in anaphylaxis, *Q. J. Exp. Physiol.* **30:**121–145.

Klickstein, L. B., Shapleigh, C., and Goetzl, E. J., 1980, Lipoxygenation of arachidonic acid as a source of polymorphonuclear leukocyte chemotactic factors in synovial fluid and tissue in rheumatoid arthritis and spondylarthritis, *J. Clin. Invest.* **66:**1166–1170.

Kreisle, R. A., and Parker, C. W., 1983, Specific binding of leukotriene, B_4 to a receptor on human polymorphonuclear leukocytes, *J. Exp. Med.* **157:**628–641.

Kreisle, R. A., Parker, C. W., Griffin, G. L., Senior, R. M., and Stenson, W. F., 1985, Studies of leukotriene B_4—specific binding and function in rat polymorphonuclear leukocytes: Absence of a chemotactic response. *J. Immunol.* **134:**3356–3363.

Leblanc, Y., Fitzsimmons, B., Adams, J., and Rokach, J., 1985, Total synthesis of lipoxin B: Assignment of stereochemistry, *Tetrahedron Lett.* **26:**1399–1402.

Lewis, R. A., Austen, K. F., Drazen, J. M., Clark, D. A., Marfat, A., and Corey, E. J., 1980, Slow reacting substances of anaphylaxis: Identification of leukotrienes C-1 and D from human and rat sources, *Proc. Natl. Acad. Sci. U.S.A.* **77:**3710–3314.

Lewis, R. A., Goetzl, E. J., Drazen, J. M., Soter, N. A., Austen, K. F., and Corey, E. J., 1981, Functional characterization of synthetic leukotriene B and its stereochemical isomers, *J. Exp. Med.* **154:**1243–1248.

Lin, A. H., Ruppel, P. L., and Gorman, R. R., 1984, Leukotriene B_4 binding to human neutrophils, *Prostaglandins* **28:**837–849.

Mannervik, B., Jensson, H., Ålin, P., Orning, L., and Hammarstrom, S., 1984, Transformation of leukotriene A_4 methyl ester to leukotriene C_4 mono methyl ester by cytosolic rat glutathione transferases, *FEBS Lett.* **175:**287–293.

Morris, H. R., and Taylor, G. W., Piper, P. J., and Tippins, J. R., 1980, Structure of slow-reacting substance of anaphylaxis from guinea-pig lung, *Nature* **285:**104–106.

Namiki, M., Igarashi, Y., Sakamoto, K., Nakamura, T., and Koga, Y., 1986, Pharmacological profiles of a potential LTB_4-antagonist, SM-9064, *Biochem. Biophys. Res. Commun.* **138:**540.

O'Flaherty, J., Kosfeld, S., and Nishihira, J., 1986, Binding and metabolism of leukotriene B_4 by neutrophils and their subcellular organelles, *J. Cell. Physiol.* **126:**359–370.

Orning, L., 1987, ω-Hydroxylation of *N*-acetylleukotriene E_4 by rat liver microsomes, *Biochem. Biophys. Res. Commun.* **143:**337–344.

Orning, L., Norin, E., Gustafsson, B., and Hammarstrom, S., 1986, *In vivo* metabolism of leukotriene C_4 in germ-free and conventional rats. Fecal excretion of *N*-acetylleukotriene E_4, *J. Biol. Chem.* **261**(2):766–771.

Rokach, J., Zamboni, R., Lau, C. K., and Guindon, Y., 1981, The stereospecific synthesis of leukotriene A_4 (LTA_4), 5-epi-LTA_4, 6-epi-LTA_4, and 5-epi, 6-epi-LTA_4, *Tetrahedron Lett.* **22:**2759–2762.

Rokach, J., Fitzsimmons, B. J., Leblanc, Y., Ueda, N., and Yamamoto, S., 1987, Recent progress in the Chemistry and Biochemistry of Lipoxygenase Products: The Lipoxins, *Adv. Prostaglandin Thromboxane Leukotriene Res.* **17:**761–767

Samuelsson, B., 1983, Leukotrienes: A new class of mediators of immediate hypersensitivity reactions and inflammations, *Adv. Prostaglandin Thromboxane Leukotriene Res.* **11:**1–13.

Serhan, C. N., Hamberg, M., and Samuelsson, B., 1984a, Trihydroxytetraenes: A novel series of compounds formed from arachidonic acid in human leukocytes, *Biochem. Biophys. Res. Commun.* **118:**943–949.

Serhan, C. N., Hamberg, M., and Samuelsson, B., 1984b, Lipoxins: A novel series of biologically active compounds formed from arachidonic acid in human leukocytes, *Proc. Natl. Acad. Sci. U.S.A.* **81:**5335–5339.

Serhan, C. N., Hamberg, M., Samuelsson, B., Morris, J., and Wishka, D. G., 1986a, On the stereochemistry and biosynthesis of lipoxin B, *Proc. Natl. Acad. Sci. U.S.A.* **83:**1983–1987.

Serhan, C. N., Nicolaou, K. C., Webber, S. E., Veale, C. A., Dahlén, S. E., Puustinen, T. J., and Samuelsson, B., 1986b, Lipoxin A stereochemistry and biosynthesis, *J. Biol. Chem.* **261:**16340–16345.

Showell, H. J., Otterness, I. G., Marfat, A., and Corey, E. J., 1982, Inhibition of leukotriene B_4-induced neutrophil degranulation by leukotriene B_4–dimethylamide, *Biochem. Biophys. Res. Commun.* **106:**741–747.

Stenson, W. F., and Lobos, E., 1982, Sulfasalazine inhibits the synthesis of chemotactic lipids by neutrophils, *J. Clin. Invest.* **69:**494–497.

IV

MOLECULAR INTERMEDIATES IN SIGNAL TRANSDUCTIONS

11

Guanine Nucleotide Regulatory Proteins in Inflammatory and Immune Responses

ALLEN M. SPIEGEL

1. Introduction

Initially discovered in studies of hormone-regulated adenylate cyclase activity (Rodbell, 1980), guanine nucleotide regulatory proteins (G proteins) are now known to couple receptors for many different types of extracellular signals to several different types of effector. Recent evidence shows that G proteins are involved in transduction of inflammatory and immune response signals in neutrophils, macrophages, and lymphocytes. With the appreciation of the diversity of receptors and effectors coupled by G proteins has come an appreciation of the diversity of the G proteins themselves. At least five different G proteins have been identified to date and more will certainly be discovered. A distinctive form of G protein appears to be particularly abundant in neutrophils and possibly other leukocytes. Notwithstanding the diversity of G proteins, each member of this family of signal transducers shares certain common features. In this chapter, I will review these common features of G proteins, highlight specific aspects of G-protein structure and function, and review information on G proteins and the transduction of inflammatory and immune response signals.

ALLEN M. SPIEGEL • Molecular Pathophysiology Section, National Institute of Diabetes, Digestive, and Kidney Diseases, National Institutes of Health, Bethesda, Maryland 20892.

2. General Aspects of G-Protein Structure and Function

2.1. Receptor–Effector Coupling and the GTPase Cycle

G proteins interact with receptors for extracellular signals including peptide and glycoprotein hormones, neurotransmitter amines, leukotrienes, chemical odorants, and photons of light. G-protein-coupled receptors appear to be transmembrane glycoproteins. Agonists bind specifically to their corresponding receptor and elicit a critical, and as yet undefined, change in receptor structure that is somehow transmitted to one or more G proteins with which each receptor interacts. Inactive G proteins contain tightly bound GDP (Ferguson *et al.*, 1986). Agonist activation of receptor leads to displacement of GDP from the receptor-associated G protein. This enables the G protein to bind GTP, which is present in the cell at higher concentration than is GDP.

The G-protein-associated receptor displays higher affinity for agonist binding than does the free receptor. Binding of GTP to the G protein dissociates it from the receptor and thereby leaves the receptor in the lower-affinity state with respect to agonist binding. This effect may be demonstrated in purified plasma membranes. Upon addition of guanine nucleotide, agonist affinity for a G-protein-coupled receptor is reduced.

G proteins interact with effectors such as enzymes involved in second messenger metabolism, as well as with certain ion channel proteins. Among the most important effectors regulated by G proteins are adenylate cyclase, responsible for cAMP formation, and phospholipase C, which cleaves phosphoinositides and forms diacylglycerol and inositol trisphosphate.

Under physiologic conditions, G-protein–effector interaction is transient and is terminated by a relatively slow GTPase activity intrinsic to the G protein. GTP hydrolysis liberates inorganic phosphate and leaves GDP tightly bound to the G protein. The GTP-bound form is the active state; the GDP-bound form is inactive. Inactive G protein is dependent on interaction with agonist-bound receptor in order to enter another round of the GTPase cycle by binding another molecule of GTP. Agonist-promoted guanine nucleotide binding results in apparent stimulation of GTPase activity. This effect of agonists can be used to monitor receptor–G-protein interaction (Cassel and Selinger, 1977).

Nonhydrolyzable analogues of GTP, for example, guanosine-3′-*O*-thiotriphosphate (GTP-γ-S), lead to persistent activation of G proteins. Fluoride, in the millimolar range, is also a general activator of G proteins. An AlF_4^- complex may mimic the γ-phosphate of GTP (Sternweis and Gilman, 1982; Bigay *et al.*, 1985). The result is that GDP-bound G protein is converted to the active "GTP-bound" state.

2.2. Subunit Structure

G proteins are composed of three distinct protein subunits. The α subunit binds guanine nucleotide, possesses GTPase activity, and serves as substrate for covalent modification by bacterial toxins (see Section 2.4). Each G protein contains a unique α subunit that may confer specificity in receptor and effector interactions. The α subunits vary in size on sodium dodecyl sulfate–polyacrylamide gel electrophoresis (SDS–PAGE) from about 39 to 52 kD.

The β and γ subunits, although not covalently linked, remain tightly associated under nondenaturing conditions. The β subunits migrate as 36-kD proteins on SDS–PAGE. A more rapidly migrating form of β subunit (35 kD) is seen in varying proportions in some tissues (Sternweis and Robishaw, 1984). Its relation to the 36-kD form has not been fully defined, but immunochemical differences (Mumby *et al.*, 1986) suggest that it may be structurally distinct. The β subunits (36-kD form) of different G proteins appear homologous by peptide mapping (Manning and Gilman, 1983) and immunochemical (Gierschik *et al.*, 1985) studies. Amino acid sequences predicted by cDNA clones show complete identity for the 340 amino acid β subunit of several distinct G proteins (Codina *et al.*, 1986).

The γ subunits are low molecular weight components (8–11 kD on SDS–PAGE). Although similar in size, the γ subunit of at least one G protein, the retinal rod outer segment membrane protein transducin, differs immunochemically (Gierschick *et al.*, 1985) from the γ subunits of other G proteins.

The holo-G protein is the form capable of interacting best with receptors (Fung, 1983; Florio and Sternweis, 1985). G-protein activation in solution leads to dissociation of α from β/γ subunits. Magnesium ion facilitates subunit dissociation (Iyengar and Birnbaumer, 1982; Katada *et al.*, 1984b). Whether subunit dissociation occurs during activation of membrane-bound G proteins is still not completely clear.

Resolved, activated α subunits are capable of activating effectors; for example, activated $G_s\alpha$ binds to and activates adenylate cyclase (May *et al.*, 1985). In general, it has been assumed that α subunits are responsible for effector interaction. The β/γ subunits are capable of inhibiting adenylate cyclase activity (Katada *et al.*, 1984b), but this effect is indirect, probably involving inhibition of G_s dissociation (Gilman, 1984). Nonetheless, β/γ subunits may also interact with effectors such as ion channels (Logothetis *et al.*, 1987), although this point is controversial (Yatani *et al.*, 1987).

2.3. Subcellular Localization

Most, if not all, G proteins involved in signal transduction appear to be membrane bound. There is no evidence that G proteins contain membrane spanning domains, nor do they appear to contain carbohydrate. In most cases, they are bound to the inner (cytoplasmic) surface of the plasma membrane. The precise details of G-protein membrane insertion have yet to be defined. Nonionic or ionic detergents such as cholate (Sternweis *et al.*, 1981) are required to release most G proteins from the plasma membrane.

The β/γ subunits may serve to anchor G proteins in the plasma membrane (Sternweis, 1985). The α subunits are relatively hydrophilic proteins, but there are hints that post-translational modifications could play a role in membrane attachment. The amino terminus of G-protein α subunits is blocked. All known α subunits begin with a methionine-glycine. Cleavage of the initiator methionine, followed by myristylation of the glycine, as in the *src* oncogene (Pellman *et al.*, 1985), is a distinct possibility.

In certain cells, including neutrophils (Okajima and Ui, 1984) and mast cells (Nakamura and Ui, 1985), G proteins are found in the cytosol. It is not clear whether this is an artifact of cell disruption or represents the native location of the G protein. Some have speculated that subunits released from the plasma membrane upon activation may interact with effectors at sites remote from the plasma membrane (Rodbell, 1985). Further work is necessary to define the extent and mode of G-protein membrane attachment to examine the possibility that G proteins are located in and function in other cellular compartments.

2.4. Effects of Bacterial Toxins

Bacterial toxins, including cholera and pertussis toxins, transfer ADP-ribose from NAD to α subunits of G proteins (Cassel and Pfeuffer, 1978; Gill and Meren, 1978; Ui, 1984). The toxins bind to cell surface receptors. After toxin binding, a subunit possessing ADP-ribosyl-transferase activity penetrates the cell. Using endogenous NAD, the toxin subunit covalently modifies G protein inside the cell.

Table I. Comparison of the Sequence (Single-Letter Amino Acid Code) Surrounding the Cholera Toxin ADP-Ribosylation Site (*) in G-Protein α Subunits

G_s192–207	P	S	D	Q	D	L	L	R	C	R	*	V	L	T	S	G	I
G_i170–185	P	T	Q	Q	D	V	L	R	T	R	*	V	K	T	T	G	I
G_o —	P	T	E	Q	D	I	L	R	T	R	*	V	K	T	T	G	I
TD165–180	P	T	E	Q	D	V	L	R	S	R	*	V	K	T	T	G	I

Table II. Comparison of the Carboxyl-Terminal Decapeptide of G-Protein α Subunits[a]

G_s385–394	R M H L R Q Y E L L
TD341–350	K E N L K D C * G L F
G_i345–355	K N N L K D C * G L F
G_o —	A N N L R G C * G L Y

[a] The cysteine (C*) is the site of pertussis toxin ADP-ribosylation.

Cholera toxin transfers ADP-ribose to an arginine residue in the α subunit of G_s and transducin (Table I). Other G-protein α subunits contain an arginine residue in the identical position (Table II) but are not very good cholera toxin substrates. Under certain conditions (no added guanine nucleotide), G-α subunits other than G_s may be modified by cholera toxin (Owens *et al.*, 1985). Cholera toxin also has functional effects in certain cells, including lymphocytes (Imboden *et al.*, 1986) and macrophages (Aksamit *et al.*, 1985), that cannot be explained by ADP-ribosylation of $G_s\alpha$. The possibility that cholera toxin acts on a unique G protein in such cases will be discussed in Section 7. Cholera toxin modification of G_s reduces GTPase activity (Cassel and Selinger, 1977). This prolongs the activity state of G_s and leads to increased activation of adenylate cyclase.

Pertussis toxin transfers ADP-ribose to a cysteine near the carboxy terminus of transducin (West *et al.*, 1985), G_i, and $G_o\alpha$ (Table II). ADP-ribosylation by pertussis toxin causes uncoupling of G protein from receptor (Ui, 1984). The result is that pertussis toxin blocks signal transduction through pathways coupled by its G-protein substrates.

Bacterial toxins have been useful as probes of both G-protein structure and function. With radioactively labeled NAD, toxin-catalyzed ADP-ribosylation can be used to identify Gα subunits in crude membranes. Toxin treatment of intact cells can be used to determine the effect of G-protein modification on signal transduction through specific pathways.

3. *Specific G Proteins and Their Functions*

3.1. *G_s and G_i*

Adenylate cyclase is under dual regulation (Gilman, 1984). A wide variety of agonists act at least in part by stimulating or inhibiting cAMP formation. Receptors for stimulatory and inhibitory agonists are coupled to adenylate cyclase by distinct G proteins, G_s and G_i, respectively.

G_s is ubiquitously distributed. At least two forms of $G_s\alpha$ have been identified in most cell types by cholera-toxin-catalyzed ADP-ribosylation

(Johnson *et al.*, 1978; Gilman, 1984). Purification of G_s confirms the heterogeneity of α subunits and shows that each α is associated with β and γ subunits as a heterotrimer (Sternweis *et al.*, 1981; Hildebrandt *et al.*, 1985). Cholera toxin stimulates cAMP formation directly and also sensitizes adenylate cyclase to stimulation by agonists. The effects of cholera toxin are due to the decrease in G_s GTPase activity discussed in Section 2.4.

Forskolin is a compound that activates adenylate cyclase in broken and intact cells (Seamon *et al.*, 1984). Forskolin appears to act directly on adenylate cyclase, an approximately 130-kD glycoprotein (Pfeuffer *et al.*, 1985), but activated Gs potentiates the effect of forskolin (Darfler *et al.*, 1982; Seamon *et al.*, 1984).

Guanine nucleotides, in addition to mediating agonist stimulation of adenylate cyclase, may under certain conditions inhibit the enzyme. Early studies involving kinetic and other indirect methods suggested that guanine nucleotide inhibition of adenylate cyclase involves a binding site distinct from that mediating stimulation (Rodbell, 1980). Pertussis toxin, which inactivates G_i without affecting G_s (Ui, 1984), led to the identification and purification of a distinct G protein mediating adenylate cyclase inhibition, G_i (Codina *et al.*, 1983; Katada *et al.*, 1984a). In S49 CYC^- mutant mouse lymphoma cells, G_s is genetically absent (Ross *et al.*, 1978). A pertussis-toxin-sensitive G protein mediates somatostatin inhibition of adenylate cyclase in CYC^-, providing further evidence that G_i is a distinct protein (Hildebrandt *et al.*, 1982). G_i is as ubiquitous as G_s, and perhaps as much as ten-fold more abundant (Gilman, 1984). Evidence to be reviewed in Sections 4 and 7 indicates that there are multiple distinct forms of G_i.

G_i functions like other G proteins with respect to receptor interactions. Receptors show high agonist affinity when bound to G_i, and inhibitory agonists stimulate G_i GTPase activity. Pertussis toxin by uncoupling G_i from receptors interferes with these interactions (Hsia *et al.*, 1984; Ui, 1984).

The mechanism of G_i–effector interaction, that is, adenylate cyclase inhibition, is unresolved. Since β/γ subunits of G_s and G_i may be identical, and since G_i concentration may exceed that of G_s in the plasma membrane, activation of G_i could lead to a substantial increase in free β/γ complex concentration. This would have the effect of retarding G_s dissociation and activation. This model of the mechanism of G_i inhibition of adenylate cyclase, indirect via action on G_s by β/γ subunits (Gilman, 1984), assumes that subunit dissociation normally accompanies G-protein activation. It also leaves open the function of $G_i\alpha$ subunits.

Experimental evidence using plasma membranes (Katada *et al.*, 1984b) and artificial phospholipid vesicles (Cerione *et al.*, 1986b) supports the potency of β/γ subunits, acting on G_s rather than directly on adenylate cyclase, as inhibitors of enzyme activation. A direct role for $G_i\alpha$ subunits in inhibition is suggested, however, by studies in CYC$^-$ cells that lack G_s. Further work is needed to resolve the mechanism of adenylate cyclase inhibition. Also unresolved is whether only one form of "$G_i\alpha$" is uniquely capable of inhibiting adenylate cyclase. Until this is clarified, the term G_i is at best imprecise (see Sections 4 and 7).

3.2. G_{pi}

Many receptor agonists stimulate phosphoinositide breakdown, leading to the generation of two second messengers—diacylglycerol (an activator of protein kinase C) and inositol trisphosphate (which mobilizes calcium from intracellular stores) (reviewed in Berridge and Irvine, 1984). One or more G proteins, here generically referred to as G_{pi}, couple receptors for such agonists to the effector enzyme, phospholipase C. Guanine nucleotides and fluoride directly stimulate phosphoinositide breakdown, and agonist stimulation of phospholipase C is dependent on guanine nucleotides in broken cell preparations (Cockroft and Gomperts, 1985; Litosch *et al.*, 1985). Agonists that act by stimulating phosphoinositide breakdown, moreover, stimulate high-affinity GTPase activity, as expected for G-protein coupled receptors (Hinkle and Phillips, 1984).

The identity of G_{pi} has not been unequivocally established. Indirect evidence suggests there may be more than one form. Evidence to be discussed in Section 7 indicates that a pertussis-toxin-sensitive G protein couples receptors to phospholipase C in cells such as neutrophils (Okajima and Ui, 1984; Smith *et al.*, 1985). In many other cell types, including pancreatic acinar cells, anterior pituitary, and liver (for review see Williamson, 1986), G-protein coupling of receptors to phospholipase C is insensitive to pertussis toxin. Although this might be due to altered accessibility to the toxin in these cell types, it might also reflect the participation of a distinct G protein.

3.3. Other G Proteins

Several other members of the G-protein family have been identified. These include two forms of transducin that couple visual pigments (rhodopsin in rods and cone opsins) to the enzyme cGMP phosphodiesterase (Shinozawa *et al.*, 1980; Nathans *et al.*, 1986; Stryer, 1986). G_o is a G protein extremely abundant in brain (Neer *et al.*, 1984; Sternweis and Robishaw, 1984). Like transducin and G_i, G_o is a pertussis toxin substrate. Specific antibodies localize G_o in brain, but it may also be present in other tissues such as heart (Huff *et al.*, 1985; Gierschik *et al.*, 1986b). The function of G_o is unclear, but recent evidence points to a possible interaction with calcium channels (Hescheler *et al.*, 1987). Other G proteins interact with potassium channels (Logothetis *et al.*, 1987; Yatani *et al.*, 1987). For both sets of channels, the involved G protein is pertussis toxin sensitive.

The *ras* and *ras*-related gene families (Gibbs *et al.*, 1984; Madaule and Axel, 1985) encode approximately 21-kD proteins that bind guanine nucleotides with high affinity, show GTPase activity, and are at least in some cases (Willumsen *et al.*, 1984) bound to the cytoplasmic surface of the cell membrane. G-protein α subunits are larger than *ras* gene products but share all the other features mentioned. Indirect evidence (reviewed in Spiegel, 1987a) suggests that *ras* gene products may function as signal transducers, coupling growth factor receptors to phosphoinositide breakdown. Further studies will clarify the function of this important class of guanine nucleotide binding proteins.

4. Molecular Biology of G Proteins

Amino acid sequencing of tryptic peptides from transducin and $G_o\alpha$ (Hurley *et al.*, 1984) permitted the synthesis of oligodeoxynucleotide probes for use in screening cDNA libraries. Multiple cDNAs encoding G-protein α subunits have now been isolated and sequenced. This has yielded a wealth of data on the primary structure of $G\alpha$ subunits, mRNA size and abundance, and distribution (reviewed in Spiegel, 1987a).

Two forms of transducin-α cDNA, corresponding to rod (Tanabe *et al.*, 1985) and cone (Lochrie *et al.*, 1985; Lerea *et al.*, 1986) forms of transducin, have been identified. Partial cDNA clones encoding G_o-α (Angus *et al.*, 1986; Itoh *et al.*, 1986) have also been obtained. Several cDNA clones coding for "G_i-related" proteins have been obtained. It is important to emphasize that the definition of these cDNAs as G_i does not derive from functional data; that is, in no case have the proteins encoded by the cDNAs been shown to be uniquely capable of inhibiting adenylate cyclase. cDNA clones found in a bovine (Nukada *et al.*, 1986b) and human (Bray *et al.*, 1987) brain library correspond exactly in amino acid sequence to the sequence of tryptic peptides derived form the 41-kD form of pertussis toxin substrate referred to as brain "$G_i\alpha$" (Neer *et al.*, 1984; Sternweis and Robishaw, 1984). $G_i\alpha$ cDNAs have also been isolated from rat C-6 glioma (Itoh *et al.*, 1986), mouse macrophage (Sullivan *et al.*, 1986), and human monocyte (Didsbury *et al.*, 1987) libraries. The latter three are highly homologous and appear to be distinct from the totally conserved human and bovine brain forms (Bray *et al.*, 1987). This suggests that at least two distinct forms of $G_i\alpha$ may exist. Since amino acid sequence differences between these two forms are clustered in regions that could confer receptor and effector interaction specificity (Bray *et al.*, 1987), the two forms may differ in function.

cDNAs encoding G_s-α have been obtained from bovine (Harris *et al.*, 1985; Nukada *et al.*, 1986a), rat (Itoh *et al.*, 1986), mouse (Sullivan *et al.*, 1986), and human (Bray *et al.*, 1986; Mattera *et al.*, 1986) libraries. These are all highly conserved. Heterogeneity is evident, however, in a limited region between amino acids 71 and 87. The variant forms correspond to at least four distinct forms of mRNA thought to be derived from a single gene via alternative splicing (Bray *et al.*, 1986; Robishaw *et al.*, 1986). The multiple forms of $G_s\alpha$ protein, in turn, are derived from the multiple mRNAs (Robishaw *et al.*, 1986). Thus, the 52-kD form of protein corresponds to an mRNA encoding 394 or 395 amino acid proteins, and the 45-kD form corresponds to mRNAs encoding 379 or 380 amino acid proteins. The 15 amino acid segment that varies in the multiple forms of $G_s\alpha$ could have functional significance, since it occurs in a putative effector interaction domain.

The Gα subunits constitute a gene family derived from a common evolutionary precursor. Transducin, G_o, and the two forms of G_i are 350–355 amino acids in length; all four forms of $G_s\alpha$ (see above) are somewhat longer. G_s is most distantly related (40% amino acid homology to the rest). Transducin, G_i, and G_o are 60–70% homologous. Invariant regions of Gα amino acid sequence correspond to regions involved in guanine nucleotide binding (Halliday, 1984; Jurnak, 1985). Regions of amino acid diversity may be involved in receptor and effector interaction (Medynski *et al.*, 1985).

5. *Specificity of Receptor–Effector Coupling by G Proteins*

The primary structures of several G-protein-coupled receptors have recently been defined. Rhodopsin and the cone opsins (Nathans *et al.*, 1986) couple to the two transducins, the β_2 (Dixon *et al.*, 1986) and β_1 (Yarden *et al.*, 1986) adrenergic receptors couple to G_s, and muscarinic cholinergic receptors (Kubo *et al.*, 1986) couple to other G proteins (G_i/G_o?). These receptors share limited amino acid homology, but more importantly they show a similar structure based on plots of hydrophobicity. All are transmembrane glycoproteins with seven membrane spanning segments. Three loops and the carboxyl terminus are exposed on the cytoplasmic surface and may interact with G proteins. Receptor–G-protein interaction would then be defined by the structure of Gα subunits and cytoplasmic domains of the receptors.

Receptor–G-protein coupling has been studied by reconstituting purified components into membranes and artificial phospholipid vesicles. Such studies show that specificity is relative rather than absolute. G_i and transducin interact well with rhodopsin, but G_s does not (Cerione *et al.*, 1985). G_i, surprisingly, interacts with the β-adrenergic receptor, but not as well as

G_s does (Cerione *et al.*, 1985). G_o and G_i both interact with the α_2-adrenergic receptor, whereas G_s does not (Cerione *et al.*, 1986a). Receptors that interact with the same G protein might be expected to show similarities in the structure of their cytoplasmic domains. This hypothesis will be evaluated as additional receptor cDNAs are sequenced.

Specificity in G-protein–effector interaction is evident in that only transducin activates cGMP phosphodiesterase and only G_s activates adenylate cyclase (Roof *et al.*, 1985). Lack of specificity, for example, the fact that both G_i and G_o activate phospholipase C in HL60 cells (Kikuchi *et al.*, 1986), must be interpreted cautiously, given the possibility of contamination with other G proteins (compare Yatani *et al.*, 1987).

6. Regulation of G-Protein Function

Appreciation of the critical role played by G proteins in signal transduction demands that one take possible quantitative and qualitative alterations in G proteins into account when studying target cell responsiveness to agonists. Genetic and developmental changes in G proteins can determine cellular response (reviewed in Spiegel, 1987b).

Covalent modifications may also be important in regulating G-protein function. The effects of bacterial toxins (Section 2.4) have obvious pathophysiologic significance. Not only is ADP-ribosylation of G_s by cholera toxin directly responsible for lethal diarrhea, but pertussis-toxin-catalyzed ADP-ribosylation of G-protein substrates may be responsible for several features of bordetella pertussis infection (Spiegel *et al.*, 1985).

Another important possibility is that G-protein subunits are substrates for phosphorylation by various types of protein kinase. *In vitro*, both transducin-α (Zick *et al.*, 1986) and $G_i\alpha$ (Katada *et al.*, 1985) are substrates for protein kinase C. Phorbol esters, activators of protein kinase C, have effects consistent with G-protein phosphorylation (see Spiegel, 1987b). Direct evidence that G proteins are phosphorylated in intact cells is needed to support this mechanism of G-protein regulation.

7. G Proteins in Inflammatory and Immune Responses

7.1. G Proteins in Neutrophils

Agonists, such as the chemotactic peptide fMet-Leu-Phe, and leukotriene B_4 bind to receptors on the surface of neutrophils and stimulate superoxide generation, lysosomal enzyme release, and other effects relevant to chemotaxis and phagocytosis. Agents such as prostaglandins of the E type and β-adrenergic agonists that activate adenylate cyclase through G_s inhibit these responses (Verghese *et al.*, 1985). The effects of fMet-Leu-Phe and other chemotactic factors are mediated by products of phosphoinositide breakdown. A pertussis-toxin-sensitive G protein couples chemotactic factor receptors to phospholipase C in neutrophils (Okajima and Ui, 1984; Goldman *et al.*, 1985; Smith *et al.*, 1985) and differentiated HL60 cells (Brandt *et al.*, 1985). Pertussis toxin blocks fMet-Leu-Phe-stimulated GTPase (Okajima and Ui, 1984), stimulation of phosphoinositide breakdown (Smith *et al.*, 1985), and distal events such as superoxide generation and lysosomal enzyme release. The effects of agents that act distal to the G protein, for example, calcium ionophores and phorbol esters, are not inhibited by pertussis toxin.

The pertussis-toxin-sensitive G protein in neutrophils was assumed to be G_i (Okajima and Ui, 1984), since pertussis-toxin-catalyzed ADP-ribosylation identified a 40-kD protein in neutrophil membranes, and ADP-ribosylation of this protein correlated with inhibition of fMet-Leu-Phe effects. The relevant G protein, however, causes activation of phospholipase C but not, as would be expected for G_i, inhibition of adenylate cyclase (Verghese *et al.*, 1985).

Pertussis toxin sensitivity does not allow discrimination among several G proteins, including G_i, G_o, and transducin, that serve as toxin substrates. Neutrophils contain a high concentration of pertussis toxin substrate, but immunochemical studies showed that neither transducin, G_o, nor the form of G_i abundant in brain could account for the neutrophil protein (Gierschik *et al.*, 1986a). An antiserum raised against a synthetic decapeptide (KENLKDCGLF) corresponding to the carboxy terminus of transducin-α strongly reacts with the pertussis toxin substrate of neutrophils. The antiserum also recognizes a protein that increases in abundance with differentiation of HL60 cells (Falloon *et al.*, 1986). This antiserum cross-reacts with both forms of G_i (see Section 4), which have the sequence KNNLKDCGLF, but not with G_o (Table II).

Purified G_i and G_o from rat brain were shown to reconstitute fMet-Leu-Phe receptor coupling to phospholipase C in pertussis-toxin-treated HL60 cells (Kikuchi *et al.*, 1986). These results must be qualified, since both purified G-protein fractions could in theory be contaminated with another form of G protein uniquely capable of coupling chemotactic peptide receptors to phospholipase C. Several lines of evidence suggest that the neutrophil pertussis toxin substrate may be distinct from G_o and the form of G_i (41 kD) abundant in brain. The pertussis toxin substrate of neutrophils and other leukocytes is susceptible to cholera toxin ADP-ribosylation, whereas brain pertussis toxin substrates are not (Verghese *et al.*, 1986). The major pertussis toxin substrate from bovine neutrophils has been purified and shown to be a 40-kD protein that differs immunochemically from the 41 kD G_i-α of brain and from the 39-kD $G_o\alpha$ (Gierschik *et al.*, 1987). An antiserum against a synthetic peptide specific for the form of $G_i\alpha$ encoded by human monocyte cDNA (Didsbury *et al.*, 1987) reacts strongly with the neutrophil protein and not with the 41-kD form of brain $G_i\alpha$ (A. M. Spiegel *et al.*, unpublished observations).

These data suggest that a specific subtype of "G_i" couples chemotactic peptide receptors to phospholipase C in neutrophils, mast cells (Nakamura and Ui, 1985), and perhaps other leukocytes (see Section 7.2), but unequivocal identification of this protein will require functional studies as well as amino acid sequencing.

Preliminary observations (Barrowman *et al.*, 1986) suggest that additional functions in neutrophils, for example, exocytosis, are also regulated by G proteins. This possibility, as well as G-protein coupling to other effector enzymes such as phospholipase A_2, deserves further study.

7.2. G Proteins in Lymphocytes and Macrophages

G proteins may also mediate signals in immune response cells. Signaling by the T-cell antigen receptor involves phosphoinositide breakdown (Imboden and Stobo, 1985). A cholera-toxin-sensitive G protein has been implicated in signal transduction by the T-cell antigen receptor (Imboden *et al.*, 1986). This protein is not G_s, since the effects of cholera toxin are not correlated with alterations in cAMP production. The protein may be similar or identical to the G protein sensitive to both cholera and pertussis toxin in neutrophils discussed in Section 7.1. Similar observations with regard to a cholera-toxin-sensitive (Aksamit *et al.*, 1985) and pertussis-toxin-sensitive (Backlund *et al.*, 1985) G protein have been made in macrophage cell lines.

Interleukin-2 stimulates specific guanine nucleotide binding and GTPase activity in T lymphocytes. These effects are blocked by pertussis toxin (Evans *et al.*, 1987). These provocative results imply that the interleukin-2 receptor may also be coupled to a pertussis-toxin-sensitive G protein. Nonhydrolyzable GTP analogues, possibly acting through the latter, are also capable of stimulating T-cell proliferation (Mustelin *et al.*, 1986). In contrast, agonists such as E type prostaglandins inhibit T-cell proliferation by increasing cAMP production through G_s (Evans *et al.*, 1987). B lymphocyte and macrophage activation by bacterial lipopolysaccharide (interleukin-1 production and mitogenesis) is blocked by pertussis toxin (Jakway and DeFranco, 1986). Further characterization of the putative G-protein substrate and comparison with pertussis-toxin-sensitive G proteins in other types of leukocyte should be interesting.

8. Conclusions

Extracellular signals may traverse the cell membrane (e.g., steroid hormones) or may bind to cell surface receptors. Cell surface receptors may have intrinsic signaling functions (e.g., tyrosine kinase activity of growth factor receptors) or may be coupled to effector mechanisms by G proteins. The functional and structural diversity of G proteins has only recently become apparent. In cells that mediate inflammatory and immune responses, several G proteins have important functions. Stimulation and inhibition of cAMP production through G_s and G_i, respectively, are important signal transduction pathways in neutrophils and other leukocytes. The stimulation of phosphoinositide breakdown, mediated by a specific form of pertussis-toxin-sensitive G protein, is equally, if not more, important. Identification of the structure and function of the relevant G proteins should allow a better understanding of the mechanisms regulating inflammatory and immune response and may offer unique targets for therapeutic intervention.

References

Aksamit, R. R., Backlund, P. S., Jr., and Cantoni, G. L., 1985, Cholera toxin inhibits chemotaxis by a cAMP-independent mechanism, *Proc. Natl. Acad. Sci. U.S.A.* **82:**7475–7479.

Angus, C. W., Van Meurs, K. P., Tsai, S.-C., Adamik, R., Miedel, M. C., Pan, Y.-C. E., Kung, H.-F., Moss, J., and Vaughan, M., 1986, Identification of the probable site of choleragen-catalyzed ADP-ribosylation in a $G_o\alpha$-like protein based on cDNA sequence, *Proc. Natl. Acad. Sci. U.S.A.* **83:**5813–5816.

Backlund, P. S., Jr., Meade, B. D., Manclark, C. R., Cantoni, G. L., and Aksamit, R. R., 1985, Pertussis toxin inhibition of chemotaxis and the ADP-ribosylation of a membrane protein in a human–mouse hybrid cell line, *Proc. Natl. Acad. Sci. U.S.A.* **82:**2637–2641.

Barrowman, M. M., Cockroft, S., and Gomperts, B. D., 1986, Two roles for guanine nucleotides in the stimulus secretion sequence of neutrophils, *Nature* **319:**504–507.

Berridge, M. J., and Irvine, R. F., 1984, Inositol trisphosphate, a novel second messenger in signal transduction, *Nature* **312:**315–321.

Bigay, J., Deterre, P., Pfister, C., and Chabre, M., 1985, Fluoroaluminates activate transducin-GDP by mimicking the γ-phosphate of GTP in its binding site, *FEBS Lett.* **191:**181–185.

Brandt, S. J., Dougherty, R. W., Lapetina, E. G., and Niedel, J. E., 1985, Pertussis toxin inhibits chemotactic peptide-stimulated generation of inositol phosphates and lysosomal enzyme secretion in human leukemic (HL60) cells, *Proc. Natl. Acad. Sci. U.S.A.* **82:**3277–3280.

Bray, P., Carter, A., Simons, C., Guo, V., Pucket, C., Kamholtz, J., Spiegel, A., and Nirenberg, M., 1986, Human cDNA clones for four species of G-α signal transduction protein, *Proc. Natl. Acad. Sci. U.S.A.* **83:** 8893–8897.

Bray, P., Carter, A., Spiegel, A., and Nirenberg, M., 1987, Human cDNA clone for brain $G_i\alpha$, *Proc. Natl. Acad. Sci. U.S.A.* **84:**5115–5119.

Cassel, D., and Pfeuffer, T., 1978, Mechanism of cholera toxin action: Covalent modification of the guanyl nucleotide-binding protein of the adenylate cyclase system, *Proc. Natl. Acad. Sci. U.S.A.* **75:**2669–2673.

Cassel, D., and Selinger, Z., 1977, Mechanism of adenylate cyclase activation by cholera toxin: Inhibition of GTP hydrolysis at the regulatory site, *Proc. Natl. Acad. Sci. U.S.A.* **74:**3307–3311.

Cerione, R. A., Staniszewski, C., Benovic, J. L., Lefkowitz, R. J., Caron, M. F., Gierschik, P., Somers, R., Spiegel, A. M., Codina, J., and Birnbaumer, L., 1985, Specificity of the functional interactions of the β-adrenergic receptor and rhodopsin with guanine nucleotide regulatory proteins reconstituted in phospholipid vesicles, *J. Biol. Chem.* **260:**1493–1500.

Cerione, R. A., Staniszewski, C., Gierschik, P., Codina, J., Somers, R. L., Birnbaumer, L., Spiegel, A. M., Caron, M. G., and Lefkowitz, R. J., 1986a, Mechanism of guanine nucleotide regulatory protein-mediated inhibition of adenylate cyclase, *J. Biol. Chem.* **261:**9514–9520.

Cerione, R. A., Regan, J. W., Nakata, H., Codina, J., Benovic, J. L., Gierschik, P., Somers, R. L., Spiegel, A. M., Birnbaumer, L., Lefkowitz, R. J., and Caron, M. G., 1986b, Functional reconstitution of the α-adrenergic receptor with guanine nucleotide regulatory proteins in phospholipid vesicles, *J. Biol. Chem.* **261:**3901–3909.

Cockcroft, S., and Gomperts, B. D., 1985, Role of guanine nucleotide binding protein in the activation of polyphosphoinositide phosphodiesterase, *Nature* **314:**534–536.

Codina, J., Hildebrandt, J., Iyengar, R., Birnbaumer, L., Sekura, R. D., and Manclark, C. R., 1983, Pertussis toxin substrate, the putative N_i component of adenylyl cyclases, is an α-β heterodimer regulated by guanine nucleotide and magnesium, *Proc. Natl. Acad. Sci. U.S.A.* **80:**4276–4280.

Codina, J., Stengel, D., Woo, S., and Birnbaumer, L., 1986, β-Subunits of the human liver G_s/G_i signal-transducing proteins and those of bovine retinal rod cell transducin are identical, *FEBS Lett.* **207:**187–193.

Darfler, F. J., Mahan, L. C., Koachman, A. M., and Insel, P. A., 1982, Stimulation by forskolin of intact S49 lymphoma cells involves the nucleotide regulatory protein of adenylate cyclase, *J. Biol. Chem.* **256:**11901–11907.

Didsbury, J. R., Ho, Y. S., and Snyderman, R., 1987, Human G_i protein α-subunit: Deduction of amino acid structure from a cloned cDNA, *FEBS Lett.* **211:**160–164.

Dixon, R. A. F., Kobilka, B. K., Strader, D. J., Benovic, J. L., Dohlman, H. G., Frielle, T., Bolanowski, M. A., Bennett, C. D., Rands, E., Diehl, R. E., Mumford, R. A., Slater, E. E., Sigal, I. S., Caron, M. G., Lefkowitz, R. J., and Strader, C. D., 1986, Cloning of the gene and cDNA for mammalian β-adrenergic receptor and homology with rhodopsin, *Nature* **321:**75–79.

Evans, S. W., Beckner, S. K., and Farrar, W. L., 1987, Stimulation of specific GTP binding and hydrolysis activities in lymphocyte membrane by interleukin-2, *Nature* **325:**166–168.

Falloon, J., Malech, H., Milligan, G., Unson, C., Kahn, R., Goldsmith, P., and Spiegel, A., 1986, Detection of the major pertussis toxin substrate of human leukocytes with antisera raised against synthetic peptides, *FEBS Lett.* **209:**352–356.

Ferguson, K. M., Higashijima, T., Smigel, M. D., and Gilman, A. G., 1986, The influence of bound GDP on the kinetics of guanine nucleotide binding to G proteins, *J. Biol. Chem.* **261:**7393–7399.

Florio, V. A., and Sternweis, P. C., 1985, Reconstitution of resolved muscarinic cholinertic receptors with purified GTP-binding proteins, *J. Biol. Chem.* **260:**3477–3483.

Fung, B. K.-K., 1983, Characterization of transducin from bovine retinal rod outer segments. I. Separation and reconstitution of the subunits, *J. Biol. Chem.* **258:**10495–10502.

Gibbs, J. B., Sigal, I. S., Poe, M., and Scolnick, E. M., 1984, Intrinsic GTPase activity distinguishes normal and oncogenic ras p21 molecules, *Proc. Natl. Acad. Sci. U.S.A.* **81:**5704–5708.

Gierschik, P., Codina, J., Simons, C., Birnbaumer, L., and Spiegel, A., 1985, Antisera against a guanine nucleotide binding protein from retina cross-react with the β subunit of the adenylyl cyclase associated guanine nucleotide binding proteins, N_s and N_i, *Proc. Natl. Acad. Sci. U.S.A.* **82:**727–731.

Gierschik, P., Falloon, J., Milligan, G., Pines, M., Gallin, J. I., and Spiegel, A., 1986a, Immunochemical evidence for a novel pertussis toxin substrate in human neutrophils, *J. Biol. Chem.* **261:**8058–8062.

Gierschik, P., Milligan, G., Pines, M., Goldsmith, P., Codina, J., Klee, W., and Spiegel, A., 1986b, Use of specific antibodies to quantitate the guanine nucleotide-binding protein G_o in brain, *Proc. Natl. Acad. Sci. U.S.A.* **83:**2258–2262.

Gierschik, P., Sidiropoulos, D., Spiegel, A., and Jakobs, K. H., 1987, Purification and characterization of the major pertussis toxin-sensitive guanine nucleotide-binding protein of bovine neutrophil membranes, *Eur. J., Biochem.* **165:**185–194.

Gill, D. M., and Meren, H., 1978, ADP-ribosylation of membrane proteins catalysed by cholera toxin: Basis of the activaiton of adenylate cyclase, *Proc. Natl. Acad. Sci. U.S.A.* **75:**305–3054.

Gilman, A. G., 1984, G proteins and dual control of adenylate cyclase, *Cell* **36:**577–579.

Goldman, D. W., Chang, F. H., Gifford, L. A., Goetzl, E. J., and Bourne, H. R., 1985, Pertussis toxin inhibition of chemotactic factor-induced calcium mobilization and function in human polymorphonuclear leukocytes, *J. Exp. Med.* **162:**145–156.

Halliday, K., 1984, Regional homology in GTP-binding proto-oncogene products and elongation factors, *J. Cyclic Nucleotide Protein Phosphor. Res.* **9:**435–448.

Harris, B. A., Robishaw, J. D., Mumby, S. M., and Gilman, A. G., 1985, Molecular cloning of complementary DNA of the α subunit of the G protein that stimulates adenylate cyclase, *Science* **229:**1274–1277.

Hescheler, J., Rosenthal, W., Trautwein, W., and Schultz, G., 1987, The GTP-binding protein, G_o, regulates neuronal calcium channels, *Nature* **325:**445–447.

Hildebrandt, J. D., Hanoune, J., and Birnbaumer, L., 1982, Guanine nucleotide inhibition of cyc-S49 mouse lymphoma cell membrane adenylyl cyclase, *J. Biol. Chem.* **257:**14723–14725.

Hildebrandt, J. D., Codina, J., Rosenthal, W., Birnbaumer, L., Neer, E. J., Yamazaki, A., and Bitensky, M. W., 1985, Characterization by two-dimensional peptide mapping of the γ subunits of N_s and N_i, the regulatory proteins of adenylate cyclase, and of transducin, the guanine nucleotide-binding protein of rod outer segments of the eye, *J. Biol. Chem.* **260:**14867–14872.

Hinkle, P. M., and Phillips, W. J., 1984, Thyrotropin-releasing hormone stimulates GTP hydrolysis by membranes from GH4Cl rat pituitary tumor cells, *Proc. Natl. Acad. Sci. U.S.A.* **81:**6183–6187.

Hsia, J. A., Moss, J., Hewlett, E. L., and Vaughan, M., 1984, ADP-ribosylation of adenylate cyclase by pertussis toxin, *J. Biol. Chem.* **259:**1086–1090.

Huff, R. M., Axton, J. M., and Neer, E. G., 1985, Physical and immunological characterization of a guanine nucleotide-binding protein purified from bovine cerebral cortex, *J. Biol. Chem.* **260:**10864–10871.

Hurley, J. B., Simon, M. I., Teplow, D. B., Robishaw, J. D., and Gilman, A. G., 1984, Homologies between signal transducing G proteins and *ras* gene products, *Science* **226:**860–862.

Imboden, J. B., and Stobo, J. D., 1985, Transmembrane signaling by the T-cell antigen receptor, *J. Exp. Med.* **161:**446–456.

Imboden, J. B., Shoback, D. M., Pattison, G., and Stobo, J. D., 1986, Cholera toxin inhibits the T-cell antigen receptor-mediated increases in inositol trisphosphate and cytoplasmic free calcium, *Proc. Natl. Acad. Sci. U.S.A.* **83:**5673–5677.

Itoh, H., Kozasa, T., Nagata, S., Nakamura, S., Katada, T., Ui. M., Iwai, S., Ohtsuka, E., Kawasaki, H., Suzuki, K., and Kaziro, Y., 1986, Molecular cloning and sequence determination of cDNAs for α subunits of the guanine nucleotide-binding proteins G_s, G_i, and G_o from rat brain, *Proc. Natl. Acad. Sci. U.S.A.* **83:**3776–3780.

Iyengar, R., and Birnbaumer, L., 1982, Hormone receptor modulates the regulatory component of adenylyl cyclase by reducing its requirement for Mg^{2+} and enhancing its extent of activation by guanine nucleotides, *Proc. Natl. Acad. Sci. U.S.A.* **79:**5179–5183.

Jakway, J. P., and DeFranco, A. L., 1986, Pertussis toxin inhibition of B cell and macrophage responses to bacterial lipopolysaccharide, *Science* **234:**743–746.

Johnson, G. L., Kaslow, H. R., and Bourne, H. R., 1978, Genetic evidence that cholera toxin substrates are regulatory components of adenylate cyclase, *J. Biol. Chem.* **253:**7120–7132.

Jurnak, F., 1985, Structure of the GDP domain of EF-Tu and location of the amino acids homologous to ras oncogene proteins, *Science* **230:**32–36.

Katada, T., Bokoch, G. M., Northup, J. K., Ui, M., and Gilman, A. G., 1984a, The inhibitory guanine nucleotide-binding regulatory component of adenylate cyclase. Properties and function of the purified protein, *J. Biol. Chem.* **259:**3568–3577.

Katada, T., Northup, J. K., Bokoch, G. M., Ui, M., and Gilman, A. G., 1984b, The inhibitory guanine nucleotide-binding regulatory component of adenylate cyclase. Subunit dissociation and guanine nucleotide-dependent hormonal inhibition, *J. Biol. Chem.* **259:**3578–3585.

Katada, T., Gilman, A. G., Watanabe, Y., Bauer, S., and Jakobs, K. H., 1985, Protein kinase C phosphorylates the inhibitory guanine nucleotide-binding regulatory component and apparently suppresses its function in hormonal inhibition of adenylate cyclase, *Eur. J. Biochem.* **151:**431–437.

Kikuchi, A., Kozawa, O., Kaibuchi, K., Katada, T., Ui, M., and Takai, Y., 1986, Direct evidence for involvement of a guanine nucleotide-binding protein in chemotactic peptide-stimulated formation of inositol bisphosphate and trisphosphate in differentiated human leukemic (HL-60) cells, *J. Biol. Chem.* **261:**11558–11562.

Kubo, T., Fukuda, K., Mikami, A., Maeda, A., Takahashi, H., Mishina, M., Haga, T., Haga, K., Ichiyama, A., Kangawa, K., Kojima, M., Matsuo, Y., Hirose, T., and Numa, S., 1986, Cloning, sequencing and expression of complementary DNA encoding the muscarinic acetylcholine receptor, *Nature* **323:**411–416.

Lerea, C. L., Somers, D. E., Hurley, J. B., Klock, I. B., and Budt-Milan, A. H., 1986, Identification of specific transducin α subunits in retinal rod and cone photoreceptors, *Science* **234:**77–80.

Litosch, I., Wallis, C., and Fain, J. N., 1985, 5-Hydroxytryptamine stimulates inositol phosphate production in a cell-free system from blowfly salivary glands. Evidence for a role of GTP in coupling receptor activation to phosphoinositide breakdown, *J. Biol. Chem.* **260:**5464–5471.

Lochrie, M.A., Hurley, J. B., and Simon, M. I., 1985, Sequence of the α subunit of photoreceptor G protein: Homologies between transducin, ras, and elongation factors, *Science* **228:**96–99.

Logothetis, D. E., Kurachi, Y., Galper, J., Neer, E. J., and Clapham, D. E., 1987, The β-γ subunits of GTP-binding proteins activate the muscarinic K^+ channel in heart, *Nature* **325:**321–326.

Madaule, P., and Axel, R., 1985, A novel ras-related gene family, *Cell* **41:**31–40.

Manning, D. R., and Gilman, A. G., 1983, The regulatory components of adenylate cyclase and transducin. A family of structure homologous guanine nucleotide binding proteins, *J. Biol. Chem.* **258:**7059–7063.

Mattera, R., Codina, J., Crozat, A., Kidd, V., Woo, S. L. C., and Birnbaumer, L., 1986, Identification by molecular cloning of two forms of the α-subunit of the human liver stimulatory (G_s) regulatory component of adenylyl cyclase, *FEBS Lett.* **206:**36–41.

May, D. C., Ross, E. M., Gilman, A. G., and Smigel, M. D., 1985., Reconstitution of catecholamine-stimulated adenylate cyclase activity using three purified proteins, *J. Biol. Chem.* **260:**15829–15833.

Medynski, D. C., Sullivan, K., Smith, D., Van Dop, C., Chang, F. H., Fung, B.K.-K., Seeburg, P. H., and Bourne, H. R., 1985, Amino acid sequence of the α subunit of transducin deduced from the cDNA sequence, *Proc. Natl. Acad. Sci. U.S.A.* **82:**4311–4315.

Mumby, S. M., Kahn, R. A., Manning, D. R., and Gilman, A. G., 1986, Antisera of designed specificity for subunits of guanine nucleotide-binding regulatory proteins, *Proc. Natl. Acad. Sci. U.S.A.* **83:**265–269.

Mustelin, T., Poso, H., and Andersson, L. C., 1986, Role of G-proteins in T-cell activation: Nonhydrolysable GTP analogues induce early ornithine decarboxylase activity in human T lymphocytes, *EMBO J.* **5:**3287–3290.

Nakamura, T., and Ui, M., 1985, Simultaneous inhibitions of inositol phospholipid breakdown, arachidonic acid release, and histamine secretion in mast cells by islet-activating protein, pertussis toxin, *J. Biol. Chem.* **260:**3584–3593.

Nathans, J., Thomas, D., and Hogness, D. S., 1986, Molecular genetics of human color vision: The genes encoding blue, green, and red pigments, *Science* **232:**193–202.

Neer, E. J., Lok, J.M., and Wolf, L. G., 1984, Purification and properties of the inhibitory guanine nucleotide regulatory unit of brain adenylate cyclase, *J. Biol. Chem.* **259:**14222–14229.

Nukada, T., Tanabe, T., Takahashi, H., Noda, M., Hirose, T., Inayama, S., and Numa, S., 1986a, Primary structure of the α-subunit of bovine adenylate cyclase-stimulating G-protein deduced from the cDNA sequence, *FEBS Lett.* **195:**220–224.

Nukada, T., Tanabe, T., Takahashi, H., Noda, M., Haga, K., Haga, T., Ichiyama, A., Kangawa, K., Hiranaga, M., Matsuo, H., and Numa, S., 1986b, Primary structure of the α-subunit of bovine adenylate cyclase-inhibiting G-protein deduced from the cDNA sequence, *FEBS Lett.* **197:**305–310.

Okajima, Fumikazu, and Ui, M., 1984, ADP ribosylation of the specific membrane protein by islet-activating protein, pertussis toxin, associated with inhibition of a chemotactic peptide-induced arachidonate release in neutrophils, *J. Biol. Chem.* **259:**13863–13871.

Owens, J. R., Frame, L. T., Ui, M., and Cooper, D. M. F., 1985, Cholera toxin ADP-ribosylates and islet-activating protein substrate in adipocyte membranes and alters its function, *J. Biol. Chem.* **260:**15946–15952.

Pellman, D., Garber, E. A., Cross, F. R., and Hanafusa, H., 1985, An N-terminal peptide from p60src can direct myristylation and plasma membrane localization when fused to heterologous proteins, *Nature* **314:**374–377.

Pfeuffer, E., Dreher, R.-M., Metzger, H., and Pfeuffer, T., 1985, Catalytic unit of adenylate cyclase: Purification and identification by affinity crosslinking, *Proc. Natl. Acad. Sci. U.S.A.* **82:**3086–3090.

Robishaw, J. D., Smigel, M. D., and Gilman, A. G., 1986, Molecular basis for two forms of the G protein that stimulates adenylate cyclase, *J. Biol. Chem.* **261:**9587–9590.

Rodbell, M., 1980, The role of hormone receptors and GTP-regulatory proteins in membrane transduction, *Nature* **284:**17–21.

Rodbell, M., 1985, Programmable messengers: A new theory of hormone action, *Trends Biochem. Sci.* **119:**461–464.

Roof, D. J., Applebury, M. L., and Sternweis, P. C., 1985, Relationships within the family of

GTP-binding proteins isolated from bovine central nervous system, *J. Biol. Chem.* **260:**16242–16249.

Ross, E. M., Howlett, A. C., Ferguson, K. M., and Gilman, A. G., 1978, Reconstitution of hormone-sensitive adenylate cyclase activity with resolved components of the enzyme, *J. Biol. Chem.* **253:**6401–6412.

Seamon, K. B., Vaillancourt, R., Edwards, M., and Daly, J. W., 1984, Binding of [^{3}H]forskolin to rat brain membranes, *Proc. Natl. Acad. Sci. U.S.A.* **81:**5081–5085.

Shinozawa, T., Uchida, S., Martin, E., Cafiso, D., Hubbell, W., and Bitensky, M., 1980, Additional Component required for activity and reconstitution of light-activated vertebrate photoreceptor GTPase, *Proc. Natl. Acad. Sci. U.S.A.* **77:**1408–1411.

Smith, C. D., Lane, B. C., Kusaka, I., Verghese, M. W., and Snyderman, R., 1985, Chemoattractant receptor-induced hydrolysis of phosphatidyl 4,5-bisphosphate in human polymorphonuclear leukocyte membranes. Requirement for a guanine nucleotide regulatory protein, *J. Biol. Chem.* **260:**5875–5878.

Spiegel, A. M., 1987a, Signal transduction by guanine nucleotide binding protiens, *Mol. Cell. Endocrinol.* **49:**1–16.

Spiegel, A. M., 1987b Guanine nucleotide binding proteins and signal transduction, *Vitam. Horm.* **44** (in press).

Spiegel, A. M., Gierschik, P., Levine, M. A., and Downs, R. W., Jr., 1985, Clinical implications of guanine nucleotide-binding proteins as receptor–effector couplers, *N. Engl. J. Med.* **312:**26–33.

Sternweis, P. C., 1985, The purified α subunits of G_o and G_i from bovine brain require β-γ for association with phospholipid vesicles, *J. Biol. Chem.* **261:**631–637.

Sternweis, P. C., and Gilman, A. G., 1982, Aluminum: A requirement for activation of the regulatory component of adenylate cyclase by fluoride, *Proc. Natl. Acad. Sci. U.S.A.* **79:**4888–4891.

Sternweis, P. C., and Robishaw, J. D., 1984, Isolation of two proteins with high affinity for guanine nucleotides from membranes of bovine brain, *J. Biol. Chem.* **259:**13806–13813.

Sternweis, P. C., Northup, J. K., Smigel, M. D., and Gilman, A. G., 1981, The regulatory component of adenylate cyclase, *J. Biol. Chem.* **256:**11517–11526.

Stryer, L., 1986, Cyclic GMP cascade of vision, *Annu. Rev. Neurosci.* **9:**87–119.

Sullivan, K. A., Liao, Y.-C, Alborzi, A., Beiderman, B., Chang, F.-H., Masters, S. B., Levinson, A. D., and Bourne, H. R., 1986, Inhibitory and stimulatory G proteins of adenylate cyclase: cDNA and amino acid sequences of the α chains, *Proc. Natl. Acad. Sci. U.S.A.* **83:**6687–6691.

Tanabe, T., Nukada, T., Nishikawa, Yl, Sugimoto, K., Suzuki, H., Takahashi, H., Noda, M., Haga, T., Ichiyama, A., Kangawa, K., Minamino, N., Matsuo, H., and Numa, S., 1985, Primary structure of the α subunit of transducin and its relationship to ras proteins, *Nature* **315:**242–245.

Ui, M., 1984, Islet-activating protein, pertussis toxin: A probe for functions of the inhibitory guanine nucleotide regulatory component of adenylate cyclase, *Trends Pharmacol. Sci.* **5:**277–279.

Verghese, M. W., Fox, K., McPhail, L. C., and Snyderman, R., 1985, Chemoattractant-elicited alterations of cAMP levels in human polymorphonuclear leukocytes require a CA^{++}-dependent mechanism which is independent of transmembrane activation of adenylate cyclase, *J. Biol. Chem.* **260:**6769–6775.

Verghese, M., Uhing, R. J., and Snyderman, R., 1986, A pertussis/cholera-toxin-sensitive N protein may mediate chemoattractant receptor signal transduction, *Biochem. Biophys. Res. Commun.* **138:**887–894.

West, R. E., Jr., Moss, J., Vaughan, M., Liu, T., and Liu, T.-Y., 1985, Pertussis toxin-catalyzed ADP ribosylation of transducin. Cysteine 347 is the ADP-ribose acceptor site, *J. Biol. Chem.* **260:**14428–14430.

Williamson, J. R., 1986, Role of inositol lipid breakdown in the generation of intracellular signals, *Hypertension* **8:**II140–II156.

Willumsen, B. M., Norris, K., Papageorge, A. G., Hubbert, N. L., and Lowy, D. R., 1984, Harvey murine sarcoma virus p21 ras protein: Biological and biochemical significance of the cysteine nearest the carboxy terminus, *EMBO J.* **3:**2581–2585.

Yarden, Y., Rodriguez, H., Wong, S. K.-F., Brandt, D. R., May, D. C., Burnier, J., Harkins, R. N., Chen, E. Y., Ramachandran, J., Ullrich, A., and Ross, E. M., 1986, The avian β-adrenergic receptor: Primary structure and membrane topology, *Proc. Natl. Acad. Sci. U.S.A.* **83:**6795–6799.

Yatani, A., Codina, J., Brown, A. M., and Birnaumer, L., 1987, Direct activation of mammalian atrial muscarinic potassium channels by GTP regulatory protein G_k, *Science* **235:**207–211.

Zick, Y., Eisenberg, R. S., Pines, M., Gierschik, P., and Spiegel, A. M., 1986, Multi-site phosphorylation of the α subunit of transducin by the insulin receptor kinase and protein kinase C, *Proc. Natl. Acad. Sci. U.S.A.* **83:**9294–9297.

12

Regulation of Phosphoinositide Breakdown

JOHN H. EXTON

1. Introduction

Many hormones, neurotransmitters, and other molecules involved in intercellular communication exert their biological actions by activating a phospholipase C that catalyzes the breakdown of a specific phospholipid (phosphatidylinositol 4,5-bisphosphate or PIP_2) in the plasma membrane of their target cells. The breakdown of PIP_2 yields inositol 1,4,5-trisphosphate (IP_3), which enters the cytosol, and 1,2-diacylglycerol (DAG), which remains in the membrane. IP_3 has a second messenger role in that it rapidly releases Ca^{2+} ions from components of the endoplasmic reticulum, causing a rise in cytosolic Ca^{2+}. This rise is responsible for many of the physiological responses observed, through the mediation of calmodulin and other Ca^{2+}-binding proteins (Fig. 1). DAG is also a second messenger because it activates the Ca^{2+}-phospholipid-dependent protein kinase C. This phosphorylates a number of cellular proteins contributing to the responses (Fig. 1).

The list of agonists that act through PIP_2 breakdown and Ca^{2+} mobilization includes epinephrine and norepinephrine, acetylcholine, vasopressin, angiotensin II, histamine, serotonin, cholecystokinin, bradykinin, substance P, chemotactic peptides, and certain hypothalamic releasing hormones and platelet activating factors. The receptors to these agonists are located on the outer surface of their target cells and are widely distributed throughout the body. In many cases, as with catecholamines, acetylcholine, histamine, and vasopressin, the agonists have more than one mechanism of action. How-

John H. Exton • Howard Hughes Medical Institute and Department of Molecular Physiology and Biophysics, Vanderbilt University School of Medicine, Nashville, Tennessee 37232.

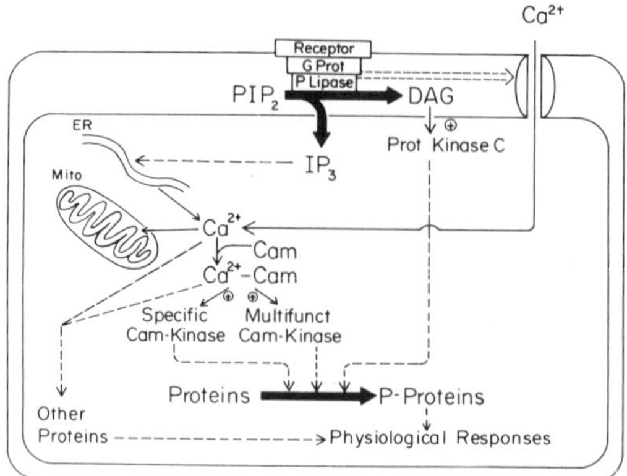

Figure 1. Generalized scheme by which Ca^{2+}-mobilizing agonists exert their physiological effects. Abbreviations: G Prot, guanine nucleotide binding protein; P lipase, phospholipase C; PIP_2, phosphatidylinositol 4,5-bisphosphate; IP_3, inositol 1,4,5-trisphosphate; DAG; 1,2-diacylglycerol; Prot Kinase C, Ca^{2+}-phospholipid-dependent protein kinase; ER, endoplasmic reticulum; Mito, mitochondrion; Cam, calmodulin; Cam-Kinase, specific or multifunctional Ca^{2+}-calmodulin-dependent protein kinase.

ever, their effects on phosphoinositide metabolism are mediated by specific receptor subtypes.

This chapter is mainly concerned with the mechanisms involved in coupling the receptors for Ca^{2+}-mobilizing agonists to the phospholipase C that hydrolyzes PIP_2. Evidence is presented to show that this involves a guanine nucleotide binding protein analogous to those (G_s and G_i) that couple stimulatory and inhibitory receptors to adenylate cyclase. The data shown are derived primarily from studies with liver, which possesses receptors for at least five different Ca^{2+}-mobilizing agonists.

2. *Phosphoinositide Metabolism*

Michell (1975, 1979) first emphasized the association between actions of certain hormones on Ca^{2+} fluxes and the turnover of phosphoinositides in a variety of tissues. In particular, this was pointed out for α-adrenergic agonists (Jones and Michell, 1978). The initial studies demonstrated that the agonists increased both the synthesis and breakdown of phosphatidylinositol (PI) as measured by labeling studies with $^{32}P_i$ (Jones and Michell, 1978). These observations were confirmed using [^{3}H]*myo*inositol (Tolbert *et al.*, 1980; Prpic *et al.*, 1982). However, the turnover of PI induced by α_1-adrenergic agonists or other Ca^{2+}-mobilizing agents, although rapid, was not fast enough to be responsible for the physiological responses (Canessa de Scarnatti and Lapetina, 1974; Kirk *et al.*, 1977, 1981; Billah and Michell, 1979; Prpic *et al.*, 1982; Uchida *et al.*, 1982).

Early observations by Schacht and Agranoff (1972) and Abdel-Latif *et al.* (1977) indicated that Ca^{2+}-mobilizing agonists also stimulated the breakdown of phosphatidylinositol 4,5-P_2 (PIP_2)in neural and smooth muscle tissue. The group of Kirk and Michell later demonstrated that the breakdown of the polyphosphoinositide induced by these agonists in liver occurred much more rapidly than that of PI (Kirk *et al.*, 1981; Michell *et al.*, 1981; Creba *et al.*, 1983). This has now been confirmed by others in liver (Litosch *et al.*, 1983; Rhodes *et al.*, 1983; Thomas *et al.*, 1983) and a variety of other tissues (Billah and Lapetina, 1982; Weiss *et al.*, 1982; Agranoff *et al.*, 1983; Downes and Wusteman, 1983; Martin, 1983; Mauco *et al.*, 1983; Putney *et al.*, 1983; Rebecchi and Gershengorn, 1983; Volpi *et al.*, 1983; Yano *et al.*, 1983; Dougherty *et al.*, 1984; Wirthensohn *et al.*, 1984).

The significance of the breakdown of PIP_2 induced by Ca^{2+}-mobilizing agonists became clear when Berridge and associates measured the changes in one of the products, *myo*inositol 1,4,5-$_3$ (IP_3), in various tissues (Berridge, 1983; Berridge *et al.*, 1983) and when Streb *et al.* (1983) showed that this compound released Ca^{2+} from nonmitochondrial stores in permeabilized pancreatic acinar cells. Since that time, IP_3 has been shown to release internal Ca^{2+} in permeabilized liver cells (Fig. 2; Burgess *et al.*, 1984a,b; Joseph *et al.*, 1984a; Joseph and Williamson, 1986), insulin-secreting cells (Biden *et al.*, 1984; Joseph *et al.*, 1984b; Prentki *et al.*, 1985; Wolf *et al.*, 1985), smooth muscle cells (Suematsu *et al.*, 1984; Somlyo *et al.*, 1985b), platelets Authi and Crawford, 1985; Brass and Joseph, 1985; O'Rourke *et al.*, 1985), neutrophils (Prentki *et al.*, 1984b), 3T3 fibroblasts (Irvine *et al.*, 1984a), macrophages (Hirata *et al.*, 1985), pituitary cells (Gershengorn *et al.*, 1984; Biden *et al.*, 1986), leukocytes (Burgess *et al.*, 1984c), NIE-115 neuronal cells (Chueh and Gill, 1986), adipocytes (Delfert *et al.*, 1986), kidney cortex cells (Thevenod *et al.*, 1986), and adrenal chromaffin cells (Stoehr *et al.*, 1986). The action of IP_3 is extremely rapid and is observed with submicromolar concentrations (Fig. 2), that is, those calculated or measured to exist intracellularly (Thomas *et al.*, 1984; Charest *et al.*, 1985; Rittenhouse and Sasson, 1985). It is transient because of the rapid metabolism of IP_3 (Prentki *et al.*, 1985; Streb *et al*). *myo*Inositol 2,4,5-P_2 and *myo*inositol 4,5-P_2 also release Ca^{2+} but are, respectively, approximately 10- and 100-fold less potent than IP_3. *myo*Inositol 1,4-P_2 and *myo*inositol 1-P are ineffective (Streb *et al.*, 1983; Burgess *et al.*, 1984b; Irvine *et al.*, 1984a; Wolf *et al.*, 1985). *myo*Inositol 1,2-(cyclic)-4,5-trisphosphate is of similar potency to IP_3, whereas *myo*inositol 1,3,4-trisphosphate is about 30-fold less potent and *myo*inositol 1,3,4,5-tetrakisphosphate is ineffective (Irvine *et al.*, 1986b).

The intracellular pool from which Ca^{2+} is released by IP_3 cosediments with mitochondria and microsomes (Dawson and Irvine, 1984; Prentki *et al.*, 1984; Streb *et al.*, 1984; Delfert *et al.*, 1986), but there is much evidence that it is not mitochondrial (Streb *et al.*, 1983, 1984; Gershengorn *et al.*, 1984; Joseph *et al.*, 1984a,b; Biden *et al.*, 1986; Thevenod *et al.*, 1986). Enzyme marker measurements in subcellular fractions of rat exocrine pancreas and

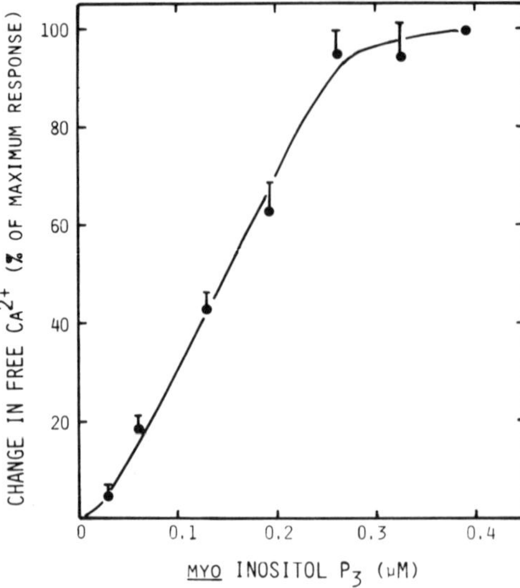

Figure 2. Effect of IP_3 to release Ca^{2+} from internal stores in permeabilized isolated rat hepatocytes. Hepatocytes were permeabilized with 0.015% digitonin and incubated in buffered medium containing the fluorescent Ca^{2+} indicator Quin 2, 110 mM KC1, 5 mM $MgCl_2$, 5 mM succinate, 5 mM pyruvate, 5 mM ATP, and an ATP-regenerating system. IP_3 was then added to give the concentration shown, and the peak fluorescence was measured over the next 10–30 s. The data are means from three experiments and are from unpublished studies by P. Thiyagarajah and P. F. Blackmore.

platelets indicate codistribution of NADPH cytochrome *c* reductase and RNA with the IP_3-sensitive pool (Streb *et al.*, 1984; Authi and Crawford, 1985). This is consistent with its location in the rough endoplasmic reticulum. However, several studies have shown that only a small fraction of the endoplasmic reticulum responds to IP_3 (Dawson and Irvine, 1984; Joseph *et al.*, 1984b; Prentki *et al.*, 1984a; Biden *et al.*, 1986).

The action of IP_3 on the Ca^{2+} store appears to be on Ca^{2+} efflux rather than uptake, but the mechanism is unknown. It is relatively insensitive to temperature (J. B. Smith, *et al.*, 1985; Chueh and Gill, 1986; Henne and Soling, 1986; Joseph and Williamson, 1986), which suggests that it involves a Ca^{2+} channel rather than a carrier. It also requires the countermovement of K^+ or another monovalent cation (Muallem *et al.*, 1985; Joseph and Williamson, 1986) and is inhibited by high concentrations of $C1^-$ or other anions. These findings indicate that the Ca^{2+} release process is electrogenic, but it is unlikely that an anion exchange mechanism or a cation/anion cotransport system is involved, since the release is insensitive to DIDS (4,4′-dithiocyanostilbene-2,2′−-disulfonic acid) or furosemide (Joseph and Williamson, 1986). Voltage-dependent Ca^{2+} channel blockers and agonists are also without effect (Henne and Soling, 1986).

There have been some recent reports of IP_3 binding to microsomal fractions from liver, adrenal cortex, and anterior pituitary (Baukal *et al.*, 1985; Spat *et al.*, 1986; Guillemette *et al.*, 1987). Some very-high-affinity (K_d,1–

10 nM), low-capacity binding sites have been identified, but it has not been convincingly demonstrated that these are responsible for Ca^{2+} mobilization.

In addition to mediating the effects of certain hormones on Ca^{2+} mobilization, IP_3 may act as a chemical messenger between transverse (T)-tubular membrane depolarization and Ca^{2+} release from sarcoplasmic reticulum in skeletal muscle (Vergara *et al.*, 1985; Volpe *et al.*, 1985; Nosek *et al.*, 1986; Thieleczek and Heilmeyer, 1986). There is also evidence that it is involved in light-induced excitation and adaptation in *Limulus* photoreceptors (Brown and Rubin, 1984; Brown *et al.*, 1984; Fein *et al.*, 1984) and in fertilization in sea urchins and *Xenopus* (Busa *et al.*, 1985; Oron *et al.*, 1985; Slack *et al.*, 1986).

Since the original reports of Berridge and associates (Berridge, 1983; Berridge *et al.*, 1983), various Ca^{2+}-mobilizing agonists have been shown to increase IP_3 in many tissues, inlcuding liver (Thomas *et al.*, 1984; Charest *et al.*, 1985), brain (Berridge *et al.*, 1983), platelets (Agranoff *et al.*, 1983; Vickers *et al.*, 1984; Rittenhouse and Sasson, 1985), salivary glands (Berridge, 1983; Berridge *et al.*, 1983; Downes and Wustemann, 1983; Aub and Putney, 1984, 1985; Irvine *et al.*, 1985), pituitary (Martin, 1983; Rebecchi and Gershengorn, 1983; Enjalbert *et al.*, 1986), exocrine pancreas (Rubin *et al.*, 1984), endocrine pancreas (Morgan *et al.*, 1985), Swiss 3T3 cells (Berridge *et al.*, 1984), adrenal cortex (Gallo-Payet *et al.*, 1986), smooth muscle cells (Akhtar and Abdel-Latif, 1984; J. B. Smith *et al.*, 1984), endothelial cells (Lambert *et al.*, 1986), lymphocytes (Imboden and Stobo, 1985), gastric mucosal cells (Baudiere *et al.*, 1986), astrocytoma cells (Masters *et al.*, 1985), PC12-phaeochromocytoma cells (Wincentini *et al.*, 1985), and adipocytes (Nanberg and Putney, 1986). In contrast to all these reports, there has been only one study indicating an inhibitory effect of an agonist (dopamine) on inositol phosphate production (Enjalbert *et al.*, 1986).

The increase in IP_3 induced by agonists occurs within a few seconds and is approximately coincident with the rise in cytosolic Ca^{2+} (Fig. 3). The concentrations of agonists that produce half-maximal changes in PIP_2 or IP_3 are similar to their K_d values for binding to their receptors (Creba *et al.*, 1983; Lynch *et al.*, 1985a). In addition, the maximum generation of IP_3 by agonists is proportional to their maximum binding sites (Lynch *et al.*, 1985a). These findings imply a close relation between receptor occupancy and phosphoinositide breakdown. However, because of the presence of spare receptors in most cells, small increases in IP_3 can elicit large physiological responses (Creba *et al.*, 1983; Rhodes *et al.*, 1983; Thomas *et al.*, 1984; Aub and Putney, 1985; Charest *et al.*, 1985; Lynch *et al.*, 1985a). This is not always the case, since in some cells the receptor reserve for α_1-adrenergic agonists may be small or nonexistent compared with that for other agonists (El-Refai *et al.*, 1979; Ambler *et al.*, 1984; Amitai *et al.*, 1984; Lynch *et al.*, 1985a).

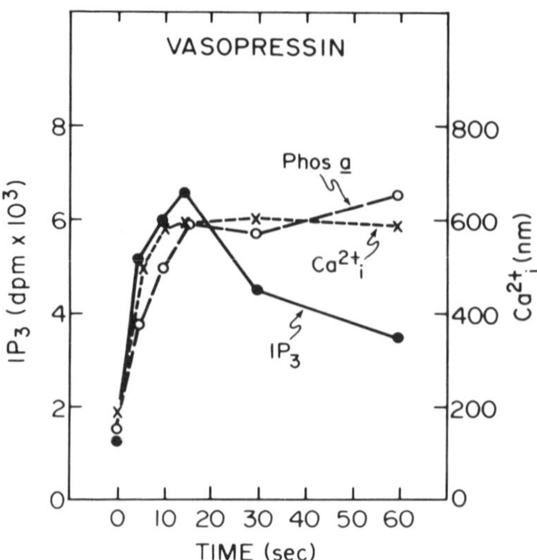

Figure 3. Time courses of increases in IP_3, cytosolic Ca^{2+}, and phosphorylase *a* in isolated rat hepatocytes incubated with vasopressin. Hepatocytes were incubated for 90 min with [^{3}H]*myo*inositol to label the inositol phospholipids and then for a further 15 min with or without the fluorescent Ca^{2+} indicator Quin 2. Aliquots were then taken for incubations with 100 nM vasopressin to measure the changes in [^{3}H]IP_3, cytosolic Ca^{2+}, and phosphorylase *a* using the methods described in Charest *et al.* (1985). The data are from unpublished studies by P. F. Blackmore.

PIP_2 is a minor phospholipid that is present mainly in the inner leaflet of the plasma membrane (Michell, 1975). The inositol phospholipids represent 2–12% of the total phospholipids of most cells, and PIP_2 represents a minor fraction of the total inositol phospholipids (Michell, 1975). The synthesis of PIP_2 involves the sequential conversion of PI to phosphatidylinositol 4-phosphate (PIP) and PIP_2 by ATP-requiring kinases (Fig. 4; Michell, 1975; Berridge, 1984). Phosphomonoesterases that reverse these reactions are also present in the plasma membrane. As alluded to already, the reaction primarily stimulated by Ca^{2+}-mobilizing agents is the breakdown of PIP_2 to inositol 1,4,5-trisphosphate (IP_3) and 1,2-diacylglycerol (DAG) under the influence of a Ca^{2+}-dependent phosphodiesterase or phospholipase C. This hormone-sensitive phospholipase C may also have some effect on phosphatidylinositol 4-P (PIP), but it does not act on PI (Aub and Putney; Downes and Wusteman, 1983; Martin, 1983; Uhing *et al.*, 1985, 1986; Taylor and Exton, 1987). The loss of PI observed in early experiments was presumably due to the accelerated conversion of PI to PIP and then PIP_2, to replace that which had been broken down by the phospholipase. To date, there is no evidence that the kinases and phosphatases involved in PIP_2 metabolism (Fig. 4) are regulated by agonists, and the increased conversion of PI to PIP_2 is thought to occur through mass action. However, the transient decrease in PIP_2 renders this rather unlikely.

IP_3 produced by cleavage of PIP_2 is metabolized by two alternative pathways involving a phosphatase and a kinase. As shown in Figs. 4 and 5, IP_3 is rapidly degraded to *myo*inositol 1,4-P_2 (IP_2) by a specific 5-phospho-

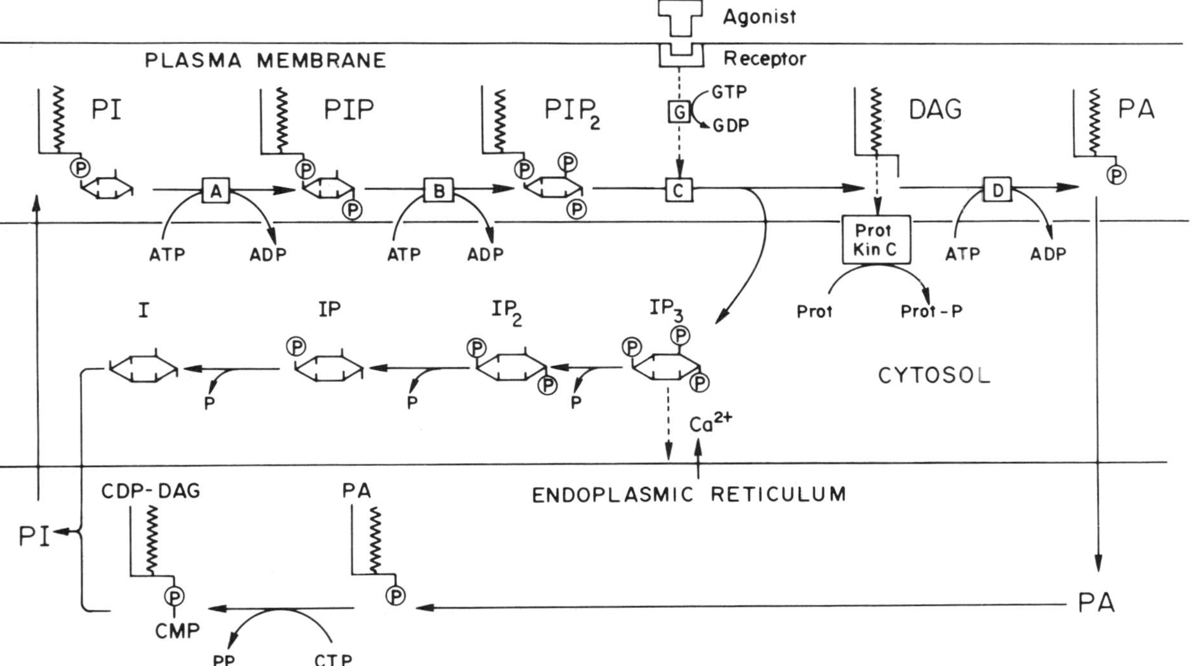

Figure 4. Generalized scheme showing the metabolism of phosphoinositides in cells. Abbreviations not defined in the text are: G, guanine nucleotide binding protein; A, phosphatidylinositol kinase; B, phosphatidylinositol 4-P kinase; C, phosphatidylinositol 4,5-P_2 phospholipase C; D, 1,2-diacylglycerol kinase; CDP-DAG, CDP-diacylglycerol; IP_2, *myo*inositol 1,4-P_2; IP, *myo*inositol 1-P; I, *myo*inositol. Recent evidence now suggests that *myo*inositol 4-P rather than *myo*inositol 1-P is the product of *myo*inositol 1, 4-P_2 metabolism.

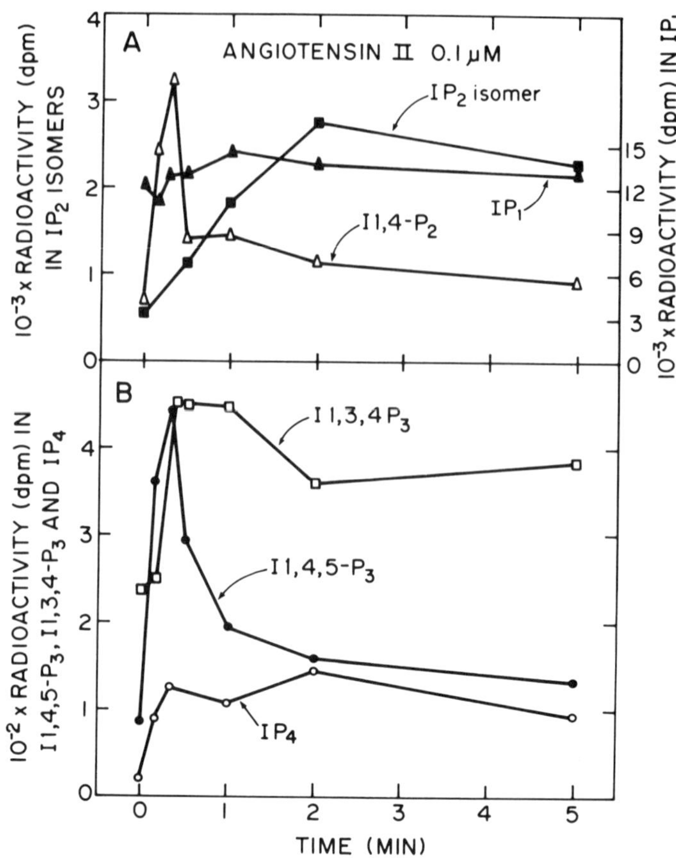

Figure 5. Effects of angiotensin II on inositol polyphosphates in isolated rat hepatocytes. Hepatocytes were initially incubated with [^{3}H]*myo*inositol for 90 min to label the inositol phospholipids and then with 0.1 μM angiotensin II for the times indicated. The radioactivity in the inositol polyphosphates was determined as described by Irvine *et al.* (1985).

monoesterase found in the plasma membrane or soluble phase (Downes *et al.*, 1982; Seyfred *et al.*, 1984; Storey *et al.*, 1984; Joseph and Williams, 1985; Connolly *et al.*, 1985). *myo*Inositol 1,4-P_2 is then sequentially degraded to *myo*inositol 4-P, and *myo*inositol by soluble phosphomonoesterases.

In addition to IP_3, *myo*inositol 1,3,4-P_3 and *myo*inositol 1,3,4,5-P_4 (IP_4) have been found in tissues exposed to Ca^{2+}-mobilizing agents (Fig. 5; Irvine *et al.*, 1984c, 1985, 1986a; Batty *et al.*, 1985; Heslop *et al.*, 1986). IP_4 arises from IP_3 through the action of a 3-kinase and is then dephosphorylated to *myo*inositol 1,3,4-P_3 by the same 5-phosphomonoesterase that acts on IP_3 (Biden and Wollheim, 1986; Downes *et al.*, 1986; Hansen *et al.*, 1986; Irvine *et al.*, 1986a; Stewart *et al.*, 1986; Connolly *et al.*, 1987). IP_4 has been reported to activate sea urchin eggs, apparently by opening a plasma membrane Ca^{2+} channel (Irvine and Moor, 1986), but a physiological role for *myo*inositol 1,3,4-P_3 has not yet been defined (Higashida and Brown, 1986). The 3-kinase that converts IP_3 to IP_4 is stimulated by Ca^{2+} (Biden and Wollheim, 1986), which may account for the transiency of the elevation in IP_3 following agonist stimulation of cells and the delayed formation of *myo*inositol 1,3,4-P_3 (Lew *et al.*, 1986). The 5-phosphomonoesterase for IP_3 has also been reported to be phosphorylated and stimulated by protein kinase C (Connolly *et al.*, 1986, 1987). Activators of this kinase have been shown to stimulate the conversion of IP_3 to IP_2 in permeabilized plateletes (Molina y Vedia and Lapetina, 1986), but this has not been observed in other cells (Orellano *et al.*, 1987). Based on the K_m values of the kinase and phosphatase for IP_3 and IP_4, it would seem that formation of IP_4 with subsequent degradation to *myo*inositol 1,3,4-P_3 would be the preferred pathway of IP_3 metabolism (Irvine *et al.*, 1986a; Connolly *et al.*, 1987), especially in the presence of elevated Ca^{2+} (Biden and Wolheim, 1986; Lew *et al.*, 1986).

Inositol 1,2-cyclic 4,5-trisphosphate is also formed during phospholipase C action on PIP_2 *in vitro* (Wilson *et al.*, 1985) and has been reported to be generated during thrombin stimulation of platelets (Ishii *et al.*, 1986) and carbachol stimulation of pancreas (Sekar *et al.*, 1987), although its presence in other tissues is uncertain. It has been reported to have equal or greater potency than IP_3 in eliciting responses in several systems (Wilson *et al.*, 1985) and is degraded to other cyclic inositol phosphates by the same enzymes that degrade IP_3 (Connolly *et al.*, 1987). However, it is not a substrate for IP_3 kinase (Connolly *et al.*, 1987).

The metabolic fate of the DAG released by PIP_2 breakdown is less well defined. The current view is that it remains in the plasma membrane and is phosphorylated to phosphatidic acid (PA) by 1,2-diacylglycerol kinase (Fig. 4) or is hydrolyzed by lipases to monoacylglycerol, glycerol, and fatty acids (Berridge, 1986). Arachidonic acid released by lipase action is a major source of eicosanoids in many tissues. It is believed that much of the PA generated by phospholipid breakdown in the plasma membrane is transferred by carrier proteins to the endoplasmic reticulum (ER), where it is utilized for the resynthesis of PI or certain other phospholipids by reaction with CTP to form CDP-diacylglycerol (CDP-DAG) under the action of CDP-diacylglycerol synthetase (Fig. 4; Michell, 1975). Alternative fates of PA are its rehydrolysis to DAG for utilization in the synthesis of triacylglycerol, phosphatidylcholine, or phosphatidylethanolamine. In general, the origins and metabolic fates of the DAG and PA species generated during agonist stimulation of most cells are poorly understood.

*myo*Inositol that is formed from the breakdown of inositol phospholipids, transported into the cell, or synthesized from glucose is incorporated into PI by two different reactions (Michell, 1975). The first is the reaction with CDP-DAG to form PI and CMP, which is catalyzed by CDP-diacylglycerol : inositol transferase (Fig. 4). The second is the exchange reaction with preformed PI or other phospholipids with release of the corresponding base. PI formed in the ER is transferred to the plasma membrane by an exchange protein (Michell, 1975).

The inositol phospholipids are enriched in stearic and arachidonic acids in the *sn*-1 and *sn*-2 positions of glycerol, respectively (Holub and Kuksis, 1978). Since the species of CDP-DAG available for their synthesis do not necessarily have this composition, unless there is compartmentation, the inositol transferase enzyme may have some specificity with regard to the fatty acid composition of the CDP-DAG utilized. However, some acyl exchange probably also occurs (Holub and Kuksis, 1978). This can occur by deacylation–reacylation reactions catalyzed by enzymes such as phospholipase A_2 and 1-acyl-*sn*-glycero-3-phosphorylinositol acyl transferase, which has a high preference for arachidonic acid in liver (Holub and Kuksis, 1978).

Although current concepts localize the site of PI synthesis to the ER, there is new evidence that some synthesis can also occur in the plasma membrane (Imai and Gershengorn, 1987). This could eliminate the need for altering the fatty acids of newly synthesized PI.

3. Receptors for Calcium-Mobilizing Agonists

The pharmacological characteristics and tissue distribution of many of the receptors for calcium-mobilizing agonists have been defined on the basis of radioligand binding studies and physiological responses, but the physicochemical features of these receptors are largely unknown. The α_1-adrenergic receptors have been the most studied. By using a variety of labeled agonists and antagonists, these receptors have been identified in brain, lung, liver, kidney, heart, uterus, adipose tissue, vas deferens, salivary glands, and certain blood vessels (Bylund and U'Prichard, 1983). Photoaffinity probes have also been employed to label specifically the α_1-adrenergic receptor of liver and other tissues (Leeb-Lundberg *et al.*, 1984; Seidman *et al.*, 1984; Lynch *et al.*, 1986a). When precautions are taken to limit proteolysis, 78–85-kD binding subunits are identified by polyacrylamide gel electrophoresis in sodium dodecylsulfate. Radiation inactivation analysis carried out in rat liver membranes also indicates that the receptor exists as a dimer with subunits of approximately 85 kD (Venter *et al.*, 1984).

The α_1-adrenergic receptor from different tissues has been purified to varying degrees using affinity chromatography (Graham *et al.*, 1982; Leeb-Lundberg *et al.*, 1985; Lomasney *et al.*, 1986). The latter group has purified the receptor from DDT_1MF-2 smooth muscle cells to apparent homogeneity using a prazosin analogue linked to agarose, wheat germ agglutinin-agarose, and high-performance liquid chromatography. The purified receptor has a subunit of 80,000 M_r containing a single ligand binding site, which has the appropriate specificity, stereoselectivity, and affinity for α_1-adrenergic agonists and antagonists. The M_r of the subunit is thus similar to those of other α_1-adrenergic receptors. Interestingly, peptide maps of the purified receptor using several proteases reveal little structural homology with the α_2-adrenergic receptor (Lomasney *et al.*, 1986).

There is little doubt that it is the α_1-subtype of the adrenergic receptor that mediates the effects of catecholamines on Ca^{2+}-mobilization and phosphoinositide metabolism. This is also true for the H-1 histamine and P_2-purinergic receptors. However, the muscarinic receptor subtype that mediates the effects of acetylcholine on these parameters is not clear. Pirenzepine, which has a higher affinity for M_1- than M_2-muscarinic receptors, has not clearly distinguished the subtype involved in phosphoinositide breakdown in several tissues (Fisher, 1986; Harden *et al.*, 1986). The serotonin receptor mediating this response is of the S-2 subtype in most, but not all, tissues (de Chaffoy de Courcelles *et al.*, 1985; Conn and Sanders-Bush, 1985), whereas the vasopressin receptor subtype linked to PIP_2 breakdown has been defined operationally as V_1 (Michell *et al.*, 1979). It is likely that the angiotensin II receptor mediating phosphoinositide hydrolysis is different from that mediating adenylate cyclase inhibition, but this has not been demonstrated.

Cross-linking and photoaffinity labeling experiments utilizing liver, uterus, adrenal and cardiac plasma membranes, and [^{125}I]angiotensin II have yielded labeled proteins of 65, 68, and 116 kD, as determined by sodium dodecylsulfate polyacrylamide gel electrophoresis (Capponi and Catt, 1980; Sen *et al.*, 1983; Rogers, 1984). The hepatic V_1 vasopressin receptor has a ligand binding subunit of 83 kD as determined by gel filtration and sucrose density gradient centrifugation (Guillon *et al.*, 1980), but photoaffinity labeling using both agonists and antagonists has yielded proteins of 30 and 38 kD on gel electrophoresis in sodium dodecylsulfate (Boer and Fahrenholz, 1985; Fahrenholz *et al.*, 1986).

4. *Role of Guanine Nucleotide Binding Proteins in Agonist Regulation of Phosphoinositide Breakdown*

Several lines of evidence have given rise to the conclusion that guanine nucleotide binding proteins (G proteins) are involved in coupling the receptors for Ca^{2+}-mobilizing agonists to the phospholipase C that hydrolyzes PIP_2. The first is that the binding of these agonists to their receptors is modulated by GTP and its poorly hydrolyzed analogues. As illustrated initially for α_1-adrenergic agonists (El-Refai *et al.*, 1979; Geynet *et al.*, 1980), the nucleotides convert the receptors from a high to a low agonist affinity state (Fig. 6; Lynch *et al.*, 1985b, 1986b; Exton, 1987). This phenomenon is well known for other receptors that are coupled to G proteins, that is, those linked to adenylate cyclase.

Further evidence for the involvement of G proteins in the actions of Ca^{2+}-mobilizing agonists is that these agonists stimulate a low K_m GTPase activity in the plasma membranes of their target cells (Hinkle and Phillips, 1984; Fitzgerald *et al.*, 1986; Grandt *et al.*, 1986; Higashida *et al.*, 1986) or increase the binding/exchange of a GTP analogue in these membranes (Lad *et al.*, 1985). Furthermore, in permeabilized mast cells and platelets, nonhydrolyzable analogues of GTP elicit Ca^{2+}-dependent exocytotic secretion (Gomperts, 1983; Haslam and Davidson 1984a, 1984b). In addition, in liver cells, NaF stimulates the breakdown of PIP_2 to IP_3 and DAG, with resultant increase in cytosolic Ca^{2+} and physiological responses (Blackmore *et al.*, 1985). Similar effects have been observed in neutrophils, GH_3 pituitary cells, WRK1 cells, platelets, fibroblasts, and astrocytoma cells (Brass *et al.*, 1986; Guillon *et al.*, 1986; Hepler and Harden, 1986; Martin *et al.*, 1986a; Strnad *et al.*, 1986; Paris and Pouyssegur, 1987). These effects are potentiated by $AlCl_3$, and AlF_4^- appears to be the active molecule. AlF_4^- is known to modulate the activity of other G proteins (Sternweis and Gilman, 1982; Katada *et al.*, 1984; Kanaho *et al.*, 1985).

More direct evidence for a role of G proteins in the regulation of the phosphoinositide phospholipase C is provided by studies showing stimulatory effects of GTP and its analogues on the breakdown of endogenous or exogenous PIP_2 and PIP, but not PI, in isolated plasma membranes from liver (Uhing *et al.*, 1985, 1986; Wallace and Fain, 1985; Guillon *et al.*, 1986; Taylor and Exton, 1987), polymorphonuclear leukocytes (Cockcroft and Gomperts, 1985), salivary glands (Litosch *et al.*, 1985), astrocytoma cells (Hepler and Harden, 1986; Orellano *et al.*, 1987), pituitary cells (Lucas *et*

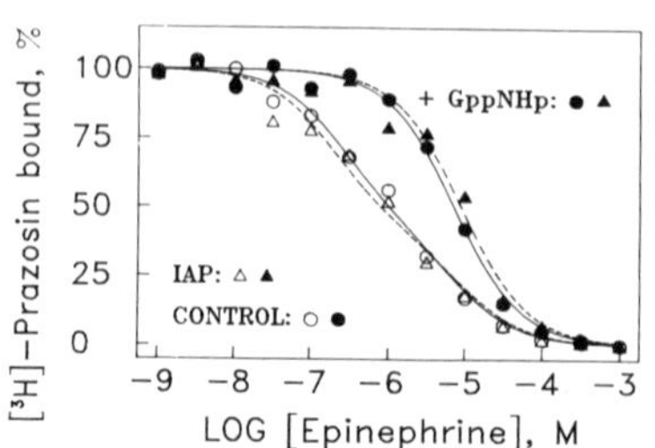

Figure 6. Effect of a GTP analogue on epinephrine binding to the α_1-adrenergic receptor in liver plasma membranes from control rats or rats injected with 25 μg of islet activating protein (IAP) 24 h previously. The ability of increasing concentrations of epinephrine to displace 1 nM [^{3}H]prazosin binding was measured as described by Lynch *et al.* (1986b). GppNHp is guanyl-5′-yl imidodiphosphate.

al., 1985; Martin *et al.*, 1986a; Straub and Gershengorn, 1986), and platelets (Baldassare and Fisher, 1986a,b). The effect is greatest with nonhydrolyzable analogues of GTP and is not seen with other nucleoside triphosphates, GDP or GMP (Fig. 7; Cockcroft and Gomperts, 1985; Litosch *et al.*, 1985; Uhing *et al.*, 1985, 1986; Wallace and Fain, 1985; Hepler and Harden, 1986; Martin *et al.*, 1986a; Taylor and Exton, 1987). Evidence that it is mediated by a G protein is provided by the facts that it is observed with micromolar concentrations of GTP analogues such as guanosine 5′-(3-*0*)-thiotriphosphate (GTPγS), is Mg^{2+} dependent, and is inhibited by guanosine 5′-(2-*0*-) thiodiphosphate (GDPβS) (Martin *et al.*, 1986a; Uhing, *et al.*, 1986; Taylor and Exton, 1987). The breakdown of PIP_2 is greater than that of PIP and requires the presence of 10 nM or higher free Ca^{2+} (Cockcroft and Gomperts, 1985; Uhing *et al.*, 1985, 1986; Taylor and Exton, 1987; see also Wallace and Fain, 1985). GTP analogues increase the phospholipase C activity at high (1 mM) Ca^{2+}, but more importantly, they increase the sensitivity of the enzyme to stimulation by Ca^{2+} (Bradford and Rubin, 1986; Uhing *et al.*, 1986; Taylor and Exton, 1987). With PIP_2 as substrate, IP_3 is the product formed initially, but this is degraded to IP_2 by the IP_3 phosphatase activity of the membranes (Uhing *et al.*, 1985; Wallace and Fain 1985; Taylor and Exton, 1987).

The direct effects of Ca^{2+}-mobilizing agonists on the breakdown of endogenous and exogenous polyphosphoinositides in membranes from a variety of cells have recently been observed (Litosch *et al.*, 1985; Lucas *et al.*, 1985; C. D. Smith *et al.*, 1985; Baldassare and Fisher, 1986a; Guillon *et al.*,

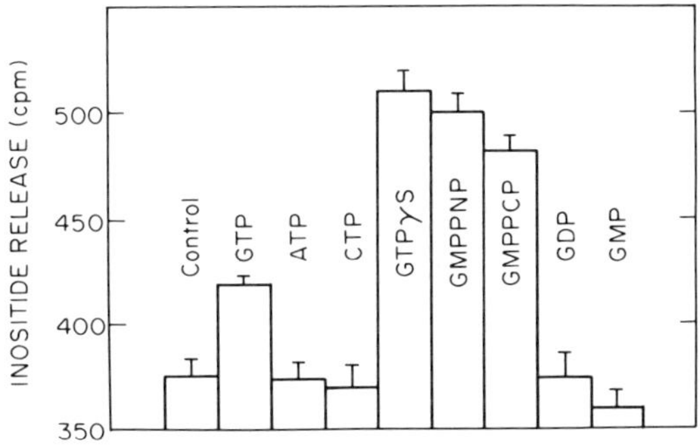

Figure 7. Effects of GTP and its analogues on the release of [^{3}H]inositol phosphates from liver plasma membranes prepared from rats injected 24 h previously with 500 μCi of [^{3}H]*myo*inositol. Membranes were incubated for 5 min with buffer containing 140 nM free Ca^{2+} and 100 μM concentration of the indicated nucleotides. A regenerating system was included with GTP. GMPPCP is guanyl-5′-yl-(β,γ-methylene) diphosphate. Other nucleotides are defined in the text or in the legend to Fig. 6. Reproduced from Uhing *et al.* (1985) by permission of the authors and publisher.

1986; Hepler and Harden, 1986; Martin *et al.*, 1986a; Straub and Gershengorn, 1986; Uhing *et al.*, 1986; Orellano *et al.*, 1987; Taylor and Exton, 1987). The effects are dependent on or potentiated by GTP or its analogues and are most evident at low concentrations of Ca^{2+} (<10 μM) (Fig. 8; Litosch and Fain, 1985; Uhing *et al.*, 1986; Martin *et al.*, 1986a; Hepler and Harden, 1986; Straub and Gershengorn, 1986; Taylor and Exton, 1987). The inositol phosphate formed initially is again IP_3, but this breaks down rapidly to IP_2 (Uhing *et al.*, 1985). The rate of formation of IP_3 is maximal within 1 min (Uhing *et al.*, 1986). The concentration dependence for agonist-induced inositide formation in liver, salivary gland, astrocytoma, or pituitary cell membranes is similar to that for IP_3 formation in the intact tissues (Litosch *et al.*, 1985; Hepler and Harden, 1986; Martin *et al.*, 1986a; Straub and Gershengorn, 1986; Uhing *et al.*, 1986). These observations differ from earlier reports of direct effects of catecholamines and vasopressin on phosphoinositide breakdown in isolated plasma membranes, which were observed in the absence of added guanine nucleotides (Lin and Fain, 1981; Wallace *et al.*, 1982, 1983; Harrington and Eichberg, 1983; Seyfred and Wells, 1984).

The formation of a high M_r hormone receptor G protein complex has been observed in liver plasma membranes treated with vasopressin and then solubilized with digitonin or another detergent and subjected to sucrose density gradient centrifugation, gel filtration, or chromatography on wheat germ lectin-agarose (Bojanic and Fain, 1986; Fitzgerald *et al.*, 1986). The complex is dissociated by GTP analogues and exhibits GTPase activity. The presence in the complex of a 35-kD β subunit common to several other G proteins has been demonstrated immunologically (Fitzgerald *et al.*, 1986).

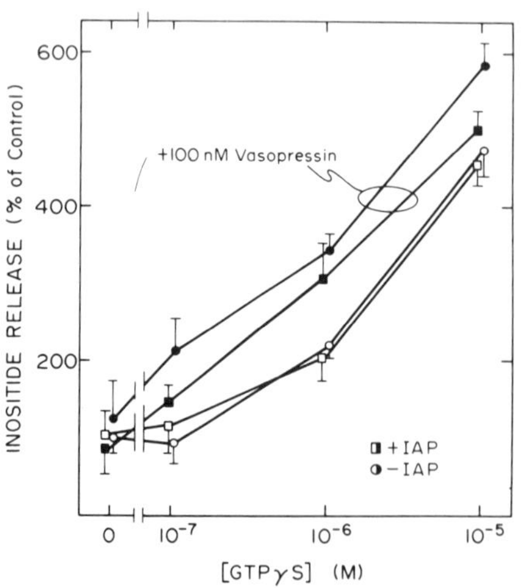

Figure 8. Effects of vasopressin and GTP analogue (GTPγS) to stimulate [^{3}H]inositol phosphate release from liver plasma membranes prepared from rats previously injected with [^{3}H]*myo*inositol with or without IAP. Plasma membranes were incubated for 10 min with buffer containing 10 mM Mg^{2+}, 300 nM free Ca^{2+}, the indicated GTPγS concentrations, and 100 nM vasopressin or no hormone. Plasma membranes were prepared from control rats or rats injected 24 h previously with 50 μg of IAP. All rats were injected with 500 μCi of [^{3}H]*myo*inositol 18 h prior to study. Reproduced from Uhing *et al.* (1986) by permission of the authors and publisher.

It is clear that more than one species of G protein is involved in polyphosphoinositide breakdown and Ca^{2+} mobilization, based on the effects of bacterial toxins. In neutrophils, mast cells, renal mesangial cells, fibroblasts, and platelets, the breakdown of PIP_2 and the associated physiological events induced by 48/80, chemotactic peptide, thrombin, or angiotensin II are inhibited by islet activating protein, a *Bordetella pertussis* toxin (Nakamura and Ui, 1983, 1985; Bokoch and Gilman, 1984; Okajima and Ui, 1984; Okajima *et al.*, 1985; Verghese *et al.*, 1985; Volpi *et al.*, 1985; Brass *et al.*, 1986; Pfeilschifter and Bauer, 1986; Paris and Pouyssegur, 1987). This toxin ADP-ribosylates G_i, transducin (a G protein involved in coupling rhodopsin to cGMP phosphodiesterase in rod outer segments of the retina), and N_o or G_o (a G protein of unknown function isolated from brain and other tissues) (Manning *et al.*, 1984; Sternweis and Robishaw, 1984; Van Dop *et al.*, 1984; Watkins *et al.*, 1984). Thus, the inhibitory effects of the toxin on neutrophils and mast cells may involve one of these G proteins or a novel protein that is also a substrate for the toxin. In NG108-15 hybrid cells, addition of purified G_o or G_i to membranes prepared from cells treated with islet activating protein restores bradykinin activation of GTPase activity (Higashida *et al.*, 1986).

In liver, islet activating protein is without effect on the stimulation of PIP_2 breakdown and Ca^{2+} mobilization induced by agonists in either intact hepatocytes or isolated plasma membranes, under conditions in which G_i is ADP-ribosylated and its functions are blocked (Lynch *et al.*, 1986b; Uhing *et al.*, 1986). Furthermore, the ability of GTP analogues to decrease high-affinity binding of epinephrine, vasopressin, or angiotensin II to liver plasma membranes is unaffected by treatment with the toxin (Fig. 6; Lynch *et al.*, 1986). Likewise, the toxin does not affect muscarinic cholinergic effects on phosphoinositide hydrolysis in cardiac myocytes and pancreatic acinar cells (Masters *et al.*, 1985; Merritt *et al.*, 1986), angiotensin II and thyrotropin-releasing hormone stimulation of IP_3 production in adrenal glomerulosa and 7315c cells, respectively (Aub *et al.*, 1986; Kojima *et al.*, 1986), thrombin action on inositol release in 3T3 fibroblasts (Murayama and Ui, 1985), α_1-adrenergic agonist binding to kidney cortex membrances (Boyer *et al.*, 1984), carbachol binding to 1321N1 astrocytoma cells (Martin *et al.*, 1985), or α_1-adrenergic agonist stimulation of respiration in brown adipocytes (Schimmel *et al.*, 1985). It also does not inhibit agonist-induced PIP_2 hydrolysis in plasma membranes from GH_3 pituitary cells and astrocytoma cells (Hepler and Harden, 1986; Martin *et al.*, 1986a,b). In liver and these tissues, therefore, it appears that the G protein involved in the actions of Ca^{2+}-mobilizing agonists is not ADP-ribosylated by islet activating protein. Since none of the G proteins regulating PIP_2 hydrolysis has yet been unequivocally identified and purified, the reasons for the differences in various cells are unknown.

5. Growth Factors and Phosphoinositide Metabolism

In addition to the well-recognized Ca^{2+}-mobilizing agonists that include hormones and neurotransmitters, it is becoming clear that certain growth factors can alter phosphoinositide metabolism in certain cells. These include epidermal growth factor (EGF) acting on A431 cells (Sawyer and Cohen, 1981; Macara, 1986; Hepler *et al.*, 1987; Pike and Eakes, 1987) and platelet-derived growth factor (PDGF) and bombesin acting on Swiss 3T3 or BALB/c 3T3 fibroblasts (Habenicht *et al.*, 1981; Berridge *et al.*, 1984; Hasegawa-Sasaki, 1985; Besterman *et al.*, 1986; Heslop *et al.*, 1986; MacDonald *et al.*, 1987). The changes in phosphoinositide metabolism involve not only the breakdown of PIP_2 and production of IP_3 (Berridge *et al.*, 1984; Hasegawa-Sasaki, 1985; Besterman *et al.*, 1986; Heslop *et al.*, 1986; Hepler *et al.*, 1987; Pike and Eakes, 1987) but also increased synthesis of polyphosphoinositides (Macara, 1986; MacDonald *et al.*, 1987; Pike and Eakes, 1987). There have also been reports that growth factors alter Ca^{2+} fluxes in certain cells (Sawyer and Cohen, 1981; Berridge *et al.*, 1984; Moolenaar *et al.*, 1984, 1986; Hasegawa-Sasaki, 1985; Bosch *et al.*, 1986; Hepler *et al.*, 1987; Johnson *et al.*, 1986; Macara, 1986) by mobilizing intracellular Ca^{2+} stores and/or opening plasma membrane Ca^{2+} channels.

The mechanisms by which growth factors alter phosphoinositide metabolism are unclear. In those situations where polyphosphoinositide synthesis is enhanced, there is evidence of increased phosphoinositide kinase activity (MacDonald *et al.*, 1987; Pike and Eakes, 1987). One possibility is that the tyrosine kinase activities associated with the receptors for growth hormones directly or indirectly increase phosphoinositide phosphorylation. Although early reports suggested that the tyrosine kinase activities of certain oncogene products ($pp60^{v\text{-}src}$, $pp68^{v\text{-}ros}$) and growth factor receptors might also have phosphoinositide kinase activity (Macara *et al.*, 1984; Machicao and Wieland, 1984; Sugimoto *et al.*, 1984), it is now clear that this is not the case (Fischer *et al.*, 1985; MacDonald *et al.*, 1985; Taylor *et al.*, 1985; Thompson *et al.*, 1985). Thus, any effect of tyrosine kinase activity on phosphoinositide phosphorylation appears to be indirect, if it occurs at all.

The mechanism by which growth factors promote the breakdown of PIP_2 to IP_3 in certain cells is even more obscure. In liver cells, the actions of epidermal growth factor to mobilize intracellular Ca^{2+} have been reported to be abolished by pertussis toxin (Johnson *et al.*, 1986), implying the involvement of a G protein. However, this would have to be different from the G protein mediating the actions of other Ca^{2+}-mobilizing agonists, since these are unaffected by the toxin (Johnson *et al.*, 1986; Lynch *et al.*, 1986b). In contrast to the findings in liver cells, the effects of the growth factor on PIP_2 breakdown and IP_3 production in A431 cells have been reported to be unaffected by the toxin (Pike and Eakes, 1987). Apart from this issue, there is the problem that in some cells, which must possess receptors for certain growth factors since they show mitogenic and other responses to these factors, no changes in phosphoinositides have been reported (Besterman *et al.*, 1986; L'Allemain and Pouyssegur, 1986), or the changes are too slow to explain the responses (Macara, 1986).

It can be speculated that the effects of growth factors on phosphoinositide metabolism and Ca^{2+} fluxes are mediated by specific subtypes of growth factor receptors present only in certain cells. The existence of receptor subtypes, which mediate diverse responses and are distributed differentially in tissues, is well known for many agonists. Alternatively, the growth factor receptors could be coupled to PIP_2 phospholipase C through a G protein in certain cells but not in others.

In contrast to other growth factors, insulin does not promote PIP_2 breakdown or synthesis, IP_3 formation, or Ca^{2+} mobilization in most of its target issues (Farese *et al.*, 1985; Taylor *et al.*, 1985; Besterman *et al.*, 1986; Bosch *et al.*, 1986; Heslop *et al.*, 1986; L'Allemain and Pouyssegur, 1986; Sakai and Wells, 1986). There has been one report that insulin increases IP_3 and other inositol phosphates in epididymal fat pads (Farese *et al.*, 1986), but this effect has not been observed in isolated fat cells (Pennington and Martin, 1985). There is also evidence that insulin increases DAG in some cell types (Farese *et al.*, 1984, 1985), but this can occur through other mechanisms besides phosphoinositide breakdown. On the other hand, there have been consistent reports of insulin stimulation of phosphoinositide and PA synthesis in adipose tissue or BC3H-1 cultured myocytes (De Torrontegui and Berthet, 1966; Stein and Hales, 1974; Farese *et al.*, 1982, 1984, 1985; Honeyman *et al.*, 1983; Pennington and Martin, 1985). The mechanism by which insulin stimulates the synthesis of these phospholipids is presently unknown.

6. *Agonist-Regulated Phosphoinositide Phospholipase C*

As described above, there is much evidence that Ca^{2+}-mobilizing agonists and GTP analogues stimulate PIP_2 breakdown by activating a specific phospholipase C through a G protein. However, this phospholipase C has not been purified to date, and its regulation by a G protein has not been demonstrated. Most investigators have studied the enzyme utilizing membranes or permeabilized cells in which the endogenous inositol phospholipids have been isotopically labeled (Cockroft and Gomperts, 1985; Litosch and Fain, 1985; Litosch *et al.*, 1985; C. D. Smith *et al.*, 1985; Uhing *et al.*, 1985, 1986; Wallace and Fain, 1985; Martin *et al.*, 1986a, 1986b; Straub and Gershengorn, 1986). Only a few studies have shown guanine nucleotide stimulation of the enzyme utilizing unilamellar vesicles containing exogenous labeled PIP_2 (Baldassare and Fisher, 1986a; Banno *et al.*, 1986b; Deckmyn *et al.*, 1986; Jackowski *et al.*, 1986; Taylor and Exton, 1987). In most of these studies, the observed nucleotide specificities and potencies were appropriate for the involvement of a G protein (Baldassare and Fisher, 1986a; Deckmyn *et al.*, 1986; Jackowski *et al.*, 1986; Taylor and Exton, 1987; see also Banno *et al.*, 1986b). Taylor and Exton (1987) further showed that stimulation of the liver plasma membrane enzyme by GTP and its analogues required the presence of phosphatidylserine or phosphatidylethanolamine in the vesicles and 2–5 mM Mg^{2+}. It was inhibited by GDPβS and by phosphatidylcholine. Other workers (Baldassare and Fisher, 1986a; Deckmyn *et al.*, 1986; Jackowski *et al.*, 1986) included other phospholipids, for example, those extracted from the relevant cells, in their PIP_2 phospholipase C assays but did not define the Mg^{2+} dependence.

Taylor and Exton (1987) were able to demonstrate hormonal stimulation of PIP_2 phospholipase C in the presence of submicromolar concentrations of GTPγS. They also observed that the phospholipase hydrolyzing PIP_2 and PIP was completely dependent on Ca^{2+} (0.1 μM–1 mM) for activity and that GTPγS increased the sensitivity of the enzyme to Ca^{2+} as well as enhancing its activity at millimolar Ca^{2+}. In contrast, the enzyme hydrolyzing PI required millimolar Ca^{2+} and was minimally stimulated by GTP analogues (Taylor and Exton, 1987). Deckmyn *et al* (1986), studying the soluble PIP_2 phospholipase C from platelets, found a Ca^{2+} dependence and GTPγS modulation that were similar to those reported by Taylor and Exton (1987). However, these workers also observed that the PI phospholipase C was sensitive to low Ca^{2+}, although it showed little response to the GTP analogue. Jackowski *et al.* (1986) and Baldassare and Fisher (1986a,b) did not study the Ca^{2+} dependence of their phospholipases in detail, but Banno *et al* (1986b) reported that GDP was as effective as GTP in stimulating cytosolic PIP_2 phospholipase C from platelets in the presence of 6 μM Ca^{2+} or 2 mM EGTA, and that the partially purified enzyme was activated by GTP in the presence of EGTA but inhibited by GTPγS under these conditions. It seems unlikely that these latter workers were studying a G-protein-regulated form of the enzyme.

The phospholipase C hydrolyzing PI has been partially purified from several tissues (for references see Shukla, 1982, and Low *et al.*, 1984). It is present mainly in the cytosol but also in particulate fractions (Shukla, 1982). The soluble enzymes from various tissues have been resolved into several forms (Chau and Tai, 1982; Hirasawa *et al.*, 1982; Hoffman and Majerus, 1982; Low *et al.*, 1984; Nakanishi *et al.*, 1985), some of which may have arisen through proteolysis (Low *et al.*, 1984). In contrast to PIP_2 phospholipase C activity, PI phospholipase C activity is minimally activated by GTP analogues (Deckmyn *et al.*, 1986; Taylor and Exton, 1987).

Phospholipase C activity that hydrolyzes PIP_2 and/or PIP has been detected in the soluble fraction of many tissues (Rittenhouse, 1983; Irvine *et al.*, 1984b; Wilson *et al.*, 1984; Nakanishi *et al.*, 1985; Baldasare and Fisher, 1986a; Banno *et al.*, 1986a; Deckmyn *et al.*, 1986; Low *et al.*, 1986; Ebstein *et al.*, 1987; Taylor and Exton, 1987). In most instances, multiple activities have been found (Wilson *et al.*, 1984; Nakanishi *et al.*, 1985; Banno *et al.*, 1986a; Low *et al.*, 1986; Ebstein *et al.*, 1987), and it has been observed that the activity toward polyphosphoinositides requires less Ca^{2+} than for PI hydrolysis (Wilson *et al.*, 1984; Nakanishi *et al.*, 1985; Banno *et al.*, 1986a). The soluble PIP_2 phospholipase C activity derived from platelets is stimulated by GTP and its analogues (Baldassare and Fisher, 1986a; Deckmyn *et al.*, 1986), but this is not the case for liver (Taylor and Exton, 1987). This presumably reflects the fact that the G protein that regulates PIP_2 phospholipase C (or its α subunit) is present in or released into the soluble phase of platelets but is not present in the soluble fraction of liver cells. In support of this idea, a pertussis-toxin-labeled substrate is detectable in the soluble phase of platelets but not of liver cells (Deckmyn *et al.*, 1986; S. Taylor, C. J. Lynch, and J. H. Exton, unpublished observations). The partially purified PIP_2 phospholipase C of calf brain shows stimulation by GTPγS, but not the highly purified enzymes from brain or seminal vesicle (Deckmyn *et al.*, 1986). To date, there have been no successful reconstitutions of a purified phosphoinositide phospholipase C with a G protein.

7. *Summary*

Ca^{2+}-mediated agonists play important roles in the regulation of many body functions. Their receptors are located on the outer surface of their target cells and are widely distributed. The specific receptor subtypes for many of the agonists have been characterized using radioligands and pharmacological approaches, but very few of the receptors have been purified even partially. The primary effect of receptor activation is to stimulate a phospholipase C that catalyzes the breakdown of phosphatidylinositol 4,5-bisphosphate (PIP_2) in the plasma membrane to yield *myo*inositol 1,4,5-trisphosphate (IP_3) and 1,2-diacylglycerol (DAG). IP_3 releases Ca^{2+} from intracellular stores whereas DAG activates a specific Ca^{2+}-phospholipid-dependent protein kinase (protein kinase C). PIP_2 is formed in the membrane by the successive phosphorylation of phosphatidylinositol and phosphatidylinositol 4-phosphate (PIP) by specific kinases.

There is much evidence that the receptors are coupled to the phospholipase C by a guanine nucleotide binding protein (G protein) and that activation occurs by a process very similar to that for the stimulatory receptors that are coupled to adenylate cyclase. Thus, micromolar concentrations of GTP and its analogues reduce the affinity of the receptors for their agonists and are required for hormonal activation of the phospholipase C in isolated plasma membranes. The hormones also activate a low K_m GTPase activity in these membranes and promote the physical interaction of their receptors with the G protein. At least two types of G protein are involved, which differ in their ability to be ADP-ribosylated by a pertussis toxin. The G protein(s) that specifically regulates the phospholipase C has not yet been identified unequivocally (e.g., by reconstitution assays).

The phospholipase C hydrolyzes PIP_2 and PIP, but not PI, and is dependent on Ca^{2+} (10^{-7}–10^{-4} M). Hormones and GTP analogues acting through the G protein activate the enzyme by increasing its sensitivity to Ca^{2+}. The activation requires millimolar concentrations of Mg^{2+} and is enhanced by phosphatidylserine and phosphatidylethanolamine but is inhibited by phosphatidylcholine or GDP analogues.

The IP_3 released by PIP_2 breakdown is either dephosphorylated successively to *myo*inositol or phosphorylated to *myo*inositol 1,3,4,5-tetrakisphosphate, which is then hydrolyzed to *myo*inositol 1,3,4-trisphosphate. No clear roles have been identified for the latter two compounds. IP_3 is believed to release Ca^{2+} from a subfraction of endoplasmic reticulum by opening a Ca^{2+} channel, but the details are uncertain. The major target of cytosolic Ca^{2+} is calmodulin, and the Ca^{2+}–calmodulin complex interacts with and modifies the activities of many enzymes and other proteins. There is very limited information concerning the specific enzymes and other proteins involved in the physiological responses of different cells to Ca^{2+}-mediated agonists.

The major, perhaps exclusive, cellular role of DAG is to activate protein kinase C by decreasing its requirement for Ca^{2+}. This action of DAG is mimicked by tumor-promoting phorbol esters, although it is unclear that protein kinase C is their sole target. Protein kinase C requires Ca^{2+} and a phospholipid such as phosphatidylserine for activity and may only be active in membranes. Few of its cellular substrates have been defined, but some of them are probably membrane proteins.

Much more work needs to be done to clarify the molecular aspects of the actions of Ca^{2+}-mediated agonists. The physicochemical characteristics and the molecular interactions of the receptors, G proteins, and PIP_2 phospholipase C need to be defined. The molecular mechanisms and site of action of IP_3 and DAG also need clarification. Finally, the specific enzymes and other proteins involved in the mechanisms underlying many physiological responses need to be identified.

References

Abdel-Latif, A. A., Akhtar, R. A., and Hawthorne, J. N., 1977, Acetylcholine increases the breakdown of triphosphoinositide of rabbit iris muscle prelabelled with [^{32}p]phosphate, *Biochem. J.* **162:**61–73.

Agranoff, B. W., Murthy, P., and Sequin, E. B., 1983, Thrombin-induced phosphodiesteratic cleavage of phosphatidylinositol bisphosphate in human platelets, *J. Biol. Chem.* **258:**2076–2078.

Akhtar, R. A., and Abdel-Latif, A. A., 1984, Carbachol causes rapid phosphodiesteratic cleavage of phosphatidylinositol 4,5-bisphosphate and accumulation of inositol phosphates in rabbit iris smooth muscle; prazosin inhibits noradrenaline- and ionophore A23187-stimulated accumulation of inositol phosphates, *Biochem. J.* **224:**291–300.

Ambler, S. K., Brown, R. D., and Taylor, P., 1984, The relationship between phosphoinositol metabolism and mobilization of intracellular calcium elicited by alpha$_1$-adrenergic receptor stimulation in BC3H-1 muscle cells, *Mol. Pharmacol.* **26:**405–413.

Amitai, G., Brown, R. D., and Taylor, P., 1984, The relationship between a$_1$-adrenergic receptor occupation and the mobilization of intracellular calcium, *J. Biol. Chem.* **259:**12519–12527.

Aub, D. L., and Putney, J. W., Jr., 1984, Metabolism of inositol phosphates in parotid cells: Implications for the pathway of the phosphoinositide effect and for the possible messenger role of inositol trisphosphate, *Life Sci.* **34:**1347–1355.

Aub, D. L., and Putney, J. W., Jr., 1985, Properties of receptor-controlled inositol trisphosphate formation in parotid acinar cells, *Biochem. J.* **225:**263–266.

Aub, D. L., Frey, E. A., Sekura, R. D., and Cote, T. E., 1986, Coupling of the thyrotropin-releasing hormone receptor to phospholipase C by a GTP-binding protein distinct from the inhibitory or stimulatory GTP-binding protein, *J. Biol. Chem.* **261:**9333–9340.

Authi, K. S., and Crawford, N., 1985, Inositol 1,4,5-trisphosphate-induced release of sequestered Ca^{2+} from highly purified human platelet intracellular membranes, *Biochem. J.* **250:**247–253.

Baldassare, J. J., and Fisher, G. J., 1986a, Regulation of membrane associated and cytosolic phospholipase C activities in human platelets by guanosine triphosphate, *J. Biol. Chem.* **261:**11942–11944.

Baldassare, J. J., and Fisher, G. J., 1986b, GTP and cytosol stimulate phosphoinositide hydrolysis in isolated platelet membranes, *Biochem. Biophys. Res. Commun.* **137:**801–805.

Banno, Y., Nakashima, S., and Nozawa, Y., 1986a, Partial purification of phosphoinositide phospholipase C from human platelet cytosol: Characterization of its three forms, *Biochem. Biophys. Res. Commun.* **136:**713–721.

Banno, Y., Nakashima, S., Tohmatsu, T., Nozawa, Y., and Lapetina, E. G., 1986b, GTP and GDP will stimulate platelet cytosolic phospholipase C independently of Ca^{2+}, *Biochem. Biophys. Res. Commun.* **140:**728–734.

Batty, I. R., Nahorski, S. R., and Irvine, R. F., 1985, Rapid formation of inositol (1,3,4,5) tetrakisphosphate following muscarinic receptor stimulation of rat cerebral corticol slices, *Biochem. J.* **232:**211–215.

Baudiere, B., Guillon, G., Bali, J.-P., and Jard, S., 1986, Muscarinic stimulation of inositol phosphate accumulation and acid secretion in gastric fundic mucosal cells, *FEBS Lett.* **198:**321–325.

Baukal, A. J., Guillemette, G., Rubin, R., Spat, A., and Catt, K. J., 1985, Binding sites for inositol trisphosphate in the bovine adrenal cortex, *Biochem. Biophys. Res. Commun.* **133:**532–538.

Berridge, M. J., 1983, Rapid accumulation of inositol trisphosphate reveals that agonists hydrolyse polyphosphoinositides instead of phosphatidylinositol, *Biochem. J.* **212:**849–858.

Berridge, M. J., 1984, Inositol trisphosphate and diacylglycerol as second messengers, *Biochem. J.* **220:**345–360.

Berridge, M. J., 1986, Intracellular signaling through inositol trisphosphate and diacylglycerol, *Hoppe Seylers Z. Physiol. Chem.* **367:**447–456.

Berridge, M. J., and Dawson, R. M. C., Downes, C. P., Heslop, J. P., and Irvine, R. F., 1983, Changes in the levels of inositol phosphates after agonist-dependent hydrolysis of membrane phosphoinositides, *Biochem. J.* **212:**473–482.

Berridge, M. J., Heslop, J. P., Irvine, R. F., and Brown, K. D., 1984, Inositol trisphosphate formation and calcium mobilization in Swiss 3T3 cells in response to platelet-derived growth factor, *Biochem. J.* **222:**195–201.

Besterman, J. M., Watson, S. P., and Cuatrecasas, P., 1986, Lack of association of epidermal growth factor-, insulin- and serum-induced mitogenesis with stimulation of phosphoinositide degradation in BALB/c 3T3 fibroblasts, *J. Biol. Chem.* **261:**723–727.

Biden, T. J., and Wollheim, C. B., 1986, Ca^{2+} regulates the inositol tris/tetrakisphosphate pathway in intact and broken preparations of insulin-secreting R1Nm5F cells, *J. Biol. Chem.* **261:**11931–11934.

Biden, T. J., Prentki, M., Irvine, R. F., Berridge, M. J., and Wollheim, C. B., 1984, Inositol 1,4,5-trisphosphate mobilizes intracellular Ca^{2+} from permeabilized insulin-secreting cells, *Biochem. J.* **223:**467–473.

Biden, T. J., Wollheim, C. B., and Schlegel, W., 1986, Inositol 1,4,5-trisphosphate and intracellular Ca^{2+} homeostasis in clonal pituitary cells, *J. Biol. Chem.* **261:**7223–7229.

Billah, M. M., and Lapetina, E. G., 1982, Rapid decrease of phosphatidylinositol 4,5-bisphosphate in thrombin-stimulated platelets, *J. Biol. Chem.* **257:**12705–12708.

Billah, M. M., and Michell, R. H., 1979, Phosphatidylinositol metabolism in rat hepatocytes stimulated by glycogenolytic hormones, *Biochem. J.* **182:**661–668.

Blackmore, P. F., Bocckino, S. B., Waynick, L. E., and Exton, J. H., 1985, Role of a guanine nucleotide-binding regulatory protein in the hydrolysis of hepatocyte phosphatidylinositol 4,5-bisphospate by calcium-mobilizing hormones and the control of cell calcium. Studies utilizing aluminum fluoride. *J. Biol. Chem.* **260:**14477–14483.

Boer, R., and Fahrenholz, F., 1985, Photoaffinity labeling of the V_1 vasopressin receptor in plasma membranes from rat liver, *J. Biol. Chem.* **260:**15051–15054.

Bojanic, D., and Fain, J. N., 1986, Guanine nucleotide regulation of [^{3}H]vasopressin binding to liver plasma membranes solubilized receptors. Evidence for the involvement of a guanine nucleotide regulatory protein, *Biochem. J.* **240:**361–365.

Bokoch, G. M., and Gilman, A. G., 1984, Inhibition of receptor-mediated release of arachidonic acid by pertussis toxin, *Cell* **39:**301–308.

Bosch, F., Bouscarel, B., Slaton, J., Blackmore, P. F., and Exton, J. H., 1986, Epidermal growth factor mimics insulin effects in rat hepatocytes, *Biochem. J.* **239:**523–530.

Boyer, J. L., Garcia, A., Posadas, C., and Garcia-Sainz, J. A., 1984, Differential effect of pertussis toxin on the affinity state for agonists of renal α_1- and α_2-adrenoceptors, *J. Biol. Chem.* **259:**8076–8079.

Brass, L. F., and Joseph, S. K., 1985, A role for inositol trisphosphate in intracellular Ca^{2+} mobilization and granule secretion in platelets, *J. Biol. Chem.* **260:**15172–15179.

Brass, L. F., Laposata, M., Banga, H. S., and Rittenhouse, S. E., 1986, Regulation of the phosphoinositide hydrolysis pathway in thrombin-stimulated platelets by a pertussis toxin-sensitive guanine nucleotide-binding protein, *J. Biol. Chem.* **261:**16838–16847.

Brown, J. E., and Rubin, L. J., 1984, A direct demonstration that inositol trisphosphate induces an increase in intracellular calcium in *Limulus* photoreceptors, *Biochem. Biophys. Res. Commun.* **125:**1137–1142.

Brown, J. E., Rubin, L. J., Ghalayini, A. J., Tarver, A. P., Irvine, R. F., Berridge, M. J., and Anderson, R. E., 1984, *myo*-Inositol polyphosphate may be a messenger for visual excitation in *Limulus* photoreceptors, *Nature* **311:**160–163.

Burgess, G. M., Godfrey, P. P., McKinney, J. S., Berridge, M. J., Irvine, R. F., and Putney, J. W., Jr., 1984a, The second messenger linking receptor activation to internal Ca release in liver, *Nature* **309:**63–66.

Burgess, G. M., Irvine, R. F., Berridge, M. J., McKinney, J. S., and Putney, J. W., Jr., 1984b, Actions of inositol phosphates on Ca^{2+} pools in guinea-pig hepatocytes, *Biochem. J.* **224:**741–746.

Burgess, G. M., McKinney, J. S., Irvine, R. F., Berridge, M. J., Hoyle, P. C., and Putney, J. W., Jr., 1984c, Inositol, 1,4,5-trisphosphate may be a signal for f-Met-Leu-Phe- induced intracellar calcium mobilization in human leucocytes (HL-60 cells), *FEBS Lett.* **176:**193–196.

Busa, W. B., Ferguson, J. E., Joseph, S. K., Williamson, J. R., and Nuccitelli, R., 1985, Activation of frog (*Xenopus laevis*) eggs by inositol trisphosphate. 1. Characterization of Ca^{2+} release from intracellular stores, *J. Cell Biol.* **101:**677–682.

Bylund, D. B., and U'Prichard, D. C., 1983, Characterization of α_1- and α_2-adrenergic receptors, *Int. Rev. Neurobiol.* **24:**343–431.

Canessa de Scarnatti, O., and Lapetina, E., 1974, Adrenergic stimulation of phosphatidylinositol labelling in rat vas deferens, *Biochim. Biophys. Acta* **360:**298–305.

Capponi, A. M., and Catt, K. J., 1980, Solubilization and characterization of adrenal and uterine angiotensin II receptors after photoaffinity labeling, *J. Biol. Chem.* **255:**12081–12086.

Charest, R., Prpic, V., Exton, J. H., and Blackmore, P. F., 1985, Stimulation of inositol trisphosphate formation in hepatocytes by vasopressin, epinephrine and angiotensin II and its relationship to changes in cytosolic free Ca^{2+}, *Biochem. J.* **227:**79–90.

Chau, L.-Y., and Tai, H.-H., 1982, Resolution into two different forms and study of the properties of phosphatidylinositol-specific phospholipase C from human platelet cytosol, *Biochim. Biophys. Acta* **713:**344–351.

Chueh, S.-H., and Gill, D. L., 1986, Inositol 1,4,5-trisphosphate and guanine nucleotides activate calcium release from endoplasmic reticulum via distinct mechanisms, *J. Biol. Chem.* **261:**13883–13886.

Cockroft, S., and Gomperts, B. D., 1985, Role of guanine nucleotide binding protein in the activation of polyphosphoinositide phosphodiesterase, *Nature* **314:**534–536.

Conn, P. J., and Sanders-Bush, E., 1985, Serotonin stimulated phosphoinositide turnover: Mediation by the S_2 binding site in rat cerebral cortex but not in subcortical regions, *J. Pharmacol. Exp. Ther.* **234:**195–203.

Connolly, T. M., Bross, T. E., and Majerus, P. W., 1985, Isolation of a phosphomonoesterase from human platelets that specifically hydrolyzes the 5-phosphate of inositol 1,4,5-trisphosphate, *J. Biol. Chem.* **260:**7868–7874.

Connolly, T. M., Lawing, W. J., Jr., and Majerus, P. W., 1986, Protein kinase C phosphorylates human platelet inositol trisphosphate 5′-phosphomonoesterase increasing the phosphatase activity, *Cell* **49:**951–958.

Connolly, T. M., Bansal, V. S., Bross, T. E., Irvine, R. F., and Majerus, P. W., 1987, The metabolism of the tris- and tetraphosphates of inositol by 5-phosphomonoesterase and 3-kinase enzymes, *J. Biol. Chem.* **262:**2146–2149.

Creba, J. A., Downes, C. P. K., Hawkins, P. T., Brewster, G., Michell, R. H., and Kirk, C. J., 1983, Rapid breakdown of phosphatidylinositol 4-phosphate and phosphatidylinositol 4,5-bisphosphate in rat hepatocytes stimulated by vasopressin and other Ca^{2+}-mobilizing hormones, *Biochem. J.* **212:**733–747.

Dawson, A. P., and Irvine, R. F., 1984, Inositol (1,4,5)trisphosphate-promoted Ca^{2+} release from microsomal fractions of rat liver, *Biochem. Biophys. Res. Commun.* **120:**858–864.

De Chaffoy de Courcelles, D., Leysen, J. E., De Clerck, F., Van Belle, H., and Janssen, P. A. J., 1985, Evidence that phospholipid turnover is the signal transducing system coupled to the serotonin-S_2 receptor sites, *J. Biol. Chem.* **260:**7603–7608.

Deckmyn, H., Tu, S.-M., and Majerus, P. W., 1986, Guanine nucleotides stimulate soluble phosphoinositide-specific phospholipase C in the absence of membranes, *J. Biol. Chem.* **261:**16553–16558.

Delfert, D. M., Hill, S., Pershadsingh, H. A., Sherman, W. R., and McDonald, J. M., 1986, *myo*Inositol 1,4,5-trisphosphate mobilizes Ca^{2+} from isolated adipocyte endoplasmic reticulum but not from plasma membranes, *Biochem. J.* **236:**37–44.

De Torrontegui, G., and Berthet, J., 1966, The action of insulin on the incorporations of [^{32}p]phosphate in the phospholipids of rat adipose tissue, *Biochim. Biophys. Acta* **116:**477–481.

Dougherty, R. W., Godfrey, P. P., Hoyle, P. C., Putney, J. W., Jr., and Freer, R. J., 1984, Secretagogue-induced phosphoinositide metabolism in human leucocytes, *Biochem. J.* **222:**307–314.

Downes, C. P., and Wusteman, M. M., 1983, Breakdown of polyphosphoinositides and not phosphatidylinositol accounts for muscarinic agonist-stimulated inositol phospholipid metabolism in rat parotid glands, *Biochem. J.* **216:**633–640.

Downes, C. P., Mussat, M. C., and Michell, R. H., 1982, The inositol trisphosphate phosphomonoesterase of the human erythrocyte membrane, *Biochem. J.* **203:**169–177.

Downes, C. P., Hawkins, P. T., and Irvine, R. F., 1986, Inositol 1,3,4,5-tetrakisphosphate is the probable precursor of inositol 1,3,4-trisphosphate in agonist-stimulated parotid gland, *Biochem. J.* **238:**501–506.

Ebstein, R. P., Bennett, E. R., Stessman, J., and Lerer, B., 1987, Isoelectric focusing of human platelet phospholipase C: Evidence for multimolecular forms, *Life Sci.* **40:**161–167.

El-Refai, M. F., Blackmore, P. F., and Exton, J. H., 1979, Evidence for two α-adrenergic binding sites in liver plasma membranes. Studies with [^{3}H]epinephrine and [^{3}H]dihydroergocryptine, *J. Biol. Chem.* **254:**4375–4386.

Enjalbert, A., Sladeczek, F., Guillon, G., Bertrand, P., Shu, C., Epelbaum, J., Garcia-Sainz, A., Jard, S., Lombard, C., Kordon, C., and Bockaert, J., 1986, Angiotensin II and dopamine modulate both cAMP and inositol phosphate production in anterior pituitary cells, *J. Biol. Chem.* **261:**4071–4075.

Exton, J. H., 1987, Mechanisms of α_1-adrenergic and related responses: Roles of calcium, phosphoinositides, guanine nucleotides, diacylglycerol, calmodulin and changes in protein phosphorylation, in: *Cell Membranes: Methods and Reviews* (E. L. Elson, W. A. Frazier, and L. Glaser, eds.), Plenum Press, New York.

Fahrenholz, F., Kojro, E., Muller, M., Boer, R., Lohr, R., and Grzonka, Z., 1986, Iodinated photoreactive vasopressin antagonists: Labelling of hepatic vasopressin receptor subunits, *Eur. J. Biochem.* **161:**321–328.

Farese, R. V., Larson, R. E., and Sabir, M. A., 1982, Insulin acutely increases phospholipids in the phosphatidate–inositide cycle in rat adipose tissue, *J. Biol. Chem.* **257:**4042–4045.

Farese, R. V., Barnes, D. E., Davis, J. S., Standaert, M. L., and Pollet, R. J., 1984, Effects of insulin and protein synthesis inhibitors on phospholipid metabolism, diacylglycerol levels, and pyruvate dehydrogenase activity in BC3-1H cultured myocytes, *J. Biol. Chem.* **259:**7094–7100.

Farese, R. V., Davis, J. S., Barnes, D. E., Standaert, M. L., Babischkin, J. S., Hoek, R., Rosie, N. K., and Pollet, R. J., 1985, The de novo phospholipid effect of insulin is associated with increases in diacylglycerol, but not inositol phosphate or cytosolic Ca^{2+}, *Biochem. J.* **231:**269–279.

Farese, R. V., Kuo, J. Y., Babischkin, J. S., and Davis, J. S., 1986, Insulin provokes a transient activation of phospholipase C in the rat epididymal fat pad, *J. Biol. Chem.* **261:**8589–8592.

Fein, A., Payne, R., Corson, D. W., Berridge, M. J., and Irvine, R. F., 1984, Photoreceptor excitation and adaptation by inositol 1,4,5-trisphosphate, *Nature* **311:**157–160.

Fischer, S., Fagard, R., Comoglio, P., and Gacon, G., 1985, Phosphoinositides are not phosphorylated by the very active tyrosine protein kinase from murine lymphoma LSTRA, *Biochem. Biophys. Res. Commun.* **132:**481–489.

Fisher, S. K., 1986, Inositol lipids and signal transduction at CNS muscarinic receptor, *Trends Pharmacol. Sci.* **7:**(Suppl.)61–65.

Fitzgerald, T. J., Uhing, R. J., and Exton, J. H., 1986, Solubilization of the vasopressin receptor from liver plasma membranes. Evidence for a receptor–GTP–binding protein complex, *J. Biol. Chem.* **261:**16871–16877.

Gallo-Payet, N., Guillon, G., Balestre, M. N., and Jard, S., 1986, Vasopressin induces breakdown of membrane phosphoinositides in adrenal glomerulosa and fasciculata cells, *Endocrinology* **119:**1042–1047.

Gershengorn, M. C., Geras, E., Purrello, V. S., and Rebecchi, M. J., 1984, Inositol trisphosphate mediates thyrotropin-releasing hormone mobilization of non-mitochondrial calcium in rat mammotropic pituitary cells, *J. Biol. Chem.* **259:**10675–10681.

Geynet, P., Borsodi, A., Ferry, N., and Hanoune, J., 1980, Proteolysis of rat liver plasma membranes cancels the guanine nucleotide sensitivity of agonist binding to the alpha-receptor, *Biochem. Biophys. Res. Commun.* **97:**947–954.

Gomperts, B. D., 1983, Involvement of guanine nucleotide-binding protein in the gating of Ca^{2+} by receptors, *Nature* **306:**64–66.

Graham, R. M., Hess, H.-J., and Homcy, C. J., 1982, Biophysical characterization of the purified α_1-adrenergic receptor and identification of the hormone binding subunit, *J. Biol. Chem.* **257:**15174–15181.

Grandt, R., Greiner, C., Zubin, P., and Jakobs, K. H., 1986, Bradykinin stimulates GTP hydrolysis in NG108-15 membranes by a high-affinity, pertussis toxin-insensitive GTPase, *FEBS Lett.* **196:**279–283.

Guillemette, G., Balla, T., Baukal, A. J., Spat, A., and Catt, K. J., 1987, Intracellular receptors for inositol 1,4,5-trisphosphate in angiotensin II target tissues, *J. Biol. Chem.* **262:**1010–1015.

Guillon, G., Courand, P.-O., Butlen, D., Cantau, B., and Jard, S., 1980, Size of vasopressin receptors from rat liver and kidney, *Eur. J. Biochem.* **111:**287–294.

Habenicht, A. J. R., Glomset, J. A., King, W. C., Nist, C., Mitchell, C. D., and Ross, R., 1981, Early changes in phosphatidylinositol and arachidonic acid metabolism in quiescent Swiss 3T3 cells stimulated to divide by platelet-derived growth factor, *J. Biol. Chem.* **256:**12329–12335.

Hansen, C. A., Mah, S., and Williamson, J. R., 1986, Formation and metabolism of inositol 1,3,4,5-tetrakisphosphate in liver, *J. Biol. Chem.* **261:**800–8103.

Harden, T. K., Tanner, L. I., Martin, M. W., Nakahata, N., Hughes, A. R., Hepler, J. R.,

Evans, T., Masters, S. B., and Brown, J. H., 1986, Characteristics of two biochemical responses to stimulation of muscarinic cholinergic receptors, *Trends Pharmacol. Sci.* **7:**(Suppl.)14–18.

Harrington, C. A., and Eichberg, J., 1983, Norepinephrine causes α_1-adrenergic receptor-mediated decrease of phosphoinositide in isolated rat liver plasma membranes supplement-eed with cytosol, *J. Biol. Chem.* **258:**2087–2090.

Hasegawa-Sasaki, H., 1985, Early changes in inositol lipids and their metabolites induced by platelet-derived growth factor in quiescent Swiss mouse 3T3 cells, *Biochem. J.* **232:**99–109.

Haslam, R. J., and Davidson, M. M. L., 1984a, Guanine nucleotides decrease the free [Ca^{2+}] required for secretion of serotonin from permeabilized blood platelets. Evidence of a role for a GTP-binding protein in platelet activation, *FEBS Lett.* **174:**90–95.

Haslam, R. J., and Davidson, M. M. L., 1984b, Receptor-induced diacylglycerol formation in permeabilized platelets; possible role for a GTP-binding protein, *J. Receptor Res.* **4:**605–629.

Henne, V., and Soling, H.-D., 1986, Guanosine 5′-triphosphate releases calcium from rat liver and guinea pig parotid gland endoplasmic reticulum independently of inositol 1,4,5-trisphosphate, *FEBS Lett.* **202:**267–273.

Hepler, J. R., and Harden, T. K., 1986, Guanine nucleotide-dependent pertussis toxin-insensitive stimulation of inositol phosphate formation by carbachol in a membrane preparation from human astrocytoma cells. *Biochem. J.* **239:**141–146.

Hepler, J. R., Nakahata, N., Lovenberg, T. W., DiGuiseppi, J., Herman, B., Earp, H. S., and Harden, T. K., 1987, Epidermal growth factor stimulates the rapid accumulation of inositol (1,4,5)-trisphosphate and a rise in cytosolic calcium mobilized from intracellular stores in A431 cells, *J. Biol. Chem.* **262:**2951–2956.

Heslop, J. P., Blakeley, D. M., Brown, K. D., Irvine, R. F., and Berridge, M. J., 1986, Effects of bombesin and insulin on inositol (1,4,5)trisphosphate and inositol (1,3,4)trisphosphate formation in Swiss 3T3 cells, *Cell* **47:**703–709.

Higashida, H., and Brown, D. A., 1986, Membrane current responses to intracellular injections of inositol 1,3,4,5-tetrakisphosphate and inositol 1,3,4-trisphosphate in GN108-15 hybrid cells, *FEBS Lett.* **208:**283–286.

Higashida, H., Streaty, R. A., Klee, W., and Nirenberg, M., 1986, Bradykinin-activated transmembrane signals are coupled via N_o or N_i to production of inoisitol, 1,4,5-trisphosphate, a second messenger in NG108-15 neuroblastoma glioma hybrid cells, *Proc. Natl. Acad. Sci. U.S.A.* **83:**942–946.

Hinkle, P. M., and Phillips, W. J., 1984, Thyrotropin-releasing hormone stimulates GTP hydrolysis by membranes from GH_4C_1 rat pitutary tumor cells, *Proc. Natl. Acad. Sci. U.S.A.* **81:**6183–6187.

Hirasawa, K., Irvine, R. F., and Dawson, R. M. C., 1982, Heterogeneity of the calcium-dependent phosphatidylinositol phosphodiesterase of rat brain, *Biochem. J.* **205:**437–442.

Hirata, M., Kukita, M., Sasaguri, T., Suematsu, E., Hashimoto, T., and Koga, T., 1985, Increase in Ca^{2+} permeabilization of intracellular Ca^{2+} store membrane of saponin-treated guinea pig peritoneal macrophages by inositol 1,4,5-trisphosphate, *J. Biochem.* **97:**1575–1582.

Hoffman, S. L., and Majerus, P. W., 1982, Identification and properties of two distinct phosphatidylinositol-specific phospholipase C enzymes from sheep seminal vesicles, *J. Biol. Chem.* **257:**6461–6469.

Holub, B. J., and Kuksis, A., 1978, Metabolism of molecular species of diacylglycerophospholipids, *Adv. Lipid Res.* **16:**1–125.

Honeyman, T. W., Strohsnitter, W., Scheid, C. R., and Schimmel, R. J., 1983, Phosphatidic acid and phosphatidylinositol labelling in adipose tissue, *Biochem. J.* **212:**489–498.

Imai, A., and Gershengorn, M. C., 1987, Independent phosphatidylinositol synthesis in pituitary plasma membrane and endoplasmic reticulum, *Nature* **325:**726–728.

Imboden, J. B., and Stobo, J. D., 1985, Transmembrane signalling by the T cell antigen receptor, *J. Exp. Med.* **161:**446–456.

Irvine, R. F., and Moor, R. M., 1986, Microinjection of inositol 1,3,4,5-tetrakisphosphate activates sea urchin eggs by a mechanism dependent upon external Ca^{2+}, *Biochem. J.* **240:**917–920.

Irvine, R. F., Brown, K. D., and Berridge, M. J., 1984a, Specificity of inositol trisphosphate-induced calcium release from permeabilized Swiss-mouse 3T3 cells, *Biochem. J.* **221:**269–272.

Irvine, R. F., Letcher, A. J., and Dawson, R. M. C., 1984b, Phosphatidylinositol 4,5-bisphosphate phosphodiesterase and phosphomonoesterase activities of rat brain, *Biochem. J.* **218:**177–185.

Irvine, R. F., Letcher, A. J., Lander, D. J., and Downes, C. P., 1984c, Inositol trisphosphates in carbachol-stimulated rat parotid glands, *Biochem. J.* **223:**237–243.

Irvine, R. J., Anggard, E. E., Letcher, A. J., and Downes, C. P., 1985, Metabolism of inositol 1,4,5-trisphosphate and inositol 1,3,4-trisphosphate in rat parotid glands, *Biochem. J.* **229:**505–511.

Irvine, R. F., Letcher, A. J., Heslop, J. P., and Berridge, M. J., 1986a, The inositol tris/tetrakisphosphate pathway—demonstration of Ins(1,4,5)P_3 3-kinase activity in animal tissues, *Nature* **320:**631–634.

Irvine, R. F., Letcher, A. J., Lander, D. J., and Berridge, M. J., 1986b, Specificity of inositol phosphate-stimulated Ca^{2+} mobilization from Swiss-mouse 3T3 cells, *Biochem. J.* **240:**301–304.

Ishii, H., Connolly, T. M., Bross, T. E., and Majerus, P. W., 1986, Inositol cyclic trisphosphate [inositol 1,2-(cyclic)-4,5-trisphosphate] is formed upon thrombin stimulation of human platelets, *Proc. Natl. Acad. Sci. U.S.A.* **83:**6397–6401.

Jackowski, S., Rettenmier, C. W., Sherr, C. J., and Rock, C. O., 1986, A guanine nucleotide-dependent phosphatidylinositol 4,5-diphosphate phospholipase C in cells transformed by the *v-fms* and *v-fes* oncogenes, *J. Biol. Chem.* **261:**4978–4985.

Johnson, R. M., Connelly, P. A., Sisk, R. B., Pobiner, B. F., Hewlett, E. L., and Garrison, J. C., 1986, Pertussis toxin or phorbol 12-myristate 13-acetate can distinguish between growth factor- and angiotensin-stimulated signals in hepatocytes, *Proc. Natl. Acad. Sci. U.S.A.* **83:**2032–2036.

Jones, L. M., and Michell, R. H., 1978, Stimulus–response coupling at α-adrenergic receptors, *Biochem. Soc. Trans.* **6:**673–688.

Joseph, S. K., and Williams R. J., 1985, Subcellular localization and some properties of the enzymes hydrolysing inositol polyphosphates in rat liver, *FEBS LETT.* **180:**150–154.

Joseph, S. K., and Williams, J. R., 1986, Characteristics of inositol trisphosphate-mediated Ca^{2+} release from permeabilized hepatocytes, *J. Biol. Chem.* **261:**14658–14664.

Joseph, S. K., Thomas, A. P., Williams, R. J., Irvine, R. F., and Williamson, J. R., 1984a *myo*-Inositol 1,4,5-trisphosphate: A second messenger for the hormonal mobilization of intracellular Ca^{2+} in liver, *J. Biol. chem.* **259:**3077–3081.

Joseph, S. K., Williamson, R. J., Corkey, B. E., Matschinsky, F. M., and Williamson, J. R., 1984b, The effect of inositol trisphosphate on Ca^{2+} fluxes in insulin-secreting tumor cells, *J. Biol. Chem.* **259:**12952–12955.

Kanaho, Y., Moss, J., and Vaughan, M., 1985, Mechanism of inhibition of transducin GTPase activity by fluoride and aluminum, *J. Biol. Chem.* **260:**11493–11497.

Katada, T., Bokoch, G. M., Northup, J. K., Ui, M., and Gilman, A. G., 1984, The inhibitory guanine nucleotide-binding regulatory component of adenylate cyclase. Properties and function of the purified protein, *J. Biol. Chem.* **259:**3568–3577.

Kirk, C. J., Verrinder, T. R., and Hems, D. A., 1977, Rapid stimulation, by vasopressin and adrenaline, or inorganic phosphate incorporation into phosphatidylinositol in isolated hepatocytes, *FEBS Lett.* **83:**267–271.

Kirk, C. J., Creba, J. A., Downes, C. P., and Michell, R. H., 1981, Hormone-stimulated metabolism of inositol lipids and its relationship to hepatic receptor function, *Biochem. Soc. Trans.* **9:**377–379.

Kojima, I., Shibata, H., and Ogata, E., 1986, Pertussis toxin blocks angiotensin II-induced calcium influx but not inositol trisphosphate production in adrenal glomerulosa cells, *FEBS Lett.* **204:**347–351.

Lad, P. M., Olson, C. V., and Smiley, P. A., 1985, Association of the *N*-formyl-Met-Leu-Phe receptor in human neutrophils with a GTP-binding protein sensitive to pertussis toxin, *Proc. Natl. Acad. Sci. U.S.A.* **82:**869–873.

L'Allemain, G., and Pouyssegur, J., 1986, EGF and insulin action in fibroblasts, *FEBS Lett.* **197:**344–348.

Lambert, T. L., Kent, R. S., and Whorton, A. R., 1986, Bradykinin stimulation of inositol polyphosphate production in porcine aortic endothelial cells, *J. Biol. Chem.* **261:**15288–15293.

Leeb-Lundberg, L. M. F., Dickinson, K. E. J., Heald, S. L., Wikberg, J. E. S., Lefkowitz, R. J., and Caron, M. G., 1984, Photoaffinity labeling of mammalian α_1-adrenergic receptors, *J. Biol. Chem.* **259:**2579–2587.

Leeb-Lundberg, L. M. F., Cotecchia, S., Lomasney, J. W., Debernadis, J. F., Lefkowitz, R. J., and Caron, M. G., 1985, Phorbol esters promote α_1-adrenergic receptor phosphorylation and receptor uncoupling from inositol phospholipid metabolism, *Proc. Natl. Acad. Sci. U.S.A.* **82:**5651–5655.

Lew, P. D., Monod, A., Krause, K.-H., Waldvogel, F. A., Biden, T. J., and Schlegel, W., 1986, The role of cytosolic free calcium in the generation of inositol 1,4,5-trisphosphate and inositol 1,3,4-trisphosphate in HL-60 cells, *J. Biol. Chem.* **261:**13121–13127.

Lin, S. H., and Fain, J. N., 1981, Vasopressin and epinephrine stimulation of phosphatidylinositol breakdown in the plasma membrane of rat hepatocytes, *Life Sci.* **29:**1905–1912.

Litosch, I., and Fain, J. N., 1985, 5-Methyltryptamine stimulates phospholipase C-mediated breakdown of exogenous phosphoinositides by blowfly salivary gland membranes, *J. Biol. Chem.* **260:**16052–16055.

Litosch, I., Lin, S. H., and Fain, J. N., 1983, Rapid changes in hepatocyte phosphoinositides induced by vasopressin, *J. Biol. Chem.* **258:**13727–13732.

Litosch, I., Wallis, C., and Fain, J. N., 1985, 5-Hydroxytryptamine stimulates inositol phosphate production in a cell-free system from blowfly salivary glands, *J. Biol. Chem.* **260:**5464–5471.

Lomasney, J. W., Leeb-Lundberg, L. M. F., Cotecchia, S., Regan, J. W., DeBernadis, J. F., Caron, M. G., and Lefkowitz, R. J., 1986, Mammalian α_1-adrenergic receptor. Purification and characterization of the native receptor ligand binding subunit. *J. Biol. Chem.* **261:**7710–7716.

Low, M. G., Carroll, R. G., and Cox, A. C., 1986, Characterization of multiple forms of phosphoinositide-specific phospholipase C, purified from human platelets, *Biochem. J.* **237:**139–145.

Low, M. G., Carroll, R. C., and Weglicki, W. B. 1984, Multiple forms of phosphoinositide-specific phospholipase C of different relative molecular masses in animal tissues. *Biochem. J.* **221:**813–820.

Lucas, D. O., Bajjalich, S. M., Kowalchyk, J. A., and Martin, T. F. J., 1985, Direct stimulation by thyrotropin-releasing hormone of polyphosphoinositide hydrolysis in GH3 cell membranes by a guanine nucleotide-modulated mechanism, *Biochem. Biophys. Res. Commun.* **132:**721–728.

Lynch, C. J., Blackmore, P. F., Charest, R., and Exton, J. H., 1985a, The relationships between receptor binding capacity for norepinephrine, angiotensin II and vasopressin and release of inositol trisphosphate, Ca^{2+} mobilization and phosphorylase activation in rat liver, *Mol. Pharmacol.* **28:**93–99.

Lynch, C. J., Charest, R., Blackmore, P. F., and Exton, J. H., 1985b, Studies on the hepatic α_1-adrenergic receptor. Modulation of guanine nucleotide effects by calcium temperature and age, *J. Biol. Chem.* **260:**1593–1600.

Lynch, C. J., Sobo, G. E., and Exton, J. H., 1986a, Studies on the hepatic α_1-adrenergic receptor. An endogenous Ca^{2+}-sensitive protease converts the α_1-adrenergic receptor to a guanine nucleotide insensitive form, *Biochim. Biophys. Acta* **885:**110–120.

Lynch, C. J., Prpic, V., Blackmore, P. F., and Exton, J. H., 1986b, Effect of islet activating pertussis toxin on the binding characteristics of Ca^{2+}-mobilizing hormones and on agonist activation of phosphorylase in hepatocytes, *Mol. Pharmacol.* **29:**196–203.

Macara, I. G., 1986, Activation of $^{45}Ca^{2+}$ influx and $^{22}Na^+/H^+$ exchange by epidermal growth factor and vanadate in A431 cells is independent of phosphatidylinositol turnover and is inhibited by phorbol ester and diacylglycerol, *J. Biol. Chem.* **261:**9321–9327.

Macara, I. G., Marinetti, G. V. and Balduzzi, P. C., 1984, Transforming protein of avian sarcoma virus UR2 is associated with phosphatidylinositol kinase activity: Possible role in tumorigenesis, *Proc. Natl. Acad. Sci. U.S.A.* **81:**2728–2732.

MacDonald, M. L., Kuenzel, E. A., Glomset, J. A., and Krebs, E. G., 1985, Evidence from two transformed cell lines that the phosphorylations of peptide tyrosine and phosphatidylinositol are catalyzed by two different proteins, *Proc. Natl. Acad. Sci. U.S.A.* **82:**3993–3997.

MacDonald, M. L. Mack, K. F., and Glomset, J. A., 1987, Regulation of phosphoinositide phosphorylation in Swiss 3T3 cells stimulated by platelet-derived growth factor, *J. Biol. Chem.* **262:**1105–1110.

Machicao, E., and Wieland, O. H., 1984, Evidence that the insulin receptor-associated protein kinase acts as a phosphatidylinositol kinase, *FEBS Lett.* **175:**113–116.

Manning, D. R., Fraser, B. A., Kahn, R. A., and Gilman, A. G., 1984, ADP-ribosylation of transducin by islet-activating protein. Identification of asparagine as the site of ADP-ribosylation, *J. Biol. Chem.* **259:**749–756.

Martin, M. W., Evans, T., and Harden, T. K., 1985, Further evidence that muscarinic cholinergic receptors of 1321N1 astrocytoma cells couple to a guanine nucleotide regulatory protein that is not N_i, *Biochem. J.* **229:**539–544.

Martin, T. F. J., 1983, Thyrotropin-releasing hormone rapidly activates the phosphodiesterase hydrolysis of polyphosphoinositides in GH3 pituitary cells *J. Biol. Chem.* **258:**14816–14822.

Martin, T. F. J., Bajjalieh, S. M., Lucas, D. O., and Kowalchyk, J. A., 1986a, Thyrotropin-releasing hormone stimulation of polyphosphoinositide hydrolysis in GH_3 cell membranes is GTP-dependent but insensitive to cholera or pertussis toxin, *J. Biol. Chem.* **261:**10041–10049.

Martin, T. F. J., Lucas, D. O., Bajjalieh, S. M., and Kowalchyk, J. A., 1986b, Thyrotropin-releasing hormone activates a Ca^{2+}-dependent polyphosphoinositide phosphodiesterase in permeable GH_3 cells, *J. Biol. Chem.* **261:**2918–2927.

Masters, S. B., Martin, M. W., Harden, T. K., and Brown, J. H., 1985, Pertussis toxin does not inhibit muscarinic receptor-mediated phosphoinositide hydrolysis or calcium mobilization, *Biochem. J.* **227:**933–937.

Mauco, G., Chap, H., and Douste-Blazy, L., 1983, Platelet activating factor (PAF-acether) promotes an early degradation of phosphatidylinositol-4,5-bisphosphate in rabbit platelets, *FEBS Lett.* **153:**361–365.

Merritt, J. E., Taylor, C. W., Rubin, R. P., and Putney, J. W., Jr., 1986, Evidence suggesting that a novel guanine nucleotide regulatory protein couples receptors to phospholipase C in exocrine pancreas, *Biochem. J.* **236:**337–343.

Michell, R. H., 1975, Inositol phospholipids and cell surface receptor function, *Biochim. Biophys. Acta* **20:**339–344.

Michell, R. H., 1979, Inositol phospholipids in membrane function, *Trends Biochem. Sci.* **June:**128–131.

Michell, R. H., Kirk, C. J., and Billah, M. M., 1979, Hormonal stimulation of phosphatidylinositol breakdown, with particular reference to the hepatic effects of vasopressin, *Biochem. Soc. Trans.* **7:**861–865.

Michell, R. H., Kirk, C. J., Jones, L. M., Downes, C. P., and Creba, J. A., 1981, Stimulation of inositol lipid metabolism that accompanies calcium mobilization in stimulated cells: Defined characteristics and unanswered questions, *Philos. Trans. R. Soc. Lond. (Biol.)* **296:**123–138.

Molina y Vedia, L. M., and Lapetina, E. G. 1986, Phorbol 12,13-dibutyrate and 1-oleyl-2-acetyldiacylglycerol stimulate inositol trisphosphate dephosphorylation in human platelets, *J. Biol. Chem.* **261:**10493–10495.

Moolenaar, W. H., Tertoolen, L. G. J., and deLaat, S. W., 1984, Growth factors immediately raise cytoplasmic free Ca^{2+} in human fibroblasts, *J. Biol. Chem.* **259:**8066–8069.

Moolenaar, W. H., Aerts, R. J., Tertoolen, L. G. J., and deLaat, S. W., 1986, The epidermal growth factor-induced calcium signal in A431 cells, *J. Biol. Chem.* **261:**279–284.

Morgan, N. G., Rumford, G. M., and Montague, W., 1985, Studies on the role of inositol trisphosphate in the regulation of insulin secretion from isolated rat islets of Langerhans, *Biochem. J.* **228:**713–18.

Muallem, S., Schoeffield, M., Pandol, S., and Sachs, G., 1985, Inositol trisphosphate modification of ion transport in rough endoplasmic reticulum, *Proc. Natl. Acad. Sci. U.S.A.* **82:**4433–4437.

Murayama, T., and Ui, M., 1985, Receptor-mediated inhibition of adenylate cyclase and stimulation of arachidonic acid release in 3T3 fibroblasts, *J. Biol. Chem.* **260:**7226–7233.

Nakamura, T., and Ui, M., 1983, Suppression of passive cutaneous anaphylaxis by pertussis toxin, an islet-activating protein, as a result of inhibition of histamine release from mast cells, *Biochem. Pharmacol.* **32:**3435–3441.

Nakamura, T., and Ui, M., 1985, Simultaneous inhibitions of inositol phospholipid breakdown, arachidonic acid release, and histamine secretion in mast cells by islet-activating protein, pertussis toxin, *J. Biol. Chem.* **260:**3584–3593.

Nakanishi, H., Nomura, H., Kikkawa, V., Kishimoto, A., and Nishizuka, Y., 1985, Rat brain and liver soluble phospholipase C: Resolution of two forms with different requirements for calcium, *Biochem. Biophys. Res. Commun.* **132:**582–590.

Nanberg, E., and Putney, J. W., Jr., 1986, α_1-Adrenergic activation of brown adipocytes leads to an increased formation of inositol polyphosphates, *FEBS Lett.* **195:**319–322.

Nosek, T. M., Williams, M. F., Zeigler, S. T., and Godt, R. E., 1986, Inositol trisphosphate enhances calcium release in skinned cardiac and skeletal muscle, *Am. J. Physiol.* **250:**C807–C811.

Okajima, F., and Ui, M., 1984, ADP-Ribosylation of the specific membrane protein by islet-activating protein, pertussis toxin, associated with inhibition of a chemotactic peptide-induced arachidonate release in neutrophils, *J. Biol. Chem.* **259:**13863–13871.

Okajima, F., Katada, T., and Ui, M., 1985, Coupling of guanine nucleotide regulatory protein to chemotactic peptide receptors in neutrophil membranes and its uncoupling by islet-activating protein, pertussis toxin, *J. Biol. Chem.* **260:**6761–6768.

Orellano, S., Solski, P. A., and Brown, J. H., 1987, Guanosine 5′-*O*-(thiotriphosphate)-

dependent inositol trisphosphate formation in membranes is inhibited by phorbol ester and protein kinase C, *J. Biol. Chem.* **262:**1638–1643.

Oron, Y., Dascal, N., Nadler, E., and Lupu, M., 1985, Inositol 1,4,5-trisphosphate mimics muscarinic response in *Xenopus* oocytes, *Nature* **313:**141–143.

O'Rourke, F. A., Halenda, S. P., Zavoico, G. B., and Feinstein, M. B., 1985, Inositol 1,4,5-trisphosphate releases Ca^{2+} from a Ca^{2+}-transporting membrane vesicle fraction derived from human platelets, *J. Biol. Chem.* **260:**956–962.

Paris, S., and Pouyssegur, J., 1987, Further evidence for a phospholipase C-coupled G protein in hamster fibroblasts, *J. Biol. Chem.* **262:**1970–1976.

Pennington, S. R., and Martin, B. R., 1985, Insulin-stimulated phosphoinositide metabolism in isolated fat cells, *J. Biol. Chem.* **260:**11039–11045.

Pfeilschifter, J., and Bauer, C., 1986, Pertussis toxin abolishes angiotensin II-induced phosphoinositide hydrolysis and prostaglandin synthesis in rat renal mesangial cells, *Biochem. J.* **236:**289–294.

Pike, L. J., and Eakes, A. T., 1987, Epidermal growth factor stimulates the production of phosphatidylinositol monophosphate and the breakdown of polyphosphoinositides in A431 cells, *J. Biol. Chem.* **262:**1644–1651.

Prentki, M. M., Biden, T. J., Janjic, D., Irvine, R. F., Berridge, M. J., and Wollheim, C. B., 1984a, Rapid mobilization of Ca^{2+} from rat insulinoma microsomes by inositol-1,4,5-trisphosphate, *Nature* **309:**562–564.

Prentki, M., Wollheim, C. B., and Lew, P. D., 1984b, Ca^{2+} homeostasis in permeabilized human neutrophils. Characterization of Ca^{2+}-sequestering pools and the action of inositol, 1,4,5-trisphosphate, *J. Biol. Chem.* **259:**13777–13782.

Prentki, M., Corkey, B. E., and Matschinsky, F. M., 1985, Inositol 1,4,5-trisphosphate and the endoplasmic reticulum Ca^{2+} cycle of a rat insulinoma cell line, *J. Biol. Chem.* **260:**9185–9190.

Prpic, V., Blackmore, P. F., and Exton, J. H., 1982, Phosphatidylinositol breakdown induced by vasopressin and epinephrine in hepatocytes is calcium-dependent, *J. Biol. Chem.* **257:**11323–11331.

Putney, J. W., Jr., Burgess, G. M., Halenda, S. P., McKinney, J. S., and Rubin, R. P., 1983, Effects of secretagogues on [^{32}p]phosphatidylinositol 4,5-bisphosphate metabolism in the exocrine pancreas, *Biochem. J.* **212:**483–488.

Rebecchi, M. J., and Gershengorn, M. C., 1983, Thyroliberin stimulates rapid hydrolysis of phosphatidylinositol 4,5-bisphosphate by a phosphodiesterase in rat mammotropic pituitary cells, *Biochem. J.* **216:**287–294.

Rhodes, D., Prpic, V., Exton, J. H., and Blackmore, P. F., 1983, Stimulation of phosphatidylinositol 4,5-bisphosphate hydrolysis in hepatocytes by vasopressin, *J. Biol. Chem.* **258:**2770–2773.

Rittenhouse, S. E., 1983, Human platelets contain phospholipase C that hydrolyzes polyphosphoinositides, *Proc. Natl. Acad. Sci. U.S.A.* **80:**5417–5420.

Rittenhouse, S. E., and Sasson, J. P., 1985, Mass changes in myoinositol trisphosphate in human platelets stimulated by thrombin: Inhibitory effects of phorbol ester, *J. Biol. Chem.* **260:**8657–8660.

Rogers, T. B., 1984, High affinity angiotensin II receptors in myocardial sarcolemmal membranes, *J. Biol. Chem.* **259:**8106–8114.

Rubin, R. P., Godfrey, P. P., Chapman, D. A., and Putney, J. W., Jr., 1984, Secretagogue-induced formation of inositol phosphates in rat exocrine pancreas, *Biochem. J.* **219:**655–659.

Sakai, M., and Wells, W. W., 1986, Action of insulin on the subcellular metabolism of polyphosphoinositides in isolated rat hepatocytes, *J. Biol. Chem.* **261:**10058–10062.

Sawyer, S. T., and Cohen, S., 1981, Enhancement of calcium uptake and phosphatidylinositol turnover by epidermal growth factor in A-431 cells, *Biochemistry* **20:**6280–6286.

Schacht, J., and Agranoff, B. W., 1972, Effects of acetylcholine on labeling of phosphatidate and phosphoinositides by [^{32}p]orthophosphate in nerve, *J. Biol. Chem.* **247:**774–777.

Schimmel, R. J., McCarthy, L., and Dzierzanowski, D., 1985, Effects of pertussis toxin treatment on metabolism in hamster brown adipocytes, *Am. J. Physiol.* **249:**C456–C463.

Seidman, C. E., Hess, H. J., Homcy, C. J., and Graham, R. M., 1984, Photoaffinity labeling of the α_1-adrenergic receptor using an ^{125}I-labeled aryl azide analogue of prazosin, *Biochemistry* **23:**3765–3770.

Sekar, M. C., Dixon, J. F., and Hokin, L. E., 1987, The formation of inositol 1,2-cyclic 4,5-trisphosphate and inositol 1,2-cyclic 4-bisphosphate on stimulation of mouse pancreatic minilobules with carbamylcholine, *J. Biol. Chem.* **262:**340–344.

Sen, I., Jim, K. F., and Soffer, R. L., 1983, Solubilization and characterization of an angiotensin II binding protein from liver, *Eur. J. Biochem.* **136:**41–49.

Seyfred, M. A., and Wells, W. W., 1984, Subcellular site and mechanism of vasopressin-stimulated hydrolysis of phosphoinositides in rat hepatocytes, *J. Biol. Chem.* **259:**7666–7672.

Seyfred, M. A., Farrell, L. E., and Wells, W. W., 1984, Characterization of D-*myo*-inositol 1,4,5-trisphosphate phosphatase in rat liver plasma membranes, *J. Biol. Chem.* **259:**13204–13208.

Shukla, S. D., 1982, Phosphatidylinositol specific phospholipases C, *Life Sci.* **30:**1323–1335.

Slack, B. E., Bell, J. E., and Benos, D. J., 1986, Inositol-1,4,5-trisphosphate injection mimics fertilization potentials in sea urchin eggs, *Am. J. Physiol.* **250:**C340–C344.

Smith, C. D., Lane, B. C., Kusaka, I., Verghese, M. W., and Snyderman, R., 1985, Chemoattractant receptor-induced hydrolysis of phosphatidylinositol 4,5-bisphosphate in human polymorphonuclear leukocyte membranes, *J. Biol. Chem.* **260:**5875–5878.

Smith, J. B., Smith, L., Brown, E. R., Barnes, D., Sabir, M. A., Davis, J. S., and Farese, R. V., 1984, Angiotensin II rapidly increases phosphatidate–phosphoinositide synthesis and phosphoinositide hydrolysis and mobilizes intracellular calcium in cultured arterial muscle cells, *Proc. Natl. Acad. Sci. U.S.A.* **81:**7812–7816.

Smith, J. B., Smith, L., and Higgins, B. L., 1985, Temperature and nucleotide dependence of calcium release by *myo*inositol 1,4,5-trisphosphate in cultured vascular smooth muscle cells, *J. Biol. Chem.* **260:**14413–14416.

Somlyo, A. V., Bond, M., Somlyo, A. P., and Scarpa, A., 1985b, Inositol trisphosphate-induced calcium release and contraction in vascular smooth muscle, *Proc. Natl. Acad. Sci. U.S.A.* **82:**5231–5235.

Spat, A., Fabiato, A., and Rubin, R. P., 1986, Binding of inositol trisphosphate by a liver microsomal fraction, *Biochem. J.* **223:**929–932.

Stein, J. M., and Hales, C. N., and 1974, The effect of insulin on $^{32}P_i$ incorporation into rat fat cell phospholipids, *Biochim. Biophys. Acta* **337:**41–49.

Sternweis, P. C., and Gilman, A. G., 1982, Aluminum: A requirement for activation of the regulatory component of adenylate cyclase by fluoride, *Proc. Natl. Acad. Sci. U.S.A.* **79:**4888–4891.

Sternweis, P. C., and Robishaw, J. D., 1984, Isolation of two proteins with high affinity for guanine nucleotides from membranes of bovine brain, *J. Biol. Chem.* **259:**13806–13813.

Stewart, S. J., Prpic, V., Powers, F. S., Bocckino, S. B., Isaacks, R. E., and Exton, J. H., 1986, Perturbation of the human T-cell antigen receptor–T3 complex leads to the production of inositol tetrakisphosphate: Evidence for conversion from inositol trisphosphate, *Proc. Natl. Acad. Sci. U.S.A.* **83:**6098–6102.

Stoehr, S. J., Smolen, J. E., Holz, R. W., and Agranoff, B. W., 1986, Inositol trisphosphate mobilizes intracellular calcium in permeabilized adrenal chromaffin cells, *J. Neurochem.* **46:**637–640.

Storey, D. J., Shears, S. B., Kirk, C. J., and Michell, R. H., 1984, Stepwise enzymatic dephosphorylation of inositol 1,4,5-trisphosphate to inositol in liver, *Nature* **312:**374–376.

Straub, R. E., and Gershengorn, M. C., 1986, Thyrotropin-releasing hormone and GTP activate inositol trisphosphate formation in membranes isolated from rat pituitary cells, *J. Biol. Chem.* **261:**2712–2717.

Streb, H., Irvine, R. F., Berridge M. J., and Schultz, I., 1983, Release of Ca^{2+} from a nonmitochondrial intracellular store in pancreatic acinar cells by inositol 1,4,5-triphosphate, *Nature* **306:**67–69.

Streb, H., Bayerdorffer, E., Haase, W., Irvine, R. F., and Schulz, I., 1984, Effect of inositol-1,4,5-trisphosphate in isolated subcellular fractions of rat pancreas, *J. Membr. Biol.* **81:**241–253.

Streb, H., Heslop, J. P., Irvine, R. F., Schulz, I., and Berridge, M. J., 1985, Relationship between secretagogue-induced Ca^{2+} release and inositol polyphosphate production is permeabilized pancreatic acinar cells, *J. Biol. Chem.* **260:**7309–7315.

Strnad, C. F., Parente, J. E., and Wong, K., 1986, Use of fluoride ion as a probe for the guanine nucleotide-binding protein involved in the phosphoinositide-dependent neutrophil transduction pathway, *FEBS Lett.* **206:**20–24.

Suematsu, E., Hirata, M., Hashimoto, T., and Kuriyama, H., 1984, Inositol 1,4,5-trisphosphate releases Ca^{2+} from intracellular store sites in skinned single cells of porcine coronary artery, *Biochem. Biophys. Res. Commun.* **120:**481–485.

Sugimoto, Y., Whitman, M., Cantley, L. C., and Erikson, R. L., 1984, Evidence that the Rous sarcoma virus transforming gene product phosphorylates phosphatidylinositol and diacylglycerol, *Proc. Natl. Acad. Sci. U.S.A.* **81:**2117–2121.

Taylor, D., Uhing, R. J., Blackmore, P. F., Prpic, V., and Exton, J. H., 1985, Insulin and epidermal growth factor do not affect phosphoinositide metabolism in rat liver plasma membranes and hepatocytes, *J. Biol. Chem.* **260:**2011–2014.

Taylor, S., and Exton, J. H., 1987, Guanine nucleotide and hormone regulation of polyphosphoinositide specific phospholipase C activity of rat liver plasma membranes: Divalent cation and phospholipid requirements, *Biochem. J.* **248:**791–799.

Thevenod, F., Streb, H., Ullrich, K. J., and Schulz, I., 1986, Inositol trisphosphate releases Ca^{2+} from a nonmitochondrial store site in permeabilized rat cortical kidney cells, *Kidney Int.* **29:**695–702.

Thieleczek, R., and Heilmeyer, L. M. G., Jr., 1986, Inositol 1,4,5-trisphosphate enhances Ca^{2+}-sensitivity of the contractile mechanism of chemically skinned rabbit skeletal muscle fibre, *Biochem. Biophys. Res. Commun.* **135:**662–669.

Thomas, A. P., Marks, J. S., Coll, K. E., and Williamson, J. R., 1983, Quantitation and early kinetics of inositol lipid changes induced by vasopressin in isolated and cultured hepatocytes, *J. Biol. Chem.* **258:**5716–5725.

Thomas, A. P., Alexander, J., and Williamson, J. R., 1984, Relationship between inositol polyphosphate production and the increase of cytosolic free Ca^{2+} induced by vasopressin in isolated hepatocytes. *J. Biol. Chem.* **259:**5574–5584.

Thompson, D. M., Cochet, C., Chambaz, E. M., and Gill, G. N., 1985, Separation and characterization of a phosphatidylinositol kinase activity that copurifies with the epidermal growth factor receptor, *J. Biol. Chem.* **260:**8824–8830.

Tolbert, M. E. M., White, A. C., Aspry, K., Cutts, J., and Pain, J. N., 1980, Stimulation by vasopressin and α-catecholamines of phosphatidylinositol formation in isolated rat liver parenchymal cells, *J. Biol. Chem.* **255:**1938–1944.

Uchida, T., Ito, H., Baum, B. J., Roth, G. S., Filburn, C. R., and Sacktor, B., 1982, $Alpha_1$-

adrenergic stimulation of phosphatidylinositol–phosphatidic acid turnover in rat parotid cells, *Mol. Pharmacol.* **21:**128–132.

Uhing, R. J., Jiang, H., Prpic, V., and Exton, J. H., 1985, Regulation of a liver plasma membrane phosphoinositide phosphodiesterase by guanine nucleotides and calcium, *FEBS Lett.* **188:**317–320.

Uhing, R. J., Prpic, V., Jiang, H., and Exton, J. H., 1986, Hormone stimulated polyphosphoinositide breakdown in rat liver plasma membranes: Roles of guanine nucleotides and calcium, *J. Biol. Chem.* **261:**2140–2146.

Van Dop, C., Yamanaka, G., Steinberg, F., Sekura, R. D., Manclark, C. R., Stryer, L., and Bourne, H. R., 1984, ADP-ribosylation of transducin by pertussis toxin blocks the light-stimulated hydrolysis of GTP and cGMP in retinal photoreceptors, *J. Biol. Chem.* **259:**23–26.

Venter, J. C., Horne, P., Eddy, B., Gregusta, R., and Fraser, C. M., 1984 Alpha$_1$-adrenergic receptor structure, *Mol. Pharmacol.* **26:**196–205.

Vergara, J., Tsien, R. Y., and Delay, M., 1985, Inositol 1,4,5-trisphosphate: A possible chemical link in excitation–contraction coupling in muscle, *Proc. Natl. Acad. Sci. U.S.A.* **82:**6352–6356.

Verghese, M. W., Smith, C. D., and Snyderman, R., 1985, Potential role for a guanine nucleotide regulatory protein in chemoattractant receptor mediated polyphosphoinositide metabolism, Ca^{2+} mobilization and cellular respiration by leukocytes, *Biochem. Biophys. Res. Commun.* **127:**450–457.

Vickers, J. D., Kinlough-Rathbone, R. L., and Mustard, J. F., 1984, Changes in the platelet phosphoinositides during the first minute after stimulation of washed rabbit platelets with thrombin, *Biochem. J.* **219:**25–31.

Vincentini, L. M., Ambrosini, A., DiVirgilio, F., Pozzan, T., and Meldolesi, J., 1985, Muscarinic receptor-induced phosphoinositide hydrolysis at resting cytosolic Ca^{2+} concentration in PC12 cells, *J. Cell. Biol.* **100:**1330–1333.

Volpe, J., Salviati, G., DiVirgilio, R., and Pozzan, T., 1985, Inositol 1,4,5-trisphosphate induces calcium release from sarcoplasmic reticulum of skeletal muscle, *Nature* **316:**347–349.

Volpi, M., Yassin, R., Naccache, P. H., and Sha'afi, R. I., 1983, Chemotactic factors cause rapid decreases in phosphatidylinositol, 4,5-bisphosphate and phosphatidylinositol 4-monophosphate in rabbit neutrophils, *Biochem. Biophys. Res. Commun.* **112:**957–964.

Volpi, M., Naccache, P. H., Molski, T. F. P., Shefcyk, J., Huang, C.-K., Marsh, M. L., Munoz, J., Becker, E. L., and Sha'afi, R. I., 1985, Pertussis toxin inhibits fMet-Leu-Phe but not phorbol ester stimulated changes in rabbit neutrophils, *Proc. Natl. Acad. Sci. U.S.A.* **82:**2708–2712.

Wallace, M. A., and Fain, J. N., 1985, Guanosine 5′-*O*-thiotriphosphate stimulates phospholipase C activity in plasma membranes of rat hepatocytes, *J. Biol. Chem.* **260:**9527–9530.

Wallace, M. A., Randazzo, P., Li, S. Y., and Fain, J. N., 1982, Direct stimulation of phosphatidylinositol degradation by addition of vasopressin to purified rat liver plasma membranes, *Endocrinology* **111:**341–343.

Wallace, M. A., Poggioli, J., Giraud, F., and Claret, M., 1983, Norepinephrine-induced loss of phosphatidylinositol from isolated rat liver plasma membrane, *FEBS Lett.* **156:**239–243.

Watkins, P. A., Moss, J., Burns, D. L., Hewlett, E. L., and Vaughan, M., 1984, Inhibition of bovine outer rod segment GTPase by *Bordetella pertussis* toxin, *J. Biol. Chem.* **259:**1378–1381.

Weiss, S. J., McKinney, J. S., and Putney, J. W., Jr., 1982, Receptor-mediated net breakdown of phosphatidylinositol 4,5-bisphosphate in parotid acinar cells, *Biochem. J.* **206:**555–560.

Wilson, D. B., Bross, T. E., Hofmann, S. L., and Majerus, P. W., 1984, Hydrolysis of polyphosphoinositides by purified sheep seminal vesicle phospholipase C enzymes, *J. Biol. Chem.* **259:**11718–11724.

Wilson, D. B., Connolly, T. M., Bross, T. E., Majerus, P. W., Sherman, W. R., Tyler, A. N., Rubin, L. J., and Brown, J. E., 1985, Isolation and characterization of the inositol cyclic phosphate products of polyphosphoinositide cleavage by phospholipase C, *J. Biol. Chem.* **260:**13496–13501.

Wirthensohn, G., Lefrank, S., and Guder, W. G., 1984, Phospholipid metabolism in rat kidney cortical tubules. II. Effects of hormones on ^{32}p incorporation, *Biochim. Biophys. Acta* **795:**401–410.

Wolf, B. A., Comens, P. G., Ackermann, K. E., Sherman, W. R., and McDaniel, M. L., 1985, The digitonin-permeabilized pancreatic islet model, *Biochem. J.* **227:**965–969.

Yano, K., Nakashima, S., and Nozawa, Y., 1983, Coupling of polyphosphoinositide breakdown with calcium efflux in formyl-methionyl-leucyl-phenylalanine-stimulated rabbit neutrophils, *FEBS Lett.* **161:**296–300.

13

Regulation of Protein Kinase C by Sphingosine/Lysosphingolipids

ROBERT M. BELL, YUSUF A. HANNUN, and CARSON R. LOOMIS

1. Introduction

1.1. Discovery of Protein Kinase C and sn-1,2-Diacylglycerol Second Messengers

Protein kinase C was discovered in rat brain cytosol as a cyclic nucleotide-independent protein kinase (Inoue *et al.*, 1977: Takai *et al.*, 1977a,b). The proenzyme form, protein kinase C, could be activated by proteolysis or by a membrane factor in the presence of Ca^{2+} (Takai *et al.*, 1979a,b). The membrane factor was initially identified as phospholipid, with phosphatidylserine (PS) being most effective (Takai *et al.*, 1979a,b). At this point, protein kinase C was referred to as the phospholipid- and Ca^{2+}-dependent protein kinase. Crude lipid extracts from brain proved to be more effective than PS and Ca^{2+} in activating the enzyme; this led to the discovery of a neutral lipid, *sn*-1,2-diacylglycerol (DAG), which increased the enzyme's affinity for both PS and Ca^{2+} (Kishimoto *et al.*, 1980). DAG could activate protein kinase C at resting levels of Ca^{2+}. DAGs were thus discovered to be "second messengers," since DAG production in response to extracellular agents, including hormones, neurotransmitters, and growth factors, had long been known to occur from an accelerated metabolism of the phosphatidylinositols

ROBERT M. BELL and CARSON R. LOOMIS • Department of Biochemistry, Duke University Medical Center, Durham, North Carolina 27710. *YUSUF A. HANNUN* • Department of Medicine, Duke University Medical Center, Durham, North Carolina 27710.

(PIs) (Kishimoto *et al.*, 1980; Berridge and Irvine, 1984; Nishizuka, 1984a,b; 1986 Bell, 1986).

1.2. Phorbol Ester Receptor and Tumor Promotion

Much interest has been focused on protein kinase C as the transducer of DAG second messengers and as the intracellular receptor of phorbol esters and other potent tumor promoters (Berridge and Irvine, 1984; Ashendel, 1985; Bell, 1986; Nishizuka, 1986). DAG and phorbol esters activate the enzyme by similar mechanisms, through interaction at a common or overlapping site (Berridge and Irvine, 1984; Ashendel, 1985; Bell, 1986; Nishizuka, 1986). Many sources describe DAGs as ''endogenous phorbol esters'' (Ebeling *et al.*, 1985; Konig *et al.*, 1985). Phorbol esters are more potent in activating protein kinase C than DAGs and unlike DAGs are not rapidly metabolized.

1.3. Signal Transduction: Growth Factors, Neurotransmitters, and Hormones

The transmembrane signaling events that are thought to give rise to DAG production and protein kinase C activation are illustrated in Fig. 1. Extracellular signals cause the turnover of PIP_2 as a consequence of interaction with specific cell surface receptors that are coupled through a GTP-dependent ((G protein) activation of phospholipase C (Bell, 1986). PIP_2 is cleaved into DAG and a second ''second messenger,'' IP_3, which functions in the mobilization of intracellular Ca^{2+} (Majerus *et al.*, 1986). This model of regulation is analogous to other transmembrane signaling pathways, for example cyclic AMP formation (Majerus *et al.*, 1986).

1.4. Oncogene Products and Signal Transduction

Several oncogene products resemble elements such as *sis,erb*B, and *ras,* which function in transmembrane signaling (depicted in Fig. 1). Other oncogene products not depicted may also function in a similar manner (e.g., *fms* and *fes*). Thus, oncogene products may function to alter the levels of DAG and IP_3, which are critical second messengers. Evidence in support of this hypothesis is beginning to mount (Fleischman *et al.*, 1986; Jackowski *et al.*, 1986; Wakelam *et al.*, 1986; Benjamin *et al.*, 1987; Roussel *et al.*, 1987; Wolfman and Marcara, 1987). The hypothesis that DAGs produced in response to growth factors stimulate cell proliferation is supported by studies using cell-permeable DAGs (Nishizuka, 1984a; Rozengurt *et al.*, 1984; Davis *et al.*, 1985; Rozengurt, 1986; Ganong *et al.*, 1987).

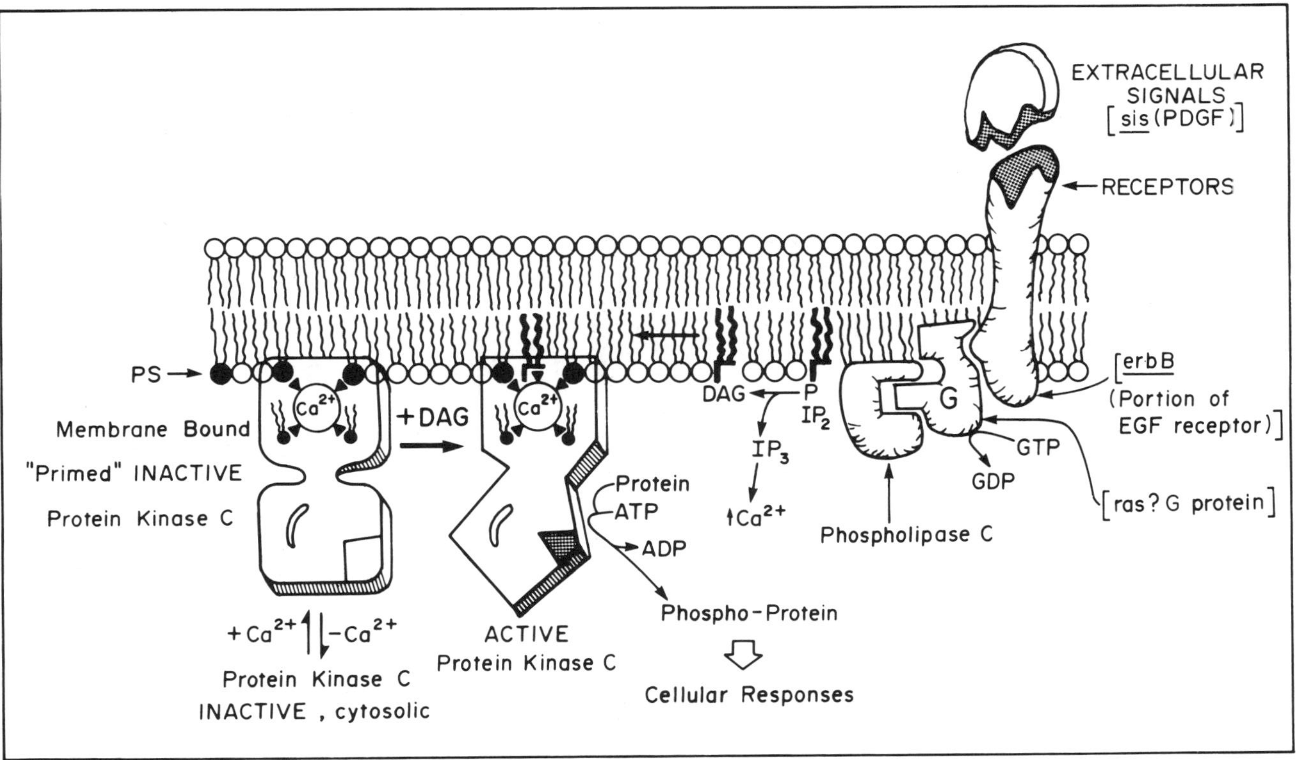

Figure 1. Model of transmembrane signaling and of protein kinase C activation by DAG second messengers. Reproduced with permission from *Cell* **45:** 631–632 (1986).

1.5. Scope and Content

Protein kinase C has become, appropriately, the target of intense investigations, because of its role in tumor promotion, oncogenesis, and signal transduction processes. Protein kinase C is under study from diverse perspectives and in diverse biological settings by numerous laboratories. The following section will briefly summarize the contributions of our laboratory toward understanding the structure, function, and regulation of protein kinase C. This background is pertinent to understanding sphingosine/lysosphingolipids as potent inhibitors of protein kinase C. This chapter will focus mainly on the mechanism of sphingosine inhibition of protein kinase C activity and of [^{3}H] PDBu binding *in vitro,* in human platelets, HL60 cells, and neutrophils (Hannun *et al.*, 1986b; Merril *et al.*, 1986; Wilson *et al.*, 1986). The data on sphingosine/lysosphingolipid inhibition of protein kinase C on a pharmacologic, pathobiologic, and physiological level will be presented, as will hypotheses to focus and guide further investigative work.

2. Structure, Function, and Regulation of Protein Kinase C

2.1. Cell-Permeable DAGs and DAG Analogues

The very limited water solubility of DAGs produced in response to extracellular agents makes their addition to cells difficult. Since protein kinase C is a peripheral membrane protein, it was inferred that its recognition of DAG occurred at the membrane interface where it interacts with PS (Bell, 1986; Ganong *et al.*, 1986). Since protein kinase C would not be deeply embedded into the hydrocarbon portion of the bilayer, the long-chain fatty acid component of the DAGs might not be recognized. DAGs containing short-chain fatty acids were prepared to ascertain whether they could activate protein kinase C *in vitro* and be of sufficient solubility to function in living cells (Ganong *et al.*, 1986). Analogues were also made to establish structure–activity relations. Among a series of DAGs containing fatty acids 3–12 carbons in length, *sn*-1,2-dioctanoylglycerol (diC$_8$) proved most effective in activating protein kinase C and displacing [^{3}H] PDBu from the enzyme in cells (Ganong *et al.*, 1986). These cell-permeable DAGs elicited numerous cellular responses as a consequence of bypassing extracellular-agent-dependent transmembrane signaling events. Moreover, these cell-permeable DAGs mimicked the biological responses of phorbol esters, and, as detailed below, precise structure–activity relations were established (Ganong *et al.*, 1986; Hannun *et al.*, 1986a).

2.2. Quantitation of DAG Second Messengers

In order to investigate changes in the levels of DAG second messengers, a sensititive assay for the quantitation of DAG mass present in crude lipid extracts of cells was developed by extension of the work of Kennerly *et al.* (1979). The assay employed *Escherichia coli* DAG kinase, which was overproduced in a plasmid-bearing strain to about 15% of the membrane protein (Loomis *et al.*, 1985). Defined mixed micellar conditions (Walsh and Bell, 1986) were used to solubilize the DAG present in crude lipid extracts. Under these conditions, DAG was quantitatively converted to [^{32}P] phosphatidic acid (Preiss *et al.*, 1986). The assay was linear over the range of 25 pmoles to 25 nmoles. With this assay, the mass of DAG in platelets and hepatocytes increased 210 and 230% over resting levels in response to thrombin and vasopressin, respectively; similar elevations of DAG in response to extracellular agents have been observed in several cell types using other methods (most frequently radiolabeling). The amounts of DAG present in NRK cells grown at 34 and 38°C were 0.47 and 0.61 nmoles/100 nmoles phospholipid, respectively. The amount of DAG increased 168 and 138% in *K-ras*-transformed NRK cells grown at 34 and 38°C, respectively. When a temperature-sensitive *K-ras*-transformed NRK cell line was studied, the amount of DAG was not higher than the NRK cell line at the restrictive temperature but was elevated at the permissive temperature. These data are consistent with the *K-ras* gene product, functioning in transmembrane signaling by activating a phospholipase C. Elevated levels of DAG were also seen in *sis*-transformed NRK cells (Preiss *et al.*, 1986). The data support the hypothesis that altered levels of DAG second messengers and protein kinase C activation (Anderson *et al.*, 1985) play important roles in cellular transformation. The methods developed should allow critical tests of the effects of other oncogenes and of extracellular agents on the mass of DAG in cells.

2.3. Cloning and Expression of Multiple Rat Brain Protein Kinase C cDNAs

Sequence data from tryptic peptides of rat brain protein kinase C permitted oligonucleotide probes to be synthetized. These probes were used to screen a rat brain cDNA library (Knopf *et al.*, 1986). Unexpectedly, three distinct cDNA clones were discovered. Two clones were sequenced in their entirety and the third was partially sequenced. The carboxy-terminal regions were homologous to the superfamily of protein kinases; three new members were added to an already long list. The N-terminal regions contained a 50 amino acid internal repeat centered around six cysteine residues (Knopf *et al.*, 1986). Homologous structures had been observed in EGF and LDL receptors. Given the unexpected multiplicity of protein kinase C cDNAs, expression studies were undertaken in COS monkey cells, using PMT-2 vectors to ensure that protein kinase C had been cloned. COS cells transfected with PKC-I and PKC-II showed a five-fold increase in [^{3}H]PDBu binding and contained increased amounts of PS, Ca^{2+}, and DAG (phorbol-ester)-dependent protein kinase activity. Northern analysis of RNAs from different tissues revealed tissue-specific expression, with PKC-I being exclusive to brain (Knopf *et al.*, 1986). Coussens *et al.* (1986) and Parker *et al.* (1986) demonstrated multiple protein kinase C cDNAs in bovine brain and in human libraries; moreover, mapping to human chromosomes revealed that the genes were located on separate chromosomes (Coussens *et al.*, 1986). Several other groups have reported similar findings (Ono *et al.*, 1986a,b; Makowske *et al.*, 1986; Ohno *et al.*, 1987).

2.4. Domain Structure of Protein Kinase C

As mentioned earlier, proteolysis of protein kinase C resulted in the formation of a catalytic fragment whose activity was independent of PS, Ca^{2+}, and DAG (phorbol esters) (Inoue *et al.*, 1977; Takai *et al.*, 1977a). When protein kinase C was treated with trypsin, 51- and 32-kD fragments were observed; the 51-kD fragment was the catalytic fragment, whereas the 32-KD fragment was found to bind [^{3}H]PDBu in a PS- and Ca^{2+}-dependent manner (Lee and Bell, 1986). Moreover, the specificity and dependencies of binding were similar to native protein kinase C (Lee and Bell, 1986). Thus, protein kinase C contains two functional domains: The lipid-binding regulatory domain, which resides at the amino-terminal region wherein the internal cysteine repeats reside, and a catalytic domain, which resides on the carboxy-terminal side (Knopf *et al.*, 1986). This two-domain structure was also supported by the studies of Hoshijima *et al.*, (1986) and Huang and Huang (1986).

2.5. Mechanism of Regulation by PS, Ca^{2+}, DAG (Phorbol Esters): Mixed Micellar Analysis

2.5.1. Stoichiometries

The physical properties of the sonic dispersion of phospholipids and diacylglycerols/phorbol esters limited efforts to determine the number of phospholipids and DAG molecules required for activating protein kinase C as well as studies to define the specificity of the phospholipid and DAG requirements. To circumvent these problems, our laboratory developed a mixed micellar assay for protein kinase C (Hannun *et al.*, 1985). This method permits the number of DAG and PS molecules present in Triton X-100 mixed micelles to be varied independently and systematically. At 8 mole % PS in the presence of Ca^{2+}, activity was strongly dependent on *sn*-1,2-dioleoylglycerol (di$C_{18:1}$); maximal activity was observed at 1 mole % di$C_{18:1}$ (Hannun *et al.*, 1985). At this point, approximately 1.4 molecules of di$C_{18:1}$ would be present per mixed micelle. Since activation occurred at levels substantially below 1 mole %, the data imply that a single molecule of diC $_{18:1}$ is sufficient to activate protein kinase C. At 12 mole % PS, maximum activity was observed at 0.1 mole % diC$_{18:1}$ (Hannun *et al.*, 1985), where only one micelle in seven would have a single molecule of diC$_{18:1}$. Plasma membranes contained 8–20 mole % PS. The activation of protein kinase C in mixed micelles established that a phospholipid bilayer was not required.

At 2.5 mole % diC$_{18:1}$, the activation of protein kinase C by PS was cooperative. A Hill number of 4.8 was calculated from these data, while a Hill number of 1 was calculated from the diC$_{18:1}$ dependence (Hannun *et al.*, 1985). Since protein kinase C activity begins to increase at 3 mole % PS, the data implied that a minimum of 4 molecules of PS were required for activation. Molecular sieve chromatography of protein Kinase C, Triton X-100 mixed micelles containing PS, and the protein kinase C–mixed micellar complex indicated that monomeric (80 kD) protein kinase C was the active species (Hannun *et al.*, 1985). Protein kinase C binding to mixed micelles was dependent on PS and Ca^{2+} and occurred in the absence of DAG. The activation of protein kinase C by PS, Ca^{2+}, and diC$_{18:1}$ proved to be interdependent (Hannun *et al.*, 1986a). Collectively, the data support the inferred stoichiometries of 4 PS molecules and 1 DAG.

The mixed micelle methods have been extended to investigate the PS dependence of the activation of protein kinase C by phorbol esters and the PS dependence of [^{3}H]phorbol-dibutyrate binding to protein kinase C (Hannun and Bell, 1986). Monomeric protein kinase C is capable of binding phorbol esters and of being activated (Hannun and Bell, 1986). These data indicate a similar mechanism of protein kinase C activation by DAG and phorbol esters; the data also support a stoichiometry of monomeric protein kinase C–4PS–Ca^{2+}–DAG (phorbol ester).

2.5.2. DAG Structure–Activity Relations

When the chain length of fatty acids was systematically varied in *sn*-1,2-DAGs, protein kinase C was activated by DAGs containing sufficient hydrophobic character to partition into phospholipid vesicles or mixed micelles (Ganong *et al.*, 1986). *sn*-1,2-Dioctanoylglycerol (diC$_8$) was as effective at activating protein kinase C *in vitro* as was diC$_{18:1}$ or 1-stearoyl-2-arachidonylglycerol (Hannun *et al.*, 1986a). Unsaturated fatty acids were not required (Ganong *et al.*, 1986). *sn*-1,2-Dibutyroylglycerol which is very soluble and lacks sufficient hydrophobic character to partition into membranes or micelles, was not effective in activating protein kinase C *in vitro* or in cells.

More than 20 DAG analogues were prepared and tested for their ability to function as activators and antagonists of protein kinase C, with the goal of determining the exact structural features required for activation (Ganong *et al.*, 1986). The synthetic work focused on the structural features required at the polar portion of the molecule, which would be expected to reside at the membrane interface where protein kinase C interacts. These analogues established that the *sn*-3-hydroxyl moiety was essential, that the position of the *sn*-3-hydroxyl was critical, that both oxygen esters were required (ethers and amides were inactive) (Ganong *et al.*, 1986), and that activation was stereospecific (Boni and Rando, 1985; Hannun *et al.*, 1986a). Three functional groups of *sn*-3-hydroxyl and two carbonyls of oxygen esters were inferred to form three bonds to the protein kinase C–PS–Ca^{2+} complex during activation (Fig. 1; Ganong *et al.*, 1986).

2.5.3. Model of Protein Kinase C Regulation

This model which is consistent with the available data (Ganong *et al.*, 1986), envisions 4 molecules of PS on the surface of a membrane (or micelle), which bind Ca^{2+} through their carboxy groups (Fig. 1). Protein kinase C binds to the 4PS–Ca^{2+} complex and thereby becomes membrane bound, "primed" but inactive (Fig. 1). DAG (phorbol ester) binds to the 4PS–Ca^{2+}–protein kinase C complex by occupying the position beneath the Ca^{2+} coordination sphere such that a direct bond forms between DAG and protein kinase C (Ganong *et al.*, 1986). This model accounts for the observed phospholipid dependencies, the increased affinity of protein kinase C for both PS an Ca^{2+} caused by DAG, and other results (Ganong *et al.*, 1986). While consistent with available data this model has not been proved.

2.5.4. Mixed Micelle Methods of Analysis of Protein Kinase C Inhibitors

The interpretation of studies of putative specific inhibitors of protein kinase C is plagued with difficulties, since many of these agents possess amphipathic properties. The mixed micellar methods developed for protein kinase C can readily be extended to amphipathic inhibitors: these methods permit the number of lipid cofactors and amphipathic inhibitors per micelle to be varied systematically and independently. The surface activity of inhibitors is easily determined. Since protein kinase C activity is independent of micelle number, activity does not change when the number of mixed micelles containing fixed mole % PS and DAG is increased (Hannun *et al.*, 1985, 1986a). The inhibitory powers of amphipathic surface-active inhibitors are decreased when the number of mixed micelles is increased. Inhibition by amphipathic molecules, like activation, is a function of the mole % (surface concentration) rather than bulk concentration.

3. Sphingosine/Lysosphingolipid Inhibition of Protein Kinase C

3.1. Inhibition in Vitro

Investigations of protein kinase C–lipid structure function relations caused us to test whether ceramide and *N*-acetylsphingosine would activate or inhibit protein kinase C, since ceramide had structural features similar to DAG. Neither compound was an effector. However, sphingosine was discovered to be a potent inhibitor (Fig. 2). When protein kinase C activity was tested using the Triton X-100 mixed micellar assay containing 6 mole % PS and 2 mole % $diC_{18:1}$, the concentration of sphingosine causing 50% inhibition occurred on a molar basis equivalent to $diC_{18:1}$ or 0.4 that of PS (Hannun *et al.*, 1986b). Inhibition occurred when only a few molecules of sphingosine were present on the micellar surface. Sphingosine did not inhibit the activity of the catalytic fragment provided that Triton X-100 micelles or phospholipid vesicles were present (Fig. 3) (Hannun *et al.*, 1986b). The potency of inhibition was strongly dependent on the concentration of mixed micelles present (Fig. 4),demonstrating that the effect of sphingosine was subject to surface dilution. When the data were expressed as mole % (sphingosine Triton X-100), identical inhibition was seen at each of the four levels of Triton X-100–PS–$diC_{18:1}$ mixed micelles employed (Fig. 4B). This demonstrated that the number of sphingosine molecules present per mixed micelle (microenvironment concentration) determined the potency. Bulk solution concentrations were meaningless. For amphipathic molecules, mole % is the appropriate concentration unit.

These results also indicated that sphingosine interacted with the surface-bound protein kinase C through the lipid binding domain, and detailed kinetic analysis supports this view (Hannun *et al.*, 1986b). Importantly, sphingosine appeared to be a competitive inhibitor of DAG (Fig. 5) and phorbol dibuty-

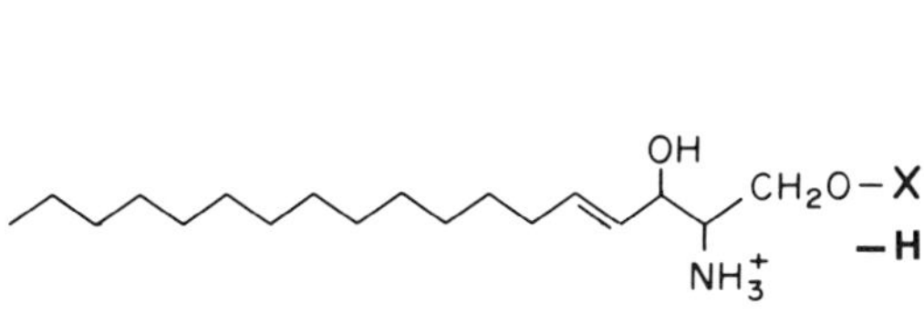

Figure 2. Structure of sphingosine and lysosphingolipids. Sphingosine possesses a hydrogen on the oxygen linked to C-1. Lysosphingolipids possess either phosphorylcholine, or carbohydrates linked to the C-1 oxygen. Lysosphingolipids are named after their parental compounds. See Fig. 8 for details of the structural diversity of X for the major sphingolipids. Many other possibilities exist as a consequence of the diversity of the glycosphingolipids. X,H–.

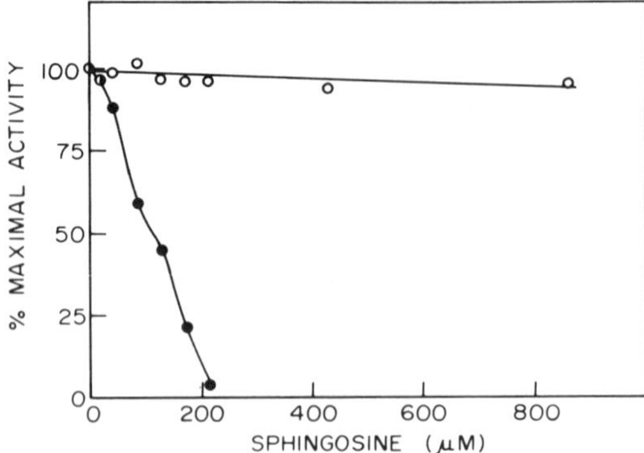

Figure 3. Inhibition of protein kinase C activity by sphingosine. Mixed micelles were formed with 3% (w/v) Triton X-100 containing PS at 6 mole %, $diC_{18:1}$ at 2 mole %, and sphingosine at tenfold the indicated concentrations. The mixed micelles were then diluted 1 : 10 into the assay mixture. One mole % of sphingosine is equivalent to 43μM. Effect of sphingosine on protein kinase C (●) and on protein kinase M (o). Identical results were obtained with protein kinase M when sphingosine was added in 0.3% Triton X-100 solution without PS and $diC_{18:1}$. Reproduced with permission from *The Journal of Biological Chemistry* **261:** 12604–12609 (1986).

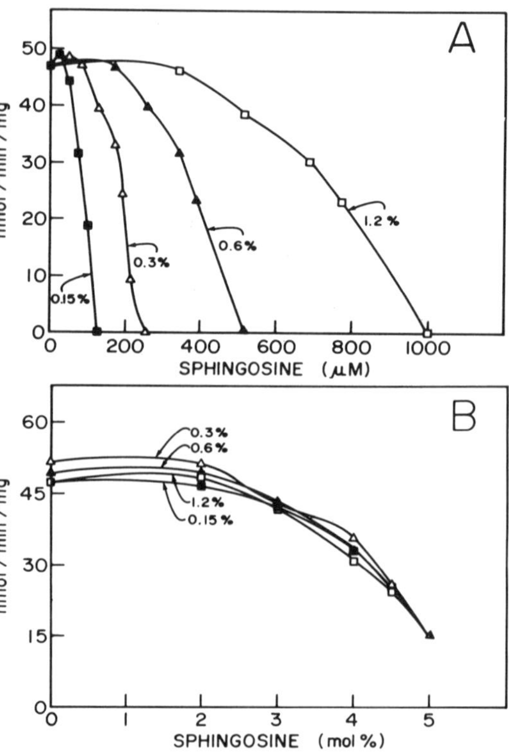

Figure 4. Effect of mixed micelle concentration on potency of sphingosine inhibition. (A) Mixed micelles were formed with Triton X-100 at 12% (w/v, □), 6% (▲), 3% (▲), and 1.5% (■) containing 7 mole % of PS, 1 mole % of $diC_{18:1}$, and sphingosine at ten fold the indicated concentrations. The mixed micelles were then diluted 1 : 10 into the assay mixture. (B)Mixed micelles were formed with 3% Triton X-100, 7 mole % of PS, 1 mole % of $diC_{18:1}$, and 2–5 mole % of sphingosine. These were then diluted 1 : 20 (■), 1 : 10 (Δ), 1 : 5 (▲), and 2 : 5 (□) into the assay mixture. In these experiments, 1 mole % of sphingosine corresponds to 21.5, 43, and 172 μM, respectively. Reproduced with permission from *The Journal of Biological Chemistry* **261:** 12604–12609 (1986).

rate (data not shown) (Hannun *et al.*, 1986b). Sphingosine was found to inhibit [^{3}H]PDBu binding to protein kinase C (Hannun *et al.*, 1986b). When the mechanism of inhibition was investigated by chromatography of protein kinase C–PS–Ca^{2+}–[^{3}H]PDBu mixed micellar complexes on Sephacryl S-200 columns, sphingosine was found to inhibit [^{3}H]PDBu binding but not to inhibit binding of protein kinase C to the mixed micelle (Fig. 6). This mechanism was consistent with a specific inhibition of [^{3}H]PDBu binding by sphingosine.

Lysosphingolipids differ from their parental sphingolipids in that they lack an amide-linked fatty acid at the 2-amino position of the sphingosine base (Fig. 2). Lysosphingolipids differ from sphingosine in that they have substituents linked through the C-1 oxygen (Fig. 2); lysosphingolipids share with sphingosine the two features noted earlier to be essential for inhibition—significant hydrophobic character and the free amino group (Hannun *et al.*, 1986b). Therefore, lysosphingolipids were prepared by base treatment of parental compounds, purified, and employed to ascertain their effects on protein kinase C activity and on [^{3}H]PPDBu binding (Hannun and Bell, 1987). All lysosphingolipids tested proved to be potent and reversible inhibitors, whereas none of the parental sphingolipids inhibited enzyme activity (Fig. 7A). As with sphingosine, lysosphingolipids were subject to surface dilution (Fig. 7B). Moreover, these compounds inhibited [^{3}H]PDBu binding

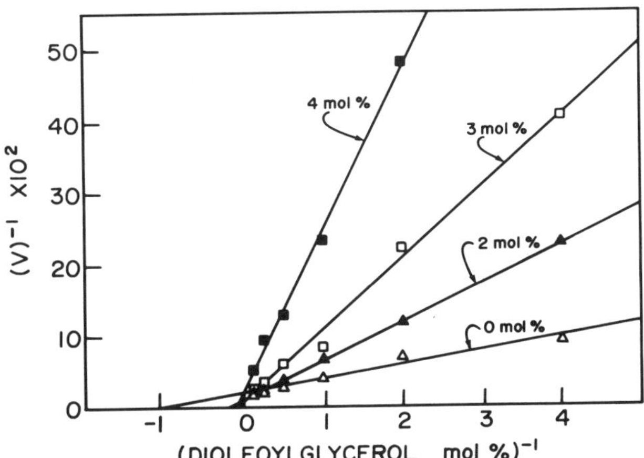

Figure 5. Interaction of sphingosine with deioleoylgycerol. Protein kinase C activity was assayed in the presence of 50 μM $CaC1_2$. Double reciprocal plots with $diC_{18:1}$ concentration in mole % are shown. Reproduced with permission from *The Journal of Biological Chemistry* **261:** 12604–12609 (1986).

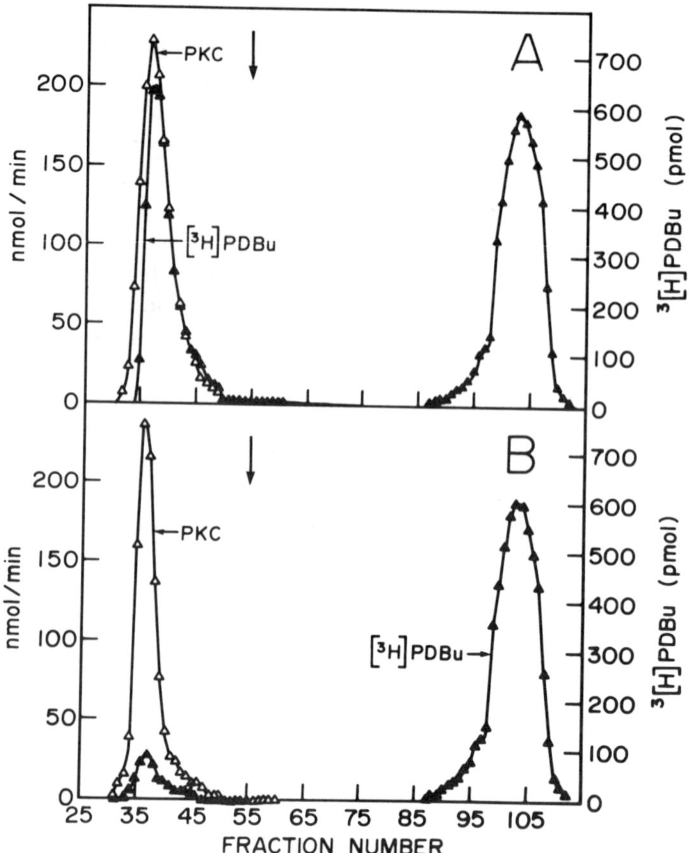

Figure 6. Interaction of PDBu and protein kinase C (PKC) with mixed micelles containing sphingosine on molecular sieves. (A) Mixed micelles contained 16 mole % of PS. A fraction of [^{3}H]PDBu (▲) coelutes with protein kinase C (Δ) at an $M_r = 200{,}000$. (B) Mixed micelles contained 16 mole % of PS and 12 mole % of sphingosine. Most of the bound [^{3}H]PDBu was displaced. Arrows indicate elution position of blue dextran protein kinase C when it is not bound to mixed micelles.

in a manner similar to sphingosine (Hannun and Bell, 1987). The lysosphingolipids found to inhibit protein kinase C are shown in Fig. 8.

3.2. *Inhibiton of Protein Kinase C in Cells by Sphingosine*

Several studies have investigated the effects of sphingosine on protein kinase C and on [^{3}H]PDBu binding in cells (Hannun *et al.*, 1986b; Merrill *et*

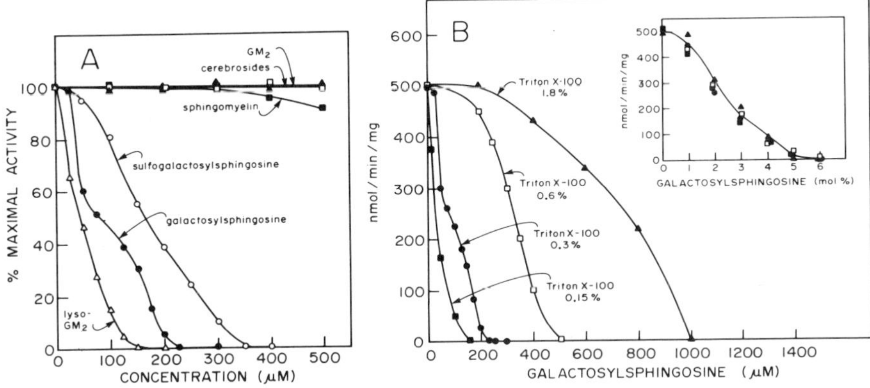

Figure 7. Inhibition of protein kinase C by lysosphingolipids. (A) Galactosylsphingosine (●), sulfogalactosylsphingosine (○), and lyso-GM_2 (Δ) were potent inhibitors of protein kinase C when activity was measured at 6 mole % phosphatidylserine, 2 mole % *sn*-1,2-dioleoyglycerol and 100 μM Ca^{2+}. These data exemplify the inhibition observed with the rest of the related lysosphingolipids (Table I). The parental sphingolipids were also evaluated for their effects on protein kinase C activity. They were dried down from chloroform–methanol solutions with the lipid cofactors of protein kinase C and solubilized in Triton X-100. None of the parent compounds inhibited protein kinase C, as shown for cerebrosides, sphingomyelin, and GM_2 (Hannun and Bell, 1987). To investigate further the specificity of inhibition of protein kinase by lysosphingolipids, the *N*-acetyl derivatives of galactosylsphingosine, sphingosylphosphorylcholine, and lyso-GM_2 were prepared as described (Hannun and Bell, 1987) for the *N*-acetyl derivative of sphingosine. The compounds were purified on a cyano high-performance liquid chromatography column and then tested for their effects on protein kinase C activity. None of these *N*-acetyl derivatives inhibited protein kinase C. (B) The lysosphingolipids displayed surface dilution kinetics. Galactosylsphingosine was less potent when protein kinase C was assayed at higher concentrations of Triton X-100 mixed micelles containing fixed amounts of phosphatidylserine (b mole %) and dioleoylglycerol (2 mole %). When the data are plotted as molar percentage of galactosphingosine Triton X-100 (0.15% w/v, ■; 0.3%, ●; 0.6%, □; and 1.8%, ▲) they appeared identical. These results indicate that the lysosphingolipids preferentially partition into the mixed micelles; therefore, effective concentrations should be expressed as the molar percentage of the lipid compounds to Triton X-100 and not as bulk concentrations. Reproduced with permission from *Science* **235:** 670–674 (1987).

al., 1986; Wilson *et al.*, 1986). Sphingosine inhibited protein kinase C activity in human platelets, as evidenced by monitoring the phosphorylation of a 40-kD polypeptide (Hannun *et al.*, 1986b). Sphingosine blocked phosphorylation when platelets were stimulated with thrombin or diC_8. Furthermore, sphingosine inhibited [^{3}H]PDBu binding to protein kinase C in human platelets (Hannun *et al.*, 1986b). N-acetylsphingosine did not inhibit activity *in vitro* or in human platelets. Certain lysosphingolipids also inhibited 40-kD polypeptide phosphorylation in human platelets (Hannun *et al.*, 1986b). Recent data indicate that sphingosine blocks platelet aggregation in response to

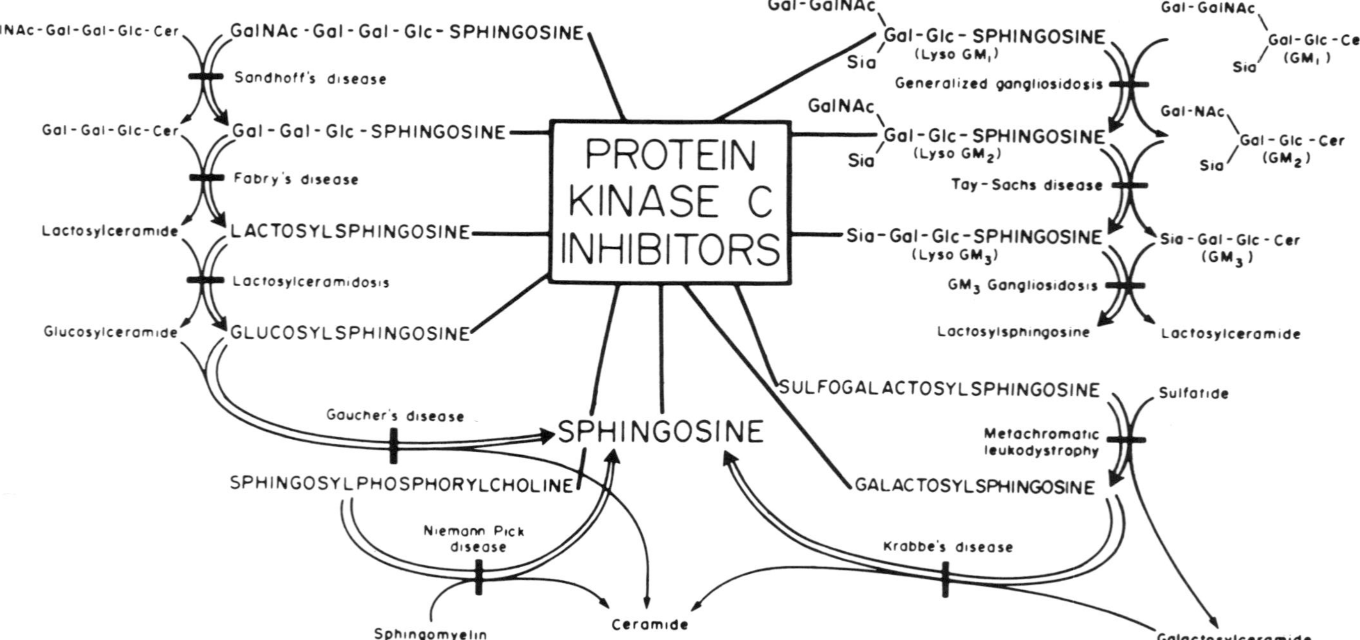

Figure 8. Proposed metabolic pathways for the degradation of lysosphingolipids. The pathways of sphingolipid catabolism are illustrated (thin arrows). Deficiency of enzymes in these pathways (solid bars) leads to the indicated diseases. It is proposed that the same enzymes also act on lysosphingolipids (double arrows); therefore, both the parent sphingolipid and the lysosphingolipid would accumulate in enzyme-deficient states. Sphingosine and lysosphingolipids differ from the parent sphingolipids in their ability to inhibit strongly protein kinase C, thus interfering in the function of important pathways of signal transduction and cell regulation. Also, the cytotoxicity of lysosphingolipids may ultimately lead to permanent tissue damage and cell death. Reproduced with permission from *Science* **235:**670–674 (1987).

numerous agonists by blocking second-wave aggregation that occurs as a consequence of secretory events; in these studies, sphingosine did not affect first-wave effects (shape change) (Y. Hannun, X. Greenberg, and R. Bell, unpublished observations). Sphingosine thus possesses antithrombotic activity.

Sphingosine inhibited the phorbol ester or DAG-dependent differentiation of the human promyelocytic leukemic HL60 cells (Merrill *et al.*, 1986). In these cells, sphingosine also inhibited [^{3}H]PFBu binding. These observations gave rise to the untested hypothesis that alteration of sphingosine levels may be the molecular basis of certain forms of human leukemias (Merrill *et al.*, 1986) (also see Section 3.3).

Sphingosine inhibited the oxidative burst of human neutrophils stimulated by opsonized zymosan, fMet-Leu-Phe, diC$_8$, phorbol esters, and arachidonate (Wilson *et al.*, 1986). Figure 9 shows the effect of sphinganine dihydrosphingosine on the oxidative burst (Wilson *et al.*, 1986). Incubation with sphinganine after PMA also led, after a lag, to complete inhibition of the oxidative burst (Fig. 9). The effects of sphinganine on oxygen consumption

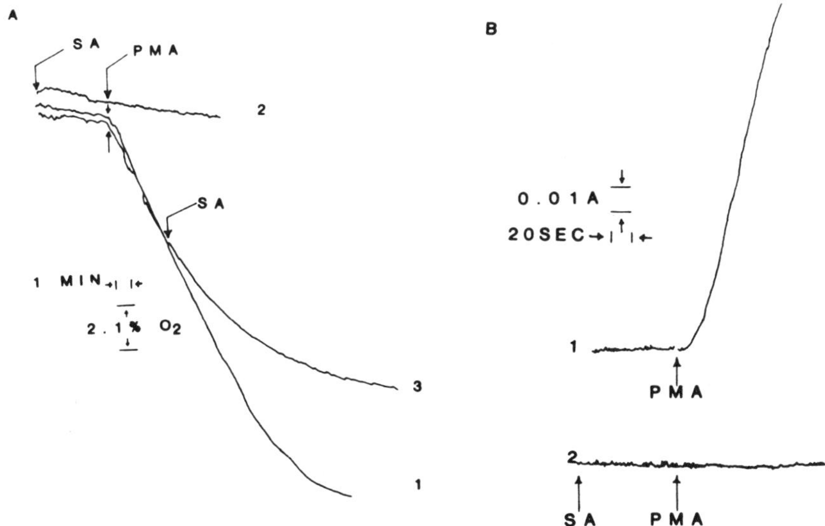

Figure 9. Effect of sphinganine (SA) on the oxidative burst. Panel (A) shows tracings of oxygraph recordings. Tracing 1 shows the effect of 1 μM PMA on oxygen consumption. In tracing 2, cells were preincubated for 3 min with sphinganine (50 μM) prior to addition of PMA. Tracing 3 shows the addition of 40 μM sphinganine after activation with PMA. Panel (B) shows recordings of reduction of cytochrome *c* (monitored at 550 nm). Tracing 1 shows PMA (1μM) induction of cytochrome *c* reduction. In control experiments, the cytochrome reduction was shown to be completely inhibited by superoxide dismutase. Tracing 2 shows effects of preincubation of cells with 10 μM sphinganine prior to addition of PMA. Reproduced with permission from *The Journal of Biological Chemistry* **261:**12616–12623 (1986).

triggered by the agents listed above are shown in Table I. Inhibition was shown to be reversible. It did not affect cell viability or normal calcium transients, and it occurred under conditions where [^{3}H]PDBu binding was inhibited (Wilson *et al.*, 1986). The data suggest that all these effectors work through a common pathway involving protein kinase C (Hannun *et al.*, 1986b; Merrill *et al.*, 1986; Wilson *et al.*, 1986). Sphingosine thus possesses anti-inflammatory activity.

3.3. Hypothesis

Sphingosine may prove to be a useful inhibitor of protein kinase C. Available data suggest that it may be a specific inhibitor; however, an insufficient number of studies have been done to demonstrate specificity. Sphingosine may exert its effects on targets other than protein kinase C.

Sphingosine and lysosphingolipids are natural components of cells and are related to a major class of cellular lipids, the sphingolipids, either as building blocks or as catabolic products. Sphingosine/lysosphingolipids may function physiologically as negative regulators of protein kinase C (Hannun and Bell, 1987). Sphingosine could be generated from ceramide by an *N*-acylsphingosine aminohydrolase; ceramide could be generated by deglycosylation or by cleavage of sphingomyelin. Lysosphingolipids, on the other hand, are a single hydrolytic step—N-deacylation—away from their parental lipids. Figure 10 presents the concept of protein kinase C regulation by positive and negative lipid second messengers.

The concept that sphingosine/lysosphingolipids function as second messengers is supported at present by little data. A specific antibody was observed to stimulate the deacylation and N-acylation of sphingomyelin in L-929 fibroblasts (Ulrich and Shearer, 1984). It is possible that a sphin-

Table I. Effect of Sphinganine on Oxygen Consumption Using Various Activators[a]

Activator	Rate[b] (nmole O_2 consumed/min/10^6 cells) No sphinganine	Sphinganine
Phorbol myristate acetate (1 μM)	4.3 ± 0.4[c]	0.2 ± 0.03
Dioctanoylglycerol (200 μM)	6.1 ± 1.0	0.3 ± 0.02
fMLP (0.5 μM) + cytochalasin B (5 μg/mL)	5.3 ± 0.4	0.3 ± 0.03
Opsonized zymosan (30 μg/mL)	5.3 ± 0.5	0.3 ± 0.02
Arachidonate (83 μM)	4.2 ± 0.6	0.2 ± 0.03

[a]Reproduced with permission of *The Journal of Biological Chemistry* **261:** 12616–12623 (1986).
[b]Oxygen consumption was measured using a Clark oxygen electrode, as described. For incubations in the presence of sphinganine, 50 μM sphinganine (final) plus 50 μM fatty acid-free BSA were added to the cells 3 min prior to addition of the activator. In control experiments, 50 μM BSA added alone prior to activation did not effect the activated rate.
[c]Data are expressed plus or minus the standard deviation of the mean.

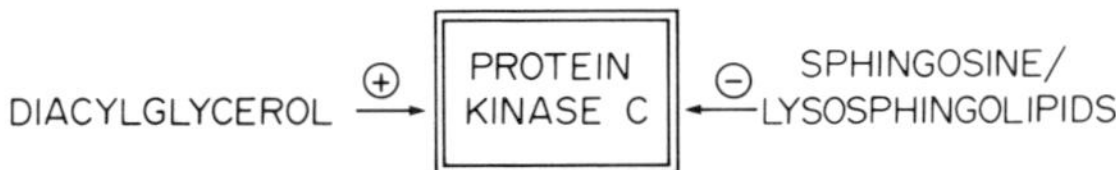

Figure 10. Positive and negative lipid effectors of protein Kinase C.

golipid cycle exists that is analogous to the phosphatidylinositol cycle (Fig. 11). [See Hannun *et al.*, (1986b) and Hannun and Bell (1987) for a more detailed discussion of the possible pathways giving rise to sphingosine/lysosphingolipid putative second messengers.] The antibody-dependent sphingomyelin N-deacylation–reacylation (Ulrich and Shearer, 1984) is one such cycle. It is unlikely that sphingosine involved in the biosynthetic pathways of sphingolipids would have such a physiological function, since the levels of sphinganine and sphingosine involved in synthesis are known to be exceedingly low (Merrill and Wang, 1986) and to turn over quickly. In this regard, sphingosine would be analogous to DAG—an ordinary metabolite in synthesis and degradation and an extraordinary metabolite in its second-messenger functions. Sphingosine is known to possess cytotoxic properties when added to CHO cells (Merrill, 1983); inhibition of protein kinase C may result in cytotoxicity.

If sphingosine/lysosphingolipids are physiological inhibitors, protein kinase C activity would be a function of the concentration of Ca^{2+}, DAG, and the negative effectors, sphingosine/lysosphingolipids. Such negative effectors could alter the sensitivity of protein kinase C to positive signals (DAG) and thereby modulate responses to extracellular agents that stimulate DAG production. Thus, a function for sphingosine/lysosphingolipids in transmembrane signaling may be emerging. It is too early to conclude that all the effects of sphingosine/lysophingolipids will occur as a consequence of modulating protein kinase C activity.

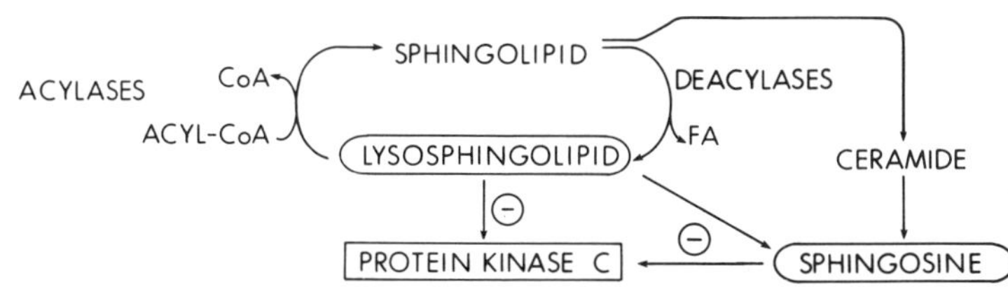

Figure 11. Sphingolipid cycle: a general depiction of breakdown and resynthesis of the sphingolipids to produce sphingosine/ lysosphingolipids.

The present data clearly indicate that sphingosine possesses interesting pharmacologic activities in human platelets (Hannun *et al.*, 1986b), neutrophils (Wilson *et al.*, 1986), and HL60 cells (Merrill *et al.*, 1986). The pharmacologic activities of sphingosine/lysosphingolipids are consistent with the possibility of physiologic inhibition. However, direct measurements of sphingosine/lysosphingolipid production in response to extracellular agents have not yet been reported. The demonstration that HL60 cells contain elevated levels of sphingosine bases (Merrill *et al.*, 1986) is consistent with *developmental trapping* occurring as a consequence of negative regulation of protein kinase C. In this speculative hypothesis, defects in the transmembrane signaling pathways giving rise to sphingosine may occur; HL60 cells can be caused to differentiate by addition of phorbol esters or diC_8, which would override the inhibitory signal.

The accumulation of lysosphingolipids in Krabbe's disease, Gaucher's disease, and other sphingolipidoses may cause the pathogenesis of these disorders (Hannun and Bell, 1987). The hypothesis that lysosphingolipid inhibition of protein kinase C represents the missing functional link between accumulation of sphingolipids and pathogenesis appears to unify existing data. The accumulation of lysosphingolipids would cause progressive dysfunction of signal transduction mechanisms vital for neural transmission, differentiation, development, and proliferation and would eventually lead to cell death. Direct evidence exists that lysosphingolipids are pathobiologic (Hannun and Bell, 1987). Thus, sphingosine/lysosphingolipids possess pathobiologic activities as well as pharmacologic activities. Direct evidence regarding the physiologic involvement of sphingosine/lysosphingolipids in the cellular regulation of protein kinase C activity is presently being sought.

The concept that the regulation of cell proliferation is a consequence of growth factors and negative growth regulators (the yin and yang) is supported by a large body of experimental data. The concept that protein kinase C regulation occurs as a consequence of positive (DAG) and negative (sphingosine/lysosphingolipid) lipid effectors could be partly an underlying chemical basis of the yin and yang. Certain oncogene products are known to cause elevated levels of DAG (the yin) (Fleischman *et al.*, 1986; Preiss *et al.*, 1986; Wolfman and Marcara, 1987). Recessive oncogenes might block sphingolipid catabolism and thus formation of sphingosine/lysosphingolipid (the yang). Therefore, carcinogenesis may occur as a consequence of elevation of DAG levels and loss of sphingosine/lysosphingolipid negative effects, the braking messenger system of negative growth regulation. Since glycosphingolipids are major tumor antigens (Hakomori, 1986; Thurin *et al.*, 1986), one cannot help but ask whether this is a direct result of failure to catabolize these glycosphingolipid tumor antigens or precursors by the pathways involved in negative growth regulation. Moreover, changes in glycosphingolipid are correlated with tumor progression (Hakomori, 1986; Thurin et al., 1986). Lysosphingolipids may be the mediators of these effects of sphingolipids. These hypotheses appear to unify existing data and may provide insight into heretofore unsuspected roles of sphingolipid metabolites in cellular regulation.

References

Anderson, W., Esterial, A., Tapiovaara, H., and Gopalakriskna, R., 1985, Altered subcellular distribution of protein kinase C (a phorbol ester receptor). Possible role in tumor promotion and the regulation of cell growth: Relationship to changes in adenylate cyclase activity, in: *Advances in Cyclic Nucleotide and Protein Phosphorylation Research,* (D. Cooper and K. Seaman, eds.) Vol. 19, pp.287–306, Raven Press, New York.

Ashendel, C., 1985, The phorbol ester receptor: A phospholipid-regulated protein kinase, *Biochim. Biophys. Acta* **822:**219–242.

Bell, R., 1986, Protein kinase C activation by diacylglycerol second messengers, *Cell* **45:**631–632.

Benjamin, C.W., Tarpley, W. G., and Gorman R. R., 1987, Loss of platelet-derived growth factor-stimulated phospholipase activity in NIH-3T3 cells expressing the EJ-*ras* oncogene, *Proc. Natl. Acad. Sci. U.S.A.* **84:**546–550.

Berridge, M., and Irvine, R., 1984, Inositol trisphosphate, a novel second messenger in cellular signal transduction, *Nature* **312:**315–321.

Boni, L., and Rando, R., 1985, The nature of protein kinase C activation by physically-defined phospholipid vesicles and diacylglycerols, *J. Biol. Chem.* **260:**10819–10825.

Coussens, L., Parker, P., Rhee, L., Yang-Feng, T., Chen, E., Watterfield, M., Francke, V., and Ullrich, A., 1986, Multiple, distinct forms of bovine and human protein kinase C suggest diversity in cellular signaling pathways, *Science* **233:**859–866.

Davis, R., Ganong, B., Bell, R., and Czech, M., 1985, *sn*-1,2-Dioctanoylglycerol. A cell-permeable diacylglycerol that mimics phorbol diester action on the epidermal growth factor receptor and mitogenesis, *J. Biol. Chem.* **260:**1562–1566.

Ebeling, J., Vandenbark, X., Kuhn, L., Ganong, B., Bell, R., and Niedel, J., 1985, Diacylglycerols mimic phorbol diester induction of leukemic cell differentiation, *Proc. Natl. Acad. Sci. U.S.A.* **82:**815–819.

Fleischman, L. F., Chahwala, S. B., and Cantley, L., 1986, *Ras*-transformed cells: Altered levels of phosphatidylinositol-4,5-bisphosphate and catabolite, *Science* **231:**407–410.

Ganong, B., Loomis, C.Hannun, Y., and Bell, R., 1986, Specificity and mechanism of protein kinase C activation by *sn*-1,2-diacylglycerols, *Proc. Natl. Acad. Sci. U.S.A.* **83:**1184–1188.

Ganong, B., Loomis, C., Hannun, Y., and Bell, R., 1987, Regulation of protein kinase C by lipid cofactors, in: *Cell Membranes: Methods and Reviews* (E., Elson, W., Frazier, and L., eds. Glaser) Plenum Press, New York.

Hannun, Y., and Bell, R., 1986, Phorbol ester binding and activation of protein kinase C on Triton X-100 mixed micelles containing phosphatidylserine, *J. Biol. Chem.* **261:**9341–9347.

Hannun, Y., and Bell, R., 1987, Lysosphingolipids inhibit protein kinase C: Implications for the sphingolipidoses, *Science* **235:**670–674.

Hannun, Y., Loomis, C., and Bell, R., 1985, Activation of protein kinase C by Triton X-100 mixed micelles containing diacylglycerol and phosphatidylserine, *J. Biol. Chem.* **260:**10039–10043.

Hannun, Y., Loomis, C., and Bell, R., 1986a, Protein kinase C activation in mixed micelles: Mechanistic implications of phospholipid, diacylglycerol and calcium dependencies, *J. Biol. Chem.* **261:**7184–7190.

Hannun, Y., Loomis, C., Merrill, A., and Bell, R., 1986b, Sphingosine inhibition of protein kinase C activity and of phorbol dibutyrate binding *in vitro* and in human platelets, *J. Biol. Chem.* **261:**12604–12609.

Hakomori, S., 1986, Glycosphingolipids: The composition of these membrane molecules changes dramatically with cell differentiation and the onset of cancer. Exploiting such changes could lead to improved diagnosis and treatment of cancer, *Sci. Am.* **254:**44–53.

Hoshijima, M., Kikuchi, A., Tanimoto, T., Kaibuchi, K., and Takai, Y., 1986, Formation of a phorbol ester binding fragment from protein kinase C by proteolytic digestion, *Cancer Res.* **46:**3000–3004.

Huang, K-P., and Huang, F., 1986, Conversion of protein kinase C from a Ca^{2+}-dependent to an independent form of phorbol ester-binding protein by digestion with trypsin, *Biochem. Biophys. Res. Commun.* **139:**320–326.

Inoue, M., Kishimoto, A., Takai, Y., and Nishizuka, Y., 1977, Studies on cyclic nucleotide-independent protein kinase and its proenzyme in mammalian tissues, *J. Biol. Chem.* **252:**7610–7616.

Jackowski, S., Rettenmier, C., Sherr, C., and Rock, C., 1986, A guanine nucleotide-dependent phosphatidylinositol 4,5-diphosphate phospholipase C in cells transformed by the v-*fms* and v-*fes* oncogenes, *J. Biol. Chem.* **261:**4978–4985.

Kennerly, D., Parker, C., and Sullivan, T., 1979, Use of diacylglycerol kinase to quantitate picomole levels of 1,2-diacylglycerol, *Anal. Biochem.* **98:**123–131.

Kishimoto, A., Takai, Y., Mori, T., Kikkawa, U., and Nishizuka, Y., 1980, Activation of calcium and phospholipid-dependent protein kinase by diacylglycerol, its possible relation to phosphatidylinositol turnover, *J. Biol. Chem.* **255:**2273–2276.

Knopf, J., Lee, M., Sultzman, L., Kritz, R., Loomis, C., Hewick, R., and Bell, R., 1986, Cloning and expression of multiple kinase C cDNAs, *Cell,* **46:**491–502.

Konig, B., Dinitto, P., and Blumberg, P., 1985, Stoichiometric binding of diacylglycerol to the phorbol ester receptor, *J. Cell Biochem.* **29:**37–44.

Lee, M., and Bell, R., 1986, The lipid binding, regulatory domain of protein kinase C, *J. Biol. Chem.* **261:**14867–14870.

Loomis, C., Walsh, J., and Bell, R., 1985, *sn*-1,2-Diacylglycerol kinase of *Escherichia coli;* purification, reconstitution and partial amino- and carboxyl-terminal analysis, *J. Biol. Chem.* **260:**4091–4097.

Majerus, P., Connolly, T., Deckmyn, H., Ross, T., Bross, T., Ishii, H., Bansal, V., and Wilson, D., 1986, The metabolism of phosphoinositide derived messenger molecules, *Science* **234:**1519–1526.

Makowske, M., Birnbaum, M., Ballester, R., and Rosen, O., 1986, A cDNA encoding protein kinase C identifies two species of mRNAs in brain and CH_3, *J. Biol. Chem.* **261:**13389–13392.

Merrill, A., 1983, Characterization of serine palmitoyltransferase activity in Chinese hamster ovary cells, *Biochim. Biophys. Acta* **754:**284–291.

Merrill, A., and Wang, E., 1986, Biosynthesis of long-chain (sphingoid) bases from serine by LM cells. Evidence for introduction of the 4-*trans*-double after *de novo* biosynthesis of *N*-acylsphinganine(s), *J. Biol. Chem.* **261:**3764–3769.

Merrill, A., Sereni, A., Stevens, V., Hannun, Y., Bell, R., and Kinkade, J., 1986, Inhibition of phorbol ester dependent differentiation of human promyelocytic leukemic (HL60) cells by sphinganine and other long-chain bases, *J. Biol. Chem.* **261:**12610–12615.

Nishizuka, Y., 1984a, The role of protein kinase C in cell surface signal transduction and tumor promotion, *Nature* **308:**693–698.

Nishizuka, Y., 1984b, Turnover of inositol phospholipids and signal transduction, *Science* **225:**1365–1370.

Nishizuka, Y., 1986, Studies and perspectives of protein kinase C, *Science* **233:**305–311.

Ohno, S., Kawasaki, H., Imajoh, S., Suzuki, K., Imagaki, M., Yokokura, H., Sakoh, T., and Hidaka, H., 1987, Tissue-specific expression of distinct types of rabbit protein kinase C, *Nature* **325:**161–166.

Ono, Y., Kurokawa, T., Kawahara, K., Nishimura, R., Igarashi, K., and Nishizuka, Y., 1986a, Cloning of rat brain protein kinase C complementary DNA, *FEBS Lett.* **203:**11–115.

Ono, Y., Kurokawa, T., Fujii, T., Kawakasa, K., Igarashi, K., Kikkawa, U., Ogita, K., and Nishizuka, Y., 1986b, Two types of complementary DNAs of rat brain protein kinase C, *FEBS Lett.* **206:**347–352.

Parker, P., Coussens, L., Totty, N., Rhee, L., Young, S., Chen, E., Stabel, S., Waterfield, M., and Ullrich, A., 1986, The complete primary structure of protein kinase C–the major phorbol ester receptor, *Science* **233:**853–859.

Preiss, J., Loomis, C., Bishop, W., Stein, R., Niedel, J., and Bell, R., 1986, Quantitative measurement of *sn*-1,2-diacylglycerols present in platelets, hepatocytes and *ras* and *sis*-transformed normal rat kidney cells, *J. Biol. Chem.* **261:**8597–8600.

Roussel, M., Dull, T., Rettenmier, C., Ralph, P., Ullrich, A., and Sherr, C., 1987, Transforming potential of the c-*fms* proto-oncogene (CSF-1 receptor), *Nature* **325:**549–552.

Rozengurt, E., 1986, Early signals in the mitogenic responses, *Science* **234:**161–166.

Rozengurt, E., Rodriquez-Pena, A., Coombs, M., and Sinnett-Smith, J., 1984, Diacylglycerol stimulates DNA synthesis and cell division in mouse 3T3 cells: Role of Ca^{2}-sensitive phospholipid-dependent protein kinase, *Proc. Natl. Acad. Sci, U.S.A.* **81:**5748–5752.

Takai, Y., Kishimoto, A., Inoue, M., and Nishizuka, Y., 1977a, Studies on a cyclic nucleotide-independent protein kinase and its proenzyme in mammalian tissues, *J. Biol. Chem.* **252:**7603–7609.

Takai, Y., Yamamoto, M., Inoue, M., Kishimoto, A., and Nishizuka, Y., 1977b, A proenzyme of cyclic nucleotide-independent protein kinase and its activation by calcium-dependent neutral protease from rat liver, *Biochem. Biophys. Res. Commun.* **77:**542–550.

Takai, Y., Kishimoto, A., Iwasa, Y., Kawahara, Y., Mori, T., and Nishizuka, Y., 1979a, Calcium-dependent activation of a multifunctional protein kinase by membrane phospholipids, *J. Biol. Chem.* **254:**3692–3695.

Takai, Y., Kishimoto, A., Iwasa, Y., Kawahara, Y., Mori, T., Nishizuka, Y., Tamura, A., and Fujii, T., 1979b, A role of membranes in the activation of a new multifunctional protein kinase system, *J. Biochem.* **86:**575–578.

Thurin, J., Thurin, M., Herlyn, M., Elder, D., Steplewski, Z., Clark, W., and Koprowski, H., 1986, GD2 ganglioside biosynthesis is a distinct biochemical event in human melanoma tumor progression, *FEBS Lett.* **209:**17–22.

Ulrich, R., and Shearer, W., 1984, Enhanced N-acylation of palmitic acid in sphingomyelin of antibody-stimulated L cells, *Biochem. Biophys. Res. Commun.* **121:**605–611.

Wakelam, M., Davies, S., Houslay, M., McKay, I., Marshall, C., and Hall, A., 1986, Normal $p21^{N-ras}$ couples bombesin and other growth factor receptors to inositol phosphate production, *Nature* **323:**172–175.

Walsh, J., an Bell, R., 1986, *sn*-1,2-Diacylglycerol kinase of *Escherichia coli:* Mixed micellar analysis of the phospholipid co-factor requirement and divalent cation dependence, *J. Biol. Chem.* **261:**6239–6247.

Wilson, E., Olcott, M., Bell, R., Merrill, A., and Lambeth, J., 1986, Inhibition of the oxidative burst in human neutrophils by sphingoid long chain bases: Role of protein kinase C in activation of the burst, *J. Biol. Chem.* **261:**12616–12623.

Wolfman, A., Marcara, I., 1987, Elevated levels of diacylglycerols and decreased phorbol ester sensitivity in *ras*-transformed fibroblasts, *Nature* **325:**359–361.

14

Regulation of Inositol Trisphosphate Formation and Action

JAMES W. PUTNEY, JR.

1. Introduction

Over 30 years have now passed since the original report by Hokin and Hokin on receptor-stimulated turnover of inositol lipids (Hokin and Hokin, 1953). Since that time, the phosphoinositides have enjoyed periods of interest, neglect, rekindled interest, controversy, and finally acceptance as important intermediaries in biological signaling processes in a wide variety of systems. The pivotal contributions that resulted in this tumultuous history came from a number of different laboratories. Michell's (1975) hypothesis that the phosphoinositides somehow served to couple receptors to cellular calcium mobilization provoked considerable research and criticism, but it was ultimately shown to be correct when Berridge, Irvine, and their collaborators demonstrated that inositol 1,4,5-trisphosphate [(1,4,5)IP_3], the water-soluble product of phospholipase C degradation of phosphatidylinositol 4,5-bisphosphate (PIP_2), was capable of releasing sequestered calcium intracellularly in a variety of biological systems (Berridge 1983, 1984; Berridge and Irvine, 1984; Streb *et al.*, 1983; Burgess *et al.*, 1984a,b; Prentki *et al.*, 1984). A parallel story evolved from the work in Nishizuka's laboratory (Nishizuka, 1983, 1984), which demonstrated that the other product of inositol lipid breakdown, diacylglycerol (DG) was also a cellular messenger; this apparently innocuous intermediate of phospholipid metabolism was shown to be a potent and specific activator of an ubiquitous protein kinase that Nishizuka designated as *C kinase*.

JAMES W. PUTNEY, JR. • Section of Calcium Regulation, Laboratory of Pharmacology, National Institute of Environmental Health Sciences, Research Triangle Park, North Carolina 27709.

This chapter summarizes recent work from the author's laboratory and from other groups that addresses fundamental questions about this important biological signaling system. These questions include the nature of the coupling of receptors to phospholipase C, the metabolism of the inositol phosphates, and the mechanism or mechanisms by which the inositol phosphates regulate cellular calcium metabolism.

2. *Control of Phospholipase C*

2.1. *Guanine Nucleotides*

The first receptor–effector system for which a role of guanine nucleotides was recognized was the adenylate cyclase system (Rodbell, 1980); Smigel *et al.*, 1984; Taylor and Merritt, 1986). It is now recognized that receptor coupling to adenylate cyclase involves an intermediate guanine-nucleotide-dependent regulatory protein (G protein) designated G_s (subscript $_s$ for "stimulation" of cyclase). In addition, a second G protein, G_i, is also known to mediate the actions of receptors that inhibit adenylate cyclase. Other G proteins have been described, including transducin, a G protein that mediates the coupling of photon excitation to the activation of a cyclic GMP phosphodiesterase, and G_o (subscript $_o$ for "other") purified from brain whose function is not yet clearly understood. All these proteins are believed to exist naturally as heterotrimers, composed of α, β and γ subunits. Receptor interaction changes the conformation of the G protein such that GTP displaces bound GDP from the α subunit and the α subunit dissociates from the β/γ complex. The dissociated GTP-bound α subunit of G_s activates adenylate cyclase. However, direct inhibition of cyclase by the α subunit of G_i has not been demonstrated. It has been suggested that inhibition of adenylate cyclase occurs through the liberation of a stoichiometric excess of β/γ subunits that associate with the free α subunits of G_s, resulting in inhibition (Smigel *et al.*, 1984). Regardless, it is apparent for the case of the receptors that regulate adenylate cyclase that the dissociation of the G protein into separate α and β/γ subunits is a necessary prerequisite for accomplishing the necessary effector function. Because receptor activation increases the affinity of the G proteins for GTP, agonists and GTP generally produce synergistic effects in activating or inhibiting adenylate cyclase. This synergism is especially apparent when nonhydrolyzable derivatives of GTP are used. This is because the reassociation of G-protein subunits is normally preceded by the hydrolysis of the bound GTP to GDP. In fact, these nonhydrolyzable derivatives of GTP (i.e., GTPγS, GppNHp) can bind to G proteins quite well in the absence of agonists; they can thus produce subunit dissociation and activation in the absence of receptor activation.

A characteristic of receptors that couple to their effectors through G proteins is that, in membrane preparations, the binding of agonists (but not antagonists) is inhibited by GTP and nonhydrolyzable guanine nucleotides through a decrease in apparent agonist affinity. The earliest indication that receptor coupling to phospholipase C might involve a G-protein mechanism similar to the adenylate cyclase system was the demonstration by a number of investigators that guanine nucleotides decrease the apparent affinity of agonists for receptors of this type (El-Refai *et al.*, 1979; Goodhardt *et al.*, 1982; Snavely and Insel, 1982; Evans *et al.*, 1985). In electrically permeabilized platelets, the effects of thrombin on secretion were potentiated by guanine nucleotides (Haslam and Davidson, 1984a,b). More recently, several investigators have demonstrated an activation by guanine nucleotides of PIP_2 phospholipase C in membranes or in permeable cells (Cockcroft and Gomperts, 1985; Gonzales and Crews, 1985; Litosch and Fain, 1985, 1986; Litosch *et al.*, 1985; Lucas *et al.*, 1985; Uhing *et al.*, 1985, 1986; Merritt *et al.*, 1986; Smith *et al.*, 1986; Taylor *et al.*, 1986a). In many cases, the appropriate synergistic interactions between agonists and guanine nucleotides have been reported. Similarly, activation of phospholipase C shows the same relative sensitivity to guanine nucleotide analogs as for regulation of adenylate cyclase: GTPγS>GppNHp>GTP (Merritt *et al.*, 1986).

Figure 1 shows results obtained when rat pancreatic acinar cells were preincubated with [^{3}H]inositol to label the endogenous inositol lipids and then electrically permeabilized (Baker and Knight, 1978) to permit experimental access to the intracellular side of the plasma membrane (Merritt *et al.*, 1986). In this preparation, GTPγS stimulated the formation of [^{3}H]IP_3. GppNHp also stimulated IP_3 formation but was less potent. GTP itself was without effect, but this was probably due to the presence of saturating levels of GTP already present in the permeable cell preparation (Merritt *et al.*, 1986). More importantly for establishing a role for this guanine-nucleotide-mediated process in the normal receptor mechanism, the effects of the nonhydrolyzable guanine nucleotides were potentiated by the phosphoinositide-linked receptor agonists, carbachol and caerulein. The effects of carbachol were blocked by the muscarinic receptor antagonist, atropine.

However, the effects on neither agonists nor guanine nucleotides were blocked by pertussis toxin in pancreatic acinar cells (Merritt *et al.*, 1986). Pertussis toxin blocks the receptor-mediated inhibition of adenylate cyclase by preventing receptor activation of G_i (Smigel *et al.*, 1984; Taylor and Merritt, 1986), and in some systems it blocks the receptor-mediated activation of phospholipase C (Ui, 1986). This latter finding suggested that G_i or a similar protein might be involved in the regulation of phospholipase C. However, in addition to the exocrine pancreas, a number of other phospholipase C-linked receptors have been shown to be insensitive to pertussis toxin. Fur-

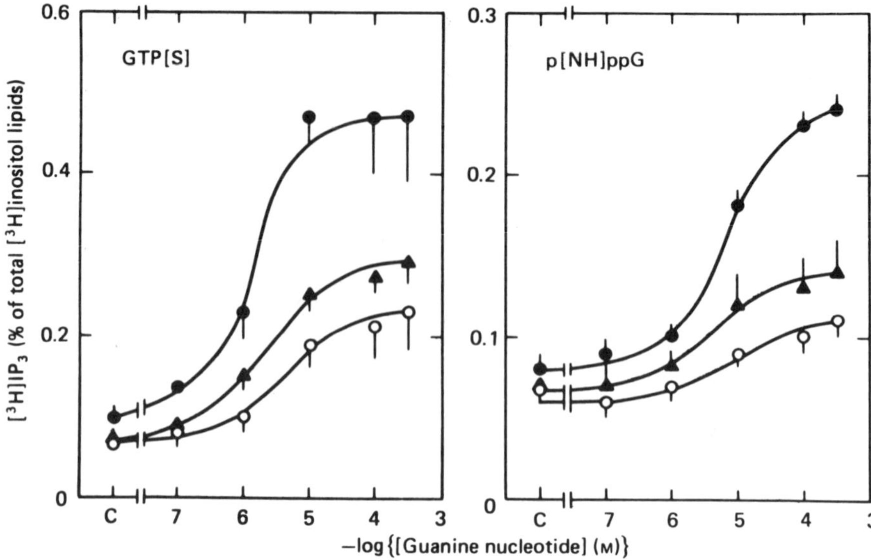

Figure 1. Concentration–response relation between stable guanine nucleotides and $[^3H]IP_3$ accumulation in the presence of receptor agonists: (o), no agonist (control); (▲), carbachol (0.1 mM); (●), caerulein (1.0 nM). For details see Merritt *et al.* (1986) from which this figure is taken with permission.

thermore, cholera toxin, which potently activates adenylate cyclase by causing the dissociation of G_s, does not activate the phospholipase C system. Collectively, these findings suggest that the coupling of receptors to phospholipase C involves a guanine-nucleotide-dependent regulatory protein that is similar, but not identical, to the proteins that regulate adenylate cyclase. The differential sensitivity to pertussis toxin may indicate that different G proteins are involved in the regulation of inositol lipid metabolism in different systems.

2.2. Calcium

For some time, it has been recognized that phospholipases, including phosphoinositide-specific phospholipase C, can be activated by Ca^{2+} (Michell, 1975). The question of whether calcium plays a role in regulation of the enzyme under physiological conditions has been somewhat controversial. Indeed, one of the main arguments raised by Michell (1975) in his proposal of the role of phosphoinositides in calcium mobilization was the relative independence of the process from changes in cytosolic $[Ca^{2+}]$. However, others have noted that inositol lipid turnover may be somewhat, or even entirely, dependent on extracellular calcium in certain model systems (Cockcroft, 1981).

The electrically permeabilized pancreatic acinar cell preparation described in Section 2.1 serves as a useful model to test the effects of [Ca^{2+}], guanine nucleotides, and agonists under controlled conditions. The data in Fig. 2 (Taylor *et al.*, 1986b) show that the peptide hormone caerulein activates IP_3 production equally well at a [Ca^{2+}] of 140 or 250 nM. However, when [Ca^{2+}] was buffered to very low levels (<10 nM), no activation occurred. This indicates a minimal requirement for calcium for agonist activation, which is apparently fully satisfied at 140 nM, the estimated free [Ca^{2+}] in the acinar cell under resting conditions (Taylor *et al.*, 1986b). When the [Ca^{2+}] was increased to 800 nM—a value close to that which occurs upon

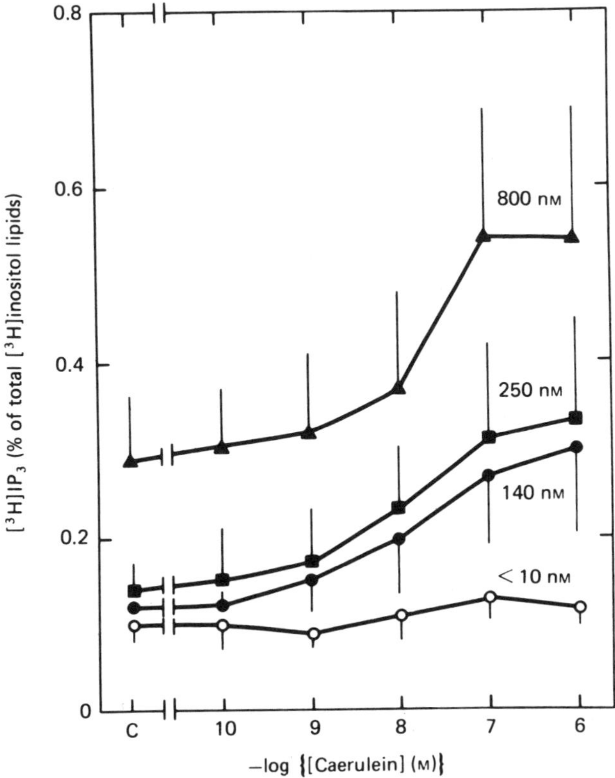

Figure 2. Effects of caerulein on [^3H]IP_3 formation at different free [Ca^{2+}] in permeabilized rat pancreatic acinar cells. For details see Taylor *et al.* (1986b) from which this figure is taken with permission.

agonist stimulation in the intact cell—the basal production of IP_3 was increased; however, caerulein still activated the enzyme, and the effect appeared to be additive with that of calcium. These data indicate that in addition to a requirement of a low level of calcium for receptor coupling to phospholipase C, high [Ca^{2+}] can directly activate the enzyme. Since the effects of high calcium and caerulein were additive, the two mechanisms of activation may be unrelated and indeed could even involve different enzymes (Taylor *et al.*, 1986b).

2.3. Receptors

Of course, the primary control of phospholipase C occurs at the level of the receptor. For receptors linked to adenylate cyclase, it has been known for some time that the receptors themselves are subject to self-imposed regulation. Most often, continuous stimulation leads to a diminution in the responsiveness to agonists, or a *desensitization*. When the signals generated by the agonist interaction with its receptor mediate the desensitization, as in the case of cyclic AMP-induced phosphorylation of the β-adrenergic receptor (Nambi *et al.*, 1985), the process is referred to as *heterologous* desensitization; this is because activation of any receptor linked to that same signaling system would induce the desensitization. In other cases, it appears that desensitization is specific for a particular receptor type; activation of the specific receptor does not generally cause desensitization of other receptors linked to the same signaling system, nor does the activation of these other receptors cause desensitization of the first. In such cases, the desensitization is referred to as *homologous* (Nambi *et al.*, 1985).

For receptors linked to the phosphoinositide cycle, there are numerous examples of heterologous desensitization. In most cases, the mechanism of heterologous desensitization appears to result from activation of the C kinase as a result of the formation of DG from inositol lipid breakdown (Leeb-Lundberg *et al.*, Lynch *et al.*, 1985; Vincenti *et al.*, 1985; Watson and Lapetina, 1985). An example of homologous desensitization of a phosphoinositide-linked receptor also has been described. In parotid acinar cells, activation of substance P receptors results in a rapid desensitization of the substance-P-induced formation of inositol phosphates (Sugiya *et al.*, 1987). Substance P receptor activation, however, does not induce a loss of responsiveness to muscarinic-cholinergic receptor stimulation with methacholine in these same cells, nor does methacholine cause any loss of responsiveness to substance P. The desensitization of the substance P response occurs within 10 seconds after the addition of agonist and is essentially complete by 1 min. The desensitization is paralleled by a loss of specific binding sites for substance P; thus the homologous desensitization apparently involves *down-regulation* of the substance P receptor (i.e., a disappearance of surface receptors). The recovery of the binding sites as well as the recovery of the responsiveness of the tissue to substance P is slow, requiring 1–2 h (Sugiya *et al.*, 1987). The mechanism underlying this homologous down regulation in the parotid gland is not known. It is interesting, and perhaps significant, that two receptor mechanisms coexisting in the same cell are linked to the same signaling system but regulated in such different ways.

3. Metabolism of Inositol 1,4,5-Trisphosphate

The inositol trisphosphate formed in stimulated cells was shown by Irvine *et al.* (1984) to be a mixture of two isomers: (1,4,5)IP_3, the expected breakdown product of PIP_2, and (1,3,4)IP_3. In most systems, (1,4,5)IP_3 increases before (1,3,4)IP_3, but on prolonged stimulation, the latter isomer predominates (Irvine *et al.*, 1984; Burgess *et al.*, 1985). The source of (1,3,4)IP_3 remained a mystery until the recent discovery by Batty *et al.* (1985) of an inositol tetrakisphosphate that was demonstrated to be (1,3,4,5)IP_4. Subsequently, Irvine *et al.* (1986a) demonstrated the presence of an inositol trisphosphate kinate that phosphorylated (1,4,5)IP_3 to (1,3,4,5)IP_4 in a variety of animal tissues. The latter compound was shown to be dephosphorylated by a 5-phosphatase [presumably the same one that degrades (1,4,5)IP_3 to (1,4)IP_2 (Tennes *et al.*, 1987)] to produce (1,3,4)IP_3 (Fig. 3). The most clearly demonstrable biological activity of an inositol phosphate is the ability of (1,4,5)IP_3 to release calcium from endoplasmic reticulum (discussed in more detail below). The general finding has been that (1,3,4,5)IP_4 does not release calcium or interfere with the ability of (1,4,5)IP_3 to do so, and so in the very least the phosphorylation of (1,4,5)IP_3 must serve as an inactivation mechanism. However, the recent findings of Irvine and Moor (1986) (discussed in detail in a later section) suggest that the phosphorylation of (1,4,5)IP_3 may serve to convert a calcium-releasing sig-

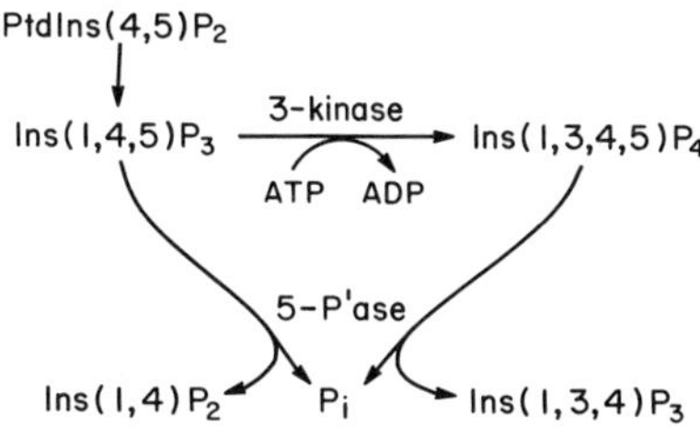

Figure 3. Initial steps in the metabolism of (1,4,5)IP_3. From Tennes *et al.* (1987) with permission.

nal into a modulator of calcium entry. Likewise, earlier indirect evidence suggested that (1,3,4)IP_3 was without biological activity (Burgess *et al.*, 1985), but a recent study (Irvine *et al.*, 1986b) demonstrated that this isomer was capable of releasing calcium, although higher concentrations were required than for (1,4,5)IP_3. Since the levels of (1,3,4)IP_3 in stimulated tissues are generally much greater than those of (1,4,5)IP_3, the possibility exists that this isomer could contribute to the intracellular calcium release signal under some conditions. What is sorely needed is a detailed characterization of the levels (not just labeling by radioactive precursors) and kinetics of changes in these various inositol phosphates, together with a detailed analysis of changes in calcium signaling and calcium pools.

4. Actions of Inositol Phosphates

4.1. Calcium Release

In virtually all systems thus far examined, receptor-activated calcium mobilization involves an initial phase of calcium release from intracellular stores, followed by a more prolonged phase of calcium entry (Putney *et al.*, 1981). The available evidence now strongly indicates that the internal release is signaled by (1,4,5)IP_3 (Berridge, 1983, 1984). Thus, resolution of the site of action of (1,4,5)IP_3 in cells would provide information on the locus of receptor-regulated calcium pools. Prior to the discovery of the second-messenger function of (1,4,5)IP_3, a number of laboratories approached this problem by examining the distribution of total and radiolabeled calcium in subcellular fractions prepared as rapidly as possible following agonist treatment, and with precautions to minimize redistribution during fractionation. This approach was employed by several laboratories in studies of calcium-mobilizing hormones in the liver. The general finding was that, following receptor activation, mitochondrial and microsomal fractions show a net loss of calcium (Exton, 1980; Williamson *et al.*, 1981). Other reports questioned the likelihood of mitochondrial involvement on the basis of data suggesting that normal, healthy cells do not generally contain appreciable quantities of calcium in their mitochondria (Burgess *et al.*, 1983).

For (1,4,5)IP_3, the results are more straightforward. Since the action of this mediator can be tested in permeable cells and subcellular fractions, specific poisons and other experimental manipulations can be used to determine unambiguously the contributions of mitochondrial and nonmitochondrial pools (Putney, 1987). Under these conditions, the results from a number of different laboratories concur that (1,4,5)IP_3 induces release of calcium from an intracellular pool that is insensitive to inhibitors of mitochondrial calcium uptake and is thus likely to be a component of the endoplasmic reticulum

(ER) (Streb *et al.*, 1983; Burgess *et al.*, 1984a,b; Joseph *et al.*, 1984; Prentki *et al.*, 1984).

However, in some cells there is evidence that all the ER stores of calcium may not be regulated by (1,4,5)IP_3 (Putney, 1987). For example, in hepatocytes, only about 30–40% of exchangeable calcium could be released by (1,4,5)IP_3 (Burgess *et al.*, 1984b). Figure 4 summarizes the results of experiments in which the effects of (1,4,5)IP_3 on the efflux of $^{45}Ca^{2+}$ from fully loaded ER were monitored following rapid depletion of ATP by glucose and hexokinase (Taylor and Putney, 1985). Under these conditions, calcium exit from the ER of the permeable cells resembles a monoexponential process with a half-time of about 5 min. This rate is dramatically increased by (1,4,5)IP_3 such that about 40% of the ^{45}Ca content is lost in 30 s. Thereafter, in the continued presence of (1,4,5)IP_3, the remaining 60% is lost at a rate similar to that observed in the absence of (1,4,5)IP_3. This remaining fraction of accumulated calcium presumably resides in a pool, or component of the ER, that is not subject to regulation by (1,4,5)IP_3.

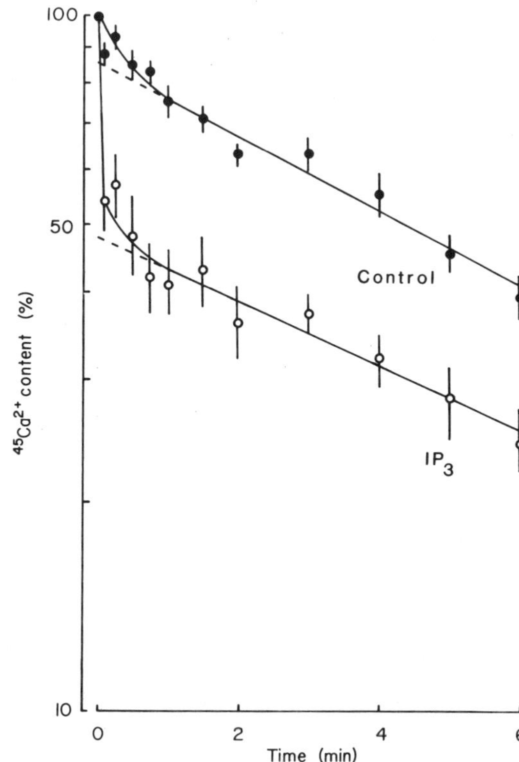

Figure 4. Effects of (1,4,5)IP_3 on ^{45}Ca efflux from the nonmitochondrial pool of permeabilized hepatocytes. Glucose (10 mM) and hexokinase (50 units/mL) were added alone (●) or with 5μM (1,4,5)IP_3 (o) to permeabilized guinea pig hepatocytes 13 min after the addition of ^{45}Ca and ATP. ^{45}Ca contents are shown as percentages of the ATP-dependent ^{45}Ca contents at the time of the additions ($t=0$). For details see Taylor and Putney (1985) from which this figure is taken with permission.

Based on the low concentration of $(1,4,5)IP_3$ necessary to release calcium and certain structural requirements for activity, it was suggested that $(1,4,5)IP_3$ acted by binding to a specific receptor on the ER (Burgess *et al.*, 1984a,b). Thus, the failure of a component of ER to respond to $(1,4,5)IP_3$ may result from the absence of the receptor in that fraction. Spat *et al.* (1986a,b) prepared ^{32}P-labeled $(1,4,5)IP_3$ of high specific activity and demonstrated the presence of specific binding sites in microsomes and permeable cells. In permeable hepatocytes, there was a good correlation between occupancy of these sites and the mobilization of calcium by both $(1,4,5)IP_3$ and $(2,4,5)IP_3$. There was also a reasonably good correlation between $(1,4,5)IP_3$ binding and calcium release in rabbit neutrophils, the latter being about tenfold more sensitive to $(1,4,5)IP_3$ than hepatocytes (Fig. 5). Despite the correlations obtained in these studies, the identity of these sites with the $(1,4,5)IP_3$ receptor is certainly not yet proved. What is sorely needed is a better selection of potent pharmacological probes to aid in the biochemical characterization of this important receptor site.

4.2. Calcium Entry

By comparison to calcium release, the regulation of calcium entry, or the second phase of calcium mobilization, is poorly understood (Putney, 1986). There is strong albeit indirect evidence that inositol lipid metabolism is important for calcium entry (Berridge and Fain, 1979), and recently, the injection of $(1,4,5)IP_3$ into sea urchin eggs produced a response pattern suggestive of an activation of calcium influx (Irvine and Moor, 1986; Slack *et al.*, 1986). However, the direct application of $(1,4,5)IP_3$ to plasma membranes does not increase their permeability to calcium (Delfert *et al.*, 1986; Ueda *et al.*, 1986). Collectively, these observations suggest that $(1,4,5)IP_3$ may activate calcium entry in cells, but not by acting directly on the plasma membrane.

Recently, a mechanism that is consistent with this idea has been proposed. According to this hypothesis, the emptying of a $(1,4,5)IP_3$-sensitive pool secondarily signals calcium entry. A detailed discussion of the circumstantial evidence supporting this idea has been presented elsewhere (Putney, 1986). The major evidence is derived from studies of the kinetics of emptying and refilling of a receptor-regulated calcium pool in the rat parotid gland. In the parotid gland under resting conditions, this intracellular calcium pool was resistent to depletion by extracellular chelating agents. However, when emptied by agonist stimulation, this pool could rapidly be filled from outside the cell, even in the absence of agonists and (presumably) second messengers, such as inositol phosphates (Aub *et al.*, 1982). These results suggest that the loss of calcium from this pool somehow activated a pathway for entry into the pool from the extracellular space. Thus, in the continued presence of agonist, when (1,4,5)IP_3 levels are being continuously maintained, the pool would presumably be held empty, the pathway from the extracellular space would be open, and calcium would enter the ER and subsequently the cytosol through the (1,4,5)IP_3-activated channels.

Recently, Irvine and Moor (1986) have presented evidence that (1,3,4,5)IP_4 could act as a specific signal for calcium entry in sea urchin eggs. (1,3,4,5)IP_4 is known to be formed by phosphorylation of (1,4,5)IP_3. In the sea urchin egg, injection of (1,4,5)IP_3 causes full activation of the eggs (i.e., raising of a fertilization envelope), and this is believed to involve intracellular calcium release as well as calcium entry. The data of Irvine and Moor (1986) show that (2,4,5)IP_3—an inositol phophate that is known to cause calcium release in other systems but which, they argue, may not be phosphorylated to an IP_4—does not fully activate eggs when injected intracellularly. Also, they found that (1,3,4,5)IP_4 did not activate eggs when injected on its own. However, when (1,3,4,5)IP_4 was injected together with the calcium-releasing inositol phosphate, (2,4,5)IP_3, a full fertilization response (i.e., envelope) was obtained. They suggest that when (1,4,5)IP_3 is

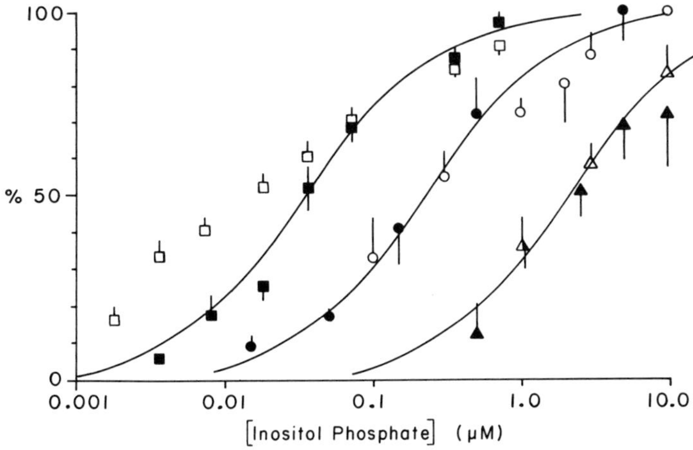

Figure 5. Comparison of inositol phosphate receptor occupancy and calcium release in guinea pig hepatocytes and rabbit neutrophils. Receptor occupancy (o,Δ,□) was measured by competition for specific binding of [^{32}P](1,4,5)IP_3, and calcium release (●,▲,■) was measured with ^{45}Ca and rapid filtration as described previously (Burgess *et al.*, 1984b). (o,●), (1,4,5)IP_3 in hepatocytes; (Δ, ▲), (2,4,5)IP_3 in hepatocytes; □, ■, (1,4,5)IP_3 in neutrophils. For details see Spat *et al.* (1986a) from which this figure is taken with permission.

injected, it causes calcium release, but that activation of calcium entry requires its phosphorylation to (1,3,4,5)IP_4. But since calcium release seemed to be a prerequisite for activation of calcium entry [i.e., (1,3,4,5)IP_4 only worked together with the calcium-releasing (2,4,5)IP_3], they concluded that the emptying of the (1,4,5)IP_3-regulated calcium pool was necessary for activation of calcium entry by (1,3,4,5)IP_4. These findings represent the first demonstrated biological action of (1,3,4,5)IP_4. Whether the specific mechanism suggested by Irvine and Moor (1986) is correct depends heavily on the assumptions made, which were based on known properties of the inositol phosphates in other biological systems. It will also be important to determine whether a mechanism similar to this can be demonstrated in a system other than the sea urchin egg. However, even within the context of the interpretation of Irvine and Moor, the data support the suggestion put forth earlier that the emptying of the (1,4,5)IP_3-regulated pool serves as a signal regulating calcium entry. (1,3,4,5)IP_4 may act in conjunction with this signal to produce a coordinated and tightly regulated biphasic pattern of calcium mobilization necessary for full expression of an appropriate physiological response.

5. *Conclusions*

In the past few years, our understanding of the relations between receptors, inositol lipids, and calcium has advanced at an incredible rate. Indeed, less than 4 years before the writing of this chapter, Streb *et al.* (1983) first demonstrated the biological activity of (1,4,5)IP_3; in the short interval since, this molecule has now gained general acceptance as an important and ubiquitous second messenger. Figure 6 summarizes some of the points emphasized in this chapter. It is apparent that this second-messenger system is exceedingly complex and possesses many potential sites of regulation. As with most important biochemical pathways, our knowledge raises numerous challenging questions: What is the nature of the G protein (or proteins) that couples receptors to phospholipase C? What is the significance of the alternate pathway (phosphorylation–dephosphorylation) of (1,4,5)IP_3 metabolism? What is the nature, mechanism, and locus of the (1,4,5)IP_3 receptor? How is calcium entry regulated? Indeed, how is the entire system regulated, and how does it interact with other signaling systems such as those for cyclic nucleotides? Hopefully, the substantial progress we have made in the past few years will continue toward the resolution of these issues and to pose new questions as well.

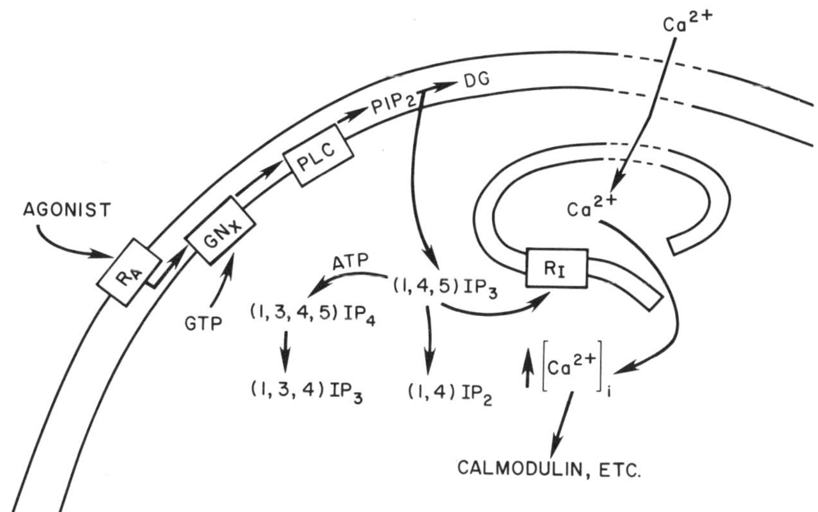

Figure 6. Calcium mobilization in nonexitable cells. The binding of agonist to its receptor (R_A) leads to activation of phospholipase C (PLC). A guanine-nucleotide-dependent regulatory protein (GN_x) mediates this coupling. PLC catalyzes the breakdown of PIP_2 into DG and $(1,4,5)IP_3$. $(1,4,5)IP_3$ can be either degraded to $(1,4)IP_2$ or phosphorylated to $(1,3,4,5)IP_4$. The latter compound can then be dephosphorylated to $(1,3,4)IP_3$. $(1,4,5)IP_3$ interacts with a specific receptor (R_I) on a subfraction of the endoplasmic reticulum, which results in the discharge of calcium to the cytosol. Calcium entry may then occur by a mechanism involving a pathway from the extracellular space into the endoplasmic reticulum, and subsequently into the cytosol.

References

Aub, D. L., McKinney, J. S., and Putney, J. W., Jr., 1982, Nature of the receptor-regulated calcium pool in the rat parotid gland, *J. Physiol. (Lond.)* **331:**557–565.

Baker, P. F., and Knight, D. E., 1978, Calcium-dependent exocytosis in bovine adrenal medullary cells with leaky plasma membranes, *Nature* **276:**620–622.

Batty, I. R., Nahorski, S. R., and Irvine, R. F., 1985, Rapid formation of inositol 1,3,4,5-tetrakisphosphate following muscarinic receptor stimulation of rat cerebral cortical slices, *Biochem. J.* **232:**211–215.

Berridge, M. J., 1983, Rapid accumulation of inositol trisphosphate reveals that agonists hydrolyze polyphosphoinositides instead of phosphatidylinositol, *Biochem J.* **212:**849–858.

Berridge, M. J., 1984, Inositol trisphosphate and diaclyglycerol as second messengers, *Biochem. J.* **220:**345–360.

Berridge, M. J., and Fain, J. N., 1979, Inhibition of phosphatidylinositol synthesis and the inactivation of calcium entry after prolonged exposure of the blowfly salivary gland to 5-hydroxytryptamine, *Biochem. J.* **178:**59–69.

Berridge, M. J., and Irvine, R. F., 1984, Inositol trisphosphate, a novel second messenger in cellular signal transduction, *Nature* **312:**315–321.

Burgess, G. M., McKinney, J. S., Fabiato, A., Leslie, B. A., and Putney, J. W., Jr., 1983, Calcium pools in saponin-permeabilized guinea-pig hepatocytes, *J. Biol. Chem.* **258:**15336–15345.

Burgess, G. M., Godfrey, P. P., McKinney, J. S., Berridge, M. J., Irvine, R. F., and Putney, J. W., Jr., 1984a, The second messenger linking receptor activation to internal Ca release in liver, *Nature* **309:**63–66.

Burgess, G. M., Irvine, R. F., Berridge, M. J., McKinney, J. S., and Putney, J. W., Jr., 1984b, Actions of inositol phosphates on Ca pools in guinea-pig hepatocytes, *Biochem.J.* **224:**741–746.

Burgess, G. M., McKinney, J. S., Irvine, R. F., and Putney, J. W., Jr., 1985, Inositol 1,4,5-trisphosphate and inositol 1,3,4-trisphosphate formation in Ca^{2+}-mobilizing hormone-activated cells, *Biochem. J.* **232:**237–243.

Cockcroft, S., 1981, Does phosphatidylinositol breakdown control the Ca^{2+}-gating mechanism?, *Trends Pharmacol. Sci.* **2:**340–342.

Cockcroft, S., and Gomperts, B. D., 1985, Role of guanine nucleotide binding protein in the activation of polyphosphoinositide phosphodiesterase, *Nature* **314:**534–536.

Delfert, D. M., Hill, S., Pershadsingh, H. H., Sherman, W. R., and McDonald, J. M., 1986, *myo*-Inositol 1,4,5-trisphosphate mobilizes Ca^{2+} from endoplasmic reticulum but not from plasma membrances, *Biochem J.* **236:**37–44.

El-Refai, M. F., Blackmore, P. F., and Exton, J. H., 1979, Evidence for two α-adrenergic binding sites in liver plasma membranes. Studies with [^{3}H]epinephrine and [^{3}H]dihydroergocryptine, *J. Biol. Chem.* **254:**4375–4386.

Evans, T., Martin, M. W., Hughes, A. R., and Harden, T. K., 1985, Guanine nucleotide-sensitive, high affinity binding of carbachol to muscarinic cholinergic receptors of 1321N1 astrocytoma cells is insensitive to pertussis toxin, *Mol. Pharmacol.* **27:**32–37.

Exton, J. H., 1980, Mechanisms involved in α-adrenergic phenomena: Role of calcium ions in actions of catecholamines in liver and other tissues, *Am. J. Physiol.* **238:**E3–E12.

Gonzales, R. A., and Crews, F. T., 1985, Guanine nucleotides stimulate production of inositol trisphosphate in rat cortical membranes, *Biochem. J.* **232:**799–804.

Goodhardt, M., Ferry, N., Geynet, P., and Hanoune, J., 1982, Hepatic α-adrenergic receptors show agonist-specific regulation by guanine nucleotides. Loss of nucleotide effect after adrenalectomy, *J. Biol. Chem.* **257:**11577–11583.

Haslam, R. J., and Davidson, M. M. L., 1984a, Potentiation by thrombin of the secretion of serotinin from permeabilized platelets equilibrated with Ca^{2+} buffers. Relationship to protein phosphorylation and diacylglycerol formation, *Biochem. J.* **222:**351–361.

Haslam, R. J., and Davidson, M. M. L., 1984b, Receptor-induced diacylglycerol formation in permeabilized platelets; possible role for a GTP-binding protein, *J. Receptor Res.* **4:**605–629.

Hokin, M. R., and Hokin, L. E., 1953, Enzyme secretion and the incorporation of ^{32}P into phospholipides of pancreas slices, *J. Biol. Chem.* **203:**967–977.

Irvine, R. F., and Moor, R. M., 1986, Micro-injection of inositol 1,3,4,5-tetrakisphosphate activates sea urchin eggs by a mechanism dependent on external Ca^{2+}, *Biochem. J.* **240:**917–920.

Irvine, R. F., Letcher, A. J., Lander, D. J., and Downes, C. P., 1984, Inositol trisphosphates in carbachol-stimulated rat parotid glands, *Biochem. J.* **223:**237–243.

Irvine, R. F., Letcher, A. J., Heslop, J. P., and Berridge, M. J., 1986a, The inositol tris/tetrakisphosphate pathway—Demonstration of Ins(1,4,5)P_3 3-kinase activity in animal tissues, *Nature* **320:**631–634.

Irvine, R. F., Letcher, A. J., Lander, D. J., and Berridge, M. J., 1986b, Specificity of inositol phosphate-stimulated Ca^{2+} mobilization from Swiss-mouse 3T3 cells, *Biochem J.* **240:**301–304.

Joseph, S. K., Thomas, A. P., Williams, R. J., Irvine, R. F., and Williamson, J. R., 1984, *myo*-Inositol 1,4,5-trisphosphate, a second messenger for the hormonal mobilization of intracellular Ca^{2+} in the liver, *J. Biol. Chem.* **259:**3077–3081.

Leeb-Lundberg, L. M., Cotecchia, S., Lomasney, J. W., DeBernardis, J. F., Lefkowitz, R. J., and Caron, M. G., 1985, Phorbol esters promote α-adrenergic receptor phosphorylation and receptor uncoupling from inositol phospholipid metabolism, *Proc. Natl. Acad. Sci. U.S.A.* **82:**5651–5655.

Litosch, I., and Fain, J. N., 1985, 5-Methyltryptamine stimulates phospholipase C-mediated breakdown of exogenous phosphoinositides by blowfly salivary gland membranes, *J. Biol. Chem.* **260:**16052–16055.

Litosch, I., and Fain, J. N., 1986, Regulation of phosphoinositide breakdown by guanine nucleotides, *Life Sci.* **39:**187–194.

Litosch, I., Wallis, C., and Fain, J. N., 1985, 5-Hydroxytryptamine stimulates inositol phosphate production in a cell-free system from blowfly salivary glands. Evidence for a role of GTP in coupling receptor activation to phosphoinositide breakdown, *J. Biol. Chem.* **260:**5464–5471.

Lucas, D. O., Bajjalieh, S. M., Kowalchyk, J. A., and Martin, T. F. J., 1985, Direct stimulation by thyrotropin-releasing hormone (TRH) of polyphosphoinositide hydrolysis in GH_3 cell membranes by a guanine nucleotide-modulated mechanism, *Biochem. Biophys. Res. Commun.* **132:**721–728.

Lynch, C. J., Charest, R., Bocckino, S. B., Exton, J. H., and Blackmore, P. F., 1985, Inhibition of hepatic α_1-adrenergic effects and binding by phorbol myristate acetate, *J. Biol. Chem.* **260:**2844–2851.

Merritt, J. E., Taylor, C. W., Rubin, R. P., and Putney, J. W., Jr., 1986, Evidence suggesting that a novel G-protein couples receptors to phospholipase C in exocrine pancreas, *Biochem. J.* **236:**337–343.

Michell, R. H., 1975, Inositol phospholipids and cell surface receptor function, *Biochim. Biophys. Acta* **415:**81–147.

Nambi, P., Stadel, J. M., Sibley, D. R., Strulovici, B., Caron, M. G., and Lefkowitz, R. J., 1985, Mechanisms of β-adrenergic receptor desensitization, *Sym. Med. Hoechst* **19:**437–451.

Nishizuka, Y., 1983, Calcium, phospholipid turnover and transmembrane signalling, *Philos.Trans.R.Soc.Lond.(Biol.)* **302:**101–112.

Nishizuka, Y., 1984, Turnover of inositol phospholipids and signal transduction, *Science* **225:**1365–1370.

Prentki, M., Biden, T. J., Janjic, D., Irvine, R. F., Berridge, M. J., and Wollheim, C. B., 1984, Rapid mobilization of Ca^{2+} from rat insulinoma microsomes by inositol 1,4,5-trisphosphate, *Nature* **309:**562–564.

Putney, J. W., Jr., 1986, A model for receptor-regulated calcium entry, *Cell Calcium,* **7:**1–12.

Putney, J. W., Jr., 1987, Formation and actions of the calcium mobilizing messenger, inositol 1,4,5-trisphosphate, *Am. J. Physiol. (Gastrointest. Liver Physiol.)* **252:**G149–G157.

Putney, J. W., Jr., Poggioli, J., and Weiss, S. J., 1981, Receptor regulation of calcium release and calcium permeability in parotid gland cells, *Philos. Trans. R. Soc. Lond. Biol.* **296:**37–45.

Rodbell, M., 1980, The role of hormone receptors and GTP-regulatory proteins in membrane transduction, *Nature* **284:**17–22.

Slack, B. E., Bell, J. E., and Benos, D. J., 1986, Inositol 1,4,5-trisphosphate injection mimics fertilization potentials in sea urchin eggs, *Am. J. Physiol. (Cell Physiol.)* **250:**C340–C344.

Smigel, M., Katada, T., Northup, J. K., Bokoch, G. M., Ui, M., and Gilman, A. G., 1984, Mechanisms of guanine nucleotide-mediated regulation of adenylate cyclase activity, *Adv. Cyclic Nucleotide Protein Phosphorylation Res.* **17:**1–18.

Smith, C. D., Cox, C. C., and Snyderman, R., 1986, Receptor-coupled activation of phosphoinositide-specific phospholipase C by an N protein, *Science* **232:**97–100.

Snavely, M. D., and Insel, P. A., 1982, Characterization of α-adrenergic receptor subtypes in the rat renal cortex. Differential regulation of α_1- and α_2-adrenergic receptors by guanyl nucleotides and Na^+, *Mol. Pharmacol.* **22:**532–546.

Spat, A., Bradford, P. G., McKinney, J. S., Rubin, R. P., and Putney, J. W., Jr., 1986a, A saturable receptor for [^{32}P]inositol-(1,4,5)trisphosphate in guinea pig hepatocytes and rabbit neutrophils, *Nature* **319:**514–516.

Spat, A., Fabiato, A., and Rubin, R. P., 1986b, Binding of inositol trisphosphate by a liver microsomal fraction, *Biochem. J.* **233:**929–932.

Streb, H., Irvine, R. F., Berridge, M. J., and Schulz, I., 1983, Release of Ca^{2+} from a nonmitochondrial intracellular store in pancreatic acinar cells by inositol-1,4,5-trisphosphate, *Nature* **306:**67–68.

Sugiya, H., Tennes, K. A., and Putney, J. W., Jr., 1987, Homologous desensitization of substance P-induced inositol polyphosphate formation in rat parotid acinar cells, *Biochem. J.* **244:**647–653.

Taylor, C. W., and Merritt, J. E., 1986, Receptor coupling to polyphosphoinositide turnover: A parallel with the adenylate cyclase system, *Trends Pharmacol. Sci.* **7:**238–242.

Taylor, C. W., and Putney, J. W., Jr., 1985, Size of the inositol 1,4,5-trisphosphate-sensitive calcium pool in guinea-pig hepatocytes, *Biochem. J.* **232:**435–438.

Taylor, C. W., Merritt, J. E., Putney, J. W., Jr., and Rubin, R. P., 1986a, A guanine nucleotide-dependent regulatory protein couples substance P receptors to phospholipase C in rat parotid gland, *Biochem. Biophys. Res. Commun.* **136:**362–368.

Taylor, C. W., Merritt, J. E., Putney, J. W., Jr., and Rubin, R. P., 1986b, Effects of Ca^{2+} on phosphoinositide breakdown in exocrine pancreas, *Biochem. J.* **238:** 765–772.

Tennes, K. A., McKinney, J. S., and Putney, J. W., Jr., 1987, Metabolism of inositol 1,4,5-trisphosphate in guinea pig hepatocytes, *Biochem. J.* **242:**797–802.

Ueda, T., Church, S. H., Noel, M. W., and Gill, D. L., 1986, Influence of inositol 1,4,5-trisphosphate and guanine nucleotides on intracellular calcium release within the N1E-115 neuronal cell line, *J. Biol. Chem.* **261:**3184–3192.

Uhing, R. J., Jiang, H., Prpic, V., and Exton, J. H., 1985, Regulation of a liver plasma membrane phosphoinositide phosphodiesterase by guanine nucleotides and calcium, *FEBS Lett.* **188:**317–320.

Uhing, R. J., Prpic, V., Jiang, H., and Exton, J. H., 1986, Hormone-stimulated polyphosphoinositide breakdown in rat liver plasma membranes, *J. Biol. Chem.* **261:**2140–2146.

Ui, M., 1986, Pertussis toxin as a probe of receptor coupling to inositol lipid metabolism, in: *Phosphoinositides and Receptor Mechanisms* (J. W. Putney, Jr., ed.) pp. 163–195, Alan R. Liss, New York.

Vincenti, L. M., Di Virgilio, F., Ambrosini, A., Pozzan, T., and Meldolesi, J., 1985, Tumor promoter phorbol 12-myristate, 13-acetate inhibits phosphoinositide hydrolysis and cytosolic Ca^{2+} rise induced by the activation of muscarinic receptors in PC12 cells, *Biochem. Biophys. Res. Commun.* **127:**310–317.

Watson, S. P., and Lapetina, E. G., 1985, 1,2-Diacylglycerol and phorbol ester inhibit agonist-induced formation of inositol phosphates in human platelets: Possible implications for negative feedback regulation of inositol phospholipid hydrolysis, *Proc. Natl. Acad. Sci. U.S.A.* **82:**2623–2626.

Williamson, J. R., Cooper, R. H., and Hoek, J. B., 1981, Role of calcium in the hormonal regulation of liver metabolism, *Biochim. Biophys. Acta* **639:**243–295.

V

RECEPTORS AND SIGNAL TRANSDUCTION PROCESSES INVOLVED IN INFLAMMATION

15

Molecular Properties of Leukocyte Receptors for Leukotrienes

CATHERINE H. KOO, LAURENT BAUD, JEFFREY W. SHERMAN, JEANNE P. HARVEY, DANIEL W. GOLDMAN, and EDWARD J. GOETZL

1. Introduction

The products of oxygenation of arachidonic acid constitute the most diverse family of inflammatory mediators, that are generated by many different pathways in almost all types of cells and are potent initiators and modulators of numerous biological functions. Distinct lipoxygenases convert arachidonic acid to an array of monohydroxy-eicosatetraenoic acids (HETEs) (see Appendix for list of abbreviations), di- and tri-HETEs, and peptide conjugates of HETEs, which participate with different functions in inflammatory and hypersensitivity reactions. The most potent of the arachidonic-acid-derived mediators of chemotaxis and other leukocyte functions is 5(S), 12(R)-dihydroxy-eicosa-6,14 *cis*-8,10-*trans*-tetraenoic acid or leukotriene B_4 (LTB_4), which is produced by the 5-lipoxygenase systems predominating in human polymorphonuclear (PMN) leukocytes, macrophages, and mast cells (Samuelsson, 1983). The C-6-sulfidopeptide leukotrienes, LTC_4, LTD_4, and LTE_4, from the 5-lipoxygenase cascade of mast cells and macrophages, are potent smooth muscle contractile and vasoactive factors but affect PMN leukocytes solely by increasing adherence to endothelial cells and other surfaces

---

CATHERINE H. KOO, LAURENT BAUD, JEFFREY W. SHERMAN, JEANNE P. HARVEY, DANIEL W. GOLDMAN, and EDWARD J. GOETZL • Howard Hughes Medical Institute and Departments of Medicine and Microbiology–Immunology, University of California Medical, Center San Francisco, California 94143-0724.

in vitro and *in vivo* (Goetzl *et al.*, 1983; Samuelsson, 1983; Hayaishi and Yamamoto, 1985). In contrast, the 15-lipoxygenase pathway is localized preferentially in epithelial cells of human skin, pulmonary airways, and the gastrointestinal tract (Burrall *et al.*, 1985; Hunter *et al.*, 1985; Krilis *et al.*, 1986) and transforms arachidonic acid into 15-HETE and multiple isomers of 8,15-di-HETE and 14,15-di-HETE. Although they act as weak agonists of some PMN leukocyte and lymphocyte functions, the most novel and potent effects of mediators from the 15-lipoxygenase pathway come from the 8,15-di-HETEs directed to the primary afferent nervous system and contributing to the hyperalgesia of inflammation and other forms of tissue injury.

The principal focus of research in this field over the past decade has been on isolating the mediators for structural characterization and synthesis and elucidating their range of functions, largely in model systems. The emphasis of current studies is on the molecular and cellular mechanisms of generation, expression of activity, and degradation of the most potent of the mediators of the cyclooxygenase and lipoxygenase pathways. The accelerated rate of recent findings of novel biochemical requirements for lipoxygenase activation, stereospecific receptors for leukotrienes, and distinctive involvement of leukotrienes in human diseases suggests that important diagnostic and therapeutic approaches will soon be available for clinical applications. This chapter describes the properties of leukocyte receptors for leukotrienes, which account for the stereospecificity and endogenous regulation of transduction of distinct functional effects of the leukotrienes.

2. *Specificity and Affinity of Human Leukocyte Receptors for Leukotrienes*

The complexity of the ligand structural determinants of binding of leukotrienes to human PMN leukocytes is similar to that characteristic of binding of the C-6-sulfidopeptide leukotrienes to smooth muscle and endothelial cells (Table I). Stereospecific binding sites for [^{3}H]LTB_4 were initially identified on human PMN leukocytes that had features consistent with a role as functionally important receptors for LTB_4 (Goetzl, 1980; Goldman and Goetzl, 1982, Kreisle and Parker, 1983). Human blood PMN leukocytes express two classes of receptors specific for LTB_4 that are clearly separate from those defined by similar techniques for the peptide chemotactic factors (Goldman and Goetzl, 1984). Of the total population of receptors detected on each intact PMN leukocyte, a mean of 4400 are of a high-affinity subset and 270,000 are of a low-affinity subset that binds [^{3}H]LTB_4 with respective mean dissociation constants (K_ds) of 0.4 and 61 nM (Goldman and Goetzl, 1984). In addition, the specificity of receptors in each subset differs signifi-

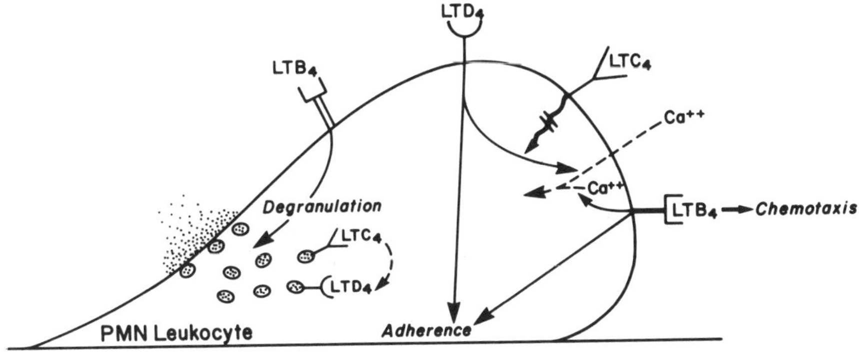

Figure 1. Diversity of specificity and function of human PMN leukocyte receptors for leukotrienes. → = chemotaxis, ⟶ = effect or action other than chemotaxis, - - → = conversion, influx, or release of a mediator or ion, ⇸ = lack of any direct effect.

cantly, as assessed by the rank order of competitive efficiency of a range of isomers in displacing the binding of [^{3}H]LTB_4. Both subsets of receptors bind native LTB_4 with the highest affinity of any of the 5,12-di-HETE isomers and exhibit lower affinity for isomers with different hydroxyl-group conformation and double bond geometry than LTB_4. The distinctive specificity of each subset is most apparent in comparisons of the affinity of LTB_4 with that of isomers with single structural changes. For example, 12(S)-LTB_4 exhibits the same differential affinity as LTB_4 in interacting with the two subsets, despite an approximately ten-fold higher K_d than LTB_4 for both types of receptors (Goldman and Goetzl, 1984). Each of the sets of receptors not only binds LTB_4 differently but is also coupled to a separate set of PMN leukocyte functions (Fig. 1). Occupancy of the high-affinity receptors at 0.2–10 nM concentrations of LTB_4 evokes a chemotactic response and increased adherence to surfaces, whereas occupancy of the low-affinity receptors at 20–200 nM LTB_4 elicits release of lysosomal enzymes and increased generation of superoxide (Goldman and Goetzl, 1984; Palmblad *et al.*, 1984).

Although characterized by substantial differences in subcellular distribution and coupling to function, the receptors for C-6-sulfidopeptide leukotrienes on intact PMN leukocytes resemble those for LTB_4 in affinity and specificity (Fig. 1). The mean affinity of the 10,000 receptors for [^{3}H]LTC_4 detected on intact human PMN leukocytes, as reflected in the K_d of approximately 30 nM, is similar to that of the low-affinity subset for LTB_4 and nearly one order of magnitude lower than that of LTC_4 receptors on endothelial and smooth muscle cells (Table I) (Krilis *et al.*, 1983; Chau *et al.*, 1986, 1987; Baud *et al.*, 1987a). The affinity of LTC_4 receptors on PMN

Table I. Characteristics of Leukotriene Receptors

Receptor specificity	LTB_4		LTC_4			LTD_4/E_4		
Target cell	Human	PMNL[a]	Human PMNL	Cultured hamster smooth muscle cell	Bovine aortic endothelial cell	Human PMNL	Human alveolar macrophage	Guinea pig lung tissue membranes
Mean binding affinity (nM)	0.4	61	26	4.4	6.8	~1	3.8	0.2–1.8
Mean number of receptors/cell ($\times 10^3$)	4.5	270	45	250	72	~40	—	—
Size of receptor binding unit (kD)		60		17,33			—	

[a]PMNL-polymorphonuclear leukocyte.

leukocytes for LTC_4 is 35- and 120-fold higher than for LTD_4 and LTE_4, respectively. The studies of binding of $[^3H]LTD_4$ to human PMN leukocytes have been compromised by the high levels of nonspecific binding and a four- to fivefold lower specific radioactivity of $[^3H]LTD_4$ than $[^3H]LTB_4$ Nonetheless, the apparent mean K_d for the LTD_4 receptors of approximately 1–5 nM reveals a more than tenfold higher affinity than LTC_4 receptors for the preferred ligand and a comparable stereospecificity (Koo *et al.*, 1987). The affinity of PMN leukocyte receptors for LTD_4 is similar to that of LTD_4 receptors on human alveolar macrophages and other types of cells (Kuehl *et al.*, 1984; Mong *et al.*, 1984; Opmeer and Hoogsteden, 1984; Sarau *et al.*, 1987), while the affinity of PMN leukocyte receptors for LTC_4 is approximately fivefold less than for that of smooth muscle and endothelial cells (Krilis *et al.*, 1983; Chau *et al.*, 1986, 1987) (Table I).

3. Subcellular Distribution of Leukotriene Receptors of Human PMN Leukocytes

The human PMN leukocyte receptors for LTB_4 are located primarily in the plasma membrane. Subcellular fractionation of PMN leukocytes revealed that the bulk of the LTB_4 binding sites are recovered with a membrane fraction that is enriched eight- to tenfold in several enzymatic markers of the plasma membrane and the endoplasmic reticulum. This membrane fraction expresses high-affinity and low-affinity binding sites for LTB_4 at respective densities of 3.4 and 27.9 pmoles/mg of protein with K_d values for LTB_4 of 0.07 and 28 nM, indicating a fivefold higher maximum affinity of the former class than the high-affinity receptors detected in intact PMN leukocytes (Sherman *et al.*, 1987). The receptors in the plasma-membrane-enriched preparation exhibited the same rank order of affinity for stereochemical isomers and analogues of LTB_4 as the subsets of receptors of intact PMN leukocytes. No specific binding activity for $[^3H]LTB_4$ was detected in the cytoplasm, and the minor amount present in granules of PMN leukocytes was attributable to plasma membrane contamination (Goldman *et al.*, 1987; Koo *et al.*, 1987).

In contrast, sonicates of human PMN leukocytes express approximately four times as many specific binding sites for $[^3H]LTC_4$ and $[^3H]LTD_4$ as intact PMN leukocytes. Scatchard analysis of the data from three studies with $[^3H]LTC_4$ revealed a single set of $44{,}536 \pm 12{,}927$ receptors per PMN leukocyte (mean $\pm$ SD) with a K_d of 25.5 ± 1.3 nM that was not statistically different from that for LTC_4 receptors on intact PMN leukocytes. The specific binding of $[^3H]LTC_4$ to separate preparations of lysosomal granules and membranes, which included fragments of the plasma membrane, again showed a single class of sites for each of the PMN leukocyte constituents, with respective mean K_d values ($\pm$SD) of 13.4 ± 1.3 and 38.3 ± 1.1 nM ($n=3$). The mean density of receptors ($\pm$SD) for [^{3}H]LTC_4 in the same preparations was 3.0 ± 1.0 pmoles/mg of protein and 15.9 ± 3.8 pmoles/mg of protein, respectively (Baud *et al.*, 1987a). The differences in receptor affinity and density between the membranes and granules were highly significant. The quantity of receptors for LTC_4 in 5×10^6 PMN leukocytes was calculated to be 312 fmoles in granules and 162 fmoles in membranes. Specificity for LTC_4 was the same for receptors in both compartments and on intact PMN leukocytes. That binding of LTC_4 was not inhibited by bilirubin and hematin was against a significant contribution from binding to glutathione *S*-transferase (Baud *et al.*, 1987a). In contrast, the intracellular receptors for LTC_4 in bovine aortic endothelial cells and guinea pig ileal smooth muscle cells were localized on mitochondrial and other membranes and exhibited an affinity equal to that of the plasma membrane subset of LTC_4 receptors (Chau *et al.*, 1986, 1987).

4. Structural Properties of the Leukotriene Receptors

A variety of affinity crosslinking techniques have been employed to radiolabel leukotriene receptor proteins prior to solubilization in different detergents, that have consistently eliminated detectable binding activity of the solubilized membrane proteins. The [^{3}H]*N*-(3-aminopropyl)-leukotriene B_4 amide ([^{3}H]LTB_4-APA) analogue of LTB_4 was affinity-crosslinked to receptors for LTB_4 in plasma-membrane-enriched preparations from human blood PMN leukocytes (Goldman *et al.*, 1985a). [^{3}H]LTB_4-APA bound to the receptors in membranes with one mean K_d of 2.3 nM at 4°C and was covalently coupled to the receptor binding unit with disuccinimidyl suberate (DSS). After solubilization in sodium dodecyl sulfate (SDS), the membrane proteins were resolved by SDS–polyacrylamide gel electrophoresis that revealed the predominance of crosslinking to one 60-kD protein (Table I). Crosslinking of the 60-kD protein was suppressed by LTB_4 and analogues of LTB_4, added prior to DSS, at concentrations of each that were similar to those capable of inhibiting the binding of [^{3}H]LTB_4 to the high-affinity receptors. The rank order of potency of $LTB_4 > 20\text{-OH } LTB_4 > LTB_4\text{-APA} = 5(S),12(R)\text{-6-}trans\text{-}LTB_4 \gg LTC_4 = LTD_4$ was identical for suppression of crosslinking of [^{3}H]LTB_4-APA and inhibition of binding of [^{3}H]LTB_4. Agents that reduced the affinity of LTB_4 receptors for [^{3}H]LTB_4, such as guanosine 5′-trisphosphate (GTP) in membrane preparations and pertussis toxin, inhibited to the same extent the crosslinking of [^{3}H]LTB_4-APA to the 60-kD membrane protein (Goldman *et al.*, 1985a). The stereospecificity and susceptibility to modulation of the crosslinking of [^{3}H]LTB$_4$-APA to the 60-kD protein support the identity of this membrane constituent as the binding unit of PMN leukocyte receptors for LTB$_4$ (Goldman *et al.*, 1987).

The LTC$_4$ receptors on bovine aortic endothelial cells and on ileal smooth muscle cells have been radiolabeled by similar affinity-crosslinking methods (Chau *et al.*, 1987). In the latter instance, a 17-and a 33-kD protein were most heavily radiolabeled with [^{3}H]LTC$_4$, but the background of other radiolabeled proteins was far greater than that observed in the pattern of [^{3}H]LTB$_4$-APA-labeled proteins. The affinity crosslinking of [^{3}H]LTC$_4$ to the ileal smooth muscle membrane receptors exhibited stereospecificity comparable to that of the primary binding interactions but was not assessed in relation to modulation of expression.

5. *Biochemical Transduction of Signals from Leukocyte Receptors for Leukotrienes*

The occupancy of PMN leukocyte receptors by LTB$_4$ and other chemotactic factors elicits cellular biochemical responses, but a relation among these events and between any of them and specific leukocyte functions has not been established unequivocally. LTB$_4$ and peptide chemotactic factors evoke rapid and transient increases in the cytosolic concentration of intracellular calcium ([Ca^{2+}]$_i$) in human PMN leukocytes (Lew *et al.*, 1984; Goldman and Goetzl, 1985; Goldman *et al.*, 1985b) and myelocytes derived from HL60 cells (Naccache *et al.*, 1984; Benjamin *et al.*, 1985; Goldman *et al.*, 1986). The increases in [Ca^{2+}]i are LTB$_4$ stereospecific and concentration dependent, related in magnitude to the concentration of extracellular calcium, unaffected by the calcium channel blocker nifedipine, and unrelated to changes in membrane potential. Decreases in the expression of LTB$_4$ receptors that result from prior exposure of PMN leukocytes to deactivating concentrations of LTB$_4$ or pertussis toxin suppress in parallel the magnitude of the responses of [Ca^{2+}]$_i$ to LTB$_4$. Stimuli of the activity of protein kinase C, such as phorbol esters, also dramatically reduce the sensitivity of the [Ca^{2+}]$_i$ response in PMN leukocytes to LTB$_4$, with a far lesser reduction in that evoked by *N*-formyl-methionyl-leucyl-phenylalanine (Goldman, 1987). Any dependence of the latter effect on phosphorylation of the 60-kD binding unit or the associated 41-kD guanine nucleotide-binding protein remains to be elucidated.

The C-6-sulfidopeptide leukotriene, LTD$_4$, also evokes increases in [Ca^{2+}]$_i$ in HL60 cell-derived myelocytes, as assessed by changes in quin-2 fluorescence (Baud *et al.*, 1987b) (Fig. 1). The increases in [Ca^{2+}]$_i$ are as rapid and transient as those elicited by LTB$_4$ and attain 50% of the maximum level at 1.2 nM LTD_4 that correlates with the apparent K_d for the LTD_4 receptors. As for LTB_4, the increases in $[Ca^{2+}]_i$ stimulated by LTD_4 are deactivated by prior exposure to LTD_4, inhibited by pertussis toxin and receptor-directed antagonists of LTD_4, unaffected by nifedipine, and unrelated to changes in membrane potential determined by 3,3′-dipentyl oxacarbocyanine (Baud *et al.*, 1987b). The increases in $[Ca^{2+}]_i$ elicited by LTD_4 were receptor dependent and related temporally to increases in myelocytic leukocyte adherence to surfaces (Fig. 1), but as for similar events stimulated by LTB_4 a direct mechanistic relationship must still be established. The receptors for LTC_4 on human PMN leukocyte membranes are apparently not coupled to any biochemical pathways. If the conversion of LTC_4 to LTD_4 is inhibited, LTC_4 does not elicit an increase in $[Ca^{2+}]_i$ or in PMN leukocyte adherence to surface (Baud *et al.*, 1987b) (Fig. 1).

LTB_4 and peptide chemotactic factors evoke in PMN leukocytes and HL60 cell-derived myelocytes a brief decrease in intracellular pH (pH_i) that is followed by a more prolonged alkalinization of the cytoplasm (Baud *et al.*, 1987c). The brief acidification, but not the subsequent alkalinization, requires a rise in $[Ca^{2+}]_i$. The capacity to inhibit the chemotactic factor-elicited alkalinization of PMN leukocytes by either prior addition of amiloride or replacement of extracellular Na^+ with organic monovalent cations suggests a role for the Na^+/H^+ antiport (Molski *et al.*, 1980; Grinstein and Furuya, 1984). LTD_4, which affects PMN leukocyte function solely by increasing adherence to surfaces, increases the pH_i of PMN leukocytes and HL60 cell-derived myelocytes with the same time course as LTB_4, but without an initial phase of acidification (Baud *et al.*, 1987c). In contrast, LTC_4 and LTE_4 have no effect on pH_i. The LTD_4-induced alkalinization is suppressed by prior exposure of myelocytic leukocytes to LTD_4 or to amiloride. The effect of the latter agent and the finding that LTD_4 enhances the rate of recovery of pH_i after introduction of an exogenous acid load support an involvement of the Na^+/H^+ antiport in the LTD_4-as well as LTB_4-induced alkalinization of myelocytic leukocytes. The lack of dependence of the alkalinization on extracellular $[Ca^{2+}]$ concentration and the failure of ionomycin-elicited increases in $[Ca^{2+}]_i$ to increase the pH_i confirm the independence of the alterations in $[Ca^{2+}]_i$ and pH_i after stimulation of myelocytic leukocytes by LTD_4.

Leukocyte receptors for LTB_4, along with a variety of hormone receptors, utilize GTP-binding proteins to regulate the coupling of receptor occupancy to biochemical transductional events. Pertussis toxin covalently modifies GTP-dependent regulatory proteins in a manner that alters the coupling of a receptor to cellular functions. Pretreatment of PMN leukocytes with pertussis toxin selectively reduces the expression of high-affinity receptors for LTB_4 and also inhibits LTB_4-induced increases in $[Ca^{2+}]_i$ and activation by LTB_4 of a range of functions. Further support for the association of the

LTB_4 receptor with a GTP-dependent regulatory protein comes from studies of the modulation of binding of [^{3}H]LTB_4 to plasma-membrane-enriched preparations from human PMN leukocytes (Sherman *et al.*, 1987). Preincubation of PMN leukocyte-derived membranes with GTP, GDP, and analogues of GTP, but not other nucleotides, converts a maximum of approximately 50% of the high-affinity receptors to low-affinity receptors for LTB_4 without changing either K_d or the total number of receptors. Pertussis toxin pretreatment of PMN leukocytes altered the membranes irreversibly, increasing the K_d of the high-affinity LTB_4 receptors to a level similar to that attained by GTP (Goldman *et al.*, 1985c).

The site of action of pertussis toxin is recognized by incubating PMN leukocyte homogenates with pertussis toxin in the presence of [^{32}P]-labeled NAD, the reactant for the ADP-ribosylation of guanine nucleotide-binding proteins. Electrophoretic analysis of the radiolabeled proteins in the PMN leukocyte homogenates reveals a single [^{32}P]-labeled component of about 41 kD. The radiolabeling is prevented by prior ADP-ribosylation with nonradioactive NAD. This 41-kD protein coelectrophoreses with the GTP-dependent protein, Ni, that is the pertussis toxin substrate of human erythrocytes but may differ significantly in primary structure.

6. Antibodies to Leukotriene Receptors

Initial studies of the immunochemical characteristics of LTB_4 receptors were directed to the preparation of rabbit polyclonal antibodies to the combining site. Rabbit anti-idiotypic IgG antibodies were raised to the combining site of a mouse monoclonal IgG_{2b} antibody to LTB_4, which binds [^{3}H]LTB_4 with a specificity similar to that of the LTB_4 receptors on human PMN leukocytes (Lee *et al.*, 1984). The rabbit IgG antibodies to the monoclonal anti-LTB_4 antibody combining site cross-reacted with the combining site of human PMN leukocyte LTB_4 receptors. IgG and Fab of the anti-idiotypic antibodies inhibited the binding of [^{3}H]LTB_4, but not chemotactic peptides, to PMN leukocytes with similar concentration–effect relations, while neither nonimmune rabbit IgG nor Fab had any inhibitory effects (Gifford *et al.*, 1987). At a concentration of anti-idiotypic IgG that inhibited by 50% the binding of [^{3}H]LTB_4 to PMN leukocytes, the high-affinity sites were recognized preferentially.

At concentrations that inhibited binding of [^{3}H]LTB_4 to PMN leukocytes, anti-idiotypic IgG and Fab suppressed PMN leukocyte chemotactic responses to LTB_4, but not chemotactic peptides, and failed to alter the release of lysosomal enzymes evoked by high concentrations of LTB_4. Increases in $[Ca^{2+}]_i$ and chemotaxis of PMN leukocytes of a magnitude equal to those responses elicited by LTB_4 were attained by the anti-idiotypic antibodies. The concentrations of anti-idiotypic IgG that maximally stimulated increases in $[Ca^{2+}]_i$ and chemotaxis, respectively, were $\frac{1}{25}$ of and equal to the level required to inhibit binding of $[^3H]LTB_4$ by 50%. The anti-idiotypic antibodies thus interacted with the combining site of LTB_4 receptors on PMN leukocytes with the specificity, preference for the high-affinity subset, and functional consequences similar to the LTB_4 ligand (Gifford *et al.*, 1987).

A panel of mouse monoclonal antibodies to framework determinants of the PMN leukocyte receptors for LTB_4 have now been prepared by immunizing mice with the purified 60-kD binding protein. The functional effects of these monoclonal antibodies are being defined in comparison with those of the polyclonal anti-idiotypic antibodies.

7. Development of Leukotriene Receptors During Differentiation of Leukocytes

The HL60 cultured line of human promyelocytic cells (Gallagher *et al.*, 1979) has served as an important *in vitro* model for analysis of the expression and transductional mechanisms of receptors for LTB_4 and other chemotactic factors (Brown *et al.*, 1984; Kitagawa *et al.*, 1984; Naccache *et al.*, 1984; Krause *et al.*, 1985). The stimulus-dependent induction of differentiation of HL60 cells into mature leukocytes with the morphological and biochemical properties of myelocytes, monocytes, or eosinophils (Collins *et al.*, 1978; Rovera *et al.*, 1979; Fischkoff *et al.*, 1984) is associated with the development of receptors for chemotactic factors (Fontana *et al.*, 1980; Niedel *et al.*, 1980; Goldman *et al.*, 1986). The undifferentiated HL60 cells express no complete receptors for chemotactic factors, but HL60 cell-derived myelocytes and monocytes have receptors for *N*-formyl-methionyl-leucyl-phenylalanine and LTB_4 that transduce increases in $[Ca^{2+}]_i$, chemotaxis, lysosomal degranulation, and bursts of oxidative metabolism (Fontana *et al.*, 1980; Niedel *et al.*, 1980; Goldman *et al.*, 1986). The exposure of HL60 cells to 1α,25-$(OH)_2$ vitamin D_3 induces the development of receptors for *N*-formyl-methionyl-leucyl-phenylalanine with a density several-fold greater than that found on dimethylsulfoxide- or retinoic acid-differentiated HL60 cells (McCarthy *et al.*, 1983).

The incubation of HL60 cells for 7 days with 100 nM 1α,25-$(OH)_2$ vitamin D_3 induced differentiation into monocyte-like cells, by standard criteria, that expressed surface receptors for LTB_4 (Goldman *et al.*, 1986). The HL60 cell-derived monocytes had a high-affinity subset of 6400 ± 3700 receptors per cell with a K_d of 2.3 ± 1.0 nM (mean $\pm$ SD) and a low-affinity subset of approximately 2.2×10^6 receptors per cell with an apparent K_d of

680 ± 410 nM. Derivatives and analogues of LTB_4, but not other chemotactic factors, inhibited the binding of [^{3}H]LTB_4 to HL60 cell-derived monocytes with a rank order of potency identical to that observed for LTB_4 receptors on PMN leukocytes. LTB_4 stimulated an increase in $[Ca^{2+}]_i$ in HL60 cell-derived monocytes that reached 50% of the maximum level at an LTB_4 concentration of 0.5 nM, defined as the EC_{50}. As for PMN leukocytes, preincubation of the monocytes with LTB_4 selectively reduced the expression of the high-affinity subset of LTB_4 receptors and increased by approximately 200-fold the EC_{50} for stimulation of increases in $[Ca^{2+}]_i$ by LTB_4, without altering the properties or functions of low-affinity receptors for LTB_4 or those for *N*-formyl-methionyl-leucyl-phenylalanine (Goldman *et al.*, 1986). Thus, the LTB_4 receptors that develop on HL60 cell-derived monocytes resemble those on human blood PMN leukocytes with respect to specificity, subset distribution, and transductional characteristics.

8. Clinical Implications

The leukotrienes have been detected at functionally relevant concentrations in the fluids and tissues of patients with a wide range of inflammatory states and hypersensitivity reactions (Goetzl *et al.*, 1986). LTB_4 has predominated quantitatively in acute inflammation, where PMN leukocytes constitute the principal inflammatory cell, such as in gout, psoriasis, and spondyloarthritis. LTC_4 and other C-6-sulfidopeptide leukotrienes have been the major mediators detected in allergic responses and chronic inflammatory reactions, where eosinophils and mononuclear leukocytes are the predominant components. However, more recent studies of intense allergic responses in human models of pollen allergy have revealed substantial concentrations of LTB_4. The repetitive introduction of antigen into chambers over denuded bases of heat-suction blisters in allergic patients for several hours led to the release of LTB_4 at high concentrations, whereas a single maximal challenge with antigen failed to generate detectable LTB_4 (Talbot *et al.*, 1985; Valone *et al.*, 1987).

The extent to which modulating the expression of leukotriene receptors on critical target cells facilitates early responses and contains and terminates them later in the course of inflammation has not been evaluated systematically. The availability of receptor-directed antagonists of leukotrienes will permit more useful studies of their role in inflammation. The rapid interruption of the effects of a leukotriene will prove involvement more conclusively than a slower subsidence after interruption of synthesis and subsequent local metabolic degradation of the fluid-phase mediators. The apparently greater diversity of receptors as compared to enzymatic pathways of generation may also permit more flexible tissue-specific suppression of the effects of the leukotrienes.

ACKNOWLEDGMENTS. Supported in part by grants AI-19784 and HL-31809 from the National Institutes of Health.

Appendix

Abbreviations used in this chapter: HETE, monohydroxy-eicosatetraenoic acid; LTB_4 5(S),12(R)-dihydroxy-eicosa-6, 14-*cis*-8, 10-*trans*-tetraenoic acid or leukotriene B_4; PMN, polymorphonuclear, K_d, dissociation constant; EC_{50}, concentration required to achieve 50% of a maximum effect; $[^3H]LTB_4$-APA, $[^3H]$*N*-(3-aminopropyl)-leukotriene B_4 amide; DSS, disuccinimidyl suberate; SDS, sodium dodecyl sulfate; GTP, guanosine 5′-triphosphate; $[Ca^{2+}]_i$, intracellular calcium; and pH_i, intracellular pH.

References

Baud, L., Koo, C. H., and Goetzl, E. J., 1987a, Specificity and cellular distribution of human polymorphonuclear leukocyte receptors for leukotriene C_4, *Immunology* **62:**53–59.

Baud, L., Goetzl, E. J., and Koo, C. H., 1987b, Stimulation by leukotriene D_4 of increases in the cytosolic concentration of calcium in dimethylsulfoxide-differentiated HL-60 cells, *J. Clin. Invest.* **80:**983–991.

Baud, L., Goetzl, E. J., and Koo, C. H., 1987c, Leukotriene D_4-induced increases in the cytoplasmic pH of dimethylsulfoxide-differentiated HL-60 cells, *J. Cell. Physiol.* (in press).

Benjamin, C. W., Rupple, P. L., and Gorman, R. R., 1985, Appearance of specific leukotriene B_4 binding sites in myeloid differentiated HL-60 cells, *J. Biol. Chem.* **260:**14208–14213.

Brown, G. E., Fischkoff, S. A., and Ordonez, J. V., 1984, Development of membrane-potential responsiveness by myeloid leukemia cells during neutrophilic differentiation, *Biochem. Biophys. Res. Commun.* **123:**937–943.

Burrall, B. A., Wintroub, B. U., and Goetzl, E. J., 1985, Selective expression of 15-lipoxygenase activity by cultured human keratinocytes, *Biochem. Biophys. Res. Commun.* **133:**208–213.

Chau, L.-Y., Hoover, R. L., Austen, K. F., and Lewis, R. A., 1986, Subcellular distribution of leukotriene C_4 binding units in cultured bovine aortic endothelial cells, *J. Immunol.* **137:**1985–1992.

Chau, L.-Y., Sun, F. F., Spur, B., Lewis, R. A., and Austen, K. F., 1987, Photoaffinity labeling of leukotriene C_4 binding sites in subcellular membranes of ileal smooth muscle, *J. Immunol.* (in press).

Collins, S. J., Ruscetti, F. W., Gallagher, R. E., and Gallo, R. C., 1978, Terminal differentiation of human promyelocytic leukemia cells induced by dimethylsulfoxide and other polar compounds, *Proc. Natl. Acad. Sci. U.S.A.* **75:**2458–2462.

Fischkoff, S. A., Pollak, A., Gleich, G. J., Testa, J. R., Misawa, S., and Reber, T. J., 1984, Eosinophilic differentiation of the human promyelocytic cell line, HL-60, *J. Exp. Med.* **160:**179–196.

Fontana, J. A., Wright, D. G., Schiffman, E., Corcoran, B. A., and Deisseroth, A. B., 1980, Development of chemotactic responsiveness in myeloid precursor cells: Studies with a human leukemia cell line, *Proc. Natl. Acad. Sci. U.S.A.***77:**3164–3166.

Gallagher, R., Collins, S., Trujillo, J., McCredie, K., Ahearn, M., Tsai, S., Metzgar, R., Aulakh, G., Ting, R., Ruscetti, F., and Gallo, R., 1979, Characterization of the continuous differentiating myeloid cell line (HL-60) from a patient with acute promyelocytic leukemia, *Blood* **54:**713–733.

Gifford, L. A., Chernov-Rogan, T., Harvey, J. P., Koo, C. H., Goldman, D. W., and Goetzl, E. J., 1987, Recognition of human polymorphonuclear leukocyte receptors for leukotriene B_4 by rabbit anti-idiotypic antibodies to a mouse monoclonal anti-leukotriene B_4, *J. Immunol.* **138:**1184–1189.

Goetzl, E. J., 1980, Mediators of immediate hypersensitivity derived from arachidonic acid, *N. Engl. J. Med.* **303:**822–825.

Goetzl, E. J., Brindley, L. L., and Goldman, D. W., 1983, Enhancement of human neutrophil adherence by synthetic leukotriene constituents of the slow-reacting substance of anaphylaxis, *Immunology* **50:**35–41.

Goetzl, E. J., Wong, M. Y. S., and Matthay, M. A., 1986, Leukotrienes in human hypersensitivity and inflammatory diseases, in: *Advances in Inflammation Research* (I. Otterness, A. Lewis, and R. Capetola, eds.), pp. 47–55, Raven Press, New York.

Goldman, D. W., 1987, Activation of protein kinase C (PKC) decreases leukotriene B_4 (LTB_4) receptor expression on human neutrophils (N), *Fed. Proc.* **46:**606.

Goldman, D. W., and Goetzl, E. J., 1982, Specific binding of leukotriene B_4 to receptors on human polymorphonuclear leukocytes, *J. Immunol.* **129:**1600–1604.

Goldman, D. W., and Goetzl, E. J., 1984, Heterogeneity of human polymorphonuclear leukocyte receptors for leukotriene B_4. Identification of a subset of high affinity receptors that transduce the chemotactic response, *J. Exp. Med.* **159:**1027–1041.

Goldman, D. W., and Goetzl, E. J., 1985, Calcium dynamics in stimulation of human polymorphonuclear leukocytes by leukotriene B_4, in: *Advances in Prostaglandin, Thromboxane, and Leukotriene Research* (O. Hayaishi, and S. Yamamoto, eds.), pp. 667–669, Raven Press, New York.

Goldman, D. W., Gifford, L. A., Young, R. N., and Goetzl, E. J., 1985a, Affinity labeling of human neutrophil (N) receptors for leukotriene B_4 (LTB4), *Fed. Proc.* **44:**781.

Goldman, D. W., Gifford, L. A., Olson, D. M., and Goetzl, E. J., 1985b, Transduction by leukotriene B_4 receptors of increases in cytosolic calcium in human polymorphonuclear leukocytes, *J. Immunol.* **135:**525–530.

Goldman, D. W., Chang, F.-H., Gifford, L.A., Goetzl, E. J., and Bourne, H. R., 1985c, Pertussis toxin inhibition of chemotactic factor-induced calcium mobilization and function in human polymorphonuclear leukocytes, *J. Exp. Med.* **162:**145–156.

Goldman, D. W., Olson, D. M., Payan, D. G., Gifford, L. A., and Goetzl, E. J., 1986, Development of receptors for leukotriene B_4 on HL-60 cells induced to differentiate by 1 alpha, 25-dihydroxyvitamin D_3, *J. Immunol.* **136:**4631–4636.

Goldman, D. W., Gifford, L. A., Marotti, T., Koo, C. H., and Goetzl, E. J., 1987, Molecular and cellular properties of human polymorphonuclear leukocyte receptors for leukotriene B_4, *Fed. Proc.* **46:**200–203.

Grinstein, S., and Furuya, W., 1984, Amiloride-sensitive N^+H^+ exchange in human neutrophils: Mechanism of activation by chemotactic factors, *Biochem. Biophys. Res. Commun.* **122:**755–762.

Hayaishi, O., and Yamamoto, S. (eds.), 1985, *Advances in Prostaglandin, Thromboxane, and Leukotriene Research*, Raven Press, New York.

Hunter, J. A., Finkbeiner, W. E., Nadel, J. A., Goetzl, E. J., and Holtzman, M. J., 1985, Predominant generation of 15-lipoxygenase metabolites of arachidonic acid by epithelial cells from human trachea, *Proc. Natl. Acad. Sci. U.S.A.* **82:**4633–4637.

Kitagawa, S., Ohta, M., Nojri, H., Kakinuma, K., Saito, M., Takaku, F., and Miura, Y., 1984, Functional maturation of membrane potential changes and superoxide-producing capacity during differentiation of human granulocytes, *J. Clin. Invest.* **73:**1062–1071.

Koo, C. H., Baud, L., Marotti, T., Cheung, M., Harvey, J. P., and Goetzl, E. J., 1987, Receptor-dependent regulation of human polymorphonuclear leukocyte responses to leukotrienes, in: *Molecular Biology of the Arterial Wall* (G. Schettler, ed.), pp. 129–131, Springer-Verlag, New York.

Krause, K.-H., Schlegel, W., Wellheim, C. B., Andersson, T., Waldvogel, F. A., and Lew, P. D., 1985, Chemotactic peptide activation of human neutrophils and HL-60 cells. Pertussis toxin reveals correlation between inositol triphosphate generation, calcium in transients, and cellular activation, *J. Clin. Invest.* **76:**1348–1354.

Kreisle, R. A., and Parker, C. W., 1983, Specific binding of leukotriene B_4 to a receptor on human polymorphonuclear leukocytes, *J. Exp. Med.* **157:**628–641.

Krilis, S. A., Lewis, R. A., Corey, E. J., and Austen, K. F., 1983, Specific receptors for leukotriene C_4 on a smooth muscle cell line, *J. Clin. Invest.* **72:**1516–1519.

Krilis, S. A., Macpherson, J. L., deCarle, D. J., Daggard, G. E., Talley, N. A., and Chesterman, C. N., 1986, Small bowel mucosa from celiac patients generates 15-hydroxyeicosatetraenoic acid (15-HETE) after *in vitro* challenge with gluten, *J. Immunol.* **137:**3768–3771.

Kuehl, F. A., DeHaven, R. N., and Pong, S.-S., 1984, Lung tissue receptors for sulfidopeptide leukotrienes, *J. Allergy Clin. Immunol.* **74:**378–381.

Lee, J. Y., Chernov, T., and Goetzl, E. J., 1984, Characteristics of the epitope of leukotriene B_4 recognized by a highly specific mouse monoclonal antibody, *Biochem. Biophys. Res. Commun.* **123:**944–950.

Lew, P. D., Dayer, J.-M., Wollheim, C. B., and Pozzan, T., 1984, Effect of leukotriene B_4, prostaglandin E_2 and arachidonic acid on cytosolic-free calcium in human neutrophils, *FEBS Lett.* **166:**44–48.

McCarthy, D. M., SanMiguel, J. F., Freake, H. C., Green, P. M., Zola, H., Catovsky, D., and Goldman, J. M., 1983, 1,25-dihydroxyvitamin D_3 inhibits proliferation of human promelocytic leukemia (HL60) cells and induces monocyte–macrophage differentiation in HL60 and normal human bone marrow cells, *Leukemia Res.* **7:**51–55.

Molski, T. F. P., Naccache, P. H., Volpi, M., Wolpert, L. M., and Sha'afi, R. T., 1980, Specific modulation of the intracellular pH of rabbit neutrophils by chemotactic factors, *Biochem. Biophys. Res. Commun.* **94:**508–514.

Mong, S., Wu, Hu.-L., Hogaboom, G. K., Clark, M. A., and Crooke, S. T., 1984, Characterization of the leukotriene D_4 receptor in guinea-pig lung, *Eur. J. Pharmacol.* **102:**1–11.

Naccache, P. H., Molski, T. F. P., Spinelli, B., Borgeat, P., and Abboud, C. N., 1984, Development of calcium and secretory responses in the human promyelocytic leukemia cell line HL60, *J. Cell. Physiol.* **119:**241–246.

Niedel, J., Kahane, I., Lachman, L., and Cuatrecasas, P., 1980, A subpopulation of cultured human promyelocytic leukemia cells (HL-60) displays the formyl peptide chemotactic receptor, *Proc. Natl. Acad. Sci. U.S.A.* **77:**1000–1004.

Opmeer, F. A., and Hoogsteden, H. C., 1984, Characterization of specific receptors for leukotriene D_4 on human alveolar macrophages, *Prostaglandins* **28:**183–194.

Palmblad, J., Gyllenhammer, H., Lindgren, J. A., and Malmsten, C. L., 1984, Effects of leukotrienes and f-Met-Leu-Phe on oxidative metabolism of neutrophils and eosinophils, *J. Immunol.* **132:**3041–3045.

Rovera, G., Santoli, D., and Damsky, C., 1979, Human promyelocytic leukemia cells in culture differentiate into macrophage-like cells when treated with a phorbol diester, *Proc. Natl. Acad. Sci. U.S.A.* **76:**2779–2783.

Samuelsson, B., 1983, Leukotrienes: Mediators of immediate hypersensitivity reactions and inflammation, *Science* 220:568–575.

Sarau, H. M., Foley, J. J., Mong, S., and Crooke, S. T., 1987, Identification of [^{3}H]LTD$_4$ receptors and associated intracellular calcium mobilization in U937 cells, *Fed. Proc.* **46:**1313A.

Sherman, J. W., Goetzl, E. J., and Koo, C. H., 1987, Guanine nucleotide modulation of human neutrophil plasma membrane receptors for leukotriene B$_4$, *Clin. Res.* **35:**618A.

Talbot, S. F., Atkins, P. C., Goetzl, E. J., and Zweiman, B., 1985, Accumulation of leukotriene C$_4$ and histamine in human allergic skin reactions, *J. Clin. Invest.* **76:**650–656.

Valone, F., Shalit, M., Atkins, P., Goetzl, E., and Zweiman, B., 1987, Platelet activating factor release in allergic skin sites in humans, *J. Allergy Clin. Immunol.* **79:**248.

16

LTD_4 Receptors and Signal Transduction Processes

STANLEY T. CROOKE, SEYMOUR MONG, MIKE CLARK, HENRY SARAU, ANGELA WONG, RAJU VEGESNA, JAMES D. WINKLER, JOANNA BALCAREK, and C. FRANK BENNETT

1. Introduction

During the past several years, substantial progress has been reported in understanding the receptors and signal transduction processes for the peptidyl leukotriene (Crooke *et al.*, 1987). Studies in our labortory have identified and characterized the receptors and the major steps in the signal transduction process. These studies have allowed us to propose a model explaining the interactions in LTD_4 with its receptors and the events that are induced by these interactions (Fig. 1).

LTD_4 interacts with highly selective specific receptors for LTD_4 that are located in the plasma membrane of target cells. The receptors are coupled to a PIP_2-specific phospholipase C (PIP_2-PLC) via guanine nucleotide binding proteins (G proteins). In some cells and tissues the G protein is a substrate for pertussis toxin (IAP), while in other cells it is not. Activation of PIP_2-PLC results in inositol phosphate metabolism, diacylglycerol (DAG) release, and calcium mobilization. Subsequently, protein kinase C is activated and the transcription of a gene or genes is enhanced. This results in the production of a protein, phospholipase A_2 activating protein (PLAP), which activates a

STANLEY T. CROOKE, SEYMOUR MONG, MIKE CLARK, HENRY SARAU, ANGELA WONG, RAJU VEGESNA, JAMES D. WINKLER, JOANNA BALCAREK, and C. FRANK BENNETT • Department of Molecular Pharmacology, Smith Kline & French Laboratories, Philadelphia, Pennsylvania 19101.

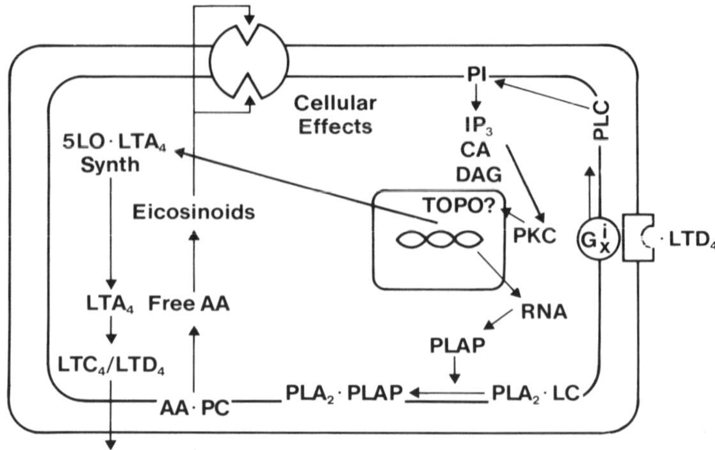

Figure 1. Current model of LTD4 receptors and signal transduction processes. Leukotrienes interact with their receptors and activate, phospholipase C (PLC) via a process mediated by inhibitory guanine nucleotide binding protein (G_i), or an unknown guanine nucleotide binding protein G_X. As a result of various mediators, phospholipase A_2 activating protein (PLAP) is induced and activates PLA_2.

phosphatidylcholine-specific PLA_2, enhancing release of arachidonic acid and metabolism to various products. In some cells, receptors for LTD_4 are present in addition to leukotriene biosynthetic enzymes.

In this chapter, we review recent progress in our laboratory in understanding the major steps in the signal transduction process and advances in understanding the genetic and epigenetic mechanisms of regulation of the processes.

2. *The LTD_4 Receptors*

Table I compares the characteristics of LTD_4 receptors in guinea pig lung, guinea pig heart, and rat basophilic leukemia cells (RBL-1) (Sarau *et al.*, 1987a). The LTD_4 receptors observed in guinea pig lung are comparable to those observed in human fetal and adult lung (Lewis *et al.*, 1984; Lewis *et al.*, 1985) and a variety of cells and tissues including RBL-1 cells (Sarau *et al.*, 1987a). Binding sites observed in guinea pig heart clearly differ, suggesting the possibility of tissue-specific subtype of LTD_4 receptors (Hogaboom *et al.*, 1985). However, much work remains to prove that the binding sites in guinea pig heart are indeed tissue-specific LTD_4 receptors and to evaluate the relevance of this observation.

LTD_4 receptors are glycoproteins located in the plasma membrane in all

Table I. Characteristics of LTD_4 Receptors in RBL-1 Cells, Guinea Pig Lung, and Guinea Pig Heart

	RBL-1	Lung[a]	Heart[a]
Affinity (K_d, nM)[b]	0.9±0.2	0.2±0.05	3.4 ±2.1
Density (B_{max}, pmoles/mg protein)	0.80±0.12	1-2	0.85±0.09
Cation modulation:			
Divalent Ca^{2+}; Mg^{2+}; Mn^{2+}	Stimulation	Stimulation	None
Monovalent (Na^+)	Inhibition	Inhibition	None
Guanine nucleotide modulation	Transition of affinity state (IC = 16±3 μM)	(IC$_{50}$ = 1.2±0.5 μM)	None
Sensitivity to –SH reagents			
Alkylating (NEM, pHMB)	Sensitive	Sensitive	Insensitive
Reducing (DTT, DTE)	Resistant	Resistant	Sensitive

[a]Monovalent, divalent, guanine nucleotide, and sulfydryl reagent data for lung and heart [^{3}H]LTD_4 specific binding were previously reported by this laboratory and others (Mong *et al.*, 1985).
[b]Affinity and density of [3H]LTD_4 binding were determined by computer-based, best fit analysis (30,31) as described (Sarau *et al.*, 1987a). Data for lung and heart membranes as previously published.

the tissues and cells we have studied (Mong *et al.*, 1988). Figure 2 shows the differential centrifugation scheme we have employed to study sheep tracheal smooth muscle (STSM). Figure 3 shows the distribution of various enzyme markers of membranes and the distribution of LTD_4 receptors. Recovery of LTD_4 receptors was greater than 90%. The characteristics of the LTD_4 receptors in the plasma membrane fractions and those in the initial membrane preparation are comparable (Mong *et al.*, 1988).

3. *The Guanine Nucleotide Binding Proteins*

Table II shows the characteristics of the G proteins involved in signal transduction for LTD_4 receptors in various cells and tissues. Obviously, the G proteins employed vary as a function of cell and tissue (Clark *et al.*, 1986). Moreover studies employing undifferentiated and differentiated U937 cells have shown that the G proteins that are employed by LTD_4 receptors may vary as a function of differentiation and the LTB_4 receptors appear to employ selectively an IAP substrate G protein (Sarau *et al.*, 1987b).

4. *The PI-Phospholipase C*

We have purified and characterized a PI-specific PLC from guinea pig uterus. In guinea pig uterus, there are at least three isotypes of this enzyme: PI-PLC I_a, I_b, and II. Table III shows the purification of PI-PLC I_a and I_b. PI-

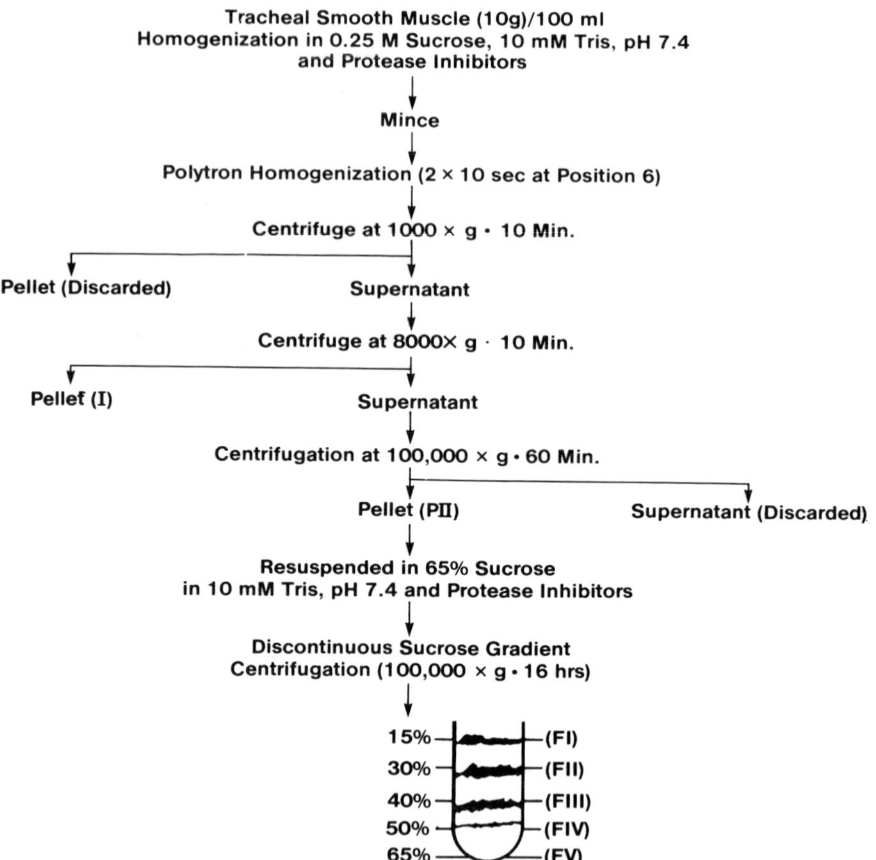

Figure 2. Fractionation of subcelluar membranes from sheep tracheal smooth muscle.

PLC I_b was purified to homogeneity and characterized relative to substrate preference, calcium and pH effects, and enzyme kinetics (Bennett and Crooke, 1987). The enzyme has all the characteristics one would expect from a PLC involved in receptor-medicated inositol signaling.

Polyclonal antibodies were prepared to PI-PLCIb and used to identify cells containing antigenically equivalent proteins. In RBL-1 cells, a protein identified by these antibodies exists (Table IV). It has enzymatic properties equivalent to the guinea pig uterus enzyme and is preciptated by the antibodies. It is a phosphoprotein that upon phosphorylation appears to translocate from the membrane to the cytosol. Phosphorylation is effectively induced by phorbol esters and when membrane bound the enzyme appears to be less active (Bennett and Crooke, 1987).

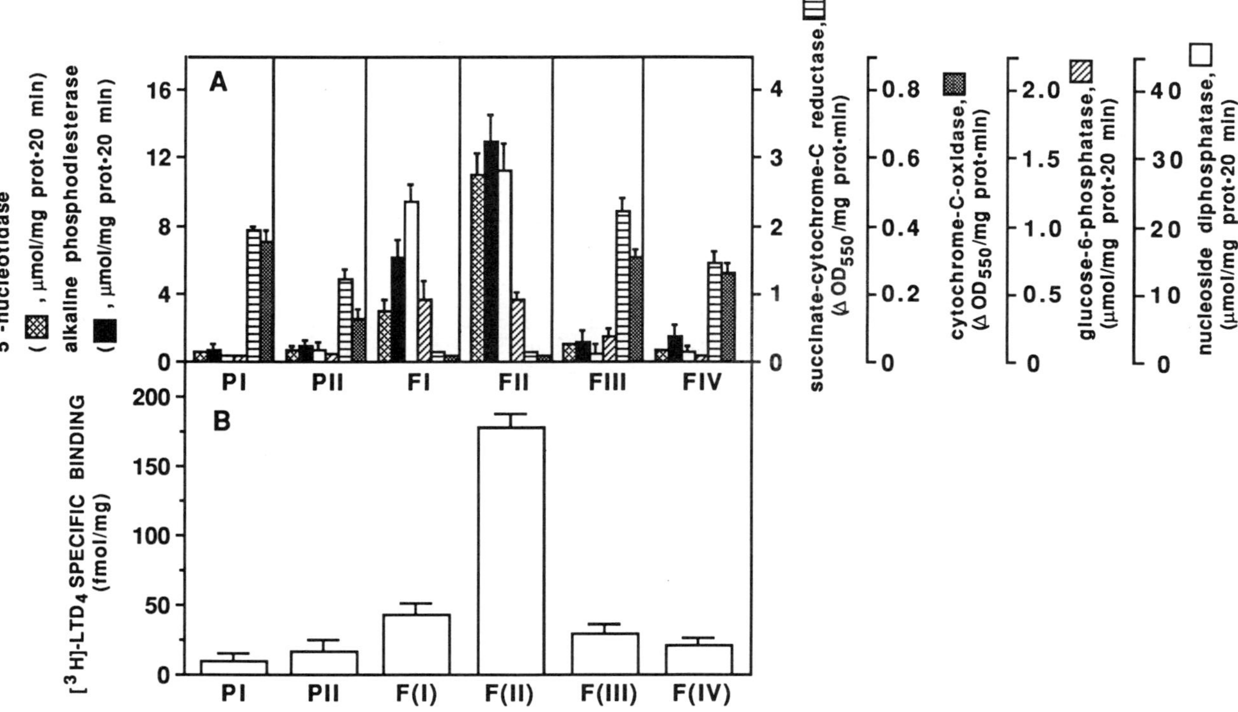

Figure 3. Distribution of marker enzymes and [3H]LTD4-specific binding sites in sheep smooth muscle subcellular membrane fractions. The values shown are mean ± SE obtained from a representative experiment of 4. The PI and PII fractions were obtained by differential centrifugation of polytron homogenized smooth muscle. The F(I), F(II), F(III), and F(IV) fractions were obtained by discontinuous sucrose density gradients as shown in Fig. 2. The specific activity of the marker enzymes and $[^3H]LTD_4$ receptor binding were as previously defined. The average recoveries ($n = 3$) of various marker enzymes were: 5′-nucleotidase, 95%; alkaline phosphodiesterase, 72%; glucose-6-phosphatase, 86%; nucleoside diphosphatase, 80%; cytochrome-C oxidase, 93%; and succinate-dependent cytochrome-C reductase, 71%.

Table II. Characteristics of Guanine Nucleotide Binding Proteins Involved in LTD_4 Signal Transduction

Cells or tissue	Guanine nucleotide binding protein	
	G_i (IAP substrate)	G_x (non-IAP substrate)
Bovine pulmonary aortic endothelial cells (CPAE)	100%	–
Rat basophilic leukemic cells (RBL-1)	–	100%
Human monocytic leukemic (U937) cells, basal state	40-80%	< 20%
Human monocytic leukemic (U937) cells (DSMO differentiated)	50%	50%
Sheep tracheal smooth muscle primary culture cells	⩾ 60%	40%
Guinea pig lung	⩾ 70%	30%

5. *Inositol Phosphate Metabolism*

Figure 4 shows the inositol phosphate metabolism induced by LTD_4 in RBL-1 cells (Sarau *et al.*, 1987a). Within 2 min after addition of LTD_4, IP_3 formation is markedly enhanced as is IP_2. In the presence of LiCl, IP_1 accumulation in response to LTD_4 is significant. The induction of inositol phosphate metabolism is clearly receptor mediated as IP_1 accumulation displays appropriate rank orders of activity for agonists and antagonists, and as a generic response employed by LTD_4 receptors (Table V). In essence, in all

Table III. Purification of Phosphatidyl-Inositol-Specific Phospholipase C from Guinea Pig Uterus

Fraction	Total protein (mg)	Specific activity[a] (nmol/mg/min)	Total activity (nmol/min)	Fold-enrichment	Yield (%)
Homogenate	2356.0	0.5	1178.0	—	100
100,000 × g supernatant	792	1.8	1425.6	3.6	121.0
DEAE Fraction I	106	7.4	784.4	14.8	66.6
AH–Sepharose Ia	21.0	9.4	197.4	18.8	16.8
Heparin–Sepharose	0.6	248.2	148.9	496.4	12.6
Sephacryl S-200	0.2	325.0	65.0	650	5.5
AH–Sepharose lb	5.9	71.0	418.9	142	35.6
Heparin–Sepharose	1.4	281.4	394.0	562.8	33.4
Sephacryl S–200	0.4	569.6	227.8	1139.2	19.3
Affi–gel Blue	0.06	717.8	43.1	1435.6	3.6

[a]Enzyme activity was determined using 10 μM phosphatidylinositol as a substrate. The reaction mixture contained 2.4 mM deoxycholate, 50 mM bis-Tris, 500 mM KC1, 1 mM $CaC1_2$; pH = 7.0.

Table IV. Characteristics of RBL-1 Cell PI-PLC I
Membrane and cytosol localized
May be inhibited in membrane
Phosphoprotein
Phorbol esters induce phosphorylation
Also present in U937 cells

cells and tissues studied to date, LTD_4 induces and increases inositol phosphate metabolism if LTD_4 receptors are present.

6. Calcium Mobilization

LTD_4 induces a rapid transient increase in intracellular free calcium and is derived from both intracellular and extracellular Ca^{2+} pools. Figure 5 shows that Ca^{2+} mobilization induced by LTD_4 is biphasic in RBL-1 cells; a rapid increase in Ca^{2+} (phase 1) is followed by a prolonged plateau (phase II). Phase I Ca^{2+} is derived from intracellular Ca^{2+} pools; phase II is derived from extracellular Ca^{2+} (Sarau *et al.*, 1987a). SK&F 104353, an LTD_4 receptor antagonist, has no effect on Ca^{2+} but blocks LTD_4-induced effects. Rank-order potencies for Ca^{2+} mobilization are as predicted by receptor

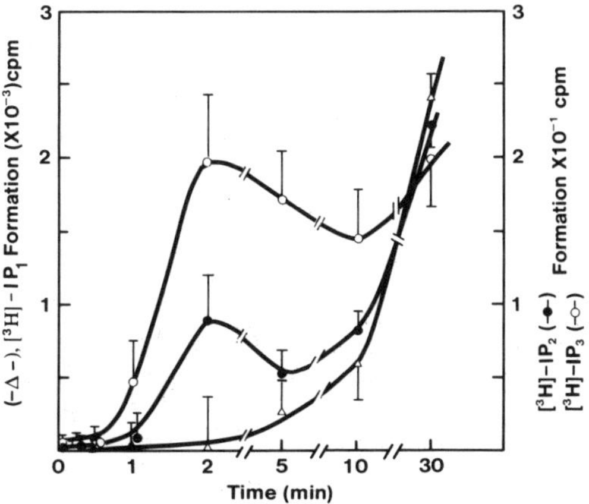

Figure 4. Kinetic responses of LTD4-induced [^{3}H]inositol phosphate formation in RBL-1 cells. RBL-1 cells were labeled with 10μCi/ml of myo-[^{3}H]inositol in the presence of 10 mM LiCl as described. These cells were incubated with 1 μM of LTD_4 from 0 to 30 min. The ^{3}H-labeled inositol phosphates were extracted and quantitated by the anion exchange method. Formation of [^{3}H]IP_1 (Δ), [^{3}H]IP_2 (●), and [^{3}H]IP_3 (0) was determined from triplicate samples of a representative experiment (of three). The difference of the LTD_4-stimulated and the control level of inositol phosphate, at each point of time, is defined as the net accumulation of [^{3}H]inositol phosphates and is shown.

Table V. Inositol Phosphate Metabolism Induced by LTD4

Sheep tracheal smooth muscle cells
Guinea pig lung
Rat basophilic leukemia cells
U937 monocytes

affinities, demonstrating that the effects are receptor mediated. In all cells and tissues with functional LTD_4 receptors studied to date, Ca^{2+} mobilization has been observed.

7. Protein Kinase C

Treatment of RBL-1 cells with phorbol esters results in activation and translocation of protein kinase C (PKC) in a fashion entirely analogous to that observed in other cells. In contrast, LTD_4 induces a rapid transient increase in PKC activity in both the supernatant and membrane with no evidence of translocation as shown in Fig. 6 (Vegesna *et al.*, 1988). PKC activation is concentration dependent and blocked by LTD_4 receptor antagonists, demonstrating that it is an LTD_4 receptor-mediated event.

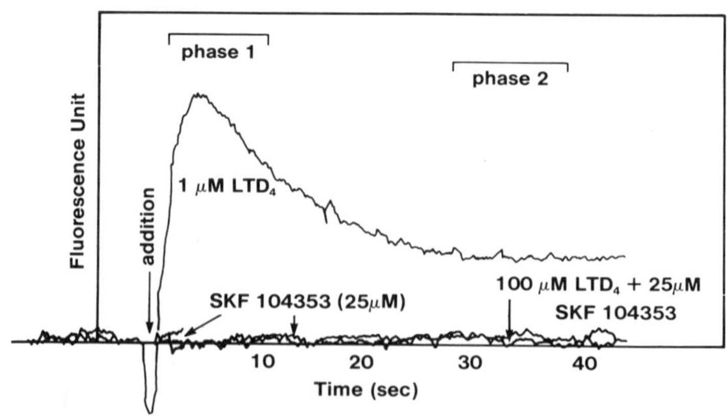

Figure 5. Calcium mobilization induced by LTD4, SK&F 104353, and LTD4 in the presence of SK&F 104353. Fura-2 loaded RBL-1 cells were resuspended in KRH buffer that contained 1 mM $CaCl_2$ and incubated with LTD_4 or SK&F 104353 and LTD_4. The fura-2/Ca^{2+} fluorescence was recorded.

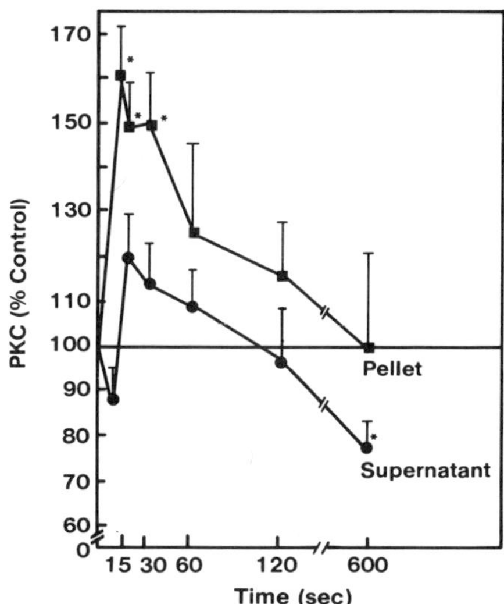

Figure 6. Kinetic response of PKC in RBL-1 cells with LTD_4. Cells were incubated with 10 nM LTD_4 for the times indicated. The cytosol (●——●) and particulate (■——■) fractions were assayed for total PKC activity. Values are expressed as percentage of control indicated by the horizontal line at 100%. Each point shows the mean ± SEM of four to six separate experiments assayed in duplicate. Values of the PKC activity for the control cells are as follows: 628 ± 132 pmoles/min for total cytosolic PKC and 84 ± 18 pmoles/min for total particulate PKC. Asterisks indicate significant difference from controls ($p<0.05$).

8. Phospholipase A_2 Activating Protein

Approximately 2 min after treatment of many cells with LTD_4, arachidonic acid release is increased in a receptor-mediated fashion (Clark *et al.*, 1986). The increase in arachidonic acid release induced by LTD_4 is dependent on RNA and protein syntheses and results from an activation of phosphatidylcholine-specific phospholipase A_2 (PC-PLA_2). We have purified a protein to homogeneity that selectively activates PC-PLA_2 (PLAP). Table VI shows that addition of PLAP to a sonicate of endothelial cells and radiolabeled substrates induces selective activation of PC-PLA_2. Although the activation is associated with an increase in the apparent V_{max} of the enzyme (Clark *et al.*, 1987b), several possible mechanisms must still be

Table VI. Enzyme and Substrate Specificity of PLAP Activity[a]

	PLC		PLA_2	
	PC	PI	PC	PE
Sonicate	4.9 ± 0.6	2.1 ± 1.1	8.2 ± 0.9	2.6 ± 0.6
Sonicate + PLAP	3.6 ± 0.7	1.6 ± 0.1	46.0 ± 0.3	2.1 ± 0.9

[a]The effects of 10 units of PLAP on phospholipase A_2 and C activity were assayed using either phosphatidylcholine (PC), phosphatidylethanolamine (PE), or phosphatidylinositol (PI) as substrate. Results shown are the mean ± SD for three separate experiments performed in triplicate and are expressed as picomoles of reaction product produced per minute per milligram of protein. PE, phosphatidylethanolamine.

entertained. PLAP may interact with substrate, induce proteolytic cleavage of PLA_2, phosphorylate PLA_2 or inhibitory proteins, or interact directly with a regulatory site on PLA_2. In any event, using a cDNA probe, we have shown induction of a mRNA by LTD_4, which we believe to be the message for PLAP in these cells, and are completing cloning and expression of PLAP (Clark *et al.*, 1988).

PLAP is induced by LTD_4 in several cell lines. In all cell lines studied to date that induce arachidonic acid release in response to LTD_4, the mechanism appears to involve PLAP. PLAP is also induced in endothelial cells by tumor necrosis factor (Clark *et al.*, 1987a) and is found in the synovial fluid of patients with rheumatoid arthritis (J.S. Bomalaski and M.A. Clark, unpublished observations).

9. 5-Lipoxygenase–LTA_4 Synthetase

In some cells, for example, rat basophilic leukemia cells (RBL-1), the enzymes required to synthesize the peptidyl leukotrienes are present as well as the receptors and signal transduction processes (Hogaboom *et al.*, 1986, Sarau *et al.*, 1987a). The purified enzyme displays both 5-lipoxygenase and LTA_4 synthetase activities. The amino acid composition and sequence of the amino-terminal portion of the molecule are shown in Table VII (Hogaboom *et al.*, 1986).

Highly specific affinity-purified polyclonal antibodies prepared against RBL-1 5LO/LTA_4 synthetase precipitate 5LO/LTA_4 synthetase activity from RBL-1 cell homogenates and have been employed to demonstrate that the protein is present in cytosol in unstimulated cells and is translocated to the plasma membrane when intracellular Ca^{2+} is increased. In the plasma membrane, the enzyme is inactive (A. Wong and S. Crooke, unpublished observations).

10. Epigenetic Regulation

We have studied homologous and heterologous desensitization in RBL-1 cells. Figure 7 shows that pretreatment of RBL-1 cells with LTD_4 results in desensitization of these cells to Ca^{2+} mobilization induced by subsequent treatment with LTD_4 (Winkler *et al.*, 1987). The maximum Ca^{2+} mobilization induced by LTD_4 is reduced by approximately 40% (Fig. 8). Studies with agonists and antagonists confirm that these effects are highly specific and mediated by receptors of LTD_4. That the B_{max} of the LTD_4 receptors was reduced by 23% while the Ca^{2+} mobilization was reduced by 40% suggests that desensitization may be mediated by effects on the receptor as well as other components of the signal transduction process. Current

Table VII. Structural Analysis of Purified RBL-1 5-Lipoxygenase[a]

Residue	Composition (mole %)	Residues/molecule
Asx	11.1	75
Glx	10.6	75
Ser	8.2	53
His	1.2	11
Gly	11.9	85
Thr	5.9	43
Arg	3.3	21
Ala	11.1	75
Tyr	4.0	32
Met	2.1	11
Val	6.5	43
Phe	3.9	32
Ile	5.4	43
Leu	9.4	64
Lys	3.8	21

N-terminal amino acid sequence

1	2	3	4	5	6	7	8	9	10	11	12	13	14	15
Pro-	Ser-	Tyr-	Thr-	Val-	Thr-	Val-	Ala-	Thr-	Gly-	Ser-	Gin-	Trp-	Phe	Ala
16	17	18	19	20	21	22	23	24	25	26	27	28	29	30
Gly-	Thr-	Asp-	Asp-	Tyr-	Ile-	Tyr-	Leu-	Ser-	Leu-	Ile-	Gly-	Glu-	Ala-	Gly

[a]The data for amino acid composition are representative values of four separate experiments. The amino acids cysteine, proline, and tryptophan were not determined by the analyses used. Results are given as mole % utilizing estimates of cysteine, proline, and tryptophan as 10% and assuming M_r = 75,000. The N-terminal amino acid sequence represents data from four separate enzyme preparations.

studies should better define the molecular mechanisms of desensitization. We have also demonstrated heterologous desensitization induced by treatment with phorbol esters (Sarau *et. al.*, 1987a).

11. Genetic Regulation

In U937 monocytes, we have shown that LTD_4 receptors are present in the plasma membrane and that when these cells are caused to differentiate the density of the LTD_4 receptors increases (Sarau *et al.*, 1987b). In the undifferentiated state, these cells display no LTB_4 receptors, but when differentiated they display a significant number of these receptors (Sarau *et al.*, 1987b). Table VIII shows the characteristics of the LTD_4 receptors and Ca^{2+} mobilization in these cells. The LTD_4 and LTB_4 receptors displayed by differentiated U937 cells are clearly distinct and the LTD_4 receptors are comparable to those found in RBL-1 cells and guinea pig and human lung.

The LTB_4 receptors are coupled entirely through a G protein that is an IAP substrate. The LTD_4 receptors, in contrast, are coupled to an IAP subs-

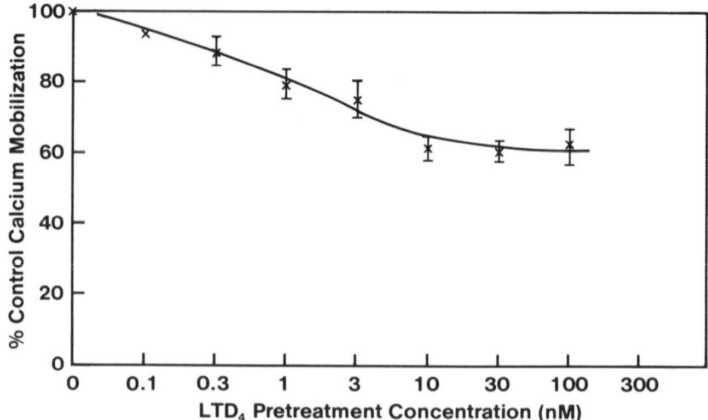

Figure 7. Effect of LTD_4 pretreatment on $[Ca^{2+}]_i$ mobilization induced by LTD_4. RBL-1 cells were incubated for 30 min at 37°C with varying concentrations of LTD_4. The cells were washed, loaded with fura-2, and the apparent $[Ca^{2+}]_i$ mobilization induced by 100 nM LTD_4 was measured. The change in the $[Ca^{2+}]_i$ between the basal value and the peak of the LTD_4-induced response was determined. The data are expressed as a percentage of the control response and each point represents the mean ± SE of two to six experiments with three determinations per experiment. The 100% level represents the $[Ca^{2+}]_i$ mobilization response in the control cells, which was 303 ± 31 nM.

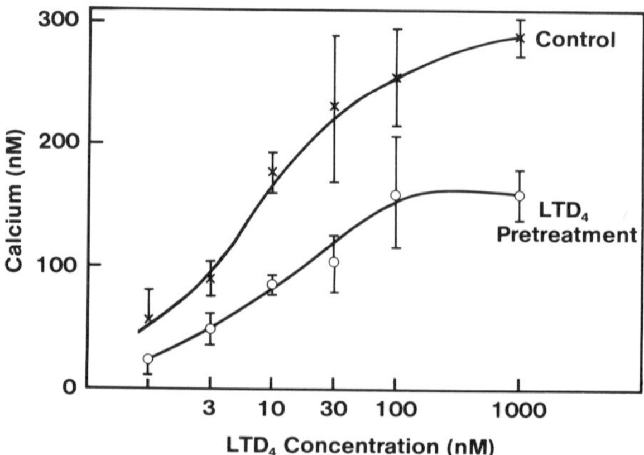

Figure 8. Concentration–response curve of LTD_4-induced $[Ca^{2+}]_i$ mobilization. RBL-1 cells were incubated with vehicle or 30 nM LTD_4 for 30 min, then washed and loaded with fura-2. $[Ca^{2+}]_i$ mobilization in response to varying concentrations of LTD_4 was determined. The change in the $[Ca^{2+}]_i$ between the baseline and the peak of the LTD_4-induced response was determined. The data points are the mean ± SE of triplicate determinations.

Table VIII. Comparison of Properties of Basal and Differentiated U937 Cells[a]

	Basal		Differentiated	
Receptor binding	LTB_4	LTD_4	LTB_4	LTD_4
K_d (nM)	N.D.[b]	0.42 ± 0.11	0.15 ± 0.06	0.35 ± 0.12
B_{max} (fmol/mg)	~8	170 ± 56	300 ± 86	440 ± 96
Divalent cations	N.D.	Stimulation	Stimulation	
Na+	N.D.	Inhibition	Inhibition	
Guanine nucleotides	N.D.	Yes	Yes	Yes
$[Ca^{2+}]_1$ mobilization				
EC_{50} (nM)	N.D.	3.0 ± 0.6	2.4 ± 0.4	3.3 ± 0.6
Max Ca^{2+} mobilization (nM)	16 ± 11	342 ± 36	334 ± 45	1154 ± 136

[a]U937 cells were treated with medium plus 1.3% DMSO for 3 days to differentiate or medium alone to maintain in the Basal (undifferentiated) state.
[b]N.D. = not determined, inusufficient signal to obtain reproducible results.

trate G protein and a G protein that is not inhibited by IAP. Moveover, when the cells are differentiated with DMSO, a greater fraction of the Ca^{2+} signal induced by LTD_4 is transduced by with a non-IAP substrate.

At similar receptor densities, the Ca^{2+} mobilized by LTD_4 receptors is significantly greater than that mobilized by LTB_4 receptors. Furthermore, the two receptors employ different Ca^{2+} pools as each was inhibited by homologous and heterologous desensitization. Finally, in addition to changes in receptors and G proteins, differentation involved changes in the PLC species present in the cells. Thus, in the U937 cells we have an excellent system to allow comparison of LTB_4 and LTD_4 receptors and signal transduction processes and to evaluate genetic and epigenetic regulatory events.

12. Conclusion

Recent progress allows us to propose a model for LTD_4 receptors and signal transduction processes that involves multiple steps and is regulated at several points via several mechanisms. We expect to define the mechanisms with increasing sophistication and to clone and characterize the genes for the key proteins involved in these processes in the near future.

References

Bennett, C. F., and Crooke, S. T., 1987, Purification and characterization of a phosphoinositide-specific phospholipase C from guinea pig uterus, *J. Bio. Chem.* **262:**13789–13797.

Clark, M.A., Cook, M., Mong, S., and Crooke, S. T., 1985, The binding of leukotriene C_4 and leukotriene D_4 to membranes of smooth muscle cell lines ($BC3H_1$) and evidence that leukotriene induced contraction in these cells is mediated by thromboxane, protein and RNA syntheses, *Eur. J. Pharmacol.* **116:**207–220.

Clark, M., Conway, T. M., Bennett, C. F., Crooke, S. T., and Stadel, J. M., 1986, Islet activating protein inhibits leukotriene D_4 and leukotriene C_4 but not bradykinin- or calcium ionophore-induced prostacyclin synthesis in bovine endothelial cells, *Proc. Nat'l. Acad. of Sci.* **83:**7320–7324.

Clark, M. A., Chen, M. J., Crooke, S. T., and Bomalaski, J. S., 1987a, Tumor necrosis factor (Cachectin) induces phospholipase A_2 activity and the synthesis of a phospholipase A_2 activating protein (PLAP) in endothelial cells, *Bioch. J.* (in press).

Clark, M., Conway, T., Shorr, R., and Crooke, S. T., 1987b, Identification and isolation of a mammalian protein which is antigenically and functionally related to the phospholipase A_2 stimulatory peptide melittin, *J. Bio. Chem.* **262:**4402–4406.

Clark, M. A., Conway, T. M., Dispoto, J. H., and Crooke, S. T., 1988, Inhibition of phospholipase A_2 activation and arachidonic acid release by phospholipase A_2 activating protein (PLAP) antisense oligodeoxynucleotide (in press).

Crooke, S. T., Mong, S., Clark, M., Hogaboom, G. K., Lewis, M., and Gleason, J. L., 1987, Leukotriene receptors and signal transduction mechanism, in: *Biochemical Actions of Hormones* (Gerald Litwack, ed.), Academic Press, New York (in press).

Hogaboom, G. K., Mong, S., Stadel, J. M., and Crooke, S. T., 1985, Characterization of [^{3}H]Leukotriene D_4 binding sites in guinea pig ventricular myocardium, *Mol. Pharmacol.* **233:**686–692.

Hogaboom, G. K., Cook, M., Newton J., Varrichio, A., Shorr, R., Sarau, H., Crooke, S. T., 1986, Purification, characterization, and structural properties of a single protein from rat basophilic leukemia (RBL-1) cells possessing 5-lipoxygenase and leukotriene A_2 synthetase activities., *Mol. Pharmacol.* **30:**510–519.

Lewis, M. A., Mong, S., Vessella, R. L., Hogaboom, G. K., Wu, H. L., and Crooke, S. T., 1984, Identification of specific binding sites for leukotriene C_4 in human fetal lung, *Prostaglandins* **27:**961–974.

Lewis, M. A., Mong, S., Vessella, R. L., and Crooke, S. T., 1985, Identification and characterization of leukotriene D_4 receptors in adult and fetal human lung, *Biochem. Pharmacol.* **34:**4311–4318.

Mong, S., Wu, H. L., Scott, M. O., Lewis, M. A., Clarke, M. A., Weichman, B. M., Kinzig, C. M., Gleason, J. G., and Crooke, S. T., 1985, Molecular heterogeneity of leukotriene receptors: Correlation of smooth muscle contraction and radioligand binding in guinea pig lung, *J. Pharmacol. Exp. Ther.* **234:**316–325.

Mong, S., Chi-Rosso, G., Clark, M. A., and Crooke, S. T., 1987, Subcellular localization of leukotriene, D_4 receptors in sheep tracheal smooth muscle, *Mol. Pharmacol.* (in press).

Sarau, H., Mong, S., Foley, J., Wu, H., and Crooke, S. T., 1987a, Identification and characterization of leukotriene D_4 receptors and signal transduction processes in rat basophilic leukemia cells, *J. Biol. Chem.* **262:**4034–4041.

Sarau, H., Foley, J. J., Wu, H.-L., Crooke, S. T., and Mong, S., 1987b, Co-expression of leukotriene B_4 and leukotriene D_4 receptors on human monocytic leukemia U-937 cells: Characterization of the receptors and signal transduction processes, (in press).

Vegesna, R. V. K., Mong, S., and Crooke, S. T., 1987, Leukotriene D_4-induced activation of protein kinases C in rat basophilic leukemia cells, *Eur. J. Pharmacol.* (in press).

Winkler, J. D., Mong, S., and Crooke, S. T., 1988, Leukotriene D_4 induced homologous desensitization of calcium mobilization in rat basophilic leukemia cells, *J. Pharmacol. Exp. Ther.* (in press).

17

Characterization of Thromboxane A_2/ Prostaglandin H_2 Receptors

PERRY V. HALUSHKA, DALE E. MAIS,
and DAVID L. SAUSSY, JR.

Introduction

In 1969, Piper and Vane reported the discovery of a labile substance capable of stimulating contraction of rabbit aorta (rabbit aorta contracting substance, RCS). RCS was later demonstrated to be cyclooxygenase metabolite of arachidonic acid (Vargaftig and Dao, 1971; Vargaftig and Zirinis, 1973), with a half-life of approximately 30 s under physiological conditions (Svensson *et al.*, 1975). Soon after the discovery of RCS, the production of a labile aggregation-stimulating substance (LASS) by platelets was reported (Willis, 1974). Like RCS, LASS was produced by the cyclooxygenase pathway of arachidonic acid metabolism; however, LASS had a longer half-life, suggesting that it was different from RCS.

The unstable cyclic endoperoxide prostaglandins PGG_2 and PGH_2 (Hamberg *et al.*, 1974) were isolated and identified. They were found to stimulate vascular smooth muscle and platelet aggregation and had half-lives of about 5 min under physiological conditions. Thus, it appeared that LASS was composed of the prostaglandin endoperoxides, but they were a minor component of RCS (Svensson *et al.*, 1975).

PERRY V. HALUSHKA • Departments of Cell and Molecular Pharmacology and Experimental Therapeutics and Medicine, Medical University of South Carolina, Charleston, South Carolina 29425. *DALE E. MAIS and DAVID L. SAUSSY, JR.* • Department of Cell and Molecular Pharmacology and Experimental Therapeutics, Medical University of South Carolina, Charleston, South Carolina 29425. *Present address for D. L. S., Jr.:* Department of Molecular Pharmacology, Smith Kline & French Laboratories, Philadelphia, Pennsylvania 19101.

Investigation of the metabolism of arachidonic acid by platelets revealed that a novel compound, which was termed PHD (now known as thromboxane B_2, TXB_2), was the major cyclooxygenase product and that the prostaglandin endoperoxides were intermediates in its formation (Hamberg and Samuelsson, 1974; Hamberg *et al.*, 1974). However, this compound proved to be biologically inactive. Trapping experiments with nucleophiles such as azide, methanol, and ethanol resulted in the formation of analogues of TXB_2 and suggested the formation of a novel, highly unstable intermediate, which was named thromboxane A_2 (TXA_2) (Hamberg *et al.*, 1975). The name was derived from thrombocyte (platelet) and the notion that the proposed structure contained an oxetane ring system. TXA_2 was given the proposed structure shown in Fig. 1 based on the structures of the derivatives of TXB_2 formed in the trapping studies. The half-life of this intermediate was similar to that of RCS (32 s), and it was found to be a very potent platelet aggregator and smooth muscle constrictor, indicating that TXA_2 was indeed the major component of RCS (Hamberg *et al.*, 1975). The enzyme responsible for the production of TXA_2 from PGG_2 and PGH_2 is thromboxanc synthase (Needleman *et al.*, 1976a; Ulrich and Graf, 1984). TXA_2 is a more potent vasoconstrictor and proaggregatory agent than PGG_2 and PGH_2 (Needleman *et al.*, 1976a,b). While natural TXA_2 has not yet been isolated in pure form and characterized, recently Bhagwat and colleagues (Bhagwat *et al.*, 1985) syn-

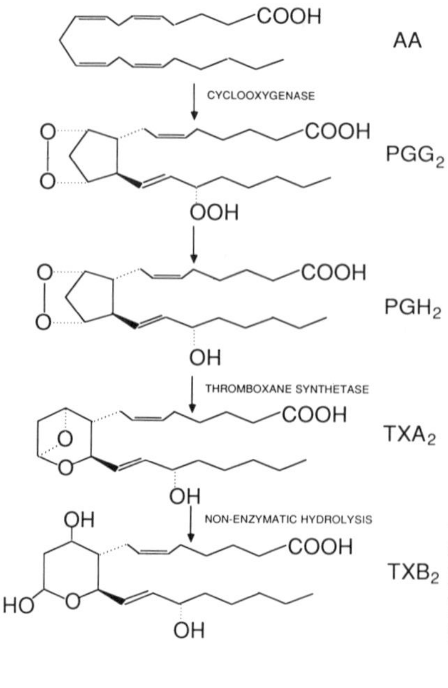

*Figure 1.*Metabolism of arachidonic acid to thromboxane A_2. Fatty acid cyclooxygenase is inhibited by aspirin and other nonsteroidal anti-inflamatory drugs. Thromboxane synthetase is inhibited by analogues of imidazole and pyridine.

thesized TXA_2, based on the structure proposed by Hamberg *et al.* (1975), and compared its properties to those of authentic, biologically generated material. They found it to be indistinguishable from platelet-derived TXA_2 in bioassays, indicating that the proposed structure was indeed correct.

Thromboxane B_2, the stablc hydrolysis product of thromboxane A_2, is often measured to quantitate the synthesis of thromboxane A_2. Two other metabolites that are also measured to estimate thromboxane A_2 synthesis are plasma 11-dehydrothromboxane B_2 (Lawson *et al.*, 1986) and urinary 2,3-dinor thromboxane B_2 (FitzGerald *et al.*, 1983).

Evidence continues to mount that the increased synthesis of thromboxane A_2 may play an importoant pathophysiologic role in a variety of cardiovascular and renal diseases (for reviews see Halushka *et al.*, 1983; FitzGerald *et al.*, 1987; Halushka and Lefer, 1987; Ogletree, 1987). In particular, increased thromboxane A_2 synthesis has been implicated as playing a role in unstable angina pectoris (Fitzgerald *et al.*, 1986), myocardial infarction (Hirsch *et al.*, 1981), enhanced platelet aggregation associated with diabetes mellitus (Halushka *et al.*, 1981; Winocour *et al.*, 1985), circulatory shock (Lefer, 1984; Ball, *et al.*, 1986), lupus nephritis (Patrono *et al.*, 1985; Kelley *et al.*, 1986), and various other renal diseases.

2. Platelet and Vascular Thromboxane A_2 Receptors

2.1. Pharmacologic Identification of Subclasses of Receptors

Shortly after its discovery, it was postulated that thromboxane A_2 was a naturally occurring calcium ionophore and therefore that one did not need to invoke the presence of a receptor for it to exert its physiologic/pharmacologic effects. However, this notion was soon discounted with the discovery of one of the first thromboxane A_2 receptor antagonists, 13-azaprostanoic acid (Le Breton *et al.*, 1979). Thus, it became possible to envision studies that would allow for the subsequent characterization of thromboxane A_2 receptors. Indeed, the synthesis of stable analogues of prostaglandin H_2 and thromboxane A_2, which acted as either agonists or antagonists of the receptor, facilitated pharmacologic studies of the receptor (for a review see Halushka and Lefer, 1987). Since it appears that prostaglandin H_2 stimulates the same receptor as thromboxane A_2, the putative receptors have been referred to as thromboxane A_2/prostaglandin H_2 (TXA_2/PGH_2) receptors but are often referred to as thromboxane A_2 receptors.

Early pharmacologic studies suggested that the platelet and vascular thromboxane A_2 receptors were different (for a review see Halushka *et al.*, 1987a). However, these earlier studies utilized a limited number of TXA_2/

PGH_2 analogues and, more importantly, used platelets and vascular tissue from different species. Since platelets and blood vessels from different species respond differently to TXA_2/PGH_2 analogues (Burke *et al.*, 1983), unequivocal interpretation of the results was not possible. Mais *et al.* (1985a) provided the most convincing evidence to date that there are subclasses of receptors. They found that the rank-order potency for a series of 13-azapinane thromboxane A_2 analogues, thromboxane receptor antagonists, were different in human platelets compared to human saphenous veins (Fig. 2). They also found that the canine and human saphenous veins had similar structure–activity relations, but canine platelets had a different structure–activity relation than the saphenous vein. In a subsequent study, Mais *et al.* (1985b) separated three of the analogues into their 15-hydroxyepimers and found that the potencies of the individual epimers were the same in platelets, but in saphenous veins they differed by a factor of 3. They have synthesized a larger series of 13-azapinane thromboxane A_2 analogues and have found that the rank-order potencies for the compounds are different in platelets com-

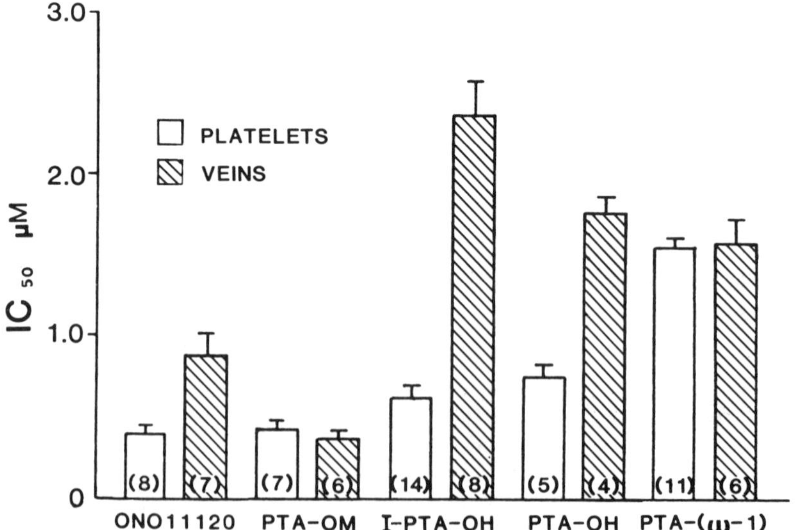

Figure 2. IC_{50} values for a series of 13-azapinane thromboxane A_2 analogues to antagonize U46619-induced platelet aggregation in platelet-rich plasma or contraction of saphenous veins obtained from human subjects. ID_{50} is the concentration required to inhibit by 50% the maximal platelet aggregation response to U46619 1 min after its addition or to produce a 50% inhibition of the maximal contraction of the saphenous vein. (See Fig. 5 for the structure of the compounds.) Adapted from the *Journal of Pharmacology and Experimental Therapeutics* **235:**799 (1985).

pared to blood vessels (Halushka *et al.*, 1986a). Of this series of 16 compounds, all were antagonists of U46619, a thromboxane A_2 mimetic (Bundy, 1975), induced platelet aggregation. In the canine saphenous vein, however, 12 were antagonists of U46619 induced contractions, but four were found to possess agonist activity. Mais *et al.* (1985a) named the vascular receptor $[TXA_2/PGH_2]_\tau$, τ for tone, and the platelet receptor $[TXA_2/PGH_2]_\alpha$, α for aggregation.

Akbar *et al.* (1985) found a different rank order of potency for a series of 13-azaprostanoic acid analogues in rat thoracic aorta and human platelets. Interestingly, however, they concluded that the receptors were not different. A series of 16-phenoxy PGE_1 analogues have been synthesized by Banerjee *et al.* (1985) that possess TXA_2/PGH_2-like activity in rabbit aortic strips and rat platelet-rich plasma. These workers noted a correlation between agonist activity in the aorta and the rat platelets. However, analysis of the same data by the Spearman rank-order correlation shows distinct rank-order differences for the compounds between the two tissues (P. V. Halushka *et al.*, personal observation). Ogletree *et al.* (1986) have also provided evidence for the existence of TXA_2/PGH_2 receptor subtypes based on relative affinities and stimulatory potencies for U46619, PGD_2, PGE_2, and $PGF_{2\alpha}$ in human platelets and rat aorta. Thus, the current evidence strongly supports the notion that the TXA_2/PGH_2 receptor in platelets is distinct from that in blood vessels.

3. Radioligand Binding Studies

The first radioligand binding studies characterizing TXA_2/PGH_2 receptors were conducted using ^3H-labeled ligands and either platelet membranes or washed platelets. Hung *et al.* (1983) described the binding characteristics of $[^3H]$13-azaprostanoic acid to a platelet membrane fraction. They found that binding was saturable, stereospecific, and displaceable by a series of compounds thought to interact with the putative TXA_2/PGH_2 receptor. A high-affinity (K_d 100 nM), low-capacity binding site (1 pmole/mg protein) and a low-affinity ($K_d = 3.5\ \mu M$), high-capacity binding site were identified for $[^3H]$13-azaprostanoic acid.

Armstrong *et al.* (1983) conducted binding studies with $[^3H]$U44069, a TXA_2/PGH_2 mimetic and an isomer of U46619 (Fig. 3), in intact washed human platelets. They reported three separate compartments for the binding of the ligand. The highest-affinity site had approximately 300 binding sites per platelet. A series of agonists and antagonists of the TXA_2/PGH_2 receptor displaced the ligand from its binding site with a rank-order potency similar to that found for their pharmacologic potencies.

Pollock *et al.* (1984) correlated receptor occupancy of $[^3H]$U44069 with

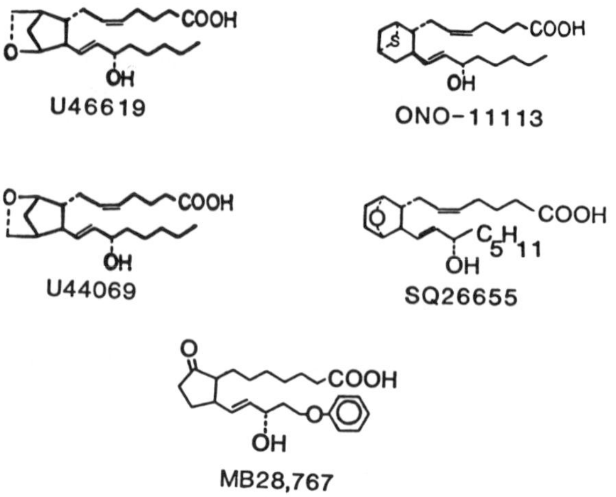

U46619

ONO-11113

U44069

SQ26655

MB28,767

Figure 3. Structures of TXA_2/PGH_2 agonists used in radioligand binding studies. The octanol–water partition coefficient for [^{3}H]U46619 was determined to assess its degree of lipophilicity and was found to be 38. For comparison, Gerrard *et al.* (1978) determined the ether–water partition coefficient for PGG_2 and reported it to be 1.7.

its ability to stimulate phosphatidate formation and elevation of cytosolic free calcium concentrations. They also reported a K_d of 70 nM for U44069. Recently, [^{3}H]U46619 has been made commercially available, and Kattelman *et al.* (1986) reported on its binding to washed human platelets. They reported a K_d for U46619 of 109 nM and a B_{max} for the high-affinity site of 350 fmoles/10^8 platelets (2000 sites/platelet). Liel *et al.* (1987) also studied the binding characteristics of [^{3}H]U46619 in washed human platelets. They reported a K_d of 20 nM and a B_{max} of 9 fmoles/10^7 platelets (550 binding sites/platelet) for the high-affinity site. The reasons for the apparent discrepancies between the binding parameters are unknown. However, these differences may be due to the different incubation conditions and methods employed for the separation of bound from free ligand.

Halushka *et al.* (1985) synthesized the first ^{125}I-labeled ligand for the study of eicosanoid receptors. It was an analogue of 13-azaprostanoic acid in which the bottom (ω) aliphatic side chain was substituted with a phenol ring system and iodinated with [^{125}I]Na using chloramine T. The analogue was named I-*cis*-APO, and its binding characteristics in human platelet membranes were evaluated. It was found to have a K_d of approximately 1 μM and a B_{max} of 18.7 pmoles/mg protein. Clearly, the affinity of the ligand for its binding site was too low to be of much further use; however, the concept of preparing ligands for eicosanoid receptors in which the aliphatic ω side chain was substituted with a phenol group and radiolabled with ^{125}I was shown to be tenable.

Mais *et al.* (1984) synthesized [^{125}I]PTA-OH, a 13-azapinane analogue of TXA_2, and utilized it for radioligand binding studies to characterize the human platelet TXA_2/PGH_2 receptor. I-PTA-OH was found to be a competitive antagonist of the human platelet TXA_2/PGH_2 receptor (Mais *et al.*, 1985c), with a pharmacologically determined K_d of 8 nM. The K_d was also determined from equilibrium binding experiments and found to be 21 nM (Fig. 4). A single class of high-affinity binding sites with a B_{max} of 0.89 pmoles/mg protein or 2500 binding sites/platelet was found. These results were confirmed by Narumiya *et al.* (1986). [^{125}I]PTA-OH binding in washed human platelets was displaced by a series of agonists (Fig. 3) with a rank-

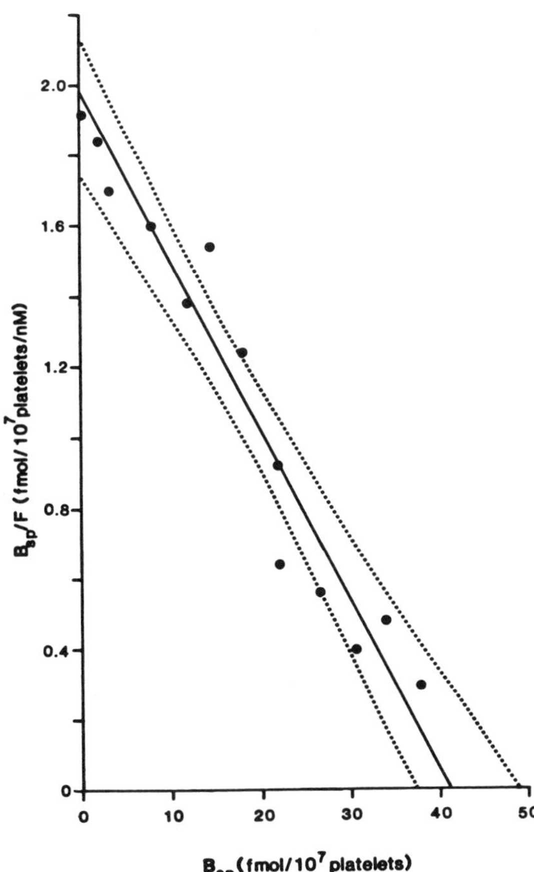

Figure 4. Scatchard plot for equilibrium binding of [^{125}I]PTA-OH to washed human platelets. The data are presented as the mean of five separate experiments and are expressed as the line of best fit and the 95% confidence band around the line. $K_d = 21$ nM; $B_{max} = 2530$ sites/platelet. Reprinted from the *Journal of Pharmacology and Experimental Therapeutics* **235:**799 (1985) by permission of the publisher.

order potency that had a high correlation with their pharmacologic potency (Table I) (Halashka *et al.*, 1987b). Similarly a series of structurally dissimilar antagonists (Fig. 5) also displaced [^{125}I]PTA-OH from its binding site, with a high degree of correlation between their pharmacologic potencies and their IC_{50} values (Table II). Binding studies have also been carried out with [^{125}I]PTA-OH in washed, intact canine (Mais *et al.*, 1985d) and guinea pig platelets (Halushka *et al.*, 1986b), with results similar to those obtained in washed human platelets. In contrast, there was no specific binding of [^{125}I]PTA-OH to washed rabbit platelets (Narumiya *et al.*, 1986).

Saussy *et al.* (1985) characterized the binding of [^{125}I]PTA-OH to a crude membrane fraction obtained from human platelets. The affinity of [^{125}I]PTA-OH for its binding site in the crude membrane fraction was similar to that found in washed platelets. In contrast, the affinity of the binding site for TXA_2/PGH_2 agonists was markedly decreased. This observation will be discussed further (*vide infra*). Treatment of the membranes with 1 mM dithiothreitol reduced displaceable binding by 58%, and treatment with 1% β-mercaptoethanol reduced binding by 97% (P. V. Halushka *et al.*, unpublished observations). These observations are consistent with the notion that disulfide bonds are present in the receptor.

Saussy *et al.* (1987) fractionated the platelet into several subcellular fractions and studied the binding of [^{125}I]PTA-OH to the various fractions. Enrichment of specific binding of [^{125}I]PTA-OH was found in the plasma membrane–dense tubular system, but not in the mitochondria or granular fractions.

The effects of the nonhydrolyzable guanine nucleotide analogue

Table I. Potencies for a Series of TXA_2,/PGH_2 Agonists in the Radioligand Binding Assay and Induction of Aggregation in Washed, Intact Human Platelets

Agonist	Aggregation[a] EC_{50} (nM)	Displacement[b] IC_{50} (nM)
ONO11,113	18 ± 2 (8)	47 ± 6 (6)
SQ26,655	20 ± 3 (10)	55 ± 6 (6)
U44069	46 ± 4 (7)	76 ± 10 (4)
U46619	64 ± 6 (11)	142 ± 9 (4)
9,11-Azo-PGH_2	61 ± 5 (9)	165 ± 57 (6)
MB28767	1300 ± 150 (6)	5380 ± 1409 (5)

[a]EC_{50} = concentration required to produce 50% of the maxium platelet aggregation response, 1 min after addition of the compound.
[b]The correlation coefficient for displacement versus aggregation is 0.99 ( ) = *n*. The data are expressed as the mean ± SEM. IC_{50} = concentration required to inhibit 50% of specifically bound [^{125}I] PTA- OH to washed, intact human platelets.

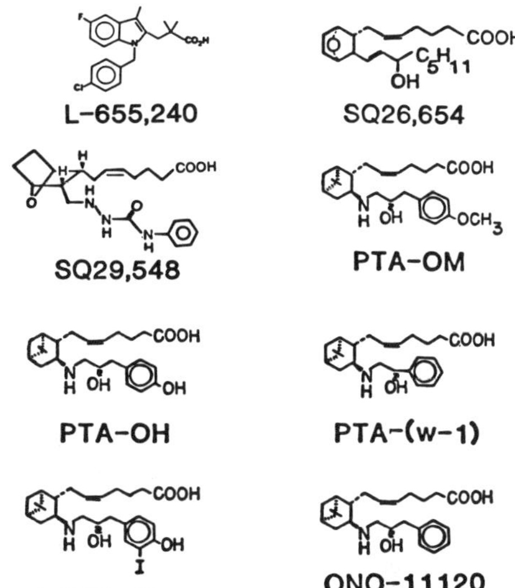

Figure 5. Structures of TXA_2/PGH_2 antagonists used in radioligand binding studies. The octanol–water partition coefficient for [^{125}I]PTA-OH was determined to be 85.

Gpp(NH)p, Na^+, K^+, Mg^{2+}, and Ca^{2+} on [^{125}I]PTA-OH and U46619 binding in platelet membranes have also been assessed. Na^+, K^+, Ca^{2+}, and Mg^{2+} had no significant effects on the binding of [^{125}I]PTA-OH (Tables III and IV). Na^+ (150 mM) also had no significant effect on the competition

Table II. Potencies for a Series of TXA_2/PGH_2 Antagonists in the Radioligand Binding Assay and Inhibition of U46619 Induced Platelet Aggregation

Antagonist	Aggregation[a] K_d (nN)	Displacement[b] IC_{50} (nM)
SQ29548	2.3 ± 0.2 (3)	4 ± 0.4 (4)
L655,240	5.5 ± 1 (4)	13 ± 1 (3)
I-PTA-OH	8.5 ± 1 (4)	21 ± 5 (5)
PTA-OM	16 ± 2 (5)	25 ± 3 (3)
ONO-11120	17 ± 2 (4)	29 ± 5 (3)
PTA-OH	40 ± 10 (5)	84 ± 34 (3)
PTA(ω-1)	55 ± 14 (5)	136 ± 20 (3)
SQ26,654	155 ± 14 (3)	298 ± 75 (5)

[a]The K_d was determined pharmacologically using a Schild analysis of the dose–response data.
[b]The IC_{50} is defined in the legend for Table I. The correlation coefficient for displacement versus aggregation was 0.99 ( ) = n. The data are expressed as the mean ± SEM.

Table III. Effects of Sodium (as NaC1) and Potassium (as KC1) on Displaceable Binding of [^{125}I]PTA-OH in a Crude Platelet Membrane Fraction[a]

Concentration (mM)	Control binding (%)	
	Na^+	K^+
0	100	100
1	87±2	101±3
5	99±8	98±5
10	106±3	98±2
50	110±2	105±3
100	109±3	103±2
150	96±7	101±4

[a]Data are normalized to control binding [determined in the presence of the sodium substitute, *N*-methyl-D-glucamine (150 mM] for the Na^+ binding data, which was arbitrarily expressed as 100%. Data are expressed as mean±SEM for three experiments performed in triplicate for each cation.

between the agonist U46619 and [^{125}I]PTA-OH for binding. In washed human platelets, neither Na^+ nor the cationophore monensin had any significant effects on the competition of U46619 with [^{125}I]PTA-OH (Halushka *et al.*, 1987a). Gpp(NH)p also had no effect on [^{125}I]PTA-OH binding to the

Table IV. Effects of Calcium (as ($CaC1_2$) and Magnesium (as $MgC1_2$) on Displaceable Binding of [^{125}I]PTA-OH in a Crude Platelet Membrane Fraction[a] EDTA (1 mM)

Concentration (mM)	Control binding (%)	
	Ca^{2+}	Mg^{2+}
0	100	100
EDTA (1 mM)	93±3	
0.1	105±4	97
0.5	98±6	
1	107±5	103
2.5	109±6	
5	115±4	107
10	103±8	101

[a]Data are normalized to control binding (determined in the absence of any addition), which was arbitrarily expressed as 100%. Data for the calcium experiments are expressed as mean±SEM for three experiments performed in triplicate. Data for the magnesium experiments are expressed as the mean of two experiments performed in triplicate.

membranes or on the competition of U46619 with [^{125}I]PTA-OH binding (data not shown).

The putative TXA_2/PGH_2 receptor has also been solubilized in active form using the detergent CHAPS (Burch *et al.*, 1985a). The binding characteristics of [^{125}I]PTA-OH to the solubilized receptor were similar to those found in the crude platelet membrane fraction (Fig. 6). The binding characteristics of several TXA_2/PGH_2 receptor agonists and antagonists to the solubilized membranes were similar to those found in the crude membrane preparation.

The molecular weight of the solubilized binding site was initially estimated using size exclusion gel chromatography and also from its hydrodynamic properties. These estimates were 200,000 and 140,000 daltons, respectively (Burch *et al.*, 1985b). These estimates were higher than what was anticipated to be the molecular weight of the receptor. Recently, we have synthesized an ^{125}I-labeled azido analogue of I-PTA-OH. Using this molecule, we have been able to show a specifically labeled band on SDS–PAGE–autoradiography with an apparent molecular weight of 58,000 daltons under nonreducing conditions and 70,000 daltons under reducing conditions (P. V. Halushka *et al.*, unpublished observations). The increase in molecular

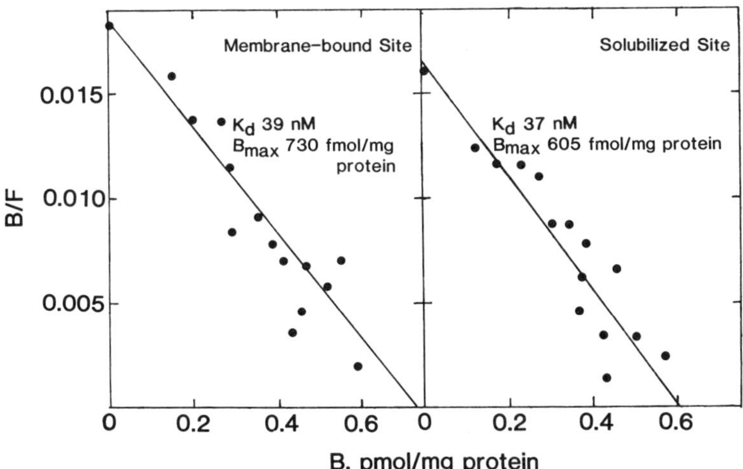

Figure 6. Scatchard plot of equilibrium binding data for [^{125}I]PTA-OH to intact and solubilized human platelet membranes from a single subject. Membrane-bound site: $K_d = 39$ nM, $B_{max} = 780$ fmoles/mg protein. Solubilized site: $K_d = 37$ nM, $B_{max} = 605$ fmoles/mg protein. Reprinted by the permission of the authors from *Proceedings of the National Academy of Sciences* **82:**7435 (1985).

weight under reducing conditions is suggestive of the presence of a disulfide bond(s) in the binding site of the receptor. This fits well with the observations that [^{125}I]PTA-OH binding to platelet membranes is decreased when disulfide bonds in the receptor are reduced (*vide supra*).

As mentioned above, in platelet membrane preparations there was a decrease in the affinity of the receptor for agonists, but not for antagonists. These observations raised the possibility that there was a substance(s) in platelets that was either released or activated during the preparation of the membranes and that decreased the affinity of the TXA_2/PGH_2 receptor for its agonists. To test this notion, Dorn *et al.* (1987a) prepared a supernatant fraction (50,000 *g*) from sonicated platelets and coincubated it with intact washed human platelets. They found that coincubation of the washed platelets with the supernatant decreased the affinity of the TXA_2/PGH_2 receptor for three different agonists but did not affect the binding affinity of [^{125}I]PTA-OH or another TXA_2/PGH_2 receptor antagonist, SQ29548. The active component of the supernatant appears to be a protein with a molecular weight of approximately 100,000 daltons. The mechanism of this effect is currently being explored.

While there have been many radioligand binding studies reported for the human platelet TXA_2/PGH_2 receptor, there have been none reported for the vascular receptor.

4. Second Messenger Systems

4.1. Platelets

The second messenger system initially proposed for the platelet TXA_2/PGH_2 receptor was adenylate cyclase. It was originally reported that PGG_2 inhibited PGE_1-stimulated adenylate cyclase in platelets (Miller and Gorman, 1976). However, this was probably secondary to the release of ADP. Subsequent work by Miller *et al.* (1977) and Miller and Gorman (1976) demonstrated that PGH_2 and TXA_2 had no effect on basal adenylate cyclase activity. Rybicki and LeBreton (1983) also reported that arachidonic acid and PGH_2 decreased platelet cAMP levels previously elevated by PGI_2. Indomethacin blocked the former effect; however, a thromboxane synthase inhibitor did not prevent this effect. The effect of TXA_2 on decreasing cAMP levels was inhibited by the calcium antagonist TMB-8 (Gorman *et al.*, 1979). Collectively, these observations led to the conclusion that the inhibition of PGE_1 stimulated adenylate cyclase by TXA_2 was not a direct effect but was secondary to an increase in intracellular free calcium, which in turn inhibited adenylate cyclase and/or release of ADP. Alternatively, these effects may be due to a direct effect on adenylate cyclase, since Avdonin *et al.* (1985) reported that U46619 and U44069 could inhibit both basal- and PGE_1-stimulated adenylate cyclase in a platelet membrane preparation. The IC_{50} for inhibition of adenylate cyclase by U46619 appeared to be similar to that found by other investigators for its effects on platelet activation and aggregation.

These results are in contrast to the findings of Best *et al.* (1979), who found that the TXA_2/PGH_2 mimetics U44069 and U46619 appeared to have no effect on basal- or PGE_1-stimulated adenylate cyclase. Thus, in the platelet the TXA_2/PGH_2 receptor does not appear to be linked directly to inhibition of adenylate cyclase.

It has previously been reported that U46619 stimulates an increase in intraplatelet free Ca^{2+} and that the size of the maximal transient rise in free Ca^{2+} in the absence of external Ca^{2+} and with 1 mM EGTA is markedly smaller than that seen in the presence of 1 mM external Ca^{2+} (Hallam *et al.*, 1984). Another TXA_2 mimetic, STA_2, has also been shown to increase intracellular free calcium concentrations in platelets (Kawahara *et al.*, 1983). Thus, agonists acting at the TXA_2/PGH_2 receptor may induce an increase in intraplatelet free Ca^{2+} partly by stimulating Ca^{2+} release from the dense tubular system, but it appears that the bulk of the Ca^{2+} movement promoted by TXA_2/PGH_2 agonists is across the plasma membrane. This Ca^{2+} influx may be through some form of a receptor-operated Ca^{2+} channel in the plasma membrane (Hallam and Rink, 1985).

Additional evidence for a TXA_2/PGH_2 receptor-operated Ca^{2+} channel in the plasma membrane is found in a series of studies by Crawford and his associates. Using a highly enriched dense tubular system preparation, they demonstrated that PG endoperoxides, TXA_2, and U46619 did not release Ca^{2+} from the dense tubular system (Carey *et al.*, 1983, 1985). In contrast to the studies cited above, they found that inositol trisphosphate was capable of releasing Ca^{2+} from this preparation (Authi and Crawford, 1985) and, in studies with saponin-permeabilized platelets, was capable of stimulating platelet aggregation (Authi *et al.*, 1986).

Stimulation of phospholipase C by a variety of agonists including PGH_2, TXA_2, and their stable mimetics results in the formation of inositol trisphosphate, cyclic inositol trisphosphate, and diacylglycerol (Pollock *et al.*, 1984; Rink and Hallam, 1984; Rittenhouse, 1984; Watson and Lapetina, 1985). The inositol polyphosphates act presumably at their receptors to cause an increase in intraplatelet free calcium, resulting in aggregation. However, this does not appear to be the entire story. Authi *et al.* (1986) recently demonstrated that when inositol trisphosphate was introduced into saponin-permeabilized platelets, aggregation resulted that could be blocked by cyclooxygenase inhibitors or TXA_2/PGH_2 receptor antagonists. They concluded that the calcium released by inositol trisphosphates, whatever its source, acts to stimulate calcium-dependent phospholipase A_2, releasing arachidonic acid from membrane phospholipids. Arachidonic acid is then converted to PGH_2 and TXA_2, which results in aggregation. Thus, the exact role of inositol polyphosphates in platelet aggregation, in particular their role as a second messenger for the TXA_2/PGH_2 receptor, remains to be determined.

Coupled to the release of inositol polyphosphates is the release of diacylglycerol with subsequent stimulation of protein kinase C. Upon activation, this enzyme catalyzes the phosphorylation of various platelet proteins. Arachidonic acid, U46619, and 9,11-azo-PGH_2 have been shown to induce the phosphorylation of at least three different proteins (Gerrard and Carroll, 1981; Seiss *et al.*, 1983). One protein was actin binding protein of MW 160 kD, and second of MW 20 kD was myosin light chain. Myosin light-chain phosphorylation precedes shape change and aggregation and probably plays an important role in the platelet response to aggregating agents. A third protein of MW 40 kD had been unidentified until very recently. This protein has been purified and shown to dephosphorylate inositol trisphosphate or cyclic inositol trisphosphate at the 5-position (Connolly *et al.*, 1986). Upon phosphorylation by protein kinase C, the enzyme becomes active. This enzyme and its activation by kinase C provide the platelet (and other cells) with a mechanism to limit the consequences of the synthesis of the inositol polyphosphates.

Guanine nucleotide regulatory proteins may play a role in coupling TXA_2/PGH_2 receptors to their second messenger(s), since U46619 and U44069 stimulate a platelet GTPase activity (Avdonin *et al.*, 1985). This GTPase activity may by due partly to stimulation of the N_i regulatory subunit of adenylate cyclase, but a large portion of the GTPase activity appears to be distinct from the N_s and N_i guanine nucleotide regulatory proteins (Houslay *et al.*, 1986). This novel GTPase may be a new guanine nucleotide regulatory protein coupled to phospholipase C hydrolysis of inositol phospholipids similar to the N_p protein described by Litosch and Fain (1986). Baldassare and Fisher (1986) reported the hydrolysis of polyphosphoinositides by platelet phospholipase C is regulated by GTP and nonhydrolyzable GTP analogues in a manner consistent with regulation of the enzyme activity by a GTP-binding regulatory protein.

Coupling of the TXA_2/PGH_2 receptor through a guanine nucleotide regulatory protein to a phosphatidylinositol-specific phospholipase C would be consistent with the observation that arachidonic acid and a variety of TXA_2/PGH_2 mimetics stimulate inositol lipid hydrolysis and the production of 1,2-diacylglycerol, phosphatidic acid, and inositol polyphosphates (Rittenhouse, 1982; Seis *et al.*, 1983; Pollock *et al.*, 1984; Rink and Hallam, 1984; Rittenhouse, 1984; Watson and Lapetina, 1985).

4.2. Vascular Smooth Muscle

In vascular smooth muscle, the second messenger(s) for the TXA_2/PGH_2 receptor remains uncertain. Both U46619 and U44069 alter $^{45}Ca^{2+}$ fluxes in canine pulmonary artery and vein (Greenberg, 1981). Loutzenhiser and van Breemen (1981) also found that U44069 enhanced $^{45}Ca^{2+}$ efflux from rabbit aorta. The cellular site(s) of calcium affected by agonists is (are) unknown. It has been postulated that mitochondria may be a possible site, since U46619 and prostaglandins have been shown to influence calcium fluxes in mitochondria (McNamara *et al.*, 1980). However, the endoplasmic reticulum or extracellular compartment has not been ruled out as a source of calcium that is released in response to TXA_2/PGH_2. It has also been shown that STA_2, a thromboxane mimetic, increases intracellular free calcium (Fukuo *et al.*, 1986) and U46619 can increase $^{45}Ca^{2+}$ efflux in vascular smooth muscle cells (Dorn *et al.*, 1987b).

It was recently shown that inositol trisphosphate could induce calcium release and vasoconstriction in permeabilized rabbit pulmonary artery (Somlyo *et al.*, 1985). Perhaps, as in the platelet, the TXA_2/PGH_2 receptor may be coupled to phosphatidylinositol polyphosphate turnover.

5. Summary

The biochemical, structural, and functional characterization of TXA_2/PGH_2 receptors is virtually in its infancy. To date, there have been no radioligand binding studies reported on TXA_2/PGH_2 receptors in any tissue other than platelets. Clearly, there are significant opportunities to enhance our understanding of the relation of structure to function for TXA_2/PGH_2 receptors in platelets and other tissues, as well as to elucidate the complexities of the second messenger systems for the receptors.

Acknowledgments. This research was supported in part by grants HL 29566 and HL 36838. Perry V. Halushka is a Burroughs Wellcome Scholar in Clinical Pharmacology. The secretarial assistance of Ms. Penny Pate, Ms. Nita Pike and Ms. Virginia Minchoff is gratefully acknowledged.

References

Akbar, H., Mukhopadhyay, A., Anderson, K., Navran, S., Ramstedt, K., Miller, D., and Feller, D., 1985, Antagonism of prostaglandin-mediated responses in platelets and vascular smooth muscle by 13-azaprostanoic acid analogs, *Biochem. Pharmacol.* **34:**641–647.

Armstrong, R. A., Jones, R. L., and Wilson, N. H., 1983, Ligand binding to thromboxane receptors on human platelets: Correlation with biological activity, *Br. J. Pharmacol.* **79:**953–964.

Authi, K. S., and Crawford, N., 1985, Inositol 1,4,5-triphosphate-induced release of sequestered Ca^{2+} from highly purified human platelet intracellular membranes, *Biochem. J.* **230:**247–253.

Authi, K., Evenden, B., and Crawford, N., 1986, Metabolic and functional consequences of introducing inositol 1,4,5-triphosphate into saponin-permeabilized human platelets, *Biochem. J.* **233:**709–718.

Avdonin, P. V., Svitina-Ulitina, I. V., Leytin, V. L., and Tkachuk, V. A., 1985, Interaction of stable prostaglandin endoperoxide analogs U46619 and U44069 with human platelet membranes and coupling of receptors with high-affinity GTPase and adenylate cyclase, *Thrombosis Res.* **40:**101–112.

Baldassare, J. J., and Fisher, G. J., 1986, Regulation of membrane-associated and cytosolic phospholipase C activities in human platelets by guanosine triphosphate, *J. Biol. Chem.* **261:**11942–11944.

Ball, H. A., Cook, J. A., Wise, W. C., and Halushka, P. V., 1986, Role of thromboxane, prostaglandins and leukotrienes in endotoxic and septic shock, *Intensive Care Med.* **12:**116–126.

Banerjee, A. K., Tuffin, D. P., and Walker, J. L., 1985, Pharmacological effects of (±)-11-deoxy,16-phenoxy-prostaglandin E_1 derivatives in the cardiovascular system, *Br. J. Pharmacol.* **84:**71–80.

Best, L. C., McGuire, M. B., Martin, T. J., Preston, F. E., and Russell, R. G. G., 1979, Effects of epoxymethano analogues of prostaglandin endoperoxides on aggregation, on release of 5-hydroxytryptamine and on the metabolism of 3′,5′-cyclic AMP and cyclic GMP in human platelets, *Biochim. Biophys. Acta* **583:**344–351.

Bhagwat, S. S., Hamann, P. R., Still, W. C., Bunting, S., and Fitzpatrick, F. A., 1985, Synthesis and structure of the platelet aggregation factor thromboxane A_2, *Nature* **315:**511–513.

Bundy, G., 1975, The synthesis of prostaglandin endoperoxide analogs, *Tetrahedron Lett.* **24:**1957–1960.

Burch, R., Mais, D. E., Suassy, D. L., and Halushka, P. V., 1985a, Solubilization of a thromboxane A_2/prostaglandin H_2 antagonists binding site from human platelets, *Proc. Natl. Acad. Sci. U.S.A.* **82:**7434–7438.

Burch, R., Mais, D., Pepkowitz, S., and Halushka, P., 1985b, Hydrodynamic properties of a thromboxane A_2/prostaglandin H_2 antagonist binding site solubilized from human platelets, *Biochem. Biophys. Res. Commun.* **132:**961–968.

Burke, S. E., Lefer, A. M., Nicolaou, K. C., Smith, G. M., and Smith, J. B., 1983, Responsiveness of platelets and coronary arteries from different species to synthetic thromboxane and prostaglandin endoperoxide analogues, *Br. J. Pharmacol.* **78:**287–292.

Carey, F., Menashi, S., and Crawford, N., 1983, Prostaglandin endoperoxides and thromboxane do not promote release of sequestered Ca^{++} from platelet intracellular membrane vesicles, *Br. J. Pharmacol.* **80:**116P.

Carey, F., Menashi, S., Authi, K. S., Hack, N., Lagarde, M., and Crawford, N., 1985, Platelet membranes, eicosanoid biosynthesis and putative endogenous calcium ionophores, *Adv. Exp. Med. Biol.* **192:**195–199.

Connolly, T., Laving, W., and Majerus, P., 1986, Protein kinase C phosphorylates human platelet inositol triphosphate 5′-phosphomonoesterase increasing the phosphatase activity, *Cell* **46:**951–958.

Dorn, G., Burch, R. M., Kochel, P., Mais, D. E., and Halushka, P. V., 1987a, Alteration of platelet thromboxane A_2/prostaglandin H_2 receptors by supernatant of platelet homogenates, *Biochem. Pharmacol.* **36:**1913–1917.

Dorn, G., Sens, D., Chaikhouni, A., Mais, D., and Halushka, P., 1987b, Cultured human vascular smooth muscle cells with functional thromboxane A_2 receptors: Measurement of U46619 induced calcium-45 efflux, *Circulation Res.* **60:**952–956.

Fitzgerald, D. J., Roy, L., Catella, F., and FitzGerald, G. A., 1986, Platelet activation in unstable coronary disease, *N. Engl. J. Med.* **315:**983–990.

FitzGerald, G. A., Pedersen, A. K., and Patrono, C., 1983, Analysis of prostacyclin and thromboxane biosynthesis in cardiovascular disease, *Circulation* **67:**1174–1177.

FitzGerald, G. A., Healy, C., and Daugherty, J., 1987, Thromboxane A_2 biosynthesis in human disease, *Fed. Proc.* **46:**154–158.

Fukuo, K., Morimoto, S., Koh, E., Yukawa, S., Tsuchiya, H., Imanaka, S., Yamamoto, H., Onishi, T., and Kumahara, Y., 1986, Effects of prostaglandins on the cytosolic free calcium concentration in vascular smooth muscle cells, *Biochem. Biophys. Res. Commun.* **136**(1)**:**247–252.

Gerrard, J. M., and Carroll, R. C., 1981, Stimulation of platlelet protein phosphorylation by arachidonic acid and endoperoxide analogs, *Prostaglandins* **22:**81–94.

Gerrard, J. M., White, J. G., and Peterson, D. A., 1978, The platelet dense tubular system: Its relationship to prostaglandin synthesis and calcium flux, *Thromb. Haemost.* **40:**224–231.

Gorman, R. R., Wierenga, W., and Miller, O. V., 1979, Independence of the cyclic AMP-lowering activity of thromboxane A_2 from the platelet release reaction, *Biochim. Biophys. Acta* **572:**95–104.

Greenberg, S., 1981, Effect of prostacyclin and 9α, 11α-epoxymethanoprostaglandin H_2 on calcium and magnesium fluxes and tension development in canine intralobar pulmonary arteries and veins, *J. Pharmacol. Exp. Ther.* **219:**326–337.

Hallam, T. J., and Rink, T. J., 1985, Agonists stimulate divalent cation channels in the plasma membrane of human platelets, *FEBS Lett.* **186:**175–179.

Hallam, T. J., Sanchez, A., and Rink, T. J., 1984, Effects of excitatory/inhibitory eicosanoids on cytoplasmic free calcium directly measured in human platelets with the fluorescent indicator quin-2, in: *Prostaglandins and Membrane Ion Transport* (P. Braquet, R. P., Garay, J. C. Frölich, and S. Nicosia, eds.) pp 157–163, Raven, New York.

Halushka, P. V., and Lefer, A. 1987, Thromboxane A_2 in health and disease, *Fed. Proc.* **46:**131–132.

Halushka, P. V., Rogers, R. C., Loadholt, C. B., and Colwell, J. A., 1981, Increased platelet thromboxane synthesis in diabetes mellitus, *J. Lab. Clin. Med.* **97:**87–96.

Halushka, P. V., Dollery, C. T., and MacDermot, J., 1983, Thromboxane and prostacyclin in disease: A review, *Q. J. Med.* **208:**461–470.

Halushka, P. V., MacDermot, J., Knapp, D. R., Eller, T., Saussy, D. L., Jr., Mais, D. B., Blair, I. A., and Dollery, C. T., 1985, A novel approach for the study of thromboxane A_2 and prostaglandin H_2 receptors using an ^{125}I-labelled ligand, *Biochem. Pharmacol.* **34:**1165–1170.

Halushka, P. V., Mais, D. E., Garvin, M., Kochel, P., and Sightler, H., 1986a, Structure–activity relationships for 13-azapinane-TXA_2 analogs in platelets and vascular TXA_2/PGH_2 receptors: Evidence for different receptors, in: *6th International Conference on Prostaglandin*, p. 166, Fondazione Giovanni Lorenzini.

Halushka, P. V., Mais, D. E., and Garvin, M., 1986b, Binding of a thromboxane A_2/prostaglandin H_2 receptor antagonist to guinea-pig platelets, *Eur. J. Pharmacol.* **131:**49–54.

Halushka, P. V., Mais, D. E., and Saussy, D. L., Jr. 1987a, Platelet and vascular smooth muscle thromboxane A_2/prostaglandin H_2 receptors, *Fed. Proc.* **46:**149–153.

Halushka, P., Kochel, P., and Mais, D., 1987b, Binding of thromboxane A_2/prostaglandin H_2

agonists to human platelets, *Br. J. Pharmacol.* **91:**223–227.

Hamberg, M., and Samuelsson, B., 1974, Prostaglandin endoperoxides. Novel transformations of arachidonic acid in human platelets, *Proc. Natl. Acad. Sci. U.S.A.* **71:**3400–3404.

Hamberg, M., Svensson, J., and Samuelsson, B., 1975, Prostaglandin endoperoxides. A new concept concerning the mode of action and release of prostaglandins, *Proc. Natl. Acad. Sci. U.S.A.* **71:**3824–3828.

Hamberg, M., Svensson, J., and Samuelsson, B., 1974, Thromboxanes: A new group of biologically active compounds derived from prostaglandin endoperoxides, *Proc. Natl. Acad. Sci. U.S.A.* **72:**2994–2998.

Hirsch, P. D., Hillis, L. D., Campbell, W. B., Firth, B. G., and Willerson, J. T., 1981, Release of prostaglandins and thromboxane into the coronary circulation in patients with ischemic heart disease, *N. Engl. J. Med.* **304:**685–691.

Houslay, M., Bojanic, D., and Wilson, A., 1986, Platelet activating factor and U44069 stimulate a GTPase activity in human platelets which is distinct from the guanine nucleotide regulatory proteins, N_s and N_i, *Biochem. J.* **234:**737–740.

Hung, S. C., Ghali, N. I., Venton, D. L., and LeBreton, G. C., 1983, Specific binding of the thromboxane A_2 antagonist 13-azaprostanoic acid to human platelet membranes, *Biochim. Biophys. Acta* **728:**171–178.

Kattelman, E., Venton, D., and LeBreton, G., 1986, Characterization of U46619 binding in unactivated, intact human platelets and determination of binding site affinities of four TXA_2/PGH_2 receptor antagonists (13-APA, BM 13.177, ONO 3708 and SQ 29,548), *Thromb. Res.* **41:**471–481.

Kawahara, Y., Yamanishi, J., Furuta, Y., Kaibuchi, K., Takai, Y., and Fukuzaki, H., 1983, Elevation of cytoplasmic free calcium concentration by stable thromboxane A_2 analogue in human platelets, *Biochem. Biophys. Res. Commun.* **117:**663–669.

Kelley, V. E., Sneve, S., and Musinski, S., 1986, Increased renal thromboxane production in murine lupus nephritis, *J. Clin. Invest.* **77:**252–259.

Lawson, J. A., Patrono, C., Ciabattoni, G., and FitzGerald, G. A., 1986, Long-lived enzymatic metabolites of thromboxane B_2 in the human circulation, *Anal. Biochem.* **15:**198–205.

LeBreton, G. C., Venton, D. L., Enke, S. E., and Halushka, P. V., 1979, 13-Azaprostanoic acid: A specific antagonist of the human blood platelets thromboxane endoperoxide receptor, *Proc. Natl. Acad. Sci. U.S.A.* **76:**4097–4101.

Lefer, A. M., 1984, Role of eicosanoids in circulatory shock, *Prostaglandins and Other Eicosanoids in the Cardiovascular System* (K. Schrör, ed.), Proceedings of the 2nd International Symposium pp. 149–159, Karger, Basel.

Liel, N., Mais, D. E., and Halushka, P. V., 1986, Binding of a thromboxane A_2/prostaglandin H_2 agonist [^{3}H]U46619 to washed human platelets, *Prostaglandins* **33:**789–797.

Litosch, I., and Fain, J., 1986, Regulation of phosphoinositide breakdown by guanine nucleotides, *Life Sci.* **39:**187–194.

Loutzenhiser, R., and van Breemen, C., 1981, Mechanism of activation of isolated rabbit aorta by PGH_2 analogue U-44069, *Am. J. Physiol.* **241:**C243–C249.

Mais, D., Knapp, D., Ballard, K., Hamanaka, N., and Halushka, P., 1984, Synthesis of thromboxane receptor antagonists with the potential to radiolabel with ^{125}I, *Tetrahedron Lett.* **25:**4207–4210.

Mais, D., Saussy, D., Chaikhouni, A., Kochel, P., Hamanaka, N., and Halushka, P., 1985a, Pharmacologic characterization of human and canine thromboxane A_2/prostaglandin H_2 receptors in platelets and blood vessels: Evidence for different receptors, *J. Pharmacol. Exp. Ther.* **233:**424–428.

Mais, D., Dunlap, C., Hamanaka, N., and Halushka, P., 1985b, Further studies on the effects of epimers of thromboxane A_2 antagonists on platelets and veins, *Eur. J., Pharmacol* **111:**125–128.

Mais, D. E., Burch, R. M., Saussy, D. L., Jr., Kochel, P. J., and Halushka, P. V., 1985c, Binding of a thromboxane A_2/prostaglandin H_2 receptor antagonist to washed human platelets, *J. Pharmacol. Exp. Ther.* **235:**729–734.

Mais, D. E., Kochel, P. J., Saussy, D. L., Jr., and Halushka, P. V., 1985d, Binding of an ^{125}I-labelled thromboxane A_2/prostaglandin H_2 receptor antagonist to washed canine platelets, *Mol. Pharmacol.* **28:**163–169.

McNamara, D. B., Roulet, M. J., Gruetter, C. A., Hymen, A. L., and Kadowitz, P. J., 1980, Correlation of prostaglandin-induced mitochondrial calcium release with contraction in bovine intrapulmonary vein, *Prostaglandins* **20:**311–320.

Miller, O. V., and Gorman, R. R., 1976, Modulation of platelet cyclic nucleotide content by PGE_1 and the prostaglandin endoperoxide PGG_2, *J. Cyclic Nucleotide Protein Phosphor. Res.* **2:**79–87.

Miller, O. V., Johnson, R. A., and Gorman, R. R., 1977, Inhibition of PGE_1-stimulated cAMP accumulation in human platelets by thromboxane A_2, *Prostaglandins* **13:**599–609.

Narumiya, S., Okuma, M., and Ushikubi, F., 1986, Binding of a radioiodinated 13-azapinane thromboxane antagonist to platelets: Correlation with antiaggregatory activity in different species, *Br. J. Pharmacol.* **88:**323–331.

Needleman, P. M., Minkes, M., and Raz, A., 1976a, Thromboxanes: Selective biosynthesis and distinct biological properties, *Science* **193:**163–165.

Needleman, P., Moncada, S., Bunting, S., Vane, J., Hamberg, M., and Samuelsson, B., 1976b, Identification of an enzyme in platelet microsomes which generates thromboxane A_2 from prostaglandin endoperoxides, *Nature* **261:**558–560.

Ogletree, M. L., 1987, Overview of physiological and pathophysiological effects of thromboxane A_2, *Fed. Proc.* **46:**133–138.

Ogletree, M., Allen, G. O'Keefe, E., Liv, K., and Hedberg, A., 1986, Activities of various prostanoids at thromboxane receptors revealed by selective receptor antagonists: Studies in human platelets and several rat and guinea-pig smooth muscles, in: *6th International Conference on Prostaglandins*, p. 350, Fondazione Giovanni Lorenzini, Florence, Italy.

Patrono, C., Ciabattoni, G., Remuzzi, G., Gotti, E., Bombardieri, S., Di Munno, O., Tartarelli, G., Cinotti, G. A., Simonetti, B. M., and Pierucci, A., 1985, Functional significance of renal prostacyclin and thromboxane A_2 production in patients with systemic lupus erythematosus, *J. Clin. Invest.* **76:**1011–1018.

Piper, P. J., and Vane, J. R., 1969, Release of additional factors in anaphylaxis and its antagonism by anti-inflammatory drugs, *Nature* **223:**29–35.

Pollock, W. K., Armstrong, R. A., Brydon, L. J., Jones, J. L., and MacIntyre, D. E., 1984, Thromboxane induced phosphatidate formation in human platelets, *Biochem. J.* **219:**833–842.

Rink, T. J., and Hallam, J. T., 1984, What turns platelets on, *Trends Biochem. Sci. Pers. Ed.* **9:**215–219.

Rittenhouse, S. E., 1984, Activaiton of human platelet phospholipase C by ionophore A23187 is totally dependent upon cyclo-oxygenase products and ADP, *Biochem. J.* **222:**103–110.

Rybicki, J. P., and LeBreton, G. C., 1983, Prostaglandin H_2 directly lowers human platelet cAMP levels, *Thromb. Res.* **30:**407–414.

Rybicki, J. P., Venton, D. L., and LeBreton, G. C., 1983, The thromboxane antagonist, 13-azaprostanoic acid, inhibits arachidonic acid-induced Ca^{2+} release from isolated platelet membrane vesicles, *Biochim. Biophys. Acta* **751:**66–73.

Saussy, D. L., Jr., Mais, D. E., Burch, R. M., and Halushka, P. V., 1985, Identification of a putative TXA_2/PGH_2 receptor in human platelet membranes, *J. Biol. Chem.* **261:**3025–3029.

Saussy, D. L., Jr., Mais, D. E., Baron, D. A., Pepkowitz, S. H., and Halushka, P. V., 1987, Subcellular localization of a thromboxane A_2/prostaglandin PGH_2 receptor antagonist binding site in human platelets *Biochem. Pharmacol.* (in press).

Seiss, W., Siegel, F. L., and Lapetina, E. G., 1983, Arachidonic acid stimulates the formation of 1,2-diacylglycerol and phosphatidic acid in human platelets, *J. Biol. Chem.* **253:**11236–11242.

Somlyo, A., Bond, M., Somlyo, A., and Scarpa, A., 1985, Inositol trisphosphate-induced calcium release and contraction in vascular smooth muscle, *Proc. Natl. Acad. Sci. U.S.A.* **82:**5231–5235.

Svensson, J., Hamberg, M., and Samuelsson, B., 1975, Prostaglandin endoperoxides IX. Characterization of rabbit aorta contracting substance (RCS) from guinea-pig lung and human platelets, *Acta Physiol. Scand.* **94:**222–228.

Ulrich, V., and Graf, H., 1984, Prostacyclin and thromboxane synthase as P-450 enzymes, *Trends Pharmacol. Sci.* **5:**352–355.

Vargafatig, B., and Dao, N., 1971, Release of vasoactive substances from guinea-pig lungs by slow-reacting substance C and arachidonic acid. It's blocked by nonsteroid anti-inflammatory agents, *Pharmacology* **6:**99–108.

Vargafatig, B., and Zirinis, P., 1973, Platelet aggregation induced by arachidonic acid is accompanied by release of potential inflammatory mediators distinct from PGE_2 and $PGF_{2\alpha}$, *Nature New Biol.* **244:**114–116.

Watson, S. P., and Lapetina, E. G., 1985, 1,2-Diacylglycerol and phorbol ester inhibit agonist-induced formation of inositol phosphates in human platelets: Possible implication for negative feedback reguatlion of inositol phospholipid hydrolysis, *Proc. Natl. Acad. Sci. U.S.A.* **82:**2623–2626.

Willis, A. L., Vane, J. M., Kuhn, D. C., Scott, C. G., and Petrin, M., 1974, An endoperoxide aggregatory (LASS) formed in platelets in response to thrombotic stimuli-purification, identification and unique biological significance, *Prostaglandins* **8:**455–507.

Winocour, P. D., Halushka, P. V., and Colwell, J. A., 1985, Platelet involvement in diabetes mellitus, in *The Platelets: Physiology and Pharmacology* (G. L., Longenecker, ed.), pp. 341–366, Academic Press, Orlando, FL.

18

Chemoattractant Receptors and Signal Transduction Processes

RONALD J. UHING, SUSAN B. DILLON, PAUL G. POLAKIS, ARTIS P. TRUETT III, and RALPH SNYDERMAN

1. Introduction

Leukocyte responses to chemotactic and other phlogistic stimuli are vital for host defense and substantial interest has thus been focused on defining the mechanisms of signal transduction in these cells. Chemoattractants initiate leukocyte activation subsequent to binding to specific receptors on the cell surface. Temporal studies show that chemoattractants elicit rapid (≤ 5 s) increases in phosphoinositide metabolism and cytosolic calcium levels followed by changes in physiological function, for example, shape change, superoxide production, or degranulation. The ability of pharmacological agents (e.g., calcium ionophores or phorbol esters) to elicit similar functions suggests the central role of calcium mobilization, phosphoinositide metabolism, and protein kinase C in leukocyte activation. Although specific receptors are present for different classes of chemoattractants, they all appear to utilize a common mechanism for increasing cytosolic calcium. Chemoattractants also increase cellular cAMP levels, in this case by a calcium-mediated inhibition of cAMP degradation. This may serve an autoregulatory role in chemoattractant-induced leukocyte activation.

RONALD J. UHING, SUSAN B. DILLON, PAUL G. POLAKIS, ARTIS P. TRUETT III, and RALPH SNYDERMAN • Howard Hughes Medical Institute and the Division of Rheumatology and Immunology, Department of Medicine, Duke University Medical Center, Durham, North Carolina 27710.

2. Chemoattractant Receptors

Specific receptors have been characterized for a number of different chemoattractants, including N-formylated peptides, C5a, and leukotriene B_4 (LTB_4). The most extensively studied chemoattractant receptor is that for N-formylated methionyl peptides, which may be analogous to bacterial-derived chemotactic factors (for reviews see Sklar *et al.*, 1984; Snyderman and Pike 1984a). Early studies with polymorphonuclear leukocytes (PMNs) showed that formylation of the amino methionyl terminus was essential for the biological activity of formylpeptide chemoattractants (Showell *et al.*, 1976). Direct binding studies with intact human PMNs using fMet-Leu-[^{3}H]Phe (Williams *et al.*, 1977; Koo *et al.*, 1982) demonstrated the presence of approximately 50,000 binding sites per cell. The binding data were best fit to a single class of binding sites with a K_d of approximately 20nM. The ability of other N-formylated peptides to inhibit fMet-Leu-[^{3}H]Phe binding paralleled that seen for eliciting biological responses, indicating that a common receptor is utilized. Other types of chemoattractants, such as C5a or LTB_4, do not compete for fMet-Leu-[^{3}H]Phe binding. Covalent crosslinking techniques have been utilized to identify the N-formylated peptide receptor. Using various labeled N-formylated peptides, a crosslinked protein of 60,000–70,000 daltons has been identified (Niedel *et al.*, 1980; Goetzl *et al.*, 1981; Painter *et al.*, 1982; Kay *et al.*, 1983). This protein is heavily glycosylated with removal of carbohydrate leaving a 32,000-dalton peptide that still exhibits agonist binding (Malech *et al.*, 1985). Partial purification of the N-formylated peptide receptor has been reported, although the binding activity of the purified preparations is poor (Goetzl *et al.*, 1981; Hoyle and Freer, 1984; Huang, 1987).

Chemoattractant receptors are not static on the cell surface. Cycling of the N-formylated peptide receptor was originally suggested from studies indicating time- and termperature-dependent endocytosis of the peptide (Niedel *et al.*, 1979; Sullivan and Zigmond, 1980). Subsequent studies have shown that the receptor is internalized following chemoattractant binding (reviewed in Sklar *et al.*, 1984). In the presence of high concentrations of N-formylated peptide or secretogues, the number of cell surface receptors can exceed that observed for unstimulated cells, indicating a reserve population of receptors. An intracellular population of receptors that comigrates with the specific granule fraction has been identified (Fletcher and Gallin, 1982; Jesaitis *et al.*, 1982). The specific granule receptor is expressed on the cell surface after exposure to secretogues and appears to be the same molecular weight as the native surface receptor (Gardner *et al.*, 1986).

In contrast to intact cells, binding of N-formylated peptides to membrane preparations from leukocytes is best fit to two classes of binding sites (Koo *et al.*, 1982; Mackin *et al.*, 1982; Snyderman *et al.*, 1984) Chemical and isotopic dilution studies exhibited identical dissociation rates, suggesting two populations of receptors (Koo *et al.*, 1982). The K_d values for the sites on human PMN membranes are approximately 0.5 and 20 nM, with the high-affinity state representing approximately 25% of the total in a crude membrane preparation (Koo *et al.*, 1982) or approximately 50% of the total in a plasma membrane preparation (R.J. Uhing, unpublished observations). The fact that high- and low-affinity binding sites were detectable in membranes, but only a single low-affinity site on intact polymorphonuclear leukocytes, suggested that the sites might be rapidly interconvertible in whole cells, allowing detection of only a single class using steady-state binding techniques (Koo *et al.*, 1983). Based on this observation, it was suspected that chemoattractant receptors might be coupled to GTP-binding proteins (G proteins), which are a group of plasma membrane components that serve a transducing function between a variety of receptors and their effector systems (for a review, see Gilman, 1984). Addition of GDP, GTP, or their nonhydrolyzable analogues (e.g., GppNHp or GTPγS), but not the corresponding adenosine compounds, causes the reversible conversion of high-affinity to low-affinity binding sites for formylpeptides on polymorphonuclear membranes (Koo *et al.*, 1983; Snyderman *et al.*, 1984). The absence of detectable high-affinity binding sites on intact cells using equilibrium binding is presumably due to the high levels of intracellular GTP and GDP, which favors a steady state that allows only the detection of the low-affinity state (Snyderman *et al.*, 1984). Using rapid (<1 min) analysis of peptide–receptor interactions, Sklar *et al.* (1987) have presented evidence for two dissociation rates on intact PMNs, with the rapid rate ($t_{1/2} \approx 10$ s) presumably being due to the receptor being either uncoupled or coupled to GTP or GDP-bound G protein. Using permeabilized cells, GTPγS was shown to increase the rate of dissociation of the entire population of receptor-bound N-formylated peptide (Sklar *et al.*, 1987). The association of N-formylated peptide receptor with a G protein is further suggested by the observation that chemoattractant binding to a solubilized high molecular weight form of the receptor was modulated by guanine nucleotides (Polakis and Snyderman, 1987) (Fig. 1). The solubilized receptor copurifies with a ~40-kD pertussis toxin substrate. Interestingly, binding to the specific granule fraction of chemoattractant receptors is not modulated by guanine nucleotides (R.J. Uhing *et al.*, unpublished observations). This is due to the apparent absence of G proteins in this fraction.

3. Receptor/GTP-Binding Protein/Phospholipase C Pathway

In a variety of cell types, the hormonal mobilization of intracellular calcium is through hormonal stimulation of inositol 1,4,5-trisphosphate (IP_3) formation (for reviews see Berridge and Irvine, 1984; Majerus *et al.*, 1986).

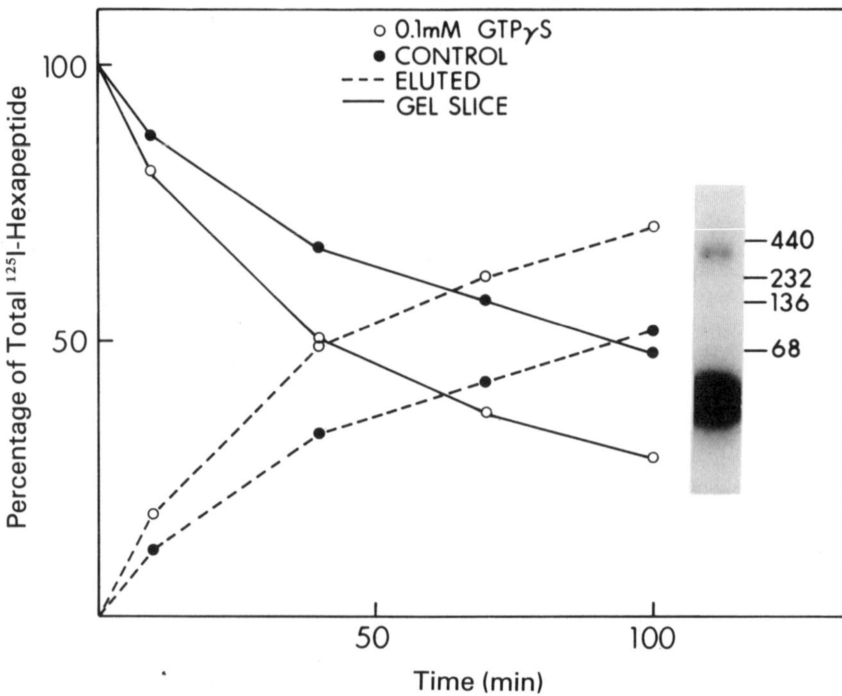

Figure 1. Elution of radioactivity following nondenaturing polyacrylamide gel electrophoresis of [^{125}I]hexapeptide–recepotor complexes. Plasma membranes from differentiated HL60 cells were noncovalently labeled with formyl-Nle-Leu-Phe-Nle-[^{125}I]Tyr-Lys, detergent extracted, and chromatographed on wheat germ agglutinin Sepharose. Eluted material was applied to a nondenaturing gradient polyacrylamide gel and the bands containing radioligand–receptor complex were visualized by autoradiography and excised from the gel; the slices were then incubated in buffer in the presence (○) or absence (●) of 0.1 mM GTPγS. Radioactivity was determined in the gel slice (solid line) or eluant (broken line) at the indicated times. Inset shows an autoradiogram of the wet gel. The following protein markers were used: ferritin, 440 kD; catalase, 232 kD; bovine serum albumin dimer, 136 kD, and monomer, 68 kD. The density appearing below 68 kD represents the dissociated, free radioligand.

The addition of chemoattractants to leukocytes results in the rapid formation of inositol phosphates and diacylglycerol derived from phosphatidylinositol 4,5-bisphosphate (PIP_2) and an increase in calcium release from intracellular stores (see section 6), suggesting that chemoattractant receptors activate a phospholipase C. Analysis of the inositol phosphates produced demonstrates that IP_3 is formed coincident with loss of PIP_2, with later formation of IP and loss of PI (Dougherty *et al.*, 1984; Bradford and Rubin, 1985, 1986; Burgess *et al.*, 1985; Lew *et al.*, 1986a; Dillon *et al.*, 1987a). These data suggest action of a phospholipase C with preference for polyphosphoinositides. Activation of this phospholipase C appears to be a common mechanism for the various chemoattractant receptors, including those for N-formyl peptides, C5a, LTB_4, and PAF (Verghese *et al.*, 1986b). Synthesis of PIP_2 occurs rapidly via the concerted action of PI and PIP kinases, leading to a decrease in the level of PI. The addition of [γ-^{32}P]ATP to leukocyte plasma membranes results in the formation of labeled PIP, PIP_2, and phosphatidic acid (Cockcroft *et al.*, 1984; Smith *et al.*, 1985). The formation of PIP and PIP_2 is accelerated by the addition of polyamines at physiological magnesium concentrations, suggesting that cellular polyamine levels may be important in maintaining the rapid flux observed in intact leukocytes (C.D. Smith *et al.*, unpublished observations). Leukocyte plasma membranes also contain phospholipase C activity with apparent specificity for the polyphosphoinositides (Cockcroft *et al.*, 1984; Smith *et al.*, 1985, 1986). Thus, leukocyte plasma membranes contain various components necessary for chemoattractant stimulation of IP_3 formation and replenishment of PIP_2 levels.

The involvement of a G protein in chemoattractant responses is suggested by the abilities of guanine nucleotides to modulate receptor affinity and of bacterial toxins to attenuate PI metabolism and leukocyte function (see Section 4). Direct evidence for the involvement of a G protein in stimulated phospholipase C activity was presented by Smith *et al.* (1985), who showed that stimulation of PIP_2 breakdown in leukocyte plasma membranes by chemoattractants required GTP. This action is mediated by a polyphosphoinositide phosphodiesterase (phospholipase C) rather than a phosphomonoesterase, since IP_2 and IP_3 accumulate concomitantly with the degradation of polyphosphoinositides (Smith *et al.*, 1986). Specificity for polyphosphoinositides is suggested, since initially only IP_2 and IP_3 accumulate subsequent to stimulation of leukocyte plasma membranes (Smith *et al.*, 1986), although breakdown of PI at later times has not been excluded. Nonhydrolyzable analogues of GTP activate phospholipase C activity, whereas adenine nucleotides, GDP, and GMP are ineffective (Smith *et al.*, 1986). Chemoattractant stimulation of phospholipase C in plasma membranes is also inhibited by pertussis toxin, indicating that the G protein involved is a substrate for ADP-ribosylation by this toxin (Smith *et al.*, 1985).

Stimulation by guanine nucleotides or the synergistic stimulation by chemoattractants plus GTP requires the presence of low concentrations of calcium, similar to resting cytosolic levels (Smith *et al.*, 1985, 1986). Stimulation by chemoattractants is due to a reduction in the calcium requirement of the phospholipase C activity (Smith *et al.*, 1986). In the absence of chemoattractants or GTP, supraphysiological concentrations of calcium ($\geq 100\ \mu M$) are needed to cause breakdown of polyphosphoinositides in PMN membranes (Cockcroft *et al.*, 1984; Smith *et al.*, 1985, 1986).

4. Characterization of the GTP-Binding Protein Involved in Leukocyte Activation

G proteins have been identified as plasma membrane components that couple a variety of receptors to their effector enzymes (for recent reviews see Gilman, 1984; Birnbaumer *et al.*, 1987; Speigel, 1987 (Table I). Common characteristics of these proteins are the ability to bind guanine nucleotides, an inherent GTPase activity, and a heterotrimeric $\alpha\beta\gamma$ structure. The α subunit confers specificity for receptor–effector system involvement, is the guanine nucleotide-binding component, and is the potential substrate for bacterial-toxin-catalyzed ADP-ribosylation. Bacterial toxins have been useful tools for the investigation of G protein roles in signal transduction (Gilman, 1984; Birnbaumer *et al.*, 1987; Spiegel, 1987). Cholera toxin catalyzes the ADP-ribosylation of G_s, reducing the GTPase activity (Cassel and Selinger, 1977), and thus causes persistent activation of adenylate cyclase. Pertussis toxin catalyzes the ADP-ribosylation of G_i and transducin, causes uncoupling with the receptor, and thus terminates the hormone signal.

As already noted, evidence has been obtained in leukocytes demonstrating involvement of a G protein in chemoattractant activation of phospholipase C. The G protein utilized shares characteristics of the previously described proteins. fMet-Leu-Phe stimulates both GTPγS binding (Smith, *et al.*, 1987) and GTPase activity (Okajima *et al.*, 1985; Feltner *et al.*, 1986). G proteins involved in calcium mobilization probably have a heterotrimeric structure, as suggested by immunochemical evidence for the vasopressin–G protein complex from liver (Fitzgerald *et al.*, 1986). Several lines of evidence suggest that the G protein involved in chemoattractant responses is a substrate for pertussis-toxin-catalyzed ADP-ribosylation. These include attenuation of chemoattractant-induced PIP_2-mediated responses (Bokoch and Gilman, 1984; Molski *et al.*, 1984; Okajima and Ui, 1984; Becker *et al.*, 1985; Brandt *et al.*, 1985; Goldman *et al.*, 1985; Lad *et al.*, 1985; Ohta *et al.*, 1985; Volpi *et al.*, 1985; Verghese *et al.*, 1985b, 1986a) and inhibition of fMet-Leu-Phe-receptor/G-protein-mediated events in membrane preparations (Okajima *et al.*, 1985; Smith *et al.*, 1985, 1986, 1987), including reduction of high-affinity fMet-Leu-Phe binding (Lad *et al.*, 1985b; Okajima *et al.*, 1985; Verghese *et al.*, 1986c).

ADP-ribosylation of leukocyte membranes in the presence of pertussis toxin shows a predominance of ~40-kD substrate. The similarity in molecular weight for the pertussis toxin substrate has led to investigations of whether the G protein involved in leukocyte activation is identical to G_i. Functional G_i has been demonstrated in intact PMNs by the ability of α_2 agonists to inhibit prostaglandin-E_1-stimulated cAMP accumulation (Verghese *et al.*, 1985a) and in membrane preparations by the ability of α_2 agonists to inhibit

Table I. Guanine Nucleotide Regulatory Proteins[a]

Characteristic	G_S	G_I	G_O	Transducin	G_P	G_C
Receptor coupling	β-Adrenergic, PGE	α_2-Adrenergic	Muscarinic	Rhodopsin	?	CTX
Subunit composition	$\alpha_s\beta\gamma$	$\alpha_i\beta\gamma$	$\alpha_o\beta\gamma$	$\alpha_t\beta\gamma$	$\alpha_p\beta\gamma$	$\alpha_c\beta\gamma$
Molecular weight of a subunit	43–50 kD	41 kD	39 kD	39 kD	21 kD	~40 kD
Ribosylated by	CT	PT	PT	CT/PT	—	CT/PT
Effector	AC^+	AC^-	?	cGMP-PDE	?	PLC

[a]Abbreviations used: AC, adenylate cyclase; CT, Cholera toxin; CTX, chemoattractant; PDE, phosphodiesterase; PLC, phospholipase C; PT, pertussis toxin. Taken from Snyderman and Uhing (1987).

cAMP production stimulated by either prostaglandins E or β agonists (Verghese *et al.*, 1985a; Bokoch, 1987). Neither report obtained any fMet-Leu-Phe-induced inhibition of agonist-stimulated cAMP formation in neutrophil membranes under similar conditions as described above (Smith *et al.*, 1985, 1986) for hormone-stimulated phospholipase C activity. These results imply that chemoattractant receptors do not couple to adenylate cyclase through G_i. Further evidence against a role for G_i in chemoattractant function is the observation that α agonists that utilize G_1 do not cause calcium mobilization in leukocytes (M. Verghese, unpublished observations). Evidence suggestive of a unique G protein in PMNs comes from studies with cholera toxin, which also ADP-ribosylates a ~40-kD protein in phagocytes (Aksamit *et al.*, 1985; Verghese *et al.*, 1986c) and myeloid cell lines (HL 60, U937) (Verghese *et al.*, 1986c). Under the same conditions, little cholera-toxin-induced ADP-ribosylation of ~40-kD proteins is evidenced in erythrocytes, liver, or brain (Verghese *et al.*, 1986c and unpublished observations). Cholera toxin treatment also resulted in decreased chemotaxis (Aksamit *et al.*, 1985), which could not be accounted for by changes in cAMP concentration. Furthermore, both pertussis and cholera toxins inhibited high-affinity fMet-Leu-Phe binding, suggesting that the same G protein may be a substrate for both (Verghese *et al.*, 1986c).

Spiegel and co-workers have utilized an immunochemical approach to identify the G protein coupled to chemoattractant receptors. Antibodies prepared against available G proteins were used to provide evidence that G_i is not present in sufficient quantities in human PMNs to account for the pertussis-toxin-catalyzed ADP-ribosylation of ~40-kD protein(s) (Giershik *et al.*, 1986). This lack of reactivity cannot be due to species differences, since G_i sequence deduced from a human monocyte-like cell line exhibits extensive homology to the G_i sequences previously deduced from other species and tissues (Didsbury *et al.*, 1987). Immunochemical evidence also suggests that other previously described pertussis toxin substrates, G_o and transducin, are low or absent in PMNs (Giershik *et al.*, 1986). The G protein utilized in leukocyte activation is nonetheless probably similar to those previously described, since antibodies to conserved regions of G protein exhibit strong reactivity with leukocyte membranes (Falloon *et al.*, 1986).

The above data suggest that leukocytes contain a unique, pertussis/cholera-toxin-sensitive, G protein that mediates chemoattractant-induced cell activation. However, Kikuchi *et al* (1986) have reported that either G_i or G_o are capable of reconstituting fMet-Leu-Phe-stimulated phospholipase C activity with similar potencies in plasma membranes prepared from pertussis-toxin-treated HL60 cells. It should be noted, however, that brain G_i/G_o preparations have been reported to contain an additional uncharacterized ~40-kD protein (Neer *et al.*, 1984. These authors did not comment on the possible presence of this protein in their preparations. Further work will be necessary to resolve the specificities of G proteins involved in phospholipase C coupling.

Work is currently in progress to purify and reconstitute the bacterial toxin substrates present in these cells. Leukocytes contain substantial amounts of high-affinity GTPγS binding (ca. 100 pmoles/mg of membrane protein; Smith *et al.*, 1987). Furthermore, quantitation of pertussis toxin labeling of PMNs suggests amounts of ~40-kD substrates similar to the large amounts found in brain (Giershik *et al.*, 1986). When the G proteins from HL60 membranes are solubilized, the major GTPγS binding activity and ~40-kD pertussis toxin substrates cochromatograph during purification (R.J. Uhing *et al.*, unpublished observations). Closely spaced doublets at both ~40 kD and ~35 kD on SDS–PAGE are evidenced (Fig. 2). Both ca. 40 kD

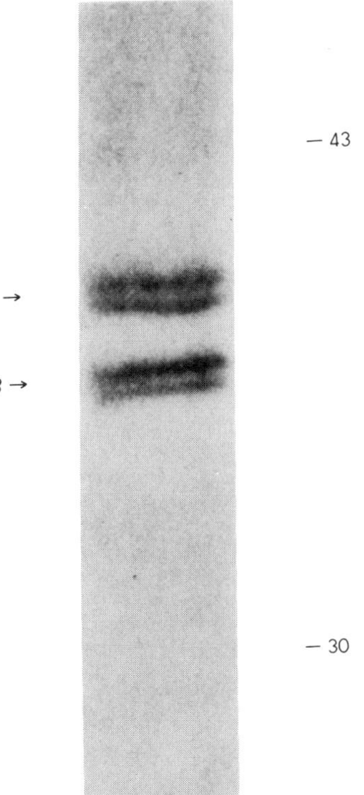

Figure 2. Pertussis toxin substrates from HL60 cells. GTP-binding proteins were purified from HL60 cells using chromatography on Ultragel AcA34, Heptylamine-Sepharose, and DEAE-Fractogel. Presented is a Coomassie blue stain of a SDS–PAGE gel of the pooled pertussis toxin substrates. Both components of the ca. 40-kD doublet are ADP-ribosylated by pertussis toxin (not shown). Taken from Snyderman and Uhing (1987).

proteins are ADP-ribosylated by pertussis toxin, but the upper protein comigrates on SDS–PAGE with brain G_i, whereas the lower one migrates more slowly than G_o, suggesting that it may represent a previously undescribed pertussis toxin substrate. This is also suggested by differences in proteolytic digestion patterns for the resolved pertussis toxin substrates (P. Polakis *et al.*, unpublished observations). Whether this protein couples to chemoattractant receptors remains to be ascertained.

Using a cDNA library prepared from dibutyrl cAMP-differentiated HL60 cells, a cDNA clone that encodes for a unique G protein has been identified (Didsbury and Snyderman, 1987). The long open reading frame is capable of coding for a potential pertussis toxin substrate that is highly homologous yet distinct from reported sequences for $G_i\alpha$ subunits. Moreover, both the 5′ and 3′ noncoding regions are clearly distinct from all reported $G\alpha$ subunit cDNAs.

5. Calcium Mobilization and Inositide Metabolism in Leukocytes

The importance of calcium mobilization in response to chemoattractants has been studied extensively in leukocytes and appears to be a primary mechanism for receptor-mediated activation. The addition of calcium ionophores in the presence of extracellular calcium stimulates superoxide production, degranulation, and aggregation (Pozzan *et al.*, 1983; Dewald *et al.*, 1984; Verghese *et al.*, 1985b; Lew *et al.*, 1986b). The addition of chemoattractants leads to a rapid increase in the cytosolic calcium concentration that precedes the cellular responses (Korchak *et al.*, 1984b; Lew *et al.*, 1984). The chemoattractant-induced increase in cytosolic calcium and stimulation of response are only partially reduced by chelation of extracellular calcium, indicating mobilization of intracellular stores (Lew *et al.*, 1984). The depletion of intracellular calcium stores by prior treatment of cells with EGTA in the presence of calcium ionophores attenuates subsequent responses invoked by chemoattractants (Lew *et al.*, 1984, 1986b; Korchak, 1984a). The fluorescent calcium indicator Quin 2, in the presence of a calcium ionophore and various extracellular calcium concentrations, has been used to examine the calcium dependency of various secretory responses (Lew *et al.*, 1984, 1986b). Increasing levels of cytosolic calcium-stimulated degranulation from secretory vesicles and specific and azurophil granules. The calcium dependencies for the various processes were each reduced approximately tenfold upon inclusion of fMet-eu-Phe. Since fMet-Leu-Phe does not further increase sytosolic calcium under these conditions, the reduction in the calcium dependency probably represents an interaction of the two second messengers (i.e., inositol 1,4,5-trisphosphate and *sn*-1,2-diacylglycerol) of the phosphoinositide pathway on expression of leukocyte function (see Section 6).

The addition of chemoattractants to leukocytes results in a rapid and transient formation of inositol trisphosphate (IP_3) (Dougherty *et al.*, 1984; Burgess *et al.*, 1985; Bradford and Rubin, 1985, 1986; Andersson *et al.*, 1986b; Lew *et al.*, 1986a; Dillon *et al.*, 1987a) (Fig. 3). The initial IP_3 formed, inosotol 1,4,5-trisphosphate, has been shown to cause the release of calcium from intracellular stores in leukocytes (Prentki *et al.*, 1984; Bradford and Rubin, 1986). Measurement of the intracellular concentration of inositol 1,4,5-trisphosphate produced in response to fMet-Leu-Phe [$\sim$1 μM at 10^{-8} M (Bradford and Rubin, 1986)] suggests sufficient quantities to provide maximum release of calcium from the IP_3-sensitive nonmitochondrial pools (Prentki *et al.*, 1984; Bradford and Rubin, 1986). Stimulated production of IP_3 occurs even when intracellular calcium stores are depleted, suggesting its importance as a primary event (DiVirgilio *et al.*, 1985; Ohta *et al.*, 1985; Dillon *et al.*, 1987a). It has recently been shown that inositol 1,4,5-trisphosphate can be metabolized via two pathways (Lew *et al.*, 1986a; Dillon *et al.*, 1987a,b) (Figs. 3 and 4). Flux through inositol 1,3,4,5-tetrakisphosphate

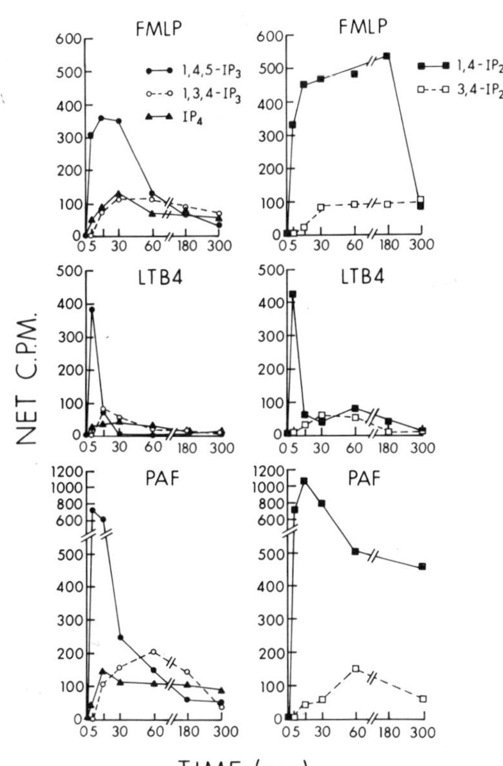

Figure 3. Kinetics of production of IP_2 and IP_3 isomers, and IP_4 in chemoattractant-stimulated PMNs. TCA-soluble material from 10^7 [^{3}H]myo-inositol labeled PMNs stimulated with 0.1 μM of the chemoattractants noted was analyzed by HPLC as described (Dillon *et al.*, 1987a,b). Values shown are from an individual donor. Net cpm indicate cpm in stimulated cells minus cpm in control (buffer only) cells.

(IP_4) and inositol 1,3,4-trisphosphate requires calcium mobilization (Lew *et al.*, 1986a; Dillon *et al.*, 1987b), although the locus of regulation is unclear (Dillon *et al.*, 1987b). Stepwise dephosphorylation patterns are distinct for the two IP_3 isomers (Dillon *et al.*, 1987a,b) (Fig. 4).

Chemoattractants cause a biphasic increase in cytosolic calcium in leukocytes (Lew *et al.*, 1984, 1986b; Andersson *et al.*, 1986a; Truett *et al.*, 1987), the second phase being dependent on extracellular calcium. Influx of $^{45}Ca^{2+}$ is stimulated by chemoattractants, a response that is slower than the initial increase in IP_3 and intracellular calcium (Korchak *et al.*, 1984a,b). A

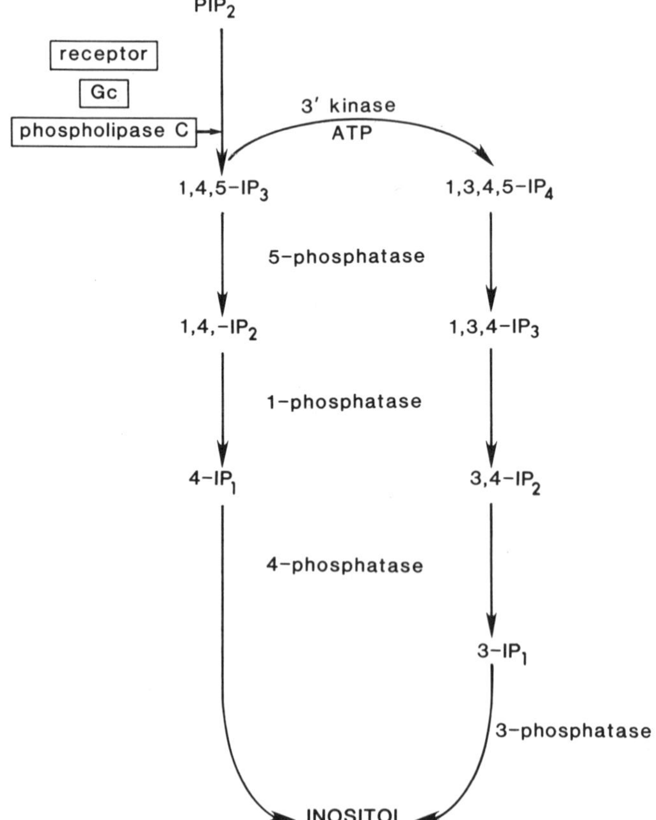

Figure 4. Pathways of 1,4,5-IP_3 metabolism in PMNs. A PIP_2-specific phospholipase C can be activated via a G protein (G_c) coupled to chemoattractant receptors. The initial products formed are 1,4,5-IP_3 and 1,2-diacylglycerol. !,4,5-IP_3 is metabolized via the indicated pathways to free inositol. Taken from Dillon *et al.* (1987b).

role for calcium influx is suggested by the fact that chelation of extracellular calcium attenuates fMet-Leu-Phe-stimulated responses in neutrophils (Lehmeyer *et al.*, 1979; Korchak *et al.*, 1984b). Leukotriene B_4, which is less potent than fMet-Leu-Phe in stimulating superoxide production (Palmblad *et al.*, 1984), degranulation (Serhan *et al.*, 1983), or protein kinase C translocation (Nishihira *et al.*, 1986), does not exhibit the slow rise in cytosolic calcium attributable to calcium influx, even though initial IP_3 levels are similar, albeit more transient, to those produced by fMet-Leu-Phe (Dillon *et al.*, 1987a). These and other results suggest a role for calcium influx in leukocyte activation and for increased diacylglycerol production independent of PIP_2 hydrolysis (see Section 6).

6. Diacylglycerol and Protein Kinase C Involvement

Chemoattractant-induced diacylglycerol formation appears to be of major importance in stimulating leukocyte activation and in feedback regulation. The addition of active phorbol esters to PMNs or to monocytes results in the redistribution of protein kinase C from the cytosol to the cells' particulate fraction (Myers *et al.*, 1985; Wolfson *et al.*, 1985; Gennaro *et al.*, 1986). The relative potencies of the various phorbol esters in causing translocation are similar to their stimulatory effects on superoxide protection (Myers *et al.*, 1985). The NADPH oxidase activity involved appears to be plasma membrane bound and consists of multiple components (for a review see McPhail and Snyderman, 1984). Protein-kinase-C-mediated activation of NADPH oxidase activity has recently been demonstrated in plasma membrane preparations (Cox *et al.*, 1985; Melloni *et al.*, 1986), suggesting that the determining step in the activation of superoxide production is translocation and activation of protein kinase C. In unstimulated PMNs, approximately 90% of the protein kinase C activity is cytosolic (Wolfson *et al.*, 1985). In the presence of cytochalasin B, fMet-Leu-Phe increases membrane-associated protein kinase C threefold at doses that stimulate a respiratory burst (McPhail *et al.*, 1984). Pharmacologic agents that inhibit chemoattractant activation of the respiratory burst also specifically inhibit chemoattractant-stimulated translocation of protein kinase C (Pike *et al.*, 1986). These data further suggest that the translocation of protein kinase C to the plasma membrane is required for activation of the respiratory burst by chemoattractants.

As noted, calcium ionophores also cause stimulation of the respiratory burst in leukocytes. The action of elevated cytosolic Ca^{2+} may also be mediated by protein kinase C. Low concentrations of phorbol esters and calcium ionophores produce a synergistic activation of the respiratory burst (White *et al.*, 1984). Elevation of cytosolic calcium also increases phorbol ester binding affinity in intact phagocytes (Dougherty and Niedel, 1986), and calcium ionophores increase endogenous diacylglycerol levels (Fig. 5).

Other chemoattractant-stimulated leukocyte functions have been suggested to be due to protein kinase C activation, using similar criteria. For example, addition of PMA or chemoattractants in the presence of cytochalasin B stimulates degranulation, primarily of specific granules (for reviews see Baggiolini and Dewald, 1984; Goldstein, 1984). The biochemical events involved in this process are not understood. Recent data also suggest that differentiation of monocytes to macrophages may be mediated by protein kinase C (Muller *et al.*, 1984; Mitchell *et al.*, 1985; Ho *et al.*, 1987). Activation of *c-fos* gene expression in monocytes is rapidly ($\leq$15 min) stimulated by chemoattractants or PMA but is not affected by calcium ionophores (Ho *et al.*, 1987).

Studies with cell-permeant diacylglycerols support a role for protein kinase C in leukocyte activation (Dewald *et al.*, 1984; Fujita *et al.*, 1984; O'Flaherty *et al.*, 1984; Cox *et al.*, 1986; Ho *et al.*, 1987). These analogues have been found to cause superoxide production, degranulation, and enhancement of *c-fos* gene expression. Interestingly, *sn*-1,2-didecanoylglycerol (diC_{10}) stimulated degranulation without increasing superoxide production (Cox *et al.*, 1986). In addition, this analogue inhibited approximately 50% of[^{3}H]PDBu binding in intact PMNs compared to the near complete inhibition observed with other analogues such as *sn*-1,2-dioctanoylglycerol, which

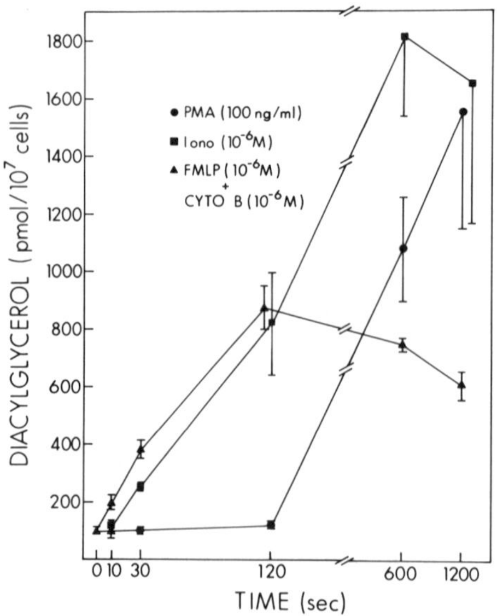

Figure 5. Stimulated diacylglycerol formation in PMNs. Cells were suspended to a concentration of 2×10^7/mL in HHBSS. For stimulation with PMA or ionomycin, the cells were prewarmed for 5 min at 37°C before addition of stimulants. For stimulation with fMet-Leu-Phe, cells were preincubated with cytochalasin B (10^{-6} M) for 5 min. Reactions were stopped and diacylglycerol quantitated by the method of Preiss *et al.* (1986). A control stimulation with an equivalent dilution of DMSO caused no increase in diacylglycerol levels over the time course.

stimulated both the respiratory burst and degranulation. These results suggest either that protein kinase C may be compartmentalized or that different isozymes may exist in PMNs.

The development of a sensitive diacylglycerol assay utilizing diacylglycerol kinase has recently been reported by Preiss *et al.* (1986) and should allow rapid advances in the understanding of the role of diacylglycerol in leukocyte function. Using this assay, Honeycutt and Niedel (1986) have shown that addition of fMet-Leu-Phe increased diacylglycerol levels. This response was potentiated six- to sevenfold by the inclusion of cytochalasin B. Chemoattractant-induced changes in phosphoinositide metabolism exhibit minimal enhancement by cytochalasin B (Serhan *et al.*, 1983; Bradford and Rubin, 1985; Truett *et al.*, 1987), suggesting that the majority of diacylglycerol is derived from sources other than phosphoinositides. Temporal studies show that stimulated diacylglycerol production is slower than calcium or IP_3 (Korchak *et al.*, 1984a,b; Honeycutt and Niedel, 1986; Dillon *et al.*, 1987a; Truett *et al.*, 1987) (compare Figs. 3 and 5), suggesting that its formation may be secondary to inositide metabolism. The source of the diacylglycerol is unknown, although other agents, such as PMA and calcium ionophores, similarly increase diacylglycerol (Fig. 5), which is apparently not derived from phosphoinositide metabolism (Truett *et al.*, 1987). Preliminary evidence from this laboratory suggests that phosphatidylcholine may provide the source for the second phase of diacylglycerol in PMNs (A. P. Truett III *et al.*, unpublished observations).

7. Termination of Chemoattractant Responses

Leukocyte activation induced by a single dose of chemoattractant is a transient phenomenon. Chemoattractant-stimulated increases in IP_3, calcium, and cAMP reach a maximum within 30 s and return to basal levels by 5 min. Subsequent superoxide production and degranulation are initiated rapidly but do not persist beyond 2–5 min. Termination of the signal can result from regulation of the chemoattractant levels, of the receptor, and/or of other steps in the transduction sequence. Recent studies suggest that multiple controls exist for termination of chemoattractant signaling. Internalization of the receptor can lead to hydrolysis of the peptide after transfer to a lysosomal fraction prior to recycling of the receptor to the cell surface (Sklar *et al.*, 1984). The uptake of N-formylated peptide can far exceed the number of cell surface receptors. The actual rate of hydrolysis of fMet-Leu-Phe far exceeds that due to cell-mediated uptake (Yuli and Snyderman, 1986). The degradation of N-formylated peptide by intact PMNs or by PMN membranes is rapid and extensive. Hydrolysis appears to be due to a membrane-associated metalloproteinase.

The addition of chemoattractants to leukocytes results in the elevation of intracellular cAMP concentrations due to calcium-dependent inhibition of cAMP degradation (Verghese *et al.*, 1985a). Studies using cell-permeable cAMP analogues or cAMP-elevating agents suggest that elevated cAMP levels may serve as an autoregulatory termination mechanism (Rivkin *et al.*, 1975; Gallin *et al.*, 1978; Simchowitz *et al.*, 1980; Fujita *et al.*, 1984). The increased intracellular cAMP concentrations inhibit chemoattractant-induced superoxide production and degranulation but do not inhibit stimulation of the same functions by PMA, indicating regulation of a step proximal to protein kinase C. Elevated cAMP levels do not alter binding of fMet-Leu-Phe (Snyderman *et al.*, 1986) but do inhibit superoxide production induced by concanavilin A (which does not utilize a pertussis-toxin-sensitive G protein) (Verghese *et al.*, 1985c), suggesting that the inhibitory action of cAMP is not due to effects on the receptor or its associated G protein. Prostaglandin-E_2-induced elevation of cAMP has been reported to inhibit the slow phase of the cytosolic calcium increase (Takenawa *et al.*, 1986). In the absence of extracellular calcium, no effects of prostaglandin E_2 on cytosolic calcium are witnessed (Takenawa *et al.*, 1986). Furthermore, diacylglycerol (A. P. Truett III *et al.*, unpublished observations) and phosphatidic acid (Takenawa *et al.*, 1986) production are inhibited by elevation of cAMP. These results suggest that the inhibitory action of cAMP may be due to effects on hormone-stimulated calcium influx or on the consequences thereof.

Although activation of protein kinase C by PMA stimulates leukocyte functon, PMA also inhibits fMet-Leu-Phe- and leukotriene B_4-stimulated degranulation and calcium mobilization (Naccache *et al.*, 1985). Platelet IP_3 5′-phosphomonoesterase has been reported to be phosphorylated and activated by protein kinase C (Connolly *et al.*, 1986; Molina *et al.*, 1986). Since concanavilin-A-stimulated IP_3 production is enhanced by PMA (Smith *et al.*, 1987), acceleration of IP_3 phosphatase does not appear to be the primary mechanism for desensitization in leukocytes. The enhancement appears to be due to increased levels of PIP and PIP_2 in PMA-treated cells. The treatment of PMNs with PMA inhibits PIP_2 hydrolysis induced by fMet-Leu-Phe, leukotrien B_4, and platelet activating factor (Smith *et al.*, 1987). The inhibition by PMA affects coupling of G_c with phospholipase C. Stimulation of phospholipase C by GTPγS, but not by high calcium, is disrupted. Receptor–G-protein interactions, as evidenced by high-affinity fMet-Leu-Phe binding and hormone-stimulated GTPγS binding, are not affected. In contrast, pertussis toxin inhibits receptor–G-protein coupling, as evidenced by reductions in high-affinity binding and hormone-stimulated GTPγS binding. The stimulation of phospholipase C by GTP plus fMet-Leu-Phe is also inhibited by pertussis toxin but not phospholipase C activity stimulated by GTPγS or calcium (Smith *et al.*, 1987).

The inhibition of leukocyte activation by cAMP, PMA, and pertussis toxin are examples of heterologous desensitization, that is, the disruption of activation for all agonists using a common transduction mechanism. Homologous desensitization (i.e., receptor specific) is also suggested to occur in leukocytes, based on the observation that exposure of leukocytes to either fMet-Leu-Phe, leukotriene B_4, or platelet activating factor desensitizes to subsequent exposure with the same agonist (M. Verghese *et al.*, unpublished observations). As noted, receptors for chemoattractants are internalized, leading to a reduction of cell surface receptors. The remaining receptors exhibit decreased high-affinity, guanine nucleotide interconvertible binding (Fig. 6), suggesting that another mechanism of desensitization occurs. This desensitization is specific for the original agonist and is not produced by either phorbol esters or calcium ionophores. The mechanisms involved for homologous desensitization in leukocytes have not been ascertained, although similar phenomena with adrenergic receptors and rhodopsin have been shown to be due to a kinase that is specific for the agonist-occupied receptor (Shichi and Somers, 1978; Benovic *et al.*, 1986).

8. Summary

The information summarized here is consistent with the model for chemoattractant-induced leukocyte activation presented in Fig. 7. Evidence indi-

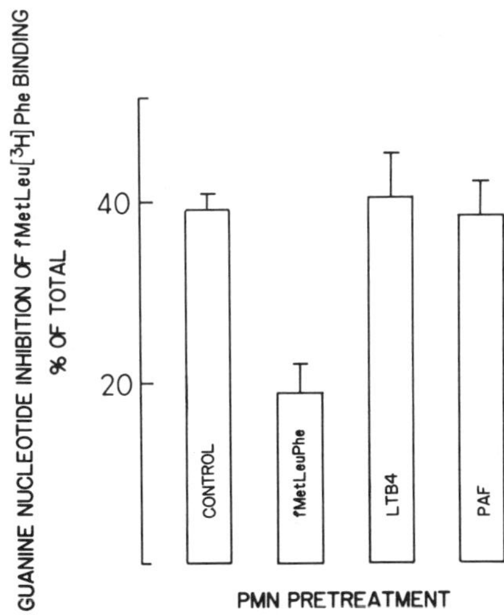

Figure 6. Desensitization of high-affinity fMet-Leu-[^{3}H]Phe binding by prior treatment of PMNs. PMNs were incubated with 1 μM of the indicated chemoattractants for 10 min at 37°C. Plasma membranes were prepared and inhibition of 1 nM fMet-Leu-[^{3}H]Phe binding by 100 μM GMPP(NH)P was determined. Values are presented as percentage of total binding (100 nM fMet-Leu-[^{3}H]Phe). Total binding (pmoles/mg) for the experiment shown was 5.48 ± 0.19 for control, 2.28 ± 0.45 for fMet-Leu-Phe, 6.75 ± 0.86 for LTB_4, and 7.08 ± 0.72 for PAF pretreatments, respectively. Pretreatment did not change affinities for either high- or low-affinity binding sites (not shown).

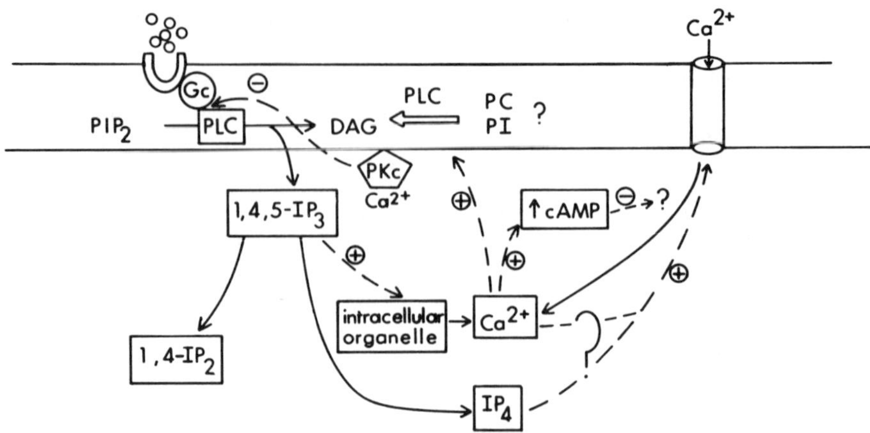

Figure 7. Model for the regulation of leukocyte responses to chemoattractants. Metabolic pathways are represented by (———) and regulatory steps by (— — —). Positive and negative regulation are represented by (+) and (−), respectively. Areas that have not been definitely established include a question mark (?). Abbreviations used are as described in the text, with PKC representing protein kinase C and PLC representing phospholipase C. Taken from Snyderman and Uhing (1987).

cates the central role of phosphoinositide metabolism, calcium mobilization, and protein kinase C activation in mediating chemoattractant-stimulated function. Chemoattractants activate a phospholipase C to produce IP_3 and diacylglycerol subsequent to binding to specific receptors. The activation of phospholipase C is mediated by a unique pertussis/cholera-toxin-sensitive GTP-binding protein termed G_c. In addition to calcium released from intracellular stores, sustained PIP_2 hydrolysis leads to enhanced membrane calcium permeability and a secondary formation of diacylglycerol from a precursor other than the phosphoinositides. These latter phenomena are well correlated with activation of the respiratory burst.

Chemoattractant-induced cellular activation can be attenuated by several mechanisms. Agonists can be hydrolyzed externally or internalized and degraded within leukocytes. The receptor can also be downregulated by internalization and an additional agonist-specific mechanism. Translocation and activation of protein kinase C disrupts the coupling of G_c and phospholipase C. Elevated cAMP levels, induced by either the calcium-dependent mechanism involved in chemoattractant-elicited responses or by hormones that act through adenylate cyclase, also attenuate the chemoattractant-induced activation of leukocytes.

Acknowledgments. This work was supported in part by grant #DE03738 from the National Institute of Dental Research and grant #CA29589 from the

National Cancer Institute. The authors wish to thank Ms. Sharon Goodwin for her excellent secretarial assistance.

References

Aksamit, R. R., Backlund, P. S., Jr., and Cantoni, G. L., 1985, Cholera toxin inhibits chemotaxis by a cAMP-independent mechanism, *Proc. Natl. Acad. Sci. U.S.A.* **82:**7475–7479.

Andersson, T., Dahlgren, C., Pozzan, T., Stendahl, O., and Lew, D. P., 1986a, Characterization of fMet-Leu-Phe receptor-mediated Ca^{2+} influx across the plasma membrane of human neutrophils, *Mol. Pharmacol.* **30:**437–443.

Andersson, T., Schlegel, W., Monod, A., Krause, K.-H., Stendahl, O., and Lew, D. P., 1986b, Leukotriene B_4 stimulation of phagocytes results in the formation of inositol 1,4,5-trisphosphate, *Biochem. J.* **240:**333–340.

Baggiolini, M., and Dewald, B., 1984, Exocytosis by neutrophils, in: *Contemporary Topics in Immunobiology* (R. Snyderman, ed), Vol. 14, pp. 221–246, Plenum Press, New York.

Becker, E. L., Kermode, J. C., Naccache, P. H., Yassin, R., Marsh, M. L., Munoz, J. J., and Sha'afi, R. I., 1985, The inhibition of neutrophil granule enzyme secretion and chemotaxis by pertussis toxin, *J. Cell Biol.* **100:**1641–1646.

Benovic, J. C., Strasser, R. H., Caron, M. G., and Lefkowitz, R. J., 1986, β-Adrenergic receptor kinase: Identification of a novel protein kinase that phosphorylates the agonist-occupied form of the receptor, *Proc. Natl. Acad. Sci. U.S.A.* **83:**2797–2081.

Berridge, M. J., and Irvine, R. F., 1984, Inositol trisphosphate, a novel second messenger in cellular signal transduction, Nature **312:**315–321.

Birnbaumer, L., Codina, J., Mattera, R., Yatani, A., Scherer, N., Toro, M.-J., and Brown, A. M., 1987, Signal transduction by G proteins, in: *Molecular Biology and the Kidney* (R.Robinson and D. K. Granner, eds.) (in press).

Bokoch, G. M., 1987, The presence of free G protein β/γ subunits in human neutrophils results in suppression of adenylate cyclase activity, *J. Biol. Chem.* **262:**589–594.

Bokoch, G. M., and Gilman, A. G., 1984, Inhibition of receptor-mediated release of arachidonic acid by pertussis toxin, *Cell* **39:**301–308.

Bradford, P. G., and Rubin, R. P., 1985, Characterization of formylmethionyl-leucyl-phenylalanine stimulation of inositol trisphosphate accumulation in rabbit neutrophils, *Mol. Pharmacol.* **27:**74–78.

Bradford, P. G., and Rubin, R. P., 1986, Quantitative changes in inositol 1,4,5-trisphosphate in chemoattractant-stimulated neutrophils, *J. Biol. Chem.* **261:**15644–15647.

Brandt, S. J., Dougherty, R. W., Lapetina, E. G., and Niedel, J. E., 1985, Pertussis toxin inhibits chemotactic peptide-stimulated generation of inositol phosphates and lysosomal enzyme secretion in human leukemic (HL-60) cells, *Proc. Natl. Acad. Sci. U.S.A.* **82:**3277–3280.

Burgess, G. M., McKinney, J. S., Irvine, R. F., and Putney, J. W., Jr., 1985, Inositol 1,4,5-trisphosphate and inositol 1,3,4-trisphosphate formation in Ca^{2+}-mobilizing-hormone-activated cells, *Biochem. J.* **232:**237–243.

Cassel, D., and Selinger, Z., 1977, Mechanism of adenylate cyclase activation by cholera toxin: Inhibition of GTP hydrolysis at the regulatory site, *Proc. Natl. Acad. Sci. U.S.A.* **74:**3307–3311.

Cockcroft, S., Baldwin, J. M., and Allan, D., 1984, The Ca^{2+}-activated polyphosphoinositide phosphodiesterase of human and rabbit neutrophil membranes, *Biochem. J.* **221:**477–482.

Connolly, T. M., Lawing, W. J., Jr., and Majerus, P. W., 1986, Protein kinase C phosphorylates human platelet inositol trisphosphate 5′-phosphomonoesterase increasing the phosphatase activity, *Cell* **46:**951–958.

Cox, C. C., Dougherty, R. W., Ganong, B. R., Bell, R. M., Niedel, J. E., and Snyderman, R., 1986, Differential stimulation of the respiratory burst and lysosomal enzyme secretion in human polymorphonuclear leukocytes by synthetic diacylglycerols, *J. Immunol.* **136:**4611–4616.

Cox, J. A., Jeng, A. Y., Sharkey, N. A., Blumberg, P. M., and Tauber, A. I., 1985, Activation of the human neutrophil nicotinamide adenine dinucleotide phosphate (NADPH)-oxidase by protein kinase C, *J. Clin. Invest.* **76:**1932–1938.

Dewald, B., Payne, T. G., and Baggiolini, M., 1984, Activation of NADPH oxidase of human neutrophils. Potentiation of chemotactic peptide by a diacylglycerol, *Biochem. Biophys. Res. Commun.* **125:**367–373.

Didsbury, J. R., and Snyderman, R., 1987, Molecular cloning of a novel human GTP-binding protein and its potential role in chemoattractant stimulus–response coupling, *Clin. Res.,* **35:**656A.

Didsbury, J. R., Ho, Y.-S., and Snyderman, R., 1987, Human G_i protein α-subunit deduction of amino acid structure from a cloned cDNA, *FEBS LETT.* **211:**160–164.

Dillon, S. B., Murray, J. J., and Snyderman, R., 1987a, Identification of a novel inositol bisphosphate isomer formed in chemoattractant stimulated human polymorphonuclear leukocytes, *Biochem. Biophys. Res. Commun.* **144:**264–270.

Dillon, S. B., Murray, J. J., Verghese, M. W., and Snyderman, R., 1987b, Regulation of inositol phosphate metabolism in chemoattractant-stimulated human polymorphonuclear leukocytes: Definition of distinct dephosphorylation pathways for IP_3 isomers, *J. Biol. Chem.* **262:**11546–11552.

DiVirgilio, F., Vicentini, L. M., Treves, S., Riz, G., and Pozzan, T., 1985, Inositol phosphate formation in fMet-Leu-Phe-stimulated human neutrophils does not require an increase in the cytosolic free Ca^{2+} concentration, *Biochem. J.* **229:**361–367.

Dougherty, R. W., and Niedel, J. E., 1986, Cytosolic calcium regulates phorbol diester binding affinity in intact phagocytes, *J. Biol. Chem.* **261:**4097–4100.

Dougherty, R. W., Godfrey, P. P., Hoyle, P. C., Putney, J. W., Jr., and Freer, R. J., 1984, Secretagogue-induced phosphoinositide metabolism in human leucocytes, *Biochem. J.* **222:**307–314.

Falloon, J., Malech, H., Milligan, G., Unson, C., Kahn, R., Goldsmith, P., and Spiegel, A., 1986, Detection of the major toxin substrate of human leukocytes with antisera raised against synthetic peptides, *FEBS Lett.* **209:**352–356.

Feltner, D. E., Smith, R. H., and Marasco, W. A., 1986, Characterization of the plasma membrane GTPase from rabbit neutrophils: I. Evidence for an N_i-like protein coupled to the formyl peptide, C5a, and leukotriene B_4 chemotaxis receptors, *J. Immunol.* **137:**1961–1970.

Fitzgerald, T. J., Uhing, R. J., and Exton, J. H., 1986, Solubilization of the vasopressin receptor from rat liver plasma membranes: Evidence for a receptor–GTP–binding protein complex, *J. Biol. Chem.* **261:**16871–16877.

Fletcher, M. P., and Gallin, J. I., 1982, Human neutrophils contain an intracellular pool of putative receptors for the chemoattractant N-formyl methionyllecylphenylalanine with a density of specific granules, *J. Cell Biol.* **95:**444a.

Fujita, I., Irita, K., Takeshige, K., and Minakami, S., 1984, Diacylglycerol, 1-oleoyl-2-acetylglycerol, stimulates superoxide-generation from human neutrophils, *Biochem. Biophys. Res. Commun.* **120:**318–324.

Gallin, J. I., Sandler, J. A., Clyman, R. I., Manganiello, V. C., and Vaughan, M., 1978,

Agents that increases cyclic AMP inhibit accumulation of cGMP and depress human monocyte locomation, *J. Immunol.* **120:**492–496.

Gardner, J. P., Melnick, D. A., and Malech, H. L., 1986, Characterization of the formyl peptide chemotactic receptor appearing at the phagocytic cell surface after exposure to phorbol myristate acetate, *J. Immunol.* **136:**1400–1405.

Gennaro, R., Florio, C., and Romeo, D., 1986, Coactivation of protein kinase C and NADPH oxidase in the plasma membrane of neutrophil cytoplasts, *Biochem. Biophys. Res. Commun.* **134:**305–312.

Giershik, P., Falloon, J., Milligan, G., Pines, M., Gallin, J. I., and Spiegel, A., 1986, Immunochemical evidence for a novel pertussis toxin substrate in human neutrophils, *J. Biol. Chem.* **261:**8058–8062.

Gilman, A. G., 1984, G proteins and dual control of adenylate cyclase, *Cell* **36:**577–579.

Goetzl, E. J., Foster, D. W., and Goldman, D. W., 1981, Isolation and partial characterization of membrane protein constituents of human neutrophil receptors for chemotactic formylmethionyl peptides, *Biochemistry* **20:**5717–5722.

Goldman, D. W., Chang, F. H., Gifford, L. A., Goetzl, E. J., and Bourne, H. R., 1985, Pertussis toxin inhibition of chemotactic factor-induced calcium mobilization and function in human polymorphonuclear leukocytes, *J. Exp. Med.* **162:**145–156.

Goldstein, I. M., 1984, Neutrophil degranulation, in: *Contemporary Topics in Immunobiology* (R. Snyderman, ed.), Vol. 14, pp. 189–220, Plenum Press, New York.

Ho, Y.-S., Lee, W. M. F., and Snyderman, R., 1987, Chemoattractant-induced activation of *c-fos* gene expression, *J. Exp. Med.* **165:**1524–1539.

Honeycutt, P. J., and Niedel, J. E., 1986, Cytochalasin B enhancement of the diacylglycerol response in formyl-stimulated neutrophils, *J. Biol. Chem.* **261:**15900–15905.

Hoyle, P. C., and Freer, R. J., 1984, Isolation and reconstitution of the N-formylpeptide receptor from HL-60 derived neutrophils, *FEBS Lett.* **167:**277–280.

Huang, C.-K., 1987, Partial purification and characterization of formylpeptide receptor from rabbit peritoneal neutrophils, *J. Leukocyte Biol.* **41:**63–69.

Jesaitis, A. J., Naemura, J. R., Painter, R. G., Sklar, L. A., and Cochrane, C. G., 1982, Intracellular localization of N-formyl chemotactic receptor and Mg^{2+} dependent ATPase in human granulocytes, *Biochim. Biophys. Acta.* **719:**556–568.

Kay, G. E., Lane, B. C., and Snyderman, R., 1983, Induction of selective biological responses to chemoattractants in a human monocyte-like cell line, *Infect. Immun.* **41:**1166–1174.

Kikuchi, A., Kozawa, O., Kaibuchi, K., Katada, T., Ui, M., and Takai, Y., 1986, Direct evidence for involvement of a guanine nucleotide-binding protein in chemotactic peptide-stimulated formation of inositol bisphosphate and trisphosphate in differentiated leukemic (HL-60) cells: Reconstitution with G_i or G_o of the plasma membranes ADP-ribosylated by pertussis toxin, *J. Biol. Chem.* **257:**11558–11562.

Koo, C., Lefkowitz, R. J., and Snyderman, R., 1982, The oligopeptide chemotactic factor receptor on human polymorphonuclear leukocyte membranes exists in two affinity states, *Biochem. Biophys. Res. Commun.* **106:**442–449.

Koo, C., Lefkowitz, R. J., and Snyderman, R., 1983, Guanine nucleotides modulate the binding affinity of the oligopeptide chemoattractant receptor on human polymorphonuclear leukocytes, *J. Clin. Invest.* **72:**748–753.

Korchak, H. M., Rutherford, L. E., and Weissmann, G., 1984a, Stimulus response coupling in the human neutrophil: I. Kinetic analysis of changes in calcium permeability, *J. Biol. Chem.* **259:**4070–4075.

Korchak, H. M., Vienne, K., Rutherford, L. E., Wilkenfeld, C., Finkelstein, M. C., and

Weissmann, G., 1984b, Stimulus response coupling in the human neutrophil: II. Temporal analysis of changes in cytosolic calcium and calcium efflux, *J. Biol. Chem.* **259:**4076–4082.

Lad, P. M., Olson, C. V., Grewal, I. S., and Scott, S. J., 1985a, A pertussis toxin-sensitive GTP-binding protein in the human neutrophil regulates multiple receptors, calcium mobilization, and lectin-induced capping, *Proc. Natl. Acad. Sci. U.S.A.* **82:**8643–8647.

Lad, P. M., Olson, C. V., and Smiley, P. A., 1985b, Association of the N-formyl-Met-Leu-Phe receptor in human neutrophils with a GTP-binding protein sensitive to pertussis toxin, *Proc. Natl. Acad. Sci. U.S.A.* **82:**869–873.

Lehmeyer, J. E., Snyderman, R., and Johnston, R. B., Jr., 1979, Stimulation of neutrophil oxidative metabolism by chemotactic peptides: Influence of calcium ion concentration and cytochalasin B and comparison with stimulation by phorbol myristate acetate, *Blood* **54:**35–45.

Lew, P. D., Wollheim, C. B., Waldvogel, F. A., and Pozzan, T., 1984, Modulation of cytosolic-free calcium transients by changes in intracellular calcium-buffering capacity: Correlation with exocytosis and O_2^- production in human neutrophils, *J. Cell Biol.* **99:**1212–1220.

Lew, P. D., Monod, A., Krause, K.-H., Waldvogel, F. A., Biden, T. J., and Schlegel, W., 1986a, The role of cytosolic calcium in the generation of inositol 1,4,5-trisphosphate and inositol 1,3,4-trisphosphate in HL-60 cells: Differential effects of chemotactic peptide receptor stimulation at distinct Ca^{2+} levels, *J. Biol. Chem.* **261:**13121–13127.

Lew, P. D., Monod, A., Waldvogel, F. A., Dewald, B., Baggiolini, M., and Pozzan, T., 1986b, Quantitative analysis of the cytosolic free calcium dependency of exocytosis from three subcellular compartments in intact neutrophils, *J. Cell Biol.* **102:**2197–2204.

Mackin, W. M., Huang, C. K., and Becker, E. L., 1982, The formyl peptide chemotactic receptor on rabbit peritoneal neutrophils, *J. Immunol.* **129:**1608–1611.

Majerus, P. W., Connolly, T. M., Deckmyn, H., Ross, T. S., Bross, T. E., Ishii, H., Bansal, V. S., and Wilson, D. B., 1986, The metabolism of phosphoinositide-derived messenger molecules, *Science* **234:**1519–1526.

Malech, H. L., Gardner, J. P., Heiman, D. F., and Rosenzweig, S. A., 1985, Asparagine-linked oligosaccharides on formyl peptide chemotactic receptors of human phagocytic cells, *J. Biol. Chem.* **260:**2509–2514.

McPhail, L. C., and Snyderman, R., 1984, Mechanisms of regulating the respiratory burst in leukocytes, in: *Contemporary Topics in Immunobiology: Regulation of Leukocyte Function* (R. Snyderman, ed.), pp. 247–281, Plenum Press, New York.

McPhail, L. C., Wolfson, M., Clayton, C., and Snyderman, R., 1984, Protein kinase C and neutrophil (PMN) activation: Differential effects of chemoattractants and phorbol myristate acetate (PMA), *Fed. Proc.* **43:**1661 (Abstract).

Melloni, E., Pontremoli, S., Salamino, F., Sparatore, B., Michetti, M., Sacco, O., and Horeker, B. L., 1986, ATP induces the release of a neutral serine proteinase and enhances the production of superoxide anion in membranes from phorbol ester-activated neutrophils, *J. Biol. Chem.* **261:**11437–11439.

Mitchell, R. L., Zokas, L., Schreiber, R. D., and Yema, I. M., 1985, Rapid induction of the expression of proto-oncogene *fos* during human monocytic differentiation, *Cell* **40:**209–217.

Molina, Y., Vedia, L., and Lapetina, E. G., 1986, Phorbol 12,13-dibutyrate and 1-oleyl-2-acetyldiacylglycerol stimulate inositol trisphosphate dephosphorylation in human platelets, *J. Biol. Chem.* **261:**10493–10495.

Molski, T. F. P., Naccache, P. H., Marsah, M. L., Kermode, J., Becker, E. L., and Sha'afi, R. I., 1984, Pertussis toxin inhibits the rise in the intracellular concentration of free calcium that is induced by chemotactic factors in rabbit neutrophils: Possible role of the "G proteins" in calcium mobilization, *Biochem. Biophys. Res. Commun.* **124:**644–650.

Muller, R., Muller, D., and Guilbert, L., 1984, Differential expression of c-fos in hematopoietic cells: Correlation with differentiation of monomyelocytic *in vitro, EMBO J.* **3:**1887–1890.

Myers, M. A., McPhail, L. C., and Snyderman, R., 1985, Redistribution of protein kinase C activity in human monocytes: Correlation with activation of the respiratory burst, *J. Immunol.* **135:**3411–3416.

Naccache, P. H., Molski, T. F. P., Borgeat, P., White, J. R., and Sha'afi, R. I., 1985, Phorbol esters inhibit the fMet-Leu-Phe- and leukotriene B_4-stimulated calcium mobilization and enzyme secretion in rabbit neutrophils, *J. Biol. Chem.* **260:**2125–2131.

Neer, E. J., Lok, J. M., and Wolf, L. G., 1984, Purification and properties of the inhibitory guanine nucleotide regulatory unit of brain adenylate cyclase, *J. Biol. Chem.* **259:**14222–14229.

Niedel, J., Kahane, I., and Cuatrecases, P., 1979, Receptor-mediated internalization of fluorescent chemotactic peptide by human neutrophils, *Science* **205:**1412–1414.

Niedel, J., David, J., and Cuatrecasas, P., 1980, Covalent affinity labeling of the formyl peptide chemotactic receptor, *J. Biol. Chem.* **255:**7063–7066.

Nishihira, J., McPhail, L. C., and O'Flaherty, J. T., 1986, Stimulus-dependent mobilization of protein kinase C, *Biochem. Biophys. Res. Commun.* **134:**587–594.

O'Flaherty, J. T., Schmitt, J. D., McCall, C. E., and Wykle, R. L., 1984, Diacylglycerols enhance human neutrophil degranulation responses: Relevancy to a multiple mediator hypothesis of cell function, *Biochem. Biophys. Res. Commun.* **123:**64–70.

Ohta, H., Okajima, F., and Ui, M., 1985, Inhibition by islet-activating protein of a chemotactic peptide-induced early breakdown of inositol phospholipids and Ca^{2+} mobilization in guinea pig neutrophils, *J. Biol. Chem.* **260:**15771–15780.

Okajima, F., and Ui, M., 1984, ADP-ribosylation of the specific membrane protein by islet-activating protein, pertussis toxin, associated with inhibition of a chemotactic peptide-induced arachidonate release in neutrophils, *J. Biol. Chem.* **259:**13863–13871.

Okajima, F., Katada, T., and Ui, M., 1985, Coupling of the guanine nucleotide regulatory protein to chemotactic peptide receptors in neutrophil membranes and its uncoupling by islet-activating protein, pertussis toxin. A possible role of the toxin substrate in Ca^{2+}-mobilizing receptor-mediated signal transduction, *J. Biol. Chem.* **260:**6761–6768.

Painter, R. G., Schmitt, M., Jesaitis, A. J., Sklar, L. A., Aissnar, K., and Cochrane, C. G., 1982, Photoaffinity labeling of the N-formyl peptide receptor on human polymorphonuclear leukocytes, *J. Cell. Biochem.* **20:**203–214.

Palmblad, J., Gyllenhammar, H., Lindgren, J. A., and Malmsten, C. L., 1984, Effects of leukotrienes and f-Met-Leu-Phe on oxidative metabolism of neutrophils and eosinophils, *J. Immunol.* **132:**3041–3045.

Pike, M. C., Jakoi, L., McPhail, L.C., and Snyderman, R., 1986, Chemoattractant-mediated stimulation of the respiratory burst in human polymorphonuclear leukocytes may require appearance of protein kinase C in the cells' particulate fraction, *Blood* **67:**909–913.

Polakis, P., and Snyderman, R., 1987, G-protein-chemoattractant receptor interaction: Copurification of the formylpeptide receptor with a guanine nucleotide binding protein, *Clin. Res.* **35:**487A.

Pozzan, T., Lew, D. P., Wollheim, C. B., and Tsien, R. Y., 1983, Is cytosolic ionized calcium regulating neutrophil activation?, *Science* **221:**1413–1415.

Preiss, J., Loomis, C. R., Bishop, W. R., Stein, R., Niedel, J. E., and Bell, R. M., 1986, Quantitative measurement of *sn*-1,2-diacylglycerols present in platelets, hepatocytes and *ras*- and *sis*-transformed normal rat kidney cells, *J. Biol. Chem.* **261:**8597–8600.

Prentki, M., Wollheim, C. G., and Lew, P. D., 1984, Ca^{2+} homeostatis in permeabilized human neutrophils: Characterization of Ca^{2+}-sequestering pools and the action of inositol 1,4,5-trisphosphate, *J. Biol. Chem.* **259:**13777–13782.

Rivkin, I., Rosenblatt, J., and Becker, E. L., 1975, The role of cyclic AMP in the chemotactic responsiveness and spontaneous motility of rabbit peritoneal neutrophils, *J. Immunol.* **115:**1126–1134.

Serhan, C. N., Radin, A., Smolen, J. E., Korchak, H., Samuelsson, B., and Weissman, G., 1982, Leukotriene, B_4 is a complete secretogogue in human neutrophils: A kinetic analysis, *Biochem. Biophys. Res. Commun.* **107:**1006–1012.

Serhan, C. N., Broekman, M. J., Korchak, H. M., Smolen, J. E., Marcus, A. J., and Weissman, G., 1983, Changes in phosphatidylinositol and phosphatidic acid in stimulated human neutrophils: Relationship to calcium mobilization, aggregation and superoxide radical generation, *Biochim. Biophys. Acta* **762:**420–428.

Shichi, H., and Somers, R. L., 1978, Light-dependent phosphorylation of rhodopsin: Purification and properties of rhodopsin kinase, *J. Biol. Chem.* **253:**7040–7046.

Showell, H. J., Freer, R. J., Zigmond, S. H., Schiffmann, E., Aswanikumar, S., Corcoran, B., and Becker, E. L., 1976, The structure–activity relations of synthetic peptides as chemotactic factors and inducers of lysosomal enzyme secretion for neutrophils, *J. Exp. Med.* **143:**1154–1169.

Simchowitz, L., Fischbein, L. C., Spilberg, I., and Atkinson, J. P., 1980, Induction of a transient elevation in intracellular levels of adenosine-3′, 5′-cyclic monophosphate by chemotactic factors: An early event in human neutrophil activation, *J. Immunol.* **124:**1482–1491.

Sklar, L. A., Jesaitis, A. J., and Painter, R. G., 1984, The neutrophil N-formyl peptide receptor: Dynamics of ligand–receptor interactions and their relationship to cellular responses, *Contemp. Topics Immunobiol.* **14:**29–82.

Sklar, L. A., Bokoch, G. M., Button, D., and Smolen, J. E., 1987, Regulation of ligand–receptor dynamics by guanine nucleotides: Real-time analysis of interconverting states for the neutrophil formylpeptide receptor, *J. Biol. Chem.* **262:**135–139.

Smith, C. D., Lane, B. C., Kusaka, I., Verghese, M. W., and Snyderman, R., 1985, Chemoattractant-receptor induced hydrolysis of phosphatidylinositol 4,5-bisphosphate in human polymorphonuclear leukoctye membranes: Requirement of a guanine nucleotide regulatory protein, *J. Biol. Chem.* **260:**5875–5878.

Smith, C. D., Cox, C. C., and Snyderman, R., 1986, Receptor-coupled activation of phosphoinositide-specific phospholipase C by an N protein, *Science* **232:**97–100.

Smith, C. D., Uhing, R. J., and Snyderman, R., 1987, Nucleotide regulatory protein-mediated activation of phospholipase C in human polymorphonuclear leukocytes is disrupted by phorbol esters, *J. Biol. Chem.* (in press).

Snyderman, R., and Pike, M. C., 1984a, Regulation of leukocyte function, in: *Contemporary Topics in Immunobiology* (R. Snyderman, ed.), pp. 1–28, Plenum Press, New York.

Snyderman, R., and Pike, M. C., 1984b, Chemoattractant receptors on phagocytic cells, in: *Annual Review of Immunology*, (W. E. Paul, ed.), Vol. 2, pp. 257–281. Annual Reviews, Inc., Palo Alto, CA.

Snyderman, R., and Uhing, R. J., 1987, Stimulus–response coupling mechanisms, in: *Inflammation: Basic Principles and Clinical Correlates* (J. I. Gallin, I. M. Goldstein, and R. Snyderman, eds.), pp. 309–323, Raven Press, New York.

Snyderman, R., Pike, M. C., Edge, S., and Lane, B., 1984, A chemoattractant receptor on macrophages exists in two affinity states regulated by guanine nucleotides, *J. Cell Biol.* **98:**444–448.

Snyderman, R., Smith, C. D., and Verghese, M. W., 1986, Model for leukocyte regulation by chemoattractant receptors: Roles of a guanine nucleotide regulatory protein and polyphosphoinositide metabolism, *J. Leukocyte Biol.* **40:**785–800.

Spiegel, A. M., 1987, Signal transduction by quanine nucleotide binding proteins, *Mol. Cell. Endocrinol.* **49:**1–16.

Sullivan, S., and Zigmond, S., 1980, Chemotactic peptide receptor modulation in polymorphonuclear leukocytes, *J. Cell Biol.* **85:**703–711.

Takenawa, T., Ishitoya, J., and Nagai, Y., 1986, Inhibitory effect of prostaglandin E_2, forskolin, and dibutyryl cAMP on arachidonic acid release and inositol phospholipid metabolism in guinea pig neutrophils, *J. Biol. Chem.* **261:**1092–1098.

Truett, A. P., III, Verghese, M. W., Dillon, S. B., and Snyderman, R., 1987, Leukocyte (PMN) activation by chemoattractants (CTX): A two phase sequential pathway mediates the respiratory burst, *Clin. Res.* **35:**618A..

Verghese, M. W., Fox, K., McPhail, L. C., and Snyderman, R., 1985a, Chemoattractant-elicited alterations of cAMP levels in human polymorphonuclear leukocytes require a Ca^{++}-dependent mechanism which is independent of transmembrane activation of adenylate cyclase, *J. Biol. Chem.* **260:**6769–6775.

Verghese, M. W., Smith, C. D., and Snyderman, R., 1985b, Potential role for a guanine nucleotide regulatory protein in chemoattractant receptor mediated polyphosphoinositide metabolism, Ca^{++} mobilization and cellular responses by leukocytes, *Biochem. Biophys. Res. Commun.* **127:**450–457.

Verghese, M. W., Smith, C. D., and Snyderman, R., 1985c, Role for a guanine nucleotide regulatory (N) protein in chemoattractant mediated Ca^{2+} mobilization and cAMP formation in human neutrophils (PMNs), *Clin. Res.* **33:**566A.

Verghese, M. W., Smith, C. D., Charles, L. A., Jakoi, L., and Snyderman, R., 1986a, A guanine nucleotide regulatory protein controls polyphosphoinositide metabolism, Ca^{2+} mobilization and cellular responses to chemoattractants in human monocytes, *J. Immunol.* **137:**271–275.

Verghese, M. W., Smith, C. D., Charles, L. A., and Snyderman, R., 1986b, A common transduction pathway for leukocyte chemoattractant receptors: Phospholipase C activation by a guanine nucleotide regulatory protein, *Clin. Res.* **34:**679A.

Verghese, M. W., Uhing, R. J., and Snyderman, R., 1986c, A pertussis/cholera toxin-sensitive N protein may mediate chemoattractant receptor signal transduction, *Biochem. Biophys. Res. Commun.* **138:**887–894.

Volpi, M., Nacchache, P. H., Molski, T. F. P., Shefcyk, J., Huang, C. K., Marsh, M. L., Munoz, J., Becker, E. L., and Sha'afi, R. I., 1985, Pertussis toxin inhibits fMet-Leu-Phe- but not phorbol ester-stimulated changes in rabbit neutrophils: Role of G proteins in excitation response coupling, *Proc. Natl. Acad. Sci. U.S.A.* **82:**2708–2712.

White, J. R., Huang, C.-K., Hill, J. M., Jr., Naccache, P. H., Becker, E. L., and Sha'afi, R. I., 1984, Effect of phorbol 12-myristate 13-acetate and its analogue 4α-phorbol 12,13-didecanoate on protein phosphorylation and lysosomal enzyme release in rabbit neutrophils, *J. Biol. Chem.* **259:**8605–8611.

Williams, L. T., Snyderman, R., Pike, M. C., and Lefkowitz, R. J., 1977, Specific receptor sites for chemotactic peptides on human polymorphonuclear leukocytes, *Proc. Natl. Acad. Sci. U.S.A.* **74:**1204–1208.

Wolfson, M., McPhail, L. C., Nasrallah, V. N., and Snyderman, R., 1985, Phorbol myristate acetate mediates redistribution of protein kinase C in human neutrophils: Potential role in the activation of the respiratory burst enzyme, *J. Immunol.* **135:**2057–2062.

Yuli, I., and Snyderman, R., 1986, Extensive hydrolysis of N-formyl-L-methionyl-L-leucyl-L-[^{3}H]phenylalanine by human polymorphonuclear leukocytes: A potential mechanism for modulation of the chemoattractant signal, *J. Biol. Chem.* **261:**4902–4908.

19

The Molecular Biology of the Human Interleukin-2 Receptor

WARNER C. GREENE

1. Introduction

The recognition of antigen by antigen-specific receptors present on the surface of resting T cells initiates a cascade of intracellular events that lead to cellular activation and growth. These events include the induction of the *de novo* synthesis and secretion of interleukin-2 (IL-2 or T cell growth factor) (Morgan *et al.*, 1976) and the transient expression of high- and low-affinity IL-2 receptors (Smith, 1980; Robb, 1984; Greene and Robb, 1985; Greene *et al.*, 1985). The interaction of IL-2 with its high-affinity membrane receptor in turn transduces the requisite intracellular signals leading to T cell proliferation and culminates in the generation of specific effector cells that mediate helper and cytotoxic T cell functions. The early increase in IL-2 receptor expression is later followed by a progressive decline in receptor number that may play an important role in terminating the T cell immune response. Thus, while specificity of this response is maintained by the antigen receptor, the magnitude and duration of the T cell immune response is controlled in part through the transient production of IL-2 and the display of IL-2 receptors.

Biochemical and molecular characterizations of the human IL-2 receptor have been facilitated by the identification of monoclonal antireceptor antibodies. The first such monoclonal antibody in the human system was anti-Tac (Uchiyama *et al.*, 1981). Through use of this antibody, the IL-2 receptor (Tac) has been characterized biochemically and purified (Leonard *et*

WARNER C. GREENE • Howard Hughes Medical Institute, Duke University School of Medicine, Durham, North Carolina 27710.

al., 1982, 1983, 1985a; Miyawaki *et al.*, 1982, Robb and Greene, 1983; Depper *et al.*, 1984; Wano *et al.*, 1984), cDNAs for the Tac receptor have been isolated (Cosman *et al.*, 1984; Leonard *et al.*, 1984; Nikaido *et al.*, 1984), and the IL-2 receptor (Tac) gene has been isolated and its organization and location within the human genome have been defined (Ishida *et al.*, 1985; Leonard *et al.*, 1985b). In addition, deregulated IL-2 receptor expression in leukemic T cells infected with the human T lymphotrophic virus types I and II (HTLV-I and -II) has been detected (Gootenberg *et al.*, 1981; Depper *et al.*, 1984b; Waldmann *et al.*, 1984; Krönke, 1985a). Emerging data suggest that these retroviruses contain similar *trans*-activator genes whose protein products activate the expression of the cellular genes encoding IL-2 and the IL-2 receptor (Tac) (Inoue *et al.*, 1986; Cross *et al.*, 1987; Maruyama *et al.*, 1987; Siekevitz *et al.*, 1987). Thus, subversion of the regulated expression of these cellular genes that control T cell growth may play a role in HTLV-I and HTLV-II-mediated leukemic transformation. In contrast to HTLV-I and HTLV-II, the human immunodeficiency virus (HIV) produces profound T cell immunosuppression and may lead to the development of the acquired immune deficiency syndrome (AIDS). Like HTLV-I, HIV infects $T4^+$ T cells; however, this infection produces cell death rather than cell transformation. Recent studies have demonstrated that HIV infection markedly impairs *IL-2* gene activation while not altering IL-2 receptor (*Tac*) gene expression (Dukovich *et al.*, 1987a). This inhibition of growth factor production may play a role, together with envelope-protein-induced cell fusion and syncytium formation (Sodroski *et al.*, 1986), in the intracellular events in HIV-infected T cells that lead to the demise of these cells. Studies of the IL-2 receptor (Tac), the biochemical basis of high- and low-affinity IL-2 receptor formation, the deregulation of IL-2 receptor expression in HTLV-I-induced adult T cell leukemia, and abnormalities of *IL-2* gene expression in HIV-infected T cells will be considered in the following sections.

2. *Identification of Anti-IL-2 Receptor Monoclonal Antibodies*

Takashi Uchiyama and Thomas Waldmann isolated a monoclonal antibody termed anti-Tac, using T cells from a patient with HTLV-I-induced adult T cell leukemia (ATL) for immunization (Uchiyama *et al.*, 1981). Anti-Tac was found to react with activated but not resting T cells. Subsequent study demonstrated that anti-Tac recognized the human IL-2 receptor or a subunit of this receptor. Evidence supporting this conclusion included the following findings: (1) anti-Tac blocked IL-2-dependent T cell proliferation (Leonard *et al.*, 1982); (2) anti-Tac inhibited the binding of radiolabeled IL-2 (Leonard *et al.*, 1982) and IL-2 inhibited the binding of radiolabeled anti-Tac

(Robb and Greene, 1983); (3) immobilized IL-2 and anti-Tac both reacted with the same 55,000-dalton membrane glycoprotein solubilized from the membranes of activated T cells (Robb and Greene, 1983); and (4) both anti-Tac and anti-IL-2 antibodies precipitated the same 65,000–70,000-dalton protein band produced by covalently crosslinking IL-2 to its membrane receptor (Leonard *et al.*, 1983).

The Tac antigen was characterized as a 50,000-dalton glycoprotein on HUT-102B2 cells and a 55,000-dalton glycoprotein on normal activated T cells (Leonard *et al.*, 1982, 1983, 1985a; Wano *et al.*, 1984). Under nonreducing conditions, the receptor migrated with a smaller apparent size, suggesting the presence of intrachain disulfide bonds. The difference in apparent M_r of the receptor in HUT-102B2 and normal cells was found to reflect differences in post-translational processing rather than an alteration in the primary amino acid sequence. Other monoclonal antireceptor antibodies have been identified, including B49.9 (Cotner *et al.*, 1983), 2A3 (Urdal *et al.*, 1984), and 7G7/B6 (Rubin *et al.*, 1985). The first two of these antibodies, B49.9 and 2A3, are similar to anti-Tac; however, 7G7/B6 identifies a distinct receptor epitope that does not interfere with IL-2 binding or action. Each of these three monoclonals reacts with the same 55,000-dalton membrane structure precipitated by anti-Tac.

3. IL-2 Receptor (Tac) Biosynthesis

Pulse-chase metabolic labeling studies performed in the presence of tunicamycin have helped to define the sequential steps involved in IL-2 receptor (Tac) biosynthesis. At early points, in the presence of tunicamycin, a M_r 33,000 protein (p33) corresponding to the unmodified polypeptide backbone was immunoprecipitated with anti-Tac (Leonard *et al.*, 1983, 1985a). This protein is approximately 1500 daltons smaller than the primary translation product (M_r 34,500) identified in a cell-free wheat germ translation system, presumably reflecting cotranslational cleavage of a signal peptide. Pulse-chase labeling performed in the absence of tunicamycin revealed two larger precursor proteins of M_r 35,000 and 37,000 (p35 and p37) (Leonard *et al.*, 1983, 1985a). These data suggested that p33 was cotranslationally processed by the addition of N-linked sugar to yield these intermediate forms (Leonard *et al.*, 1983, 1985a). This hypothesis was confirmed in studies using endoglycosidase F, an enzyme that cleaves N-linked sugar moieties. Endoglycosidase F treatment produced conversion of the p35/p37 doublet to p33. With longer periods of chase, further modifications of p35 and p37 proteins were revealed. After 30–60 min. the p35/p37 doublet increased sharply in size, yielding a broad band of M_r 55,000. This saltatory increase in size appeared to be due to Golgi-associated processing, since it was inhibited by monensin. Enzymatic digestion studies revealed the addition of O-linked sugar and sialic acid during this stage of receptor biosynthesis (Leonard *et al.*, 1985a).

The IL-2 receptor protein is also modified by the post-translation addition of sulfate (Leonard *et al.*, 1985a). At present, it is unknown whether the sulfate groups are associated with carbohydrate (usually N-linked sugar) or protein (occasionally tyrosine residues). Interestingly, the difference in apparent size of the HUT-102B2 and normal activated T cell IL-2 receptor (M_r 50,000 versus 55,000) is at least in part related to the diminished addition of sulfate to the leukemic receptor protein. Both IL-2 and anti-Tac immobilized on bead supports react with the p33, p35, and p37 precursor forms of the receptor (Leonard *et al.*, 1985a). Thus, post-translational processing of the Tac protein does not appear to be required for IL-2 binding; however, it is possible that these processing events may modify the resultant affinity of the receptor for its ligand.

Following post-translational processing, the mature receptor protein (M_r 55,000 in normal activated T cells) is inserted into the cell surface. Constitutive serine-based phosphorylation of the receptor protein may then occur (Shackelford and Trowbridge, 1984; Leonard *et al.*, 1985a). Convincing evidence for IL-2-induced phosphorylation of the Tac receptor has not thus far been obtained; however, this possibility cannot be excluded entirely.

4. Molecular Cloning of IL-2 Receptor (Tac) cDNAs

IL-2 receptor protein was purified from the HUT-102B2 cell line by immunoaffinity chromatography using the anti-Tac antibody. This cell line was previously shown to express large numbers of IL-2 receptors (150,000–300,000 sites/cell) (Depper *et al.*, 1984b). The sequence of the N-terminal 29 residues of the protein was determined by gas-phase microsequencing and was as follows:NH_2-Glu-Leu-*Cys-Asp-Asp-Asp-Pro-Pro* Glu-Ile-Pro-His-Ala-Thr-Phe-Lys-Ala-Met-Ala-Tyr-Lys-Glu-Gly-Thr-Met-Leu-Asn-Cys-Glu. A pool of 64 oligonucleotide probes, 17 bases in length, were then synthesized based on the sequence of residues 3–8 and used to screen a λgt10 cDNA library prepared with HUT-102B2 mRNA. Serial screening permitted isolation of several candidate phage clones, of which three were selected for further study. These cDNAs were subsequently shown to correspond to the Tac protein based on selective hybridization of mRNA, DNA sequence analysis, and expression in COS-1 and L cells (Leonard *et al.*, 1984; Greene *et al.*, 1985).

Two of the longest clones, denoted pIL2R3 (2.4kb) and pIL2R4 (1.7 kb) (Leonard *et al.*, 1984), displayed an interesting difference in sequence.

Clone 3 contained a region of 216 bases, located within the protein coding region, that was absent in clone 4. In order to examine the proteins encoded by these respective cDNAs, each was transiently expressed in COS-1 cells. Transfection of clone 3, which retained the 216-base segment, led to the display of M_r 50,000 receptors that reacted with both IL-2 and anti-Tac. In contract, expression of clone 4 did not direct the membrane display of an IL-2 binding protein. However, clone 4 was not the result of a cloning artifact, since aberrantly spliced mRNA corresponding to this cDNA has been detected in normal and leukemic activated T cells (Leonard *et al.*, 1985b).

The primary structure of the IL-2 receptor (Tac) was deduced by sequence analysis of the full-length receptor cDNA (Leonard *et al.*, 1984). This receptor was found to be composed of 272 amino acids, including a 21-residue hydrophobic signal peptide. Multiple cysteine residues were also identified, some of which participate in the formation of intramolecular disulfide bonds that are required for maintenance of the IL-2 and anti-Tac binding sites. Consensus sequences for two N-linked glycosylation sites (Asn-X-Ser/Thr) were also defined. A markedly hydrophobic region composed of 19 residues was present near the carboxy terminus of the protein. Given the location of the signal peptide and N-linked glycosylation sites, the receptor appeared to be oriented as an "NH_2 terminus out, COOH terminus in" transmembrane structure. Unexpectedly, this finding indicated that the intracytoplasmic domain of this receptor was composed of only 13 residues. Thus, in contrast to virtually all other growth factor receptors, this intracytoplasmic domain did not appear to be associated with an enzymatic activity, notably, tyrosine kinase. The intracytoplasmic domain does, however, contain six positively charged amino acids that may play a role in anchoring the receptor protein within the plasma membrane through interactions with the negatively charged membrane phospholipids. This possibility, as well as the general orientation of the Tac protein, was tested by introducing a translation termination codon into the cDNA sequence following the fourth residue of the transmembrane domain. When expressed in mouse fibroblasts, this truncated "anchor minus" IL-2 receptor cDNA was found to direct exclusively the synthesis of a secreted form of the IL-2 receptor, which retained the capacity to bind IL-2 and anti-Tac (Treiger *et al.*, 1986).

5. *The IL-2 Receptor (Tac) Gene*

A genomic phage library was screened with the *Tac* cDNA to characterize the Tac gene. Sequence analysis of these phage clones demonstrated that this receptor gene is organized within eight exons and seven introns and spans a minimum distance of 25 kb (Leonard *et al.*, 1985b). *In situ* hybridization studies permitted localization of the IL-2 receptor (Tac) gene to chromosome 10p band 14→15 (Leonard *et al.*, 1985d). The first exon of the receptor gene encodes the 5′ untranslated region as well as the 21 amino acid signal peptide. The first intron is quite large, encompassing at least 15 kb. Phage clones completely bridging this intron have not been isolated, thus the precise size of the Tac receptor gene remains undefined and may be considerably larger than the minimal estimate of 25 kb. The second exon encodes the first 66 amino acids of the mature receptor protein, including one of the two N-glycosylation sites. The third exon contains the next 36 amino acids as well as the second-N-glycosylation site. The fourth exon encodes 72 amino acids, which share unexpected homology with the second exon. These data suggest that these exons were derived from an internal gene duplication event. The mouse IL-2 receptor contains a similar duplicated region (Miller *et al.*, 1985), suggesting that this duplication occurred at least 50 million years ago, prior to the genetic radiation of mouse and human. The location and presence of these duplicated domains also raise the interesting, although presently unproven, possibility that the IL-2 receptor may contain two ligand-binding sites. The fourth exon is also homologous to the recognition domain of human complement factor B (Ba fragment), indicating an ancestral relation between these genes. The fourth exon also corresponds to the 216 bases that are infrequently removed by aberrant post-transcriptional splicing, resulting in the joining of exon 3 and exon 5. As noted above, expression of a cDNA corresponding to this aberrantly spliced mRNA (e.g., pIL-2R4) in COS-1 or L cells did not result in the display of functional membrane IL-2 receptors (Leonard *et al.*, 1984). These data suggest that the fourth exon is either importantly involved in the normal transport of the receptor and/or is required for IL-2 and anti-Tac binding. The sixth exon contains multiple potential sites for O-linked glycosylation and, topographically, is located immediately exterior to the plasma membrane. A similarly positioned exon rich in O-linked glycosylation sites has been detected in the low-density lipoprotein receptor (Yamamoto *et al.*, 1986). The seventh exon encodes the majority of the 19-residue hydrophobic transmembrane domain. The eighth exon encodes part of the 13 amino acid region comprising the intracytoplasmic domain as well as the 3′ untranslated region of the IL-2 receptor. This noncoding region of the IL-2 receptor may vary in length owing to the presence of three functional polyadenylation signal sequences (Leonard *et al.*, 1985b,c).

The 5′ end of the IL-2 receptor mRNAs may also vary. Both primer extension and S1 nuclease protection analyses have revealed two principal capping sites for receptor mRNA. These sites are separated by 58 nucleotides (Leonard *et al.*, 1985b) and flanked upstream by TATA-like promoter sequences. In addition, other minor capping sites also appear to exist. This 5′ flanking region of the Tac gene contains functional promoter activity, demonstrated by its capacity to direct the expression of the chloramphenicol-acetyl transferase (CAT) gene following transfection into JURKAT T cells. Each of the two promoters controlling the two transcription initiation sites appears to share nearly equivalent strength in normal activated T cells. Ongoing studies are characterizing the sequences within this region, which control mitogen inducibility.

6. Expression of Multiple IL-2 Receptor (Tac) mRNAs

Northern blotting studies of activated T cell RNA revealed three discrete size classes of IL-2 receptor mRNA (Leonard *et al.*, 1984). S1 nuclease protection studies demonstrated that these differently sized mRNAs were produced by the utilization of at least three different 3′ polyadenylation sequences (Krönke *et al.*, 1985a; Leonard *et al.*, 1985c). Suboptimal stimulation of normal T cells appears to generate preferentially mRNAs utilizing the most distal poly A site, while full stimulation produces a more equivalent use of each of the three sites (Leonard *et al.*, 1985c). At present, it is unknown whether certain of these receptor mRNA species are more stable or more efficiently translated than others. The combination of two principal transcription initiation sites, the variable post-transcriptional splicing of exon 4, and the use of three poly A sites suggest that as many as 12 different IL-2 receptor (Tac) mRNAs may exist in activated normal T cells.

7. High- and Low-Affinity Receptors for IL-2

The initial IL-2 binding studies utilizing purified radiolabeled IL-2 indicated the presence of 2000–4000 IL-2 receptors per PHA-activated T lymphoblast (Robb *et al.*, 1981). In addition, Scatchard plots suggested a single high-affinity class of receptors with an apparent K_d of 2–10 pM. In contrast, binding studies performed with the radiolabeled anti-Tac antibody demonstrated the presence of 30,000–60,000 IL-2 receptors per cell (Deppler *et al.*, 1984a). Scatchard analysis of these binding data again suggested a single affinity class of receptors. This discrepancy in receptor number in the two assays was proved not to be the result of inadvertent Fc receptor binding by anti-Tac, nor was it due to partial occupancy of receptor sites by endogenously produced IL-2. These conflicting data were resolved when the IL-2 binding studies were repeated in the presence of much greater quantities of ligand (Robb *et al.*, 1984; Lowenthal *et al.*, 1985). These assays revealed a previously undetected large pool of IL-2 binding sites displaying a 1000–10,000-fold lower apparent affinity for IL-2. The total number of high- and low-affinity binding sites detected in the IL-2 binding assays proved to be nearly equivalent (within a factor of 2) to the number of sites measured with anti-Tac. These data indicate that the anti-Tac antibody binds equivalently to the high- and low-affinity forms of the IL-2 receptor. In most cell types analyzed, the high-affinity receptor component comprises only 2–10% of the total number of membrane IL-2 receptors; however, titration of these high-affinity sites is well correlated with IL-2-induced growth. The biological function of the far more numerous low-affinity IL-2 receptors presently remains unresolved. Studies of the post binding fate of IL-2 have also demonstrated rapid internalization of ligand bound to the high-affinity receptors, while IL-2 bound to low-affinity sites remains on the membrane (Fujii *et al.*, 1986; Weissman *et al.*, 1986). These differences in receptor-mediated endocytosis further underscore a basic structural difference in these two receptor affinity classes.

To define the role of Tac antigen in the high- and low-affinity receptor display, the Tac cDNA was stably expressed in murine fibroblasts (Sabe *et al.*, 1984; Greene *et al.*, 1985; Hatakeyama *et al.*, 1985). Interestingly, these transfected fibroblasts were found to display only low-affinity IL-2 binding sites. These results raised the possibility that the Tac cDNA encoded only the low-affinity form of the receptor, while a second gene encoded the high-affinity receptor. However, this two-gene model for receptor affinity classes became untenable with the demonstration that the Tac cDNA encoded both high- and low-affinity forms of the receptor when expressed in mouse or human T cells (Hatakeyama *et al.*, 1985; Kondo *et al.*, 1986; Wano *et al.*, 1987). Furthermore, Robb (1986) demonstrated that a fraction of the low-affinity mouse IL-2 receptors present on fibroblasts transfected with the murine Tac cDNA could be converted to high-affinity receptors through fusion with human T cell lines displaying high-affinity human IL-2 receptors. These studies took advantage of the ability of anti-Tac to block IL-2 binding to human but not mouse receptors.

Together, these expression and fusion studies suggested that an additional component(s) present within T cells but missing in fibroblasts was involved in the formation of high-affinity IL-2 receptors. These data are perhaps best explained by a receptor complex model. The potential assembly of such a high-affinity receptor complex was also attractive in terms of providing a mechanism for growth signal transduction internalization of ligand by the high-affinity IL-2 receptor in view of the short intracytoplasmic domain of the Tac antigen. Recently, evidence supporting the existence of such a high-affinity IL-2 receptor membrane complex has been obtained in several laboratories (Kuo *et al.*, 1986; Sharon *et al.*, 1986; Tsudo *et al.*, 1986; Dukovich *et al.*, 1987b; Robb *et al.*, 1987; Teshigawara *et al.*, 1987). Utilizing [^{125}I]IL-2 for chemical crosslinking to high-affinity IL-2 receptors, two different proteins have been detected. One of these proteins corresponds to the Tac antigen, while the second protein is a 70,000–75,000-dalton polypeptide (p70) [the crosslinked molecular complex including IL-2 (15,500) migrates as 85,000–90,000 daltons, hence this indirect estimate of size]. Endoglycosidase cleavage and metabolic labeling with [^{3}H]mannose have demonstrated that the p70 polypeptide is a glycoprotein (Sharon *et al.*, 1986; Robb *et al.*, 1987). Furthermore, peptide mapping following proteolytic digestion has clearly demonstrated a difference in the p70 and Tac proteins (Sharon *et al.*, 1986; Robb *et al.*, 1987). Tsudo *et al.*, (1986) first demonstrated expression of the p70 protein in the absence of the Tac antigen on gibbon MLA-144 T cells and found that this protein, like Tac, was able to bind IL-2. This observation was subsequently confirmed in many other laboratories (Dukovich *et al.*, 1987b; Robb *et al.*, 1987; Teshigawara *et al.*, 1987). This previously unrecognized p70 IL-2 receptor appears to bind IL-2 with intermediate affinity (0.6–1.2 nM) compared with the recognized high- and low-affinity forms of the IL-2 receptor. The p70 protein, when expressed alone, also fails to react with the anti-Tac antibody. This protein has also been identified, in the absence of Tac antigen, on the surface of resting T cells, large granular lymphocytes (natural killer cells), and some B cells, including the SKW6.4 B cell line (Dukovich *et al.*, 1987b).

As noted above, the selective crosslinking of radiolabeled IL-2 to high-affinity IL-2 receptors produced labeling of both the Tac and p70 proteins. Furthermore, the crosslinking of IL-2 to both of these proteins was blocked by the anti-Tac antibody. As noted, however, anti-Tac did not react with the p70 protein. These findings suggested that the high-affinity IL-2 receptor might correspond to a receptor complex involving the Tac antigen and the p70 protein. Such a high-affinity receptor complex model has recently been strengthened by studies involving the expression of the Tac cDNA in the MLA-144 T cells, which display only the p70 protein (Dukovich *et al.*, 1987b). High-affinity IL-2 receptors ($K_d = 35$–70 pM) were reconstituted by transfection of plasmids containing the Tac cDNA but not by transfection of control plasmids. Furthermore, the reconstituted high-affinity receptors appeared to involve the Tac antigen, since the addition of anti-Tac blocked high-affinity, but not intermediate-affinity, IL-2 binding. Similarly, high-affinity IL-2 receptors were reconstituted in the natural killer-like YT cell line by induction of endogenous Tac gene expression (Robb *et al.*, 1987; Teshigawara *et al.*, 1987). These cells express the p70 protein in the virtual absence of the Tac antigen; however, stimulation with forskolin, IL-1 or adult T-cell-leukemia-derived factor (ADF) activates Tac gene expression (Yodoi *et al.*, 1985) and leads to the display of large numbers of high-affinity IL-2 receptors concomitant with the disappearance of the intermediate-affinity receptors. Addition of anti-Tac results in the blockade of the high-affinity component of binding and produces the reappearance of intermediate-affinity ($K_d = 810$ pM) p70 binding sites (Robb *et al.*, 1987).

Recent antibody blocking studies have also provided evidence suggesting that the p70 and Tac proteins may interact with different epitopes on the IL-2 molecule (Robb *et al.*, 1987). As noted earlier, Kuo *et al.*, (1986) had demonstrated that residues 8–27 and 33–54 within the primary sequence of IL-2 appeared importantly involved in high-affinity receptor interactions. Utilizing antibodies specific for these peptide regions, evidence was found suggesting that amino acids 8–27 interact with the p70 protein, whereas residues 33–54 interact with the Tac antigen but also may play a role in IL-2 binding to the p70. Certainly, the identification of a receptor complex that interacts with IL-2 at two different sites provides a plausible mechanism to explain the observed high-affinity binding of ligand.

The ability of IL-2 to augment cytolytic activity in natural killer cells and immunoglobulin production in SKW6.4 B cells suggests that the p70 protein may be capable of signal transduction. This supposition is further supported by the recent finding that the p70 protein alone is able to mediate rapid endocytosis of cell-bound IL-2 (Robb and Greene, 1987). Using lymphoid cell lines that express p70 but lack Tac antigen and high-affinity receptors, it has been demonstrated that IL-2 bound to the p70 protein is internalized with kinetics essentially identical to those observed for the high-affinity form of the receptor ($t_{1/2} = 10–15$ min). In contrast, the Tac antigen alone failed to promote internalization of IL-2. This finding predicts that the intracytoplasmic domain of the p70 protein may be larger than that characteristic of the Tac protein and that the endocytotic properties of the high-affinity IL-2 receptor reflect properties imparted by the presence of the p70 protein.

Other studies have suggested a role for the p70 protein in the direct activation by IL-2 of resting T cell proliferation occurring in the absence of other mitogenic stimuli (Le Thi Bich Thuy *et al.*, 1987). This response appears to involve two discrete stages mediated by the interaction of IL-2 with two different forms of the IL-2 receptor. Some, and perhaps all, resting T cells constitutively display the p70 protein (400–600 sites/cell assuming uniform distribution within the resting T cells). Currently, data suggest a model for IL-2 activation of resting T cells involving the initial binding of IL-2 to the intermediate-affinity p70 receptors, which then transduce signals leading to the generation of *competence*. These changes involve the rapid induction of several T cell activation genes, including c-*myc*, c-*myb*, and *Tac*. The production and display of Tac antigen in the presence of the p70 protein then appears to permit the assembly of high-affinity IL-2 receptors that bind IL-2 a second time and generate the second-stage *progression* signals leading to T cell proliferation. This two-step model for IL-2 activation of resting T cell growth, involving the intermediate- and high-affinity forms of the IL-2 receptor, is supported by the finding that the anti-Tac antibody markedly inhibits events associated with progression but not with competence.

At present, the p70 protein remains poorly characterized. The protein has not yet been purified, nor have specific monoclonal antibodies been prepared. In addition, formal proof for the high-affinity receptor as a ternary complex of p70, Tac, and IL-2 has not been assembled, perhaps because of the inherent inefficiency of the crosslinking reactions and the requirement for two internal crosslinks to stabilize this putative tripartite complex. As a more complete biochemical and molecular description of the p70 protein becomes available, several of the critical questions surrounding the structure and function of the high-affinity IL-2 receptor may be answered. It remains possible that other, yet unidentified, proteins also participate with Tac and p70 in the formation of the high-affinity IL-2 receptor complex. It seems likely that the total number of high-affinity IL-2 receptors expressed during T cell activation normally are limited by the availability of the p70 receptors. In contrast, the constitutive nature of p70 expression on many lymphoid cells also underscores the central regulatory role played by the inducible expression of the Tac gene in the control of high-affinity receptor display.

8. HTLV-I, IL-2 Receptors, and the Adult T Cell Leukemia

Infection of human T4$^+$ lymphocytes by the human T lymphotrophic virus type I (HTLV-I) may lead to development of adult T cell leukemia (ATL) (Poiesz *et al.*, 1980, 1981; Yoshida *et al.*, 1982). Clinically, this leukemia is often fatal and may be associated with hypercalcemia, dermal or epidermal leukemic infiltrates, and increased susceptibility of the patients to opportunistic infections (Uchiyama *et al.*, 1977; Bunn *et al.*, 1983). Cases of ATL are geographically clustered in areas where the virus is endemic. These regions include southwestern Japan (Kyushu), the Caribbean basin, the southeastern United States, and sub-Saharan Africa. To date, drug regimens capable of producing consistent and durable remissions in the acute form of ATL have not been defined. A hallmark of cultured HTLV-I-infected ATL cell lines is the uniform expression of large numbers of membrane IL-2 receptors (Gootenberg *et al.*, 1981; Depper *et al.*, 1984b; Waldmann *et al.*, 1984). Both high- and low-affinity binding sites are present; however, most of these leukemic cell lines do not produce IL-2 (Arya *et al.*, 1984), nor do they require IL-2 for growth. Notwithstanding, as discussed below, an autocrine or paracrine mechanism of leukemic cell growth may play an important role perhaps during the early stages of T cell transformation induced by this oncogenic retrovirus.

The constitutive expression of Tac receptors in the ATL cell lines does not appear to be the result of chromosomal translocation, rearrangement, or amplification of the Tac gene (Krönke *et al.*, 1985a). Furthermore, based on DNA sequencing, the primary amino acid sequence of the normal and HUT-102B2 leukemic receptor protein is identical (Ishida *et al.*, 1985; Leonard *et al.*, 1985b). ATL cell lines constitutively produce large quantities of Tac mRNA, and each of the three known poly A addition sites are used. As with normal T cells, post-transcriptional splicing of the fourth exon also occurs in HTLV-I-infected T cells. Nuclear runoff assays have confirmed high-level *Tac* gene transcription in the ATL cells (Krönke *et al.*, 1985a).

The HTLV-I-retrovirus is able to acutely transform and immortalize human cord blood T cells in culture (Wong-Staal and Gallo, 1985). However, this virus does not contain a recognized oncogene as found with many other acutely transforming animal retroviruses. In addition, these viruses do not appear to transform target T cells by promoter–enhancer insertion with *cis*-activation of an endogenous cellular oncogene, since the sites of proviral integration differ from tumor to tumor (Seiki *et al.*, 1984).

The first clues to the mechanism of HTLV-I transformation emerged with the sequencing of an HTLV-I proviral isolate (Seiki *et al.*, 1983). Like other retroviruses, HTLV-I was found to contain a 5′ and 3′ flanking long terminal repeat (LTR). In addition, *gag, pol*, and *env* genes were present. However, an extra region, termed pX, was also detected near the 3′ end of the virus containing at least three open reading frames (Haseltine *et al.*, 1984; Kiyokawa *et al.*, 1984; Seiki *et al.*, 1985; Shimotohno *et al.*, 1985). Sodroski *et al.*, (1984) subsequently demonstrated that this pX region was involved in the production of a subgenomic mRNA and corresponding protein that was capable of activating in *trans* the transcription of genes controlled by the viral LTR. The transacting responsive (TAR) sequences have also been localized to the 21-bp enhancer-like repeats within the viral LTR (Paskalis *et al.*, 1986; Shimotohno *et al.*, 1986). This transactivator gene was originally termed LOR, X-LOR, or p40, but more recently it has been termed *tat* (*t*rans*a*ctivator of *t*ranscription), given its function. Considerable speculation has focused on the possibility that the *tat* protein may, in addition to augmenting viral transcription and viral replication, also activate the expression of select cellular genes involved in T cell growth.

In this regard, Inoue and colleagues (1986) have demonstrated that the *tat*-I cDNA cloned into an expression vector is able to stimulate the transient expression of the cellular genes encoding both Tac and IL-2 following transfection into JURKAT or HSB-2 T cells. In parallel studies, Siekevitz *et al.*, (1987) have prepared a *tat*-I cDNA expression vector containing the HTLV-I

LTR as the promoter. This vector transactivates its own production of the *tat*-I gene product, thus leading to high-level protein production. This plasmid was shown to encode a functional transactivator gene product in cotransfection assays in JURKAT T cells, as evidenced by marked stimulation of the HTLV-I LTR linked to the reporter gene, chloramphenicol acetyl transferase (CAT). (The activity of a promoter can be quantitated by examining the degree of acetylation of radiolabeled chloramphenicol catalyzed by the CAT enzyme.) In contrast, a control cDNA corresponding to a 21,000-dalton pro tein from the pX region (Δ*tat*-I) had no transactivating properties.

In order to examine potential effects on cellular gene expression, the *tat*-I and Δ*tat*-I cDNA expression vectors were cotransfected separately into JURKAT T cells with a plasmid containing the upstream regulatory sequences of the IL-2 receptor linked to the CAT gene. CAT assays performed on these transfectants revealed that the *tat*-I, but not the Δ-*tat*-I gene product, activated the IL-2 receptor promoter region (Siekevitz *et al.*, 1987). In contrast, the expression of several other cellular or viral promoters was not altered by *tat*-I (Siekevitz *et al.*, 1987). These findings thus provide a mechanistic explanation for the deregulated display of IL-2 receptors encountered in HTLV-I-infected T cells. Similar results have been reported by Maruyama *et al.*, (1987) and Cross *et al.*, (1987).

Potential effects of the *tat*-I gene on IL-2 gene expression were also evaluated. In contrast to the IL-2 receptor promoter, the *tat*-I gene product alone appears to have little or no effect on the IL-2 promoter (Siekevitz *et al.*, 1987). However, in the JURKAT T cell model, two mitogenic signals (PHA and PMA) are required for the activation of IL-2 production and IL-2 promoter CAT expression. In contrast, only a single signal (PHA or PMA) is required for IL-2 receptor (Tac) gene expression. In view of this finding, combinations of signals were investigated. Cotransfection studies revealed that the *tat*-I gene product markedly synergized with either PHA or PMA, leading to high-level IL-2 promoter activity. Similar results were observed with other T-cell mitogens including monoclonal antibodies specific for T3 (Maruyama *et al.*, 1987). Furthermore, in contrast to mitogen stimulation, *tat*-I-coinduced IL-2 promoter activity was found to be largely resistant to the inhibitory effects of the immunosuppressive agent cyclosporin A (Siekevitz *et al.*, 1987). These findings suggest that the *tat*-I gene product may exert its effects on *IL-2* gene expression at a point distal to the site of inhibition produced by cyclosporin A.

The mechanism by which the *tat* proteins induce expression of the IL-2R and IL-2 cellular genes remains unclear. Immunofluorescent staining studies have suggested that the *tat*-I protein is largely localized in the nucleus (Goh *et al.*, 1985; Slamon *et al.*, 1985). However, it remains unclear whether this protein possesses DNA binding properties. A direct interaction of the *tat*-I protein with specific DNA sequences seems unlikely, since the promoters of the *IL-2* and *IL-2R* genes do not share strong sequence homology with the 21-bp enhancer-like region of the HTLV-I LTR, which has been shown to be involved in *tat*-I-induced transactivation (Paskales *et al.*, 1986; Shimotohno *et al.*, 1986). Alternatively, as found with the adenovirus Ela protein, it is possible that the *tat*-I protein alters cellular gene expression by interacting with another protein or family of proteins, which in turn bind DNA. In this regard, transfection studies with the natural-killer-like YT cell line have strongly implicated an additional cellular factor(s) in the stimulation of Tac gene and gene expression by *tat*-I (Siekevitz *et al.*, 1987). Although the Tac promoter is readily activated in these cells by addition of PMA (Yodoi *et al.*, 1985; Siekevitz *et al.*, 1987), it is completely unresponsive to activation by *tat*-I. This failure is not explained by a lack of the *tat*-I protein production in the transfected YT cells, since the LTR of HTLV-I is markedly transactivated. These results suggest the requirement for a cellular cofactor(s) that is present in JURKAT T cells, but absent in YT cells, for *tat*-I activation of the IL-2 receptor gene. In this regard, it is of interest that HTLV-I-induced leukemia is virtually confined to $T4^+$ T lymphocytes. It is possible that the apparent restriction of HTLV-I-induced leukemic transformation to this subset of T cells reflects the cellular distribution of the putative cofactor required for *tat*-I-induced changes in cellular gene expression.

Retroviral-induced tumors have now been associated with the introduction of oncogenes resembling a growth factor (e.g., v-*sis*→B chain of platelet-derived growth factor) or growth factor receptors (v-*erb*-B→truncated epidermal growth factor receptor, v-*fms*→CSF-I receptor). Alternatively, retroviruses may induce leukemic transformation by *cis*-activation of an endogenus cellular oncogene via promoter–enhancer insertion (e.g., avian leukosis virus activation of c-*myc* expression). The findings presented above suggest that the HTLV-I retrovirus employs a third strategy involving its *trans*-activator gene product, which acts to induce the expression of the Tac gene and to partially activate the *IL-2* gene. The coinduced expression of these genes, which participate in the regulation of T cell proliferation, could lead to a period of uncontrolled autocrine polyclonal T cell growth. The *in vitro* transformation of T cells by HTLV-I is in fact characterized by an early stage of vigorous polyclonal T cell proliferation (Popovic *et al.*, 1983). Such early autocrine growth might facilitate the occurrence of later second-stage events required for the completion of the T cell transformation process and lead to the emergence of a clonal growth-factor-independent population of leukemic T cells. Support for such an autocrine growth model has recently been assembled by Arima and colleagues (1986). These investigators have identified primary ATL tumor cells that simultaneously display IL-2 receptors and produce IL-2. Furthermore, the growth of these leukemic cells is inhibited by antibodies specific for either IL-2 or the Tac protein component of the IL-2 receptor. Thus, it seems likely that the *tat*-I protein may play a central role in HTLV-I-induced malignancy, not only by amplifying viral replication but also by deregulating the early expression of the cellular genes that normally control T cell growth, including IL-2 and the IL-2 receptor.

As noted earlier, no effective treatment for patients with ATL has yet been identified. The presence of the Tac antigen on the surface of these cells serves as a relatively tumor-specific marker, since most resting T and B cells do not express this gene product. Krönke and colleagues (1985b) and FitzGerald *et al.*, (1984) have demonstrated that anti-Tac antibody conjugated to the A chain of the ricin toxin or to the *Pseudomanas* exotoxin, respectively, forms an effective cytotoxic agent capable of killing HTLV-I-infected T cells *in vitro*. Furthermore, immunotoxin killing of ATL cells can be accentuated by the addition of lysosomotrophic agents (e.g., chloroquine, monensin), which presumably facilitate entry of the immunotoxin into the cytoplasm (Krönke *et al.*, 1986). While *in vitro* successes with immunotoxins do not necessarily predict *in vivo* utility, the clinical efficacy of the anti-Tac–*Pseudomonas* exotoxin conjugates is presently under study at the National Cancer Institute.

9. *Human Immunodeficiency Virus, Interleukin-2, and the Acquired Immune Deficiency Syndrome*

While HTLV-I infection of T4+ human lymphocytes may produce leukemic transformation, infection of this subset of T cells by the human immunodeficiency virus (HIV) often leads to cell death and clinically to the development of a profound acquired immunodeficiency termed AIDS (Wong-Staal and Gallo, 1985). The mechanism by which this pathogenic retrovirus, HIV, induces T4+ cell death has remained controversial. At least in part, cell death is produced by cell fusion and syncytia formation mediated by the interaction of the HIV major envelope protein with the T4+ surface antigen (Sodroski *et al.*, 1986). However, it seems likely that these fusion events may be complemented by other HIV-induced intracellular changes that lead to deranged T4+ cell function.

In this regard, we have investigated whether HIV infection of T4+ lymphocytes produces alternations in IL-2 or IL-2 receptor (*Tac*) gene expression (Dukovich *et al.*, in preparation). For these studies, T4+ lymphocytes from normal donors were purified and activated with PHA and PMA and then infected at high multiplicity with HIV. These cells, as well as uninfected cells, were then cultured for varying periods of time in the presence of IL-2 and restimulated with PHA and PMA, and 12–18 h later IL-2 and IL-2 receptor (Tac) gene expression was measured. Virtually no viral gene expression was observed in these cultures until day 6. In addition, the induction of IL-2 and IL-2 receptor (*Tac*) gene expression was entirely normal during these initial 6 days. However, on day 8 of culture a significant decline in the level of IL-2 mRNA was noted following mitogen activation. In contrast, IL-2 receptor (*Tac*) gene expression was normally induced by these mitogenic agents. Sequential measurements from day 8 to day 21 demonstrated a progressive inhibition of *IL-2* gene expression in the absence of alterations in IL-2 receptor (*Tac*) gene induction. Northern blotting studies performed with cells harvested on day 14 confirmed these marked changes in the level of IL-2 mRNA Furthermore, in contrast to the changes observed for IL-2, only modest inhibition of mitogen-induced interferon-γ mRNA production was present.

Cell surface phenotype analysis indicated that these changes in IL-2 production were not the artifactual result of the overgrowth of the cultures by T8+ cells. However, selective destruction of IL-2-producing cells within the original T4+ lymphocyte population or outgrowth of T4−/T8− cells, rather than a noncytolytic inhibition of gene expression, remained a possibility. To address these concerns T4+ JURKAT leukemic T cells were infected with HIV and studied for altered *IL-2* gene expression following mitogen activation. By using this cloned T cell line it is possible to infect virtually 100% of the cell population yet maintain a continuous viable culture. Northern blotting studies with PHA- and PMA-activated HIV-infected JURKAT cells revealed identical inhibitory effects on *IL-2* gene expression as obtained with the T4+ peripheral blood lymphocytes. In contrast, mitogen stimulation produced normal activation of the IL-2 receptor (*Tac*) gene.

Next, studies with the HIV-infected peripheral blood T4+ lymphocyte populations were initiated to investigate whether the observed inhibition of *IL-2* gene expression could be circumvented. In the prior studies, the infected cells were activated with PHA and PMA. The mitogenic lectin, PHA, is believed to act by binding to the T3–Ti complex; thus, signal transduction through this complex is required for cellular activation. At least one of the early signals generated by this receptor–ligand interaction is an increase in intracellular levels of Ca^{2+} (Weiss *et al.*, 1984). Interestingly, when the T3–Ti complex was bypassed by stimulation of the HIV-infected cells with the calcium ionophore, ionomycin, and PMA, considerable IL-2 mRNA was detected. These findings demonstrate that *IL-2* gene expression can be induced in these cell populations if an appropriate stimulus is employed. These findings also raise the possibility that the defect in IL-2 production in the HIV-infected T4+ cells may involve an altered ability of the T3–Ti complex to transduce effectively the required intracellular activation signals.

Since T4+ lymphocytes require IL-2 for growth, it is possible that the blockade of antigen- or mitogen-induced IL-2 production produced by HIV infection may contribute to the cytopathic effect of this virus on these cells. Thus, while the HTLV-I retrovirus appears to activate IL-2 receptor and *IL-2* gene expression and produce leukemic transformation, the HIV retrovirus appears to produce marked inhibition of *IL-2* gene expression, which may contribute to the multifactorial development of the profound immunodeficiency.

10. Summary

The critical role of IL-2 and IL-2 receptors in normal T cell immune response is now well established. The genes encoding this growth factor and one subunit of its high-affinity cellular receptor have been cloned, sequenced, and expressed in eukaryotic and prokaryotic hosts. Several studies are presently underway to assemble a more complete biochemical and molecular description of the recently recognized second subunit of the high-affinity IL-2 receptor. These investigations of the p70 protein should provide important clues to the elusive mechanism of signal transduction employed by this receptor. It is also possible that these studies will reveal that yet other proteins are required for the formation of functional high-affinity receptors. The oncogenic human retrovirus HTLV-I has been shown to be etiologically associated with the adult T cell leukemia. This virus has also been found to encode a transacting protein (*tat*) that not only participates in the regulation of viral gene expression but is also able to produce activation of the genes encoding the Tac antigen and IL-2. These findings suggest that *tat*-I induced changes in the expression of these cellular genes that control T cell growth and may play a role in the leukemic transformation induced by this virus. Finally, the human immunodeficiency virus cells have been shown to produce marked inhibition of *IL-2* gene expression following infection of T4+ lymphocytes. While the molecular mechanism for this inhibition remains unknown, the lack of IL-2 protein production may contribute to the T cell immunodeficiency characteristic of AIDS.

References

Arima, N., Daitoku, Y., Ohgaki, S., Fukumori, J., Tanaka, H., Yamamoto, Y., Fujimoto, K., and Onoue, K., 1986, Autocrine growth of interleukin 2-producing leukemic cells in a patient with adult T cell leukemia, *Blood* **68:**779–782.

Arya, S. K., Wong-Staal, F., and Gallo, R. C., 1984, T-cell growth factor gene: Lack of expression on human T-cell leukemia lymphoma virus infected cells, *Science* **223:**1086–1087.

Bunn, P. A., Schechter, G. P., Jaffe, E., Blayney, D., Young, R. C., Matthews, M. J., Blattner, W., Broder, S., Robert-Guroff, M., and Gallo, R. C., 1983, Clinical course of retrovirus associated adult T cell lymphoma in the United States, *N. Engl. J. Med.* **309:**257–262.

Cosman, D., Cerretti, D. P., Larsen, A., Park, L., March, C., Dower, S., Gillis, S., and Urdal, D., 1984, Cloning, sequence and expressions of human interleukin-2 receptor, *Nature* **312:**768–771.

Cotner, T., Williams, J. M., Christenson, L., Shapiro, H. M., Strom, T. B., and Strominger, J., 1983, Simultaneous flow cytometric analysis of human T cell activation antigen expression and DNA content, *J. Exp. Med.* **157:**461–472.

Depper, J. M., Leonard, W. J., Krönke, M., Noguchi, P. D., Cunningham, R. E., Waldmann, T. A., and Greene, W. C., 1984a, Regulation of interleukin-2 receptor expression: Effects of phorbol diester, phospholipases, and reexposure to lectin and antigen, *J. Immunol.* **133:**3054–3061.

Depper, J. M., Leonard, W. J., Krönke, M., Waldmann, T. A., and Greene, W. C., 1984b, Augmented T-cell growth-factor receptor expression in HTLV-I infected human leukemic T-cells, *J. Immunol.* **133:**1691–1695.

Dukovich, M., Wano, Y., Le Thi Bich Thuy, Katz, P., Cullen, B. R., Kehrl, J. H., and Greene, W. C., 1987b, Identification of a second human IL-2 binding protein and its possible role in the high affinity IL-2 receptor, *Nature* **327:**518–522.

FitzGerald, D. J. P., Waldmann, T. A., Willingham, M. C., and Pastan, I., 1984, *Pseudomonas* exotoxin–anti-Tac cell specific immunotoxin active against cells expressing the human T-cell growth factor receptor, *J. Clin. Invest.* **74:**966–971.

Fujii, M., Sugamura, K., Sano, K., Nakai, M., Sugita, K., and Hinuma, Y., 1986, High-affinity receptor-mediated internalization and degradation of interleukin 2 in human T cells, *J. Exp. Med.* **163:**550–562.

Goh, W. C., Sodroski, J., Rosen, C., Essex, M., and Haseltine, W. A., 1985, Subcellular localization of the product of the long open reading frame of human T-cell leukemia virus type I, *Science* **227:**1227–1228.

Gootenberg, J. E., Ruscetti, F. W., Mier, J. W., Gazdar, A., and Gallo, R. C., 1981, Human cutaneous T-cell lymphoma and leukemia-cell lines produce and respond to T-cell growth factor, *J. Exp. Med.* **154:**1403–1417.

Greene, W. C. and Robb, R. J., 1985, Receptors for T-cell growth factor: Structure, function and expression on normal and neoplastic cells, in: *Contemporary Topics in Molecular Immunology* (S. Gillis and F. P. Inman, eds.), Vol. 10, pp. 1–34, Plenum Press, New York.

Greene, W. C., Robb, R. J., Svetlik, P. B., Rusk, C. M., Depper, J. M., and Leonard, W. J., 1985, Stable expression of cDNA encoding the human interleukin receptor in eukaryotic cells, *J. Exp. Med.* **162:**363–368.

Haseltine, W. A., Sodroski, J., Patarca, R., Briggs, D., Perkins, D., and Wong-Staal, F., 1984, Structure of the 3′ terminal repeat region of type-II human T-lymphotropic virus: Evidence for a new coding region, *Science* **225:**419–421.

Hatakeyama, M., Minamoto, S., Uchiyama, T., Hardy, R. R., Yamada, G., and Taniguchi, T., 1985, Reconstitution of functional receptor for human interleukin-2 in mouse cells, *Nature* **318:**467–470.

Inoue, J.,Seiki, M., Taniguchi, T., Tsuru, S., and Yoshida, M., 1986, Induction of interleukin-2 receptor gene expression by $p40^x$ encoded by human T cell leukemia virus-type I, *Embo J.* **5:**2883–2888.

Ishida, N., Kanamori, H., Noma, T., Nikaido, T., Sabe, H., Suzuki, N., Shimizu, A., and Honjo, T., 1985, Molecular cloning and structure of the human interleukin-2 gene, *Nucleic Acids Res.* **13:**7579–7589.

Kiyokawa, T., Seiki, M., Imagawa, K., Shimizu, F., and Yoshida, M., 1984, Identification of a protein (p40^x^) encoded by a unique sequence of pX of human T-cell leukemia-virus type-I, *Gann* **75:**747–751.

Kondo, S., Shimizu, A., Maeda, M., Tagaya, Y., Yodoi, J., and Honjo, T., 1986, Expression of functional human interleukin-2 receptor in mouse T cells by cDNA transfection, *Nature* **320:**75–77.

Kovesdi, I., Reichel, R., and Nevins, J. R., 1986, E1A transcription induction: Enhanced binding of a factor to upstream promoter sequences, *Science* **231:**719–722.

Krönke, M., Leonard, W. J., Depper, J. M., and Greene, W. C., 1985a, Deregulation of interleukin receptor gene expression in HTLV-I induced adult T cell leukemia, *Science* **228:**1215–1217.

Krönke, M., Depper, J. M., Leonard, W. J., Vitetta, E. S., Waldmann, T. A., and Greene, W. C., 1985b, Adult T cell leukemia: A potential target for ricin A chain immunotoxins, *Blood* **65:**1416–1421.

Krönke, M., Schlick, E., Waldmann, T. A., Vitetta, E. S., and Greene, W. C., 1986, Selective killing of human T lymphotropic virus-I infected leukemic T cells by monoclonal anti-interleukin-2 receptor antibody-ricin A chain conjugates: Potentiation by ammonium chloride and monensin, *Cancer Res.* **46:**3695–13698.

Kuo, L. M., Rusk, C. M., and Robb, R. J., 1986, Structure–function relationships for the IL 2 receptor system. II. Localization of an IL 2 binding site on high and low affinity receptors, *J. Immunol.* **137:**1544–1551.

Le Thi Bich Thuy, Dukovich, M., Peffer, N. J., Fauci, A. S., Kehrl, J. H., and Greene, W. C., 1987, Direct activation of human T cells by IL-2: The role of an IL-2 receptor distinct from the Tac protein, *J. Immunol.* **139:**1550–1559.

Leonard, W. J., Depper, J. M., Uchiyama, T., Smith, K. A., Waldmann, T. A., and Greene, W. C., 1982, A monoclonal antibody that appears to recognize the receptor for human T cell growth factor: Partial characterization of the receptor, *Nature* **300:**267–269.

Leonard, W. J., Depper, J. M., Robb, R. J., Waldmann, T. A., and Greene, W. C., 1983, Characterization of the human receptor for T cell growth factor, *Proc. Natl. Acad. Sci. U.S.A.* **80:**6957–6961.

Leonard, W. J., Depper, J. M., Crabtree, G. R., Rudikoff, S., Pumphrey, J., Robb, R. J., Krönke, M., Svetlik, P. B., Peffer, N. J., Waldmann, T. A., and Greene, W. C., 1984, Molecular cloning and expression of cDNAs for the human interleukin-2 receptor, *Nature* **311:**626–631.

Leonard, W. J., Depper, J. M., Krönke, M., Robb, R. J., Waldmann, T. A., and Greene, W. C., 1985a, The human receptor for T-cell growth factor: Evidence for variable post-translational processing. phosphorylation, sulfation, and the ability of precursor forms of the receptor to bind TCGF, *J. Biol. Chem.* **260:**1872–1880.

Leonard, W. J., Depper, J. M., Krönke, M., Peffer, N. J., Svetlik, P. B., Kanehisa, M., Sullivan, M., and Greene, W. C., 1985b, Structure of the human interleukin-2 receptor gene, *Science* **230:**633–639.

Leonard, W. J., Krönke, M., Peffer, N. J., Depper, J. M., and Greene, W. C., 1985c, Interleukin-2 receptor gene expression in normal human T lymphocytes, *Proc. Natl. Acad. Sci. U.S.A.* **82:**6281–6285.

Leonard, W. J., Donlon, T. A., Lebo, R. V., and Greene, W. C., 1985d, Localization of the gene encoding the human interleukin-2 receptor on chromosome 10, *Science* **228:**1547–1549.

Lowenthal, J. W., Zubler, R. H., Nabholz, M., and MacDonald, H. R., 1985, Similarities between interleukin-2 receptor number and affinity on activated B and T lymphocytes, *Nature* **315:**669–672.

Maruyama, M., Shibuya, H., Harada, H., Hatakeyama, M., Seiki, T., Fujita, T., Inoue, J.,

Yoshida, M., and Taniguchi, T., 1987, Evidence for aberrant activation of the interleukin-2 autocrine loop by HTLV-I encoded $p40^x$ and T3–Ti complex triggering, *Cell* **48:**343–350.

Miller, J., Malek, T. R., Leonard, W. J., Greene, W. C., Shevach, E. M., and Germain, R., 1985, Nucleotide sequence and expression of a mouse interleukin-2 receptor cDNA, *J. Immunol.* **134:**4212–4217.

Miyawaki, T. A., Yachie, A., Uwandana, N., Ohzeki, S., Nagaoki, T., and Taniguchi, N., 1982, Functional significance of Tac antigen expressed on activated human T lymphocytes: Tac antigen interacts with T cell growth factor in cellular proliferation, *J. Immunol.* **129:**2474–2478.

Morgan, D. A., Ruscetti, F. W., and Gallo, R. C., 1976, Selective *in vitro* growth of T lymphocytes from normal human bone marrows, *Science* **193:**1007–1008.

Nikaido, T., Shimizu, N., Ishida, N., Sabe, H., Teshigawara, K., Maeda, M., Uchiyama, T., Yodoi, J., and Honjo, T., 1984, Molecular cloning of cDNA encoding human interleukin-2 receptor, *Nature* **311:**631–635.

Paskalis, H., Felber, B. K., and Pavlakis, G. N., 1986, Cis-acting sequences responsible for the transcriptional activation of human T-cell leukemia virus type I constitute a conditional enhancer, *Proc. Natl. Acad. Sci. U.S.A.* **83:**6558–6562.

Poiesz, B. J., Ruscetti, F. W., Gazdar, A. F., Bunn, P. A., Minna, J. D., and Gallo, R. C., 1980, Detection and isolation of type-C retrovirus particles from fresh and cultured lymphocytes of a patient with cutaneous T-cell lymphoma, *Proc. Natl. Acad. Sci. U.S.A.* **77:**7415–7419.

Poiesz, B. J., Ruscetti, F. W., Reitz, M. S., Kalyanaraman, V. S., and Gallo, R. C., 1981, Isolation of a new type-C retrovirus (HTLV) in primary uncultured cells of a patient with Sezary T-cell leukemia, *Nature* **294:**268–271.

Popovic, M., Sarin, P. S., Robert-Guroff, M., *et al* 1983, Isolation and transmission of human retrovirus (human T cell leukemia virus), *Science* **219:**856–859.

Robb, R. J., 1984, Interleukin-2: The molecule and its function, *Immunol. Today* **5:**203–209.

Robb, R. J., 1986, Conversion of low-affinity interleukin 2 receptors to a high-affinity state following fusion of cell membranes, *Proc. Natl. Acad. Sci. U.S.A.* **83:**3992–3996.

Robb, R. J., and Greene, W. C., 1983, Direct demonstration of the identity of T cell growth factor binding protein and the Tac antigen, *J. Exp. Med.* **158:**1332–1337.

Robb, R. J., and Greene, W. C. 1987b, Internalization of IL-2 is mediated by the beta chain of the high affinity IL-2 receptor, *J. Exp. Med.* **165:**1201–1206.

Robb, R. J., Munck, A., and Smith, K. A., 1981, T-cell growth factor receptors: Quantification, specificity, and biological relevance, *J. Exp. Med.* **154:**1455–1474.

Robb, R. J., Greene, W. C., and Rusk, C. M., 1984, Low and high affinity cellular receptors for interleukin 2. Implications for the level of Tac antigen, *J. Exp. Med.* **160:**1126–1146.

Robb, R. J., Rusk, C. M., Yodoi, J., and Greene, W. C., 1987, An interleukin-2 binding molecule distinct from the Tac protein: Analysis of its role in formation of high affinity receptors, *Proc. Natl. Acad. Sci. U.S.A.* **84:**2002–2006.

Rubin, L. A., Kurman, C. L., Biddison, W. E., Goldman, N. D., and Nelson, D. L., 1985, A monoclonal antibody, 7G7/B6, binds to an epitope on the human IL-2 receptor that is distinct from that recognized by IL-2 or anti-Tac, *Hybridoma* **4:**91–102.

Sabe, H., Kondo, S., Shimizu, A., Tagaya, Y., Yodoi, J., Kobayashi, N., Hatanaka, M., Matsunami, N., Maeda, M., Noma, T., and Honjo, T., 1984, Properties of human interleukin-2 receptors expressed on nonlymphoid cells by cDNA transfection, *Mol. Biol. Med.* **2:**379–396.

Seiki, M., Hattori, S., Hirayama, Y., and Yoshida, M., 1983, Human adult T-cell leukemia virus: Complete nucleotide sequence of the provirus genome integrated in leukemia cell DNA, *Proc. Natl. Acad. Sci. U.S.A.* **80:**3618–3622.

Seiki, M., Eddy, R., Shows, T., and Yoshida, M., 1984, Nonspecific integration of HTLV provirus genome into adult T-cell leukemia cells, *Nature* **309:**640–642.

Seiki, M., Hikikoshi, A., Taniguchi, T., and Yoshida, M., 1985, Expression of the pX gene of HTLV-I: General splicing mechanisms in the HTLV family, *Science* **228:**1532–1534.

Shackelford, D. A., and Trowbridge, I. S., 1984, Induction of expression and phosphorylation of the human interleukin 2 receptor by a phorbol diester, *J. Biol. Chem.* **259:**11706–11712.

Sharon, M., Klausner, R. D., Cullen, B. R., Chizzonite, R., and Leonard, W. J., 1986, Novel interleukin-2 receptor subunit detected by crosslinking under high affinity conditions, *Science* **234:**859–863.

Shimotohno, K., Miwa, M., Slamon, D. J., Chen, I.S.Y., Hoshino, H. O., Takano, M., Fujino, M., and Sugimura, T., 1985, Identification of new gene-products encoded from X-regions of human T-cell leukemia viruses, *Proc. Natl. Acad. Sci. U.S.A.* **82:**302–306.

Shimotohno, K., Takano, M., Teruuchi, T., and Miwa, M., 1986, Requirement of multiple copies of a 21-nucleotide sequence in the U3 regions of human T-cell leukemia virus type I and type II long terminal repeats for trans-acting activation of transcription, *Proc. Natl. Acad. Sci. U.S.A.* **83:**8112–8116.

Siekevitz, M., Feinberg, M. B., Holbrook, N., Yodoi, J., Wong-Staal, F., Gallo, R. C. and Greene, W. C., 1987, Activation of interleukin-2 and interleukin-2 receptor promoter expression by the transactivator (*tat*) gene product of HTLV-I, *Proc. Natl. Acad. Sci. U.S.A.* **84:**5389–5394.

Slamon, D. J., Press, M. F., Souza, L. M., Murdock, D. C., Cline, M. J., Golde, D. W., Gasson, J. C., and Chen, I. S., 1985, Studies of the putative transforming protein of the type I human T-cell leukemia virus, *Science* **228:**1427–1430.

Smith, K. A., 1980, T-cell growth factor, *Immunol. Rev.* **51:**337–357.

Sodroski, J. G., Rosen, C. A., and Haseltine, W. A., 1984, Trans-acting transcriptional activation of the long terminal repeat of human T-lymphotropic viruses in infected cells, *Science* **225:**381–385.

Sodroski, J., Goh, W. C., Rosen, C., Campbell, K., and Haseltine, W. A., 1986, Role of the HTLV-III/LAV envelope in syncytium formation and cytopathicity, *Nature* **322:**470–474.

Teshigawara, K., Wang, H. M., Kato, K., and Smith, K. A., 1987, Interleukin-2 high affinity receptor expression requires two distinct binding proteins, *J. Exp. Med.* **165:**223–238.

Treiger, B. F., Leonard, W. J., Svetlik, P., Rubin, L., Nelson, D. L., and Greene, W. C., 1986, A secreted form of the human interleukin-2 receptor encoded by an "anchor-minus" cDNA, *J. Immunol.* **136:**4099–4105.

Tsudo, M., Kozak, R. W., Goldman, C. K., and Waldmann, T. A., 1986, Demonstration of a non-Tac peptide that binds interleukin-2: A potential participant in a multichain interleukin-2 receptor complex, *Proc. Natl. Acad. Sci. U.S.A.* **83:**9694–9698.

Uchiyama, T., Yodoi, J., Sagawa, K., Takatsuki, K., and Uchino, H., 1977, Adult T cell leukemia: Clinical and hematologic features of 16 cases, *Blood* **50:**481–489.

Uchiyama, T., Broder, S., and Waldmann, T. A., 1981, A monoclonal antibody (anti-Tac) reactive with activated and functionally mature human T cells, *J. Immunol.* **126:**1293–1297.

Urdal, D. L., March, C. J., Gillis, S., Larsen, A., and Dower, S. K., 1984, Purification and chemical characterization of the receptor for interleukin-2 from activated human T lymphocytes and from a human T-cell lymphoma cell-line, *Proc. Natl. Acad. Sci. U.S.A.* **81:**6481–6485.

Waldmann, T. A., Greene, W. C., Sarin, P. S., Saxinger, C., Blayney, D. W., Blattner, W. A., Goldman, C. K., Bongiovanni, K., Sharrow, S., Depper, J. M., Leonard, W. J., and Gallo, R. C., 1984, Functional and phenotypic comparison of human T cell leukemia/lymphoma virus positive adult T cell leukemia with human T cell leukemia/lymphoma virus negative Sezary leukemia, *J. Clin. Invest.* **73:**1711–1718.

Wano, Y., Uchiyama, T., Fukui, K., Maeda, M., Uchino, H., and Yodoi, J., 1984, Characterization of human interleukin 2 receptor (Tac expression) in normal and leukemic T-cells: Coexpression of normal and aberrant receptors in HUT 102 cells, *J. Immunol.* **132:**3005–3010.

Wano, Y., Cullen, B. R., Svetlik, P., Peffer, N. J., and Greene, W. C., 1987, Reconstitution of high affinity IL-2 receptor expression in a human T cell line using a retroviral cDNA expression vector, *Mol. Biol. Med.* **4:**95–109.

Weiss, A., Imboden, J., Shoback., D., and Stobo, J., 1984, Role of T3 surface molecules in human T-cell activation: T3-dependent activation results in an increase in cytoplasmic free calcium, *Proc. Natl. Acad. Sci. U.S.A.* **81:**4169–4173.

Weissman, A. M., Harford, J. B., Svetlik, P. B., Leonard, W. J., Depper, J. M., Waldmann, T. A., Greene, W. C., and Klausner, R. D., 1986, Only high affinity receptors for interleukin-2 mediate internalization of ligand, *Proc. Natl. Acad. Sci. U.S.A.* **83:**1463–1466.

Wong-Staal, F., and Gallo, R. C., 1985, Human T lymphotrophic retroviruses, *Nature* **317:**395–403.

Yamamoto, T., Bishop, R. W., Brown, M. S., Goldstein, J. L., and Russell, D. W., 1986, Deletion in cysteine-rich region of LDL receptor impedes transport to cell surface in WHHL rabbit, *Science* **232:**1230–1237.

Yodoi, J., Teshigawara, K., Nikaido, T., Fukui, K., Noma, T., Honjo, T., Takigawa, M., Sasaki, M., Minato, N., Tsudo, M., *et al.*, 1985, TCGF (IL 2)-receptor inducing factor(s). I. Regulation of IL 2 receptor on a natural killer-like cell line (YT cells), *J. Immunol.* **134:**1623–1630.

Yoshida, M., Miyoshi, I., and Hinuma, Y., 1982, Isolation and characterization of retrovirus from cell lines of human adult T cell leukemia and its implication in disease, *Proc. Natl. Acad. Sci. U.S.A.* **78:**6476–6480.

VI

THE ROLE OF PHOSPHOLIPASES IN INFLAMMATION

20

Some Novel Phospholipase C Activities

Actions on Phosphatidylcholine and on Phosphatidylinositol-Glycans as Anchors for Membrane Proteins and as Precursors for Possible Insulin Mediators

PEDRO CUATRECASAS

During the past few years, as illustrated by this and other recent volumes, there has been considerable interest in a hormone-sensitive phospholipase C that acts specifically on polyphosphoinositides. Upon receptor activation by certain agonists (e.g., adrenergic agonists, acetylcholine, serotonin, vasopressin, and platelet-derived growth factor), this membrane-localized enzyme(s) hydrolyzes primarily phosphatidylinositol 4, 5-bisphosphate into 1, 2-diacylglycerol and inositol-1,4,5-trisphosphate. These substances, which often act synergistically in certain biological systems, activate protein kinase C(s) and release calcium from intracellular stores, respectively. Many interesting complexities of this important regulatory system continue to evolve.

Much less attention has been directed to the study of other membrane-localized phospholipases of the C type that are now also emerging as potentially significant pathways in cellular regulation. It is the purpose of this chapter to briefly review some of the recently described developments in these areas and to speculate provocatively on possible mechanisms and interpretations.

PEDRO CUATRECASAS • Glaxo Research Laboratories, Glaxo, Inc., Research Triangle Park, North Carolina 27709.

A case in point may be an enzyme that selectively hydrolyzes phosphatidylcholine into diacylglycerol and phosphocholine. Such an enzyme has been purified from dog heart (Wolf and Gross, 1985). Recent studies (Besterman *et al.*, 1986) have shown that the putative activity of such an enzyme may occur as the result of treating cells with phorbol esters (i.e., the activation of protein kinase C) (Fig. 1). Certain agonists (e.g., serum) that activate the PI-specific phospholipase C may thus act indirectly on this enzyme or perhaps even independently by the same receptor (but a different G or N

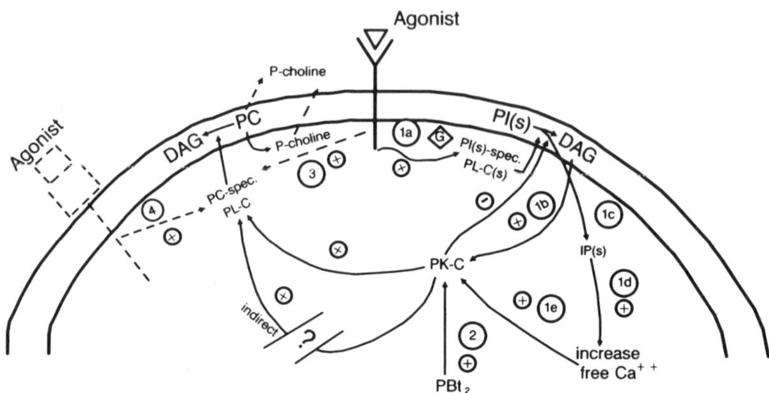

Figure 1. Schematic representation of possible pathways involved in agonist-induced formation of DAG. This model depicts documented and implied routes (mechanisms) involved in the process of agonist (i.e., PBt$_2$, serum, and PDGF) induced generation of DAG. Solid lines depict documented pathways, lightly dashed lines and symbols represent structures and pathways proposed to exist. Step a of pathway 1 represents agonist-induced, receptor-mediated activation of phosphoinositide-specific phospholipase C [PI(s)-spec. PL-C(s)]. Activation of this phospholipase(s) is thought to be coupled through a GTP-binding protein (g) and results in the formation of DAG and the inositol phosphates [IP(s), step 1c]. At least one of the inositol phosphates, InsP$_3$, causes the release of stored calcium from endoplasmic reticulum (step 1d). The resulting increase in free cytosolic calcium combined with the increased level of membrane DAG synergizes to activate protein kinase C (PK-C, steps 1b and 1e). Addition of an active phorbol ester, such as PBt$_2$, directly activates protein kinase C (pathway 2). Protein kinase C stimulates, either directly or indirectly, a phosphatidylcholine-specific phospholipase C (PC-spec. PL-C), resulting in the formation of DAG and PCho. Increased levels of PCho have been detected both intracellularly and extracellularly, but it is not yet known whether the two pools arise in parallel or sequentially (heavily dashed arrows). It is proposed that phosphatidylcholine-specific phospholipase C may be activated by at least two mechanisms, which are independent of protein kinase C. Pathway 3 represents agonist–receptor pairs also capable of activating phospholinositide-specific phospholipase C. Pathway 4 offers the possibility that there may exist agonist–receptor systems preferentially coupled to phosphatidylcholine-specific phospholipase C. From Besterman, *et al.*, (1986).

protein?). Alternatively, certain agonist receptors may act exclusively on this PC-specific enzyme. Regardless of the mechanism that may be involved, an important biological consequence might be to allow production of diacylglycerol (for kinase C activation) without affecting calcium mobilization from intracellular stores. Of course, it is also possible that the diacylglycerol produced, either because of its fatty acid composition or its localization, may act on a totally different protein kinase C. Furthermore, the phosphocholine (or rapidly produced choline) may also be of biological relevance.

Although in the early 1960s Slein and Logan (1965) described the release of alkaline phosphatase from mammalian tissues after incubation with a partially purified preparation of phospholipase C from *Bacillus cereus,* these observations remained virtually dormant until very recently. During the past year or two, these results have been confirmed and extended to several other systems. It is now clear that certain phospholipase C enzymes (e.g., from *B. cereus, Staphylococcus aureus, Clostridium novyi* type A, and *B. thuringiensis)* are very specific for phosphatidylinositol mannosides but not for PI 4-phosphate or PI 4,5-bisphosphate (for reviews see Low *et al.,* 1986; Low, 1987). Furthermore, these enzymes can release from intact cells or membranes a variety of proteins of differing function, in widely varying species and in all tissues (Table I). It is now clear that these proteins, normally localized on the external surfaces of cells, have a covalently attached, complex glycosyl phosphatidylinositol moiety at their carboxyl-terminal end.

Table I. Proteins Anchored to Membranes Covalent Attachment to Glycosyl-Phosphatidylinositol[a]

Alkaline phosphatase
5′-Nucleotidase
Acetylcholinesterase
Alkaline phosphodiesterase I
Variant surface glycoprotein
Thy-1
Trehalase
Decay accelerating factor
p63 Protease
RT-6 antigen
Qa antigen
ThB antigen
T-cell activating protein
N-CAM 120
Heparan sulfate proteoglycan

[a]Adapted from Low *et al.* (1987).

This structure, now believed to be closely related to that illustrated in Fig. 2, appears to anchor those proteins tightly into cell membranes. It is not yet known whether any proteins directed toward the cytoplasmic phase of the membrane may also be anchored in this fashion; this is obviously a more difficult problem to study.

Given the known structural mechanisms for attaching proteins to membranes (e.g., the transmembrane hydrophobic domains of the proteins themselves and covalently attached lipid), what function could be served by such an unusual mechanism? It is notable that these proteins are all of a highly diverse nature and function. The most obvious explanation is that the anchor serves to permit the selective release of these proteins into the extracellular milieu, where their function is perhaps more appropriately exercised. It is also possible that their enzymatic release is a way of freeing the cells of enzymes or proteins that are not needed for the cell's new functions. Such a mechansim would obviously require the presence of very specific PI-glycan phospholipase C enzymes, the activities of which would, in turn, probably be under careful regulatory control.

Indeed, such specific enzymes appear to exist. One such enzyme, *Trypanosoma brucei,* has been purified, and its activity is relatively specific for the variant surface glycoprotein (VSG) linked to the PI-glycan (for references see Low *et al.,* 1986; Low, 1987). A very similar calcium independent enzyme has recently been purified from rat liver (Fox *et al.,* 1987). It is notable that many of these released proteins are normally present in the extracellular medium as well as on cell surfaces, and certain types of physiologic stimuli

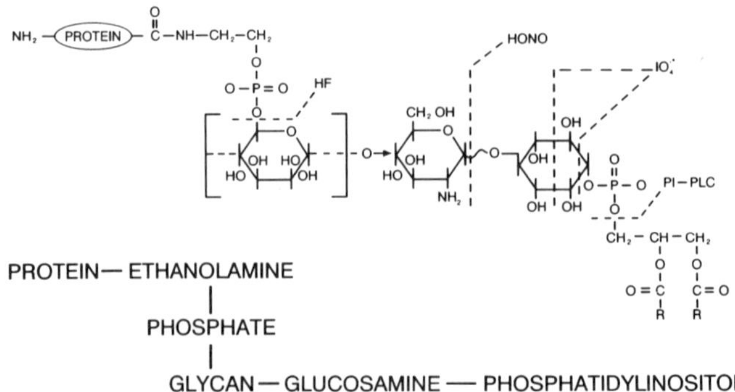

Figure 2. Putative structure of the phosphatidylinositol glycosyl structure attached covalently to various cell surface proteins, serving to anchor them into the hydrophobic domain of membranes. From Low (1987).

release some of these proteins into the medium (reviewed in Low, 1987). Also, the VSG coat is completely and rapidly removed from trypanosomes during antigenic variation or during differentation while in the insect host.

Of considerable interest, but not yet understood, is the mechanism of biosynthesis, glycolipid coupling, and membrane insertion of these proteins, or their orientation within the bilayer. Furthermore, it is not known if the diacylglycerol moiety released upon enzymatic hydrolysis has any biological function, although the fatty acid composition appears to be largely myristate in some case (e.g., *T. brucei)* and saturated in others (Low *et al.*, 1986; Low, 1987). Another potential explanation for this unusual protein anchor is that it endows these surface proteins with greater lateral mobility within the plane of the membrane, since in this way they would not be constrained by interactions with cytoplasmic or cytoskeletal proteins (Low 1987).

A remarkable coincidence appears to be emerging between these PI-glycan protein systems and the possible mechanism of action of insulin. Recent investigations reveal that the treatment of cell membranes with insulin releases into the medium a small molecular weight substance that can mimic some of the actions of insulin, and which chemically resembles a similarly active substance that can be released by PI-specific phospholipase C (from *S. aureus* or *T. brucei)* (Saltiel and Cuatrecasas, 1986; Saltiel *et al.*, 1986). Furthermore, this inositol-containing substance has at least some of the chemical features of the inositol-glycan moiety present in the proteins described above. The substance can activate various insulin-sensitive enzymes, such as pyruvate dehydrogenase and cyclic AMP phosphodiesterase, *in vitro* (Saltiel and Cuatrecasas, 1986; Saltiel *et al.*, 1986). Thus, it is possible that a phosphoinositol-glycan from a membrane precursor, releasing both an active some of the actions of insulin.

The data suggest that the insulin receptor may, perhaps by way of an intermediary N or G protein, activate a phospholipase C that can release a phosphoinositol-glycan from a membrane precursor, releasing both an active mediator and diacylglycerol (Fig. 3). Could other hormones, growth factors, or regulators for which no well-defined chemical messengers are yet known be acting by analogous pathways?

An apparent anomaly in this scheme is the inwardly oriented representation of the insulin-target glycolipid. However, the only reasons for such a proposition is to rationalize the bias (in analogy with cyclic AMP) that an insulin mediator should be released directly into the interior of the cell. There may be good reasons to propose an opposite orientation, based on an analogy with the PI-glycan anchored proteins described above. A general scheme, which could perhaps be applicable to all these systems, is suggested in Fig. 4. This could explain and rationalize a surface localization for the unique

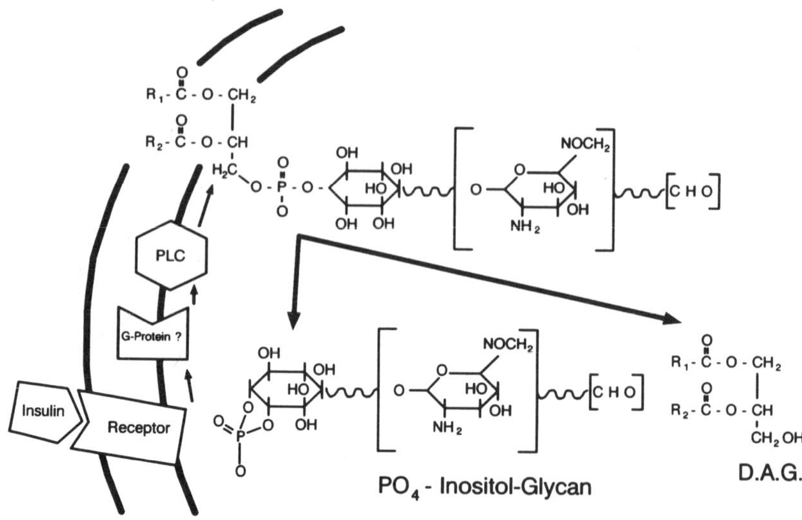

Figure 3. Hypothetical mechanism by which the insulin receptor may activate a phosphatidylinositol-glycan membrane precursor to produce a general chemical mediator (PO_4-inositol-glycan) and diacylglycerol (DAG).

phospholipase C(s) involved. Furthermore, the ease with which purified PI-specific phospholipase C(s) releases the "mediator" into the medium, without obvious injury to cells, is also consistent with this model. In addition, purified PI-specific phospholipase C generates insulin-like responses in intact cells (A.R. Saltiel, unpublished observations).

If such a proposal is to have merit, it is obvious that the substances released must be active extracellularly or on the surface of cells. For proteins and enzymes, this is not difficult to visualize. For an intracellular mediator, however, this would require a specific receptor on the cell surface, acting either directly or through internalization of the ligand–receptor complex. It is notable that the purified phosphoinositol glycan can induce insulin-like effects when added externally to cells (A.R. Saltiel, unpublished observations).

Furthermore, extremely provocative experiments by Ishihara *et al.*, (1987) demonstrate that insulin treatment of liver cells releases heparan sulfate proteoglycan into the medium and that there appears to be a receptor-mediated process by which this protein is internalized and delivered to the nucleus, where it may induce DNA synthesis. This protein has a phosphoinositol glycan group on its C-terminal end. Thus, this precedent is suppor-

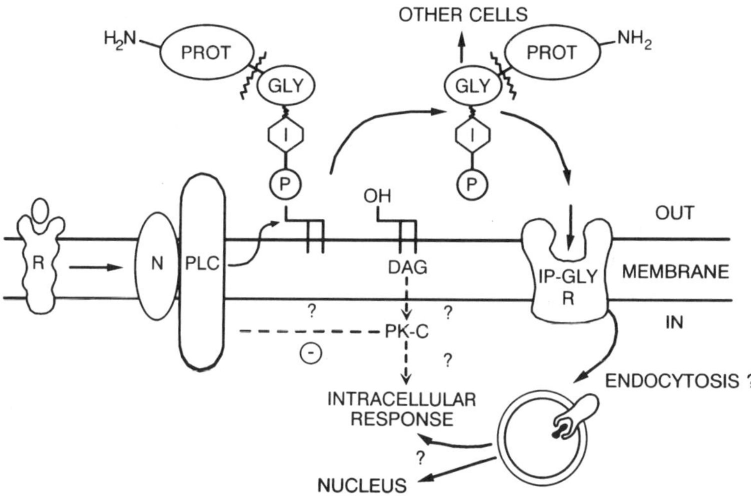

Figure 4. Speculative model proposing receptor-mediated activation of a phosphatidylinositol-glycan phospholipase C that releases phosphoinositol-glycan substances (with or without attached proteins) into the medium. The proteins so released (e.g., enzymes) could act at appropriate extracellular targets or could be internalized for actions inside the cells. The non-protein substances (e.g., insulin mediators) might act on cell surface receptors directly or might also be subject to internalization for direct intracellular actions. R, receptor; N, N- or G-protein; PLC, phospholipase C; IP-GLY R, phosphoinositol-glycan receptor; P, phosphate; I, inositol; GLY, glycan; PROT, protein; DAG, diacylglycerol.

tive of a mechanism for the external generation of signal molecules. Are other specific proteins with a similar anchor also released by insulin? Is it possible that a unique yet general receptor exists (Fig. 4) that might recognize (and internalize) molecules containing a phosphoinositol-glucosamine-glycan moiety?

References

Besterman, J. M., Duronio, V., and Cuatrecasas, P., 1986, Rapid formation of diacylglycerol from phosphatidylcholine: A pathway for generation of a second messenger, *Proc. Natl. Acad. Sci. U.S.A.* **83:**6785–6789.

Fox, J. A., Soliz, N. M., and Saltiel, A. R., 1987, Purification of a phosphatidylinositol-glycan-specific phospholipase-C from liver plasma-membranes—a possible target of insulin action, *Proc. Natl. Acad. Sci. U.S.A.* **84:** 2663–2667.

Ishihara, M., Fedarko, N. S., and Conrad, H. E., 1987, Involvement of phosphatidylinositol and insulin in the coordinate regulation of proteoheparan sulfate metabolism and hepatocyte growth, *J. Bio. Chem.* **262**(10): 4708–4716.

Low, M. G., 1987, Biochemistry of the glycosyl-phosphatidylinositol membrane-protein anchors, *Biochem. J.* **244**(1): 1–13.

Low, M. G., Ferguson, M. A., Futerman, I., and Silman, I., 1986, Covalently attached phosphatidylinositol as a hydrophobic anchor for membrane proteins, *Trends Biochem. Sci.* **11:** 212–215.

Saltiel, A. R., and Cuatrecasas, P., 1986, Insulin stimulates the generation from hepatic plasma membranes of modulators derived from an inositol glycolipid, *Proc. Natl. Acad. Sci. U.S.A.* **83:** 5793–5797.

Saltiel, A. R., Fox, J. A., Sherline, P., and Cuatrecasas, P., 1986, Insulin-stimulated hydrolysis of a novel glycolipid generates modulators of cAMP phosphodiesterase, *Science* **233:** 967–972.

Slein, M. W., and Logan, G. F., 1965, Characterization of the phospholipases of *Bacillus cereus* and their effects on erythrocytes, bone, and kidney cells, *J. Bacteriol.* **90:** 69–81.

Wolf, R. A., and Gross, R. W., 1985, Identification of neutral active phospholipase C which hydrolyzes choline glycerophospholipids and plasmalogen selective phospholipase A_2 in canine myocardium, *J. Biol. Chem.* **260:** 7295–7303.

21

Enzymatic Mechanisms and Inhibition of Phospholipase A_2

From Manoalide to the Lipocortins

EDWARD A. DENNIS, FLORENCE F. DAVIDSON, and RAYMOND A. DEEMS

1. Introduction

The control of eicosanoid production in inflammatory and other cells depends on the release of free arachidonic acid from the *sn*-2 position of membrane phospholipids (summarized in Fig. 1). Phospholipase A_2 is the simplest and most obvious candidate for the responsible enzyme, which makes its regulation, including both its activation and inhibition, crucial determinants of eicosanoid production (Dennis, 1987). Our laboratory has developed the *dual phospholipid model* (Roberts *et al.*, 1977) to explain the mechanism of action of an extracellular phospholipase A_2; our current understanding of the enzyme is summarized elsewhere (Dennis, 1983; Dennis and Plückthun, 1986). Our studies on the enzymes from snake venom and mammalian pancreas provide a paradigm for the phospholipases responsible for arachidonic acid release. In addition, they provide the best sources of readily available, pure, stable phospholipase A_2 for testing potential inhibitors of phospholipase A_2. This chapter discusses three inhibitor systems that we have recently characterized. These include amide ether analogues of phospholipids, which comprise an example of a reversible competitive inhibitor (Davidson *et al.*, 1986), manoalide, which is an example of an irreversible inhibitor (Lom-

EDWARD A. DENNIS, FLORENCE F. DAVIDSON, and RAYMOND A. DEEMS • Department of Chemistry, University of California, San Diego, La Jolla, California 92093.

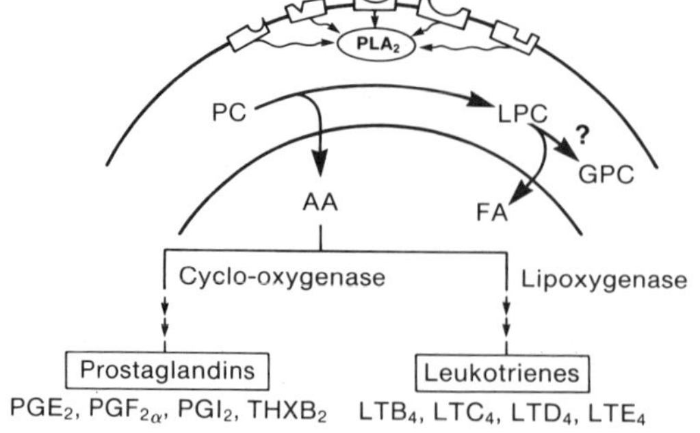

Figure 1. Possible role for a membrane-bound phospholipase A_2 (PLA_2) in the release of free arachidonic acid (AA) from the *sn*-2 position of membrane phospholipids such as phosphatidylcholine (PC). The AA is converted to the various prostaglandins and leukotrienes by a series of reactions. Whether the phospholipase is stimulated directly by a receptor or via one or more mediators is unclear at this time. Reproduced with permission from Dennis (1987).

bardo and Dennis, 1985) that is perhaps mechanism based and for which analogues can be studied (Deems *et al.*, 1987), and the lipocortins or calpactins, where calpactin I is an example of inhibition by a *substrate depletion model (Davidson et al.*, 1987). Meaningful interpretation of such studies can only be accomplished within the framework of the enzyme's kinetics and mechanism of action at the lipid–water interface, which are discussed.

2. Competitive Inhibitors

There are many kinds of inhibitors and potential inhibitors of phospholipase A_2. Their interpretation is complicated by the lipid–water interface. A competitive inhibitor is the simplest sort to analyze conceptually, but as we shall see below, it is not necessarily straightforward.

The ether amides are representative of the general class of substrate analogues (Davidson *et al.*, 1986). Compound (1), shown below, is an analogue of phosphatidylcholine (PC) with a nonhydrolyzable amide at the *sn*-2 position and an ether at the *sn*-1 position. Its synthesis is described elsewhere (Chandrakumar and Hajdu, 1983).

$$CH_3(CH_2)_{16}\overset{\overset{\displaystyle O}{\|}}{C}{-}H\ \ N{-}\overset{\overset{\displaystyle CH_2O(CH_2)_{17}CH_3}{|}}{\underset{\underset{\displaystyle CH_2{-}O{-}\overset{\overset{\displaystyle O}{\|}}{\underset{\underset{\displaystyle O^-}{|}}{P}}{-}OCH_2CH_2{-}\overset{\overset{\displaystyle CH_3}{|}}{\underset{\underset{\displaystyle CH_3^+}{|}}{N}}{-}CH_3}{|}}{C}}{-}H \qquad (1)$$

If one looks at this compound as an inhibitor with a substrate, such as dipalmitoyl phosphatidylcholine, it inhibits in a normal dose response as shown in Fig. 2, with half inhibition at about 40μM. However, it is desirable to have a more quantitative interpretation of the inhibition. With soluble substrates and soluble enzymes, one would normally determine the velocity as a function of substrate. The results would be plotted in a Lineweaver–Burke manner (1/V versus 1/S) at various constant concentrations of inhibitor.

In the case of the phospholipases, however, one does not know what kind of substrate concentration dependency to examine—that is, bulk concentration or surface concentration. In this case, we accepted the fact that the substrate is in a micelle, as well as the inhibitor, and varied substrate concentration in surface concentration units (1/X), keeping the total amount of surface the same in all cases. This kept both the bulk and the surface concentrations of the inhibitor constant for each of the lines shown in Fig. 3. The kinetic data can be interpreted as fitting an intersecting velocity pattern consistent with a competitive inhibitor. This amide ether is an inhibitor that would be expected to be competitive and to bind more tightly than dipalmitoyl PC, because the ether linkage gives approximately a tenfold tighter binding than the acyl group, as can be seen from a comparison with the acyl analogue (Chandrakumar and Hajdu, 1983). The amide is also a little tighter-binding than an ester linkage.

In order to examine the kinetics in greater detail, we studied a thio analogue of the phospholipid that serves as a substrate. The enzyme specifically hydrolyzes the thio ester group at the *sn*-2 position, producing a free thiol as product. When the assay is carried out in the presence of dithiobispyridine, this product undergoes a ready chemical reaction to produce a chromophore; the kinetics of the substrate can be followed spectrophotometrically. If one looks at the inhibition by the ether amide, however, with thio phosphatidylcholine (PC) and thio phosphatidylethanolamine

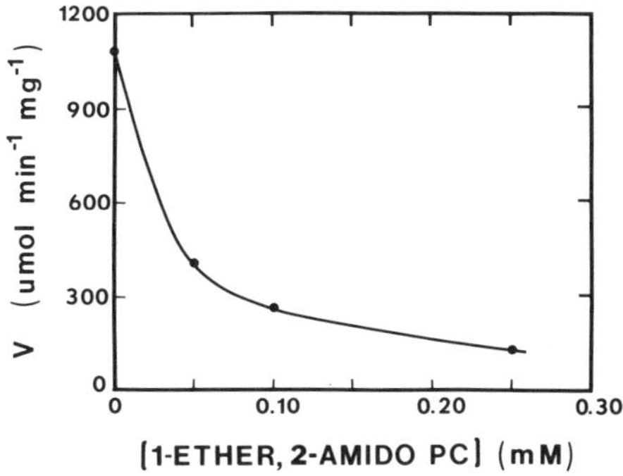

Figure 2. Inhibition of phospholipase A_2 activity toward dipalmitoyl PC–Triton X-100 mixed micelles by 1-stearyl,2-stearoylaminodeoxyphosphatidylcholine (1-ether, 2-amido PC). The standard pH-stat assay was employed with 5 mM dipalmitoyl PC and 40 mM Triton X-100. Reproduced with permission from Davidson *et al.* (1986).

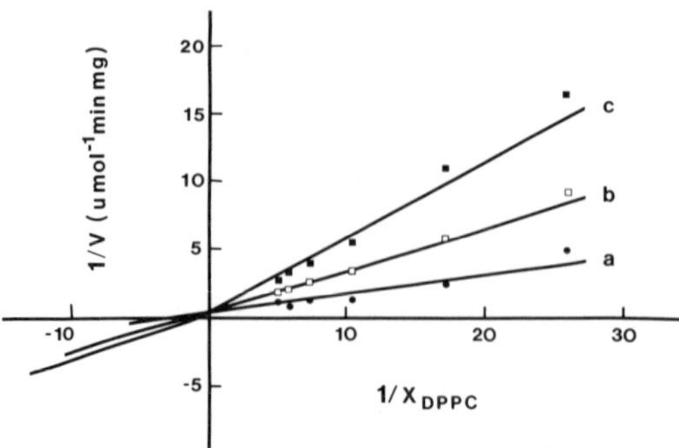

Figure 3. Phospholipase A_2 activity toward dipalmitoyl PC–Triton X-100 mixed micelles as a function of varying surface concentration of substrate and fixed surface concentrations of 1-stearyl, 2-stearoylaminodeoxyphosphatidylcholine: (a) no inhibitor (●), (b) 0.05 mM inhibitor, 0.96×10^{-3} mol fraction (□), (c) 0.10 mM inhibitor, 1.92×10^{-3} mol fraction (■). Variations in dipalmitoyl PC concentration from 2 to 10 mM were compensated by changes in Triton X-100 concentration such that the total concentration of substrate, inhibitor, and Triton X-100 was constant at 52 mM. Reproduced with permission from Davidson *et al.* (1986) with a minor correction in the ordinate.

(PE) substrates, one obtains very different results (see Fig. 4). Using thio PC as substrate, one observes a dose–response curve similar to that with dipalmitoyl PC as substrate, with half-inhibition at about 20μM. The thio PC binds about tenfold tighter than a diacyl PC, but it is also a poorer substrate, so the catalytic efficiency toward this substrate is not very different than toward a normal phospholipid.

Interestingly, if one uses thio PE as a substrate, one observes a very different result; rather than looking like an inhibitor, the ether amide looks like an activator. We believe that this is related to the fact that thio PE is a poor substrate. The presumed inhibitor is actually an activator because it

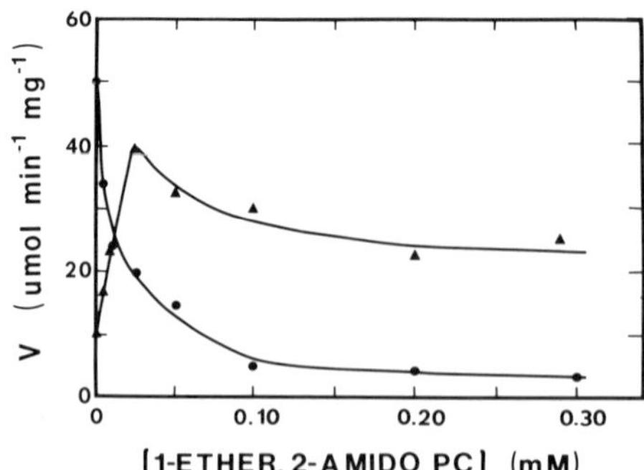

Figure 4. Effect of increasing concentration of 1-stearyl, 2-stearoylaminodeoxyphosphatidylcholine (1-ether, 2-amido PC) on phospholipase A_2 activity toward thio PC–Triton X-100 (●) and thio PE–Triton X-100 (▲) mixed micelles. Substrate concentrations were 0.5 mM and Triton X-100 was 4.0 mM. Reproduced with permission from Davidson *et al.* (1986).

contains a phosphorylcholine polar head group, which has previously been shown to activate PE hydrolysis. Thus, one gets a dramatic activation coupled with a competitive inhibition at higher concentrations giving rise to the observed curve. This demonstrates how critical it is to know the substrate used to analyze a potential inhibitor since one might otherwise interpret the amide ether kinetics as those of an activator rather than an inhibitor.

3. *Irreversible Inhibitors*

Manoalide is an example of an irreversible inhibitor. Manoalide is a natural product from sponge which Jacobs and co-workers (Jacobs *et al.*, 1985) have shown has anti-inflammatory activity *in vivo* . We have shown that it inhibits the extracellular phospholipase A_2 *in vivo* (Lombardo and Dennis, 1985). The structure of manoalide [See (2) below] reveals the presence of two rings, a lactone and a hemiacetal, either of which, under the correct conditions, may be converted to an open form as shown. In the open form, α, β unsaturated aldehydes are present.

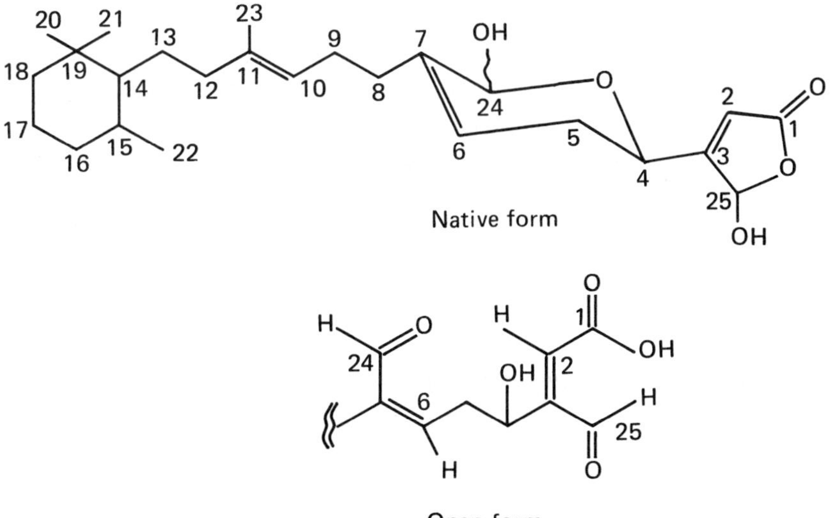

Native form

(2)

Open form

Manoalide inhibits cobra venom phospholipase A_2 in a time-dependent manner; its inhibition is enhanced by Ca^{2+} and protected by lyso PC, which is a reaction product (Lombardo and Dennis, 1985). The inhibition of the enzyme by manoalide is concentration dependent. If a simple dose–response curve is made, half-inhibition occurs at about 2μM, so manoalide appears at first, to be an exceedingly good inhibitor. However, since this inhibition is time dependent, the apparent inhibition constant depends on how the assay is conducted, and the meaning of this kind of inhibition thus becomes quite complicated. If the inhibition assay is carried out for a longer time, one gets a higher percentage of inhibition, making it hard to know the significance of $2 \mu M IC_{50}$.

Inhibition by manoalide is time dependent and irreversible. Analysis of inhibited enzyme shows the addition of carbonyl groups to the enzyme, and amino acid analysis shows loss of some lysine residues (Lombardo and Dennis, 1985). Since there do not appear to be lysines in the catalytic site of the enzyme, we would like to suggest that manoalide irreversibly interacts with lysine residues at or near the surface of the enzyme. This may occur in the activator site rather than the catalytic site or, it may lie elsewhere on the surface, where the presence of the bulky manoalide interferes in some manner with the specificity or activation of the enzyme.

In more recently reported experiments (Deems *et al.*, 1987), we noted that in the presence of enzyme the manoalide itself appears to be converted in a time-dependent manner to some form in which there are more free carbonyls. Thus, in the presence of phospholipase A_2, one gets a time-dependent production of carbonyls that is enhanced by Ca^{2+}, which is required by the enzyme and also by lyso PC. Lyso PC serves as an activator lipid when one has a non-phosphorylcholine-containing substrate. This suggests that the enzyme may somehow be involved in a conversion of the manoalide and raises the possibility that it is a mechanism-based or suicide type of inhibitor (Deems *et al.*, 1987).

To explore this possibility in more detail, we examined what happens to manoalide as a function of pH. Initially, our studies on manoalide employed ^{1}H-NMR to follow changes of the carbonyl group. When the pH of manoalide was raised from neutral pH to 9, new peaks appeared in the 7 –10-ppm region in the NMR spectrum. The 10-ppm peak is indicative of an aldehyde; the 7-ppm peak is indicative of the proton on the β position of the double bond. Because two aldehyde peaks were being produced and there were two double bond peaks at this pH, we thought at first that both the hemiacetal and the lactone ring were being opened. Upon further consideration however, we (Lombardo and Dennis, 1985) suggested that only the lactone ring was being opened, but that some isomerization was also occurring, producing another form of manoalide that represented the second pair of peaks. To explore that further, we have now studied an analogue of manoalide, 3(*cis, cis*-7, 10)-hexadecadienyl-4-hydroxy-2-butenolide (HDHB) as shown in structure (3) below (Deems *et al.*, 1987).

This compound contains only the lactone ring of manoalide and some hydrophobicity. The ^{1}H-NMR spectrum for the proton on the ring that be-

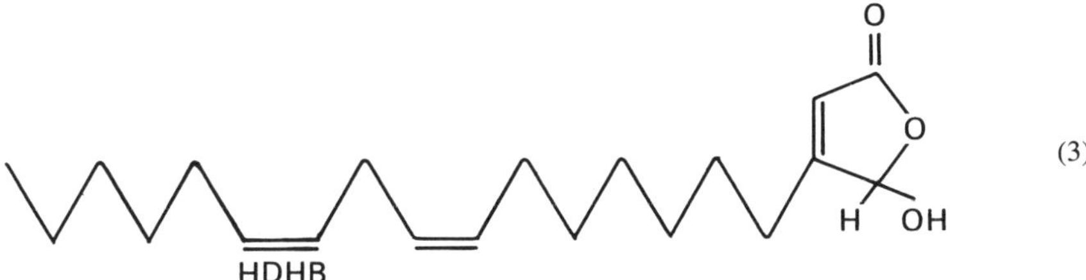

(3)

comes the aldehyde is at 6.1 ppm and the proton on the β position of the double bond is at 5.9 ppm, as shown in Fig. 5 (panel A). Interestingly, when we examine the NMR spectrum as a function of pH, we see that these two peaks move downfield as one raises the pH, until at about pH 9.5 two new distinct peaks exist at 6.8 and 10.2 ppm. This corresponds to the open form of HDHB. When we then lowered the pH, the effect on the spectrum was completely reversible to reproduce the closed form. At neutral pH, there is a reversible opening and closing of the lactone ring to produce the *cis*-HDHB

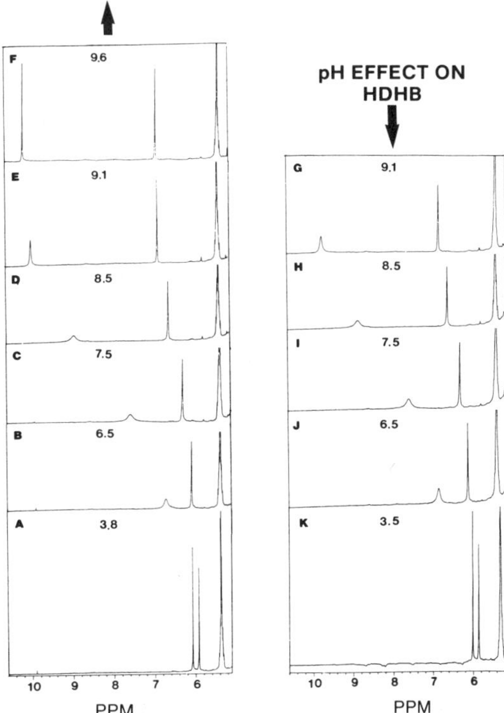

Figure 5. The effect of pH on the ^{1}H-NMR spectrum of HDHB between 5 and 10.5 ppm. HDHB was dissolved in 20% D_2O–80% deuteromethanol to a final concentration of 43 mM. The pH was measured by pH paper and adjusted by addition of NaOD or DCl. The initial pH of 3.8 was raised in steps to pH 9.6 (panels A–F) and then lowered to pH 3.5 (panels G–K). Reproduced with permission from Deems *et al.* (1987).

product. When this *cis*-open form at pH 9.5 is taken up to higher pH, there are further changes in the NMR spectrum, until by pH 13 two new peaks are formed that are chemically shifted from those at pH 9.5 (data not shown). When we take the spectrum at high pH and lower it back down, there are no further changes. It does not go back to the original pH 9.5 spectrum or to that at neutral pH. We believe that at high pH there is a base-catalyzed isomerization of the double bond, which converts the *cis*-HDHB form to the *trans*-HDHB form, which is more stable (Deems *et al.*, 1987). When the pH is lowered after equilibrium is reached, this form is trapped.

HDHB itself is an inhibitor of phospholipase A_2, but it does not appear to irreversibly inhibit the enzyme and the nature of the inhibition is complicated, as shown in Fig. 6; one cannot describe it as either purely competitive or purely noncompetitive. The *trans*-HDHB is also inhibitory and it is several-fold more potent than the *cis*-HDHB; again, it is not simply competitive or noncompetitive.

In summary, we believe that HDHB undergoes a reversible ring opening to the *cis*-HDHB form at neutral pH, and at high pH a further isomerization occurs, converting it to the *trans* form as shown in Fig. 7. These same

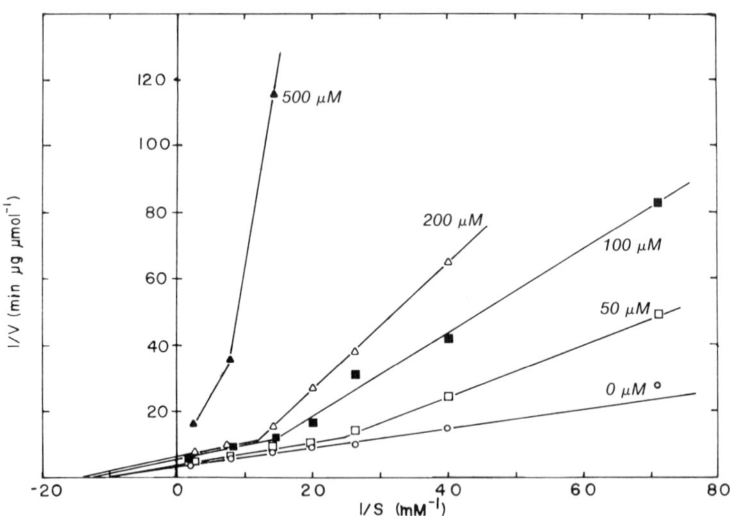

Figure 6. Double reciprocal plot of the inhibition of cobra venom phospholipase A_2 by HDHB. The assays were conducted using thio PC as substrate. All lines, except for 500 μM inhibitor, are linear least-squares fits. The inhibitor concentrations were (○) none, (□) 50 μM, (■) 100 μM, (Δ) 200 μM, and (▲) 500 μM Reproduced with permission from Deems *et al.* (1987).

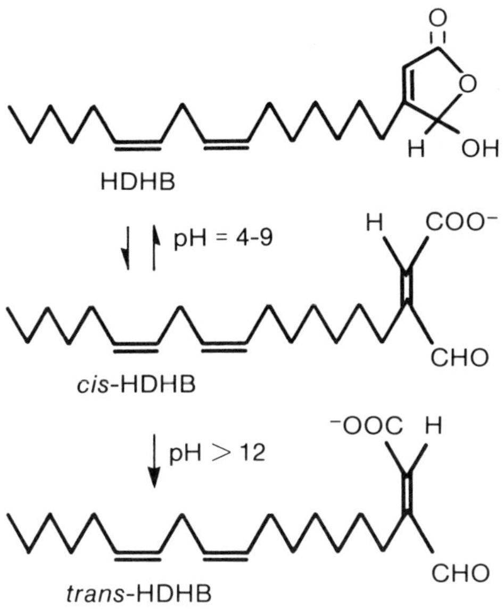

Figure 7. Proposed conversions of 3 (*cis,cis*-7,10)-hexadecadienyl-4-hydroxy-2-butenolide (HDHB) to *cis*-HDHB and *trans*-HDHB.

conversions also occur in manoalide, but at a lower pH. We are currently investigating just how the enzyme might participate in these events and are trying to determine whether the enzyme itself catalyzes either the ring opening or the isomerization, thereby qualifying manoalide as a suicide substrate or a mechanism-based inhibitor.

4. *Lipocortins*

The lipocortins have been defined as comprising a family of proteins that are induced by steroids and are inhibitors of the phospholipases, specifically phospholipase A_2. It has been suggested (Flower *et al.*, 1984) that the lipocortins are produced in inflammatory cells in response to cortisone and that they act in their anti-inflammatory role as inhibitors of phospholipase A_2. Independent of the lipocortin literature, there exists a group of abundant cytoskeletal proteins known as the calpactins. They have the properties of binding calcium, phospholipid, and actin, hence the name *cal* for calcium, *p* for phospholipid, and *actin* for actin-binding proteins. In the last few years, it has been discovered that the calpactins are substrates of the tyrosine kinases, thereby linking them to oncogenes. Recently, it was discovered from the sequence of cloned calpactins that they are homologous to the lipocortins. Specifically, the protein known as calpactin I (Kristensen *et al.*, 1986) is approximately 50% homologous to the cloned human lipocortin (Huang *et al.*, 1986). Other calpactins have subsequently been isolated, and calpactin II has been found by sequence homology to be the same as lipocortin (now called lipocortin I) (Glenney, 1986). We have studied inhibition of phospholipase A_2 by a number of calpactins isolated from bovine lung, including calpactin I, which is also known as p36 and is the same as lipocortin II. It exists as a tetramer with two 38-kD subunits and two 11-kD subunits or it can be prepared in a monomeric form that is 38 kD. We have also studied calpactin II, which is known as p35. It is actually the EGF receptor substrate and is believed to be the same as lipocortin.

When we initially examined the calpactins in our normal assays with the extracellular phospholipase A_2, we could not find any inhibition (Davidson *et al.*, 1987). We therefore studied the inhibition in an *E. coli* assay, which is very similar to those used in the literature for lipocortin. *E. coli* are grown in $[^3H]$oleic acid; it is a very simple assay for phospholipase A_2. One can stop the reaction, centrifuge the *E. coli* in the assay mix, and count the radioactive fatty acid in the supernatant. Indeed, when we used this assay for $[^3H]$ release in the supernatant, we did find inhibition of phospholipase A_2. Both calpactin I and II gave a similar inhibition, and a related 73-kD protein also inhibits (Davidson *et al.*, 1987).

Because we were concerned that one could be misled in the inhibition by some artifact, we looked in detail at the labeled lipids using a two-dimensional TLC system (Davidson *et al.*, 1987). The major phospholipids in *E. coli* are PE and PG. There is no PC or sphingomyelin, which are major components in mammalian cell membranes. We found that phospholipase A_2 produces free fatty acid at the expense of PE and that calpactin does indeed inhibit that production. In fact, the *E. coli* assay does adequately measure phospholipase A_2 activity and the inhibition thereof.

Binding studies suggest that calpactin I will bind to phosphatidylserine (PS). We examined the activity of the pancreatic phospholipase A_2 in the *E. coli* assay (Davidson *et al.*, 1987). Under experimental conditions where 100% activity was obtained in the absence of calpactin but the was inhibited in the presence of calpactin. Interestingly, if we added PS in the form of vesicles during this assay, the inhibition by calpactin was overcome; that is, the activity increased with the addition of a very small amount of PS. On the other hand, if PS was added in the absence of calpactin, there was a decrease in the apparent activity of the enzyme. This is really a *surface dilution effect*, because the unlabeled PS serves as a substrate and is not measured in an assay that only detects labeled *E. coli* products.

Because of the potential complications of the *E. coli* assay, we wanted to examine the phospholipase activity toward a pure lipid system. We used extracted lipid from *E. coli* in lipid vesicles as substrate. Calpactin I gives a dose-dependent inhibition with this pure lipid (shown in Fig. 8). However, we discovered an important aspect of this inhibition by examining the lipid vesicle assay in detail (Davidson *et al.*, 1987); inhibition depends strongly on the substrate concentration at a set calpactin and phospholipase concentration (shown in Fig. 9). In the presence of calpactin, there is a dramatic inhibition, approaching 100%, at low substrate concentrations. However, this falls off considerably as the substrate concentration is increased. Roughly, there is no inhibition above 8 μM substrate. We would like to suggest that this is because calpactin is actually binding to the substrate phospholipid in the presence of Ca^{2+}, thereby blocking it as a substrate to the enzyme. If one increases the substrate concentration high enough so not all of it is blocked, the enzyme can act on available substrate. Thus, the calpactin is actually serving to deplete the substrate for the enzyme.

There is also a substrate-dependent inhibition with the whole *E. coli* assay, as shown in Fig. 10. The point at which the inhibition is abolished is higher with *E. coli* than with vesicles because the total phospholipid in *E. coli* is not available to the inhibitor on the outside of the membrane. We do not know how much is accessible, but it is certainly much less than in the isolated vesicles.

In summary, we would like to suggest a *substrate depletion model* (Davidson *et al.*, 1987) to explain the inhibition by calpactin I of the phospholipase A_2:

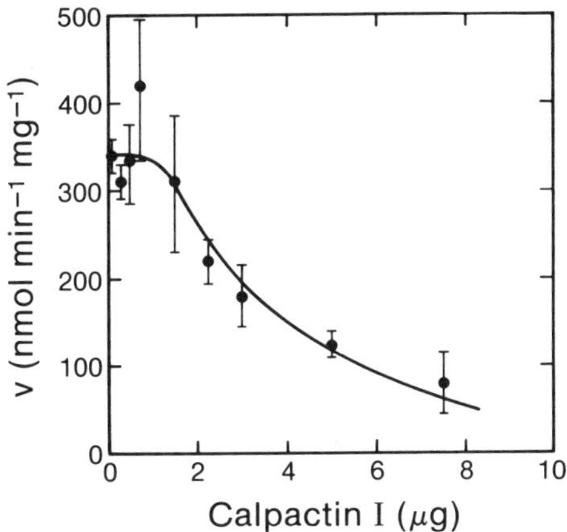

Figure 8. Dose response of calpactin I inhibition of phospholipase A_2 hydrolysis of phospholipid vesicles. Substrate phospholipid, which had been extracted from *E. coli* and reconstituted into sonicated vesicles, was kept constant at 5 μM. The reaction was started by the addition of phospholipase A_2 (5.1×10^{-10} M). Note that half-maximal inhibition was achieved with approximately 3 μg calpactin I, or 5.3×10^{-7} M. Reproduced with permission from Davidson *et al.* (1987).

$$\begin{array}{ccc} & K_s & k_p \\ E + S & \rightleftarrows & ES \rightarrow E + P \\ + & & \\ I & & \\ K_0 \upharpoonleft\!\downharpoonright & & \\ SI & & \end{array}$$

While the enzyme and phospholipid normally form an ES complex, the inhibitor in this case competes with the enzyme for binding to the substrate. There is a very tight binding of inhibitor to substrate to form an SI complex, which is not available to the enzyme; this is a complex kinetic situation. A dose-dependent inhibition of the enzyme is observed only at very low substrate concentrations where all the substrate can be bound or blocked by the inhibitor. These results are not consistent with the lipocortins acting as specific noncompetitive or competitive inhibitors of the enzyme.

While substrate depletion can explain the inhibition of phospholipase A_2 observed for the lipocortin examined so far, it is of course possible that other lipocortins will be found that are specific inhibitors of phospholipase A_2. However, such a potential inhibitor would have to be shown by the stringent tests described herein to be a specific inhibitor, rather than acting through the substrate depletion model.

Acknowledgments. We wish to acknowledge financial assistance from the National Institutes of Health (GM-20,501) and the National Science Foundation (DMB 85-18684), which supported this work in our laboratory.

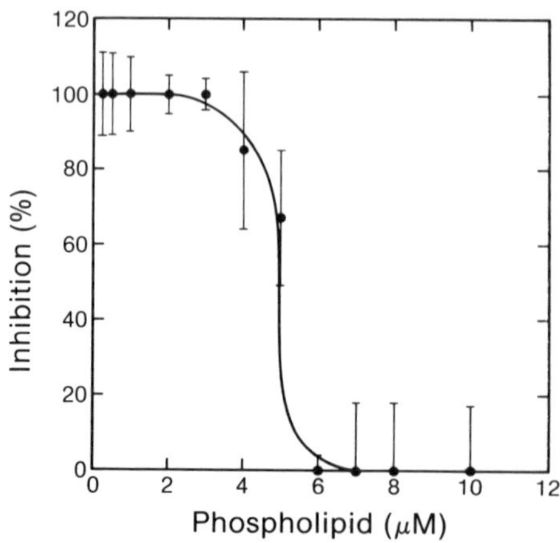

Figure 9. Dependence of calpactin I inhibition on the concentration of extracted *E. coli* phospholipid vesicles. The concentrations of phospholipase A_2 (5.1×10^{-10} M) and calpactin I (1.1×10^{-7} M) were kept constant as the concentration of *E. coli*-derived phospholipid vesicles was increased. Percent inhibition is shown as a function of the phospholipid substrate concentration. Reproduced with permission from Davidson *et al.* (1987).

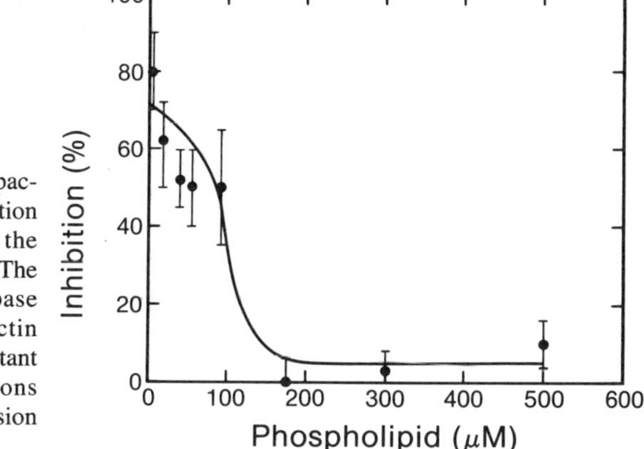

Figure 10. Dependence of calpactin I inhibition on the concentration of phospholipid substrate in the form of whole *E. coli* cells. The concentrations of phospholipase A_2 (8.9×10^{-8} M) and calpactin (3.5×10^{-7} M) were kept constant over all substrate concentrations used. Reproduced with permission from Davidson *et al.* (1987).

References

Chandrakumar, N. S., and Hajdu, J., 1983, Syntheses of enzyme-inhibitory phospholipid analogues. Stereospecific synthesis of 2-amidophosphatidylcholines and related derivatives, *J. Org. Chem.* **8:**1197–1202.

Davidson, F. F., Hajdu, J., and Dennis, E. A., 1986, 1-Stearyl, 2-stearoylaminodeoxy phosphatidylcholine, a potent reversible inhibitor of phospholipase A_2, *Biochem. Biophys. Res. Commun.* **137:**587–592.

Davidson, F. F., Dennis, E. A., Powell, M., and Glenney, J., 1987, Inhibition of phospholipase A_2 by lipocortins: An effect of binding to phospholipids, *J. Biol. Chem.* **262:**1698–1705.

Deems, R. A., Lombardo, D., Morgan, B. P., Mihelich, E. D., and Dennis, E. A., 1987, The inhibition of phospholipase A_2 by manoalide and manoalide analogues, *Biochim. Biophys. Acta* **917:**258–263.

Dennis, E. A., 1987, Phospholipase A_2 mechanism: inhibition and role in arachidonic acid release, *Drug Dev. Res.* **10:**205–220.

Dennis, E. A., 1987, Phospholipase A_2 mechanism, inhibition and role in arachidonic acid release, *Drug Dev. Res.* (in press).

Dennis, E. A., and Plückthun, A., 1986, Mechanism of interaction of phospholipase A_2 with phospholipid substrates and activators, in: *Enzymes of Lipid Metabolism* (L. Freysz and S. Gatt, eds.), pp. 121–132, Plenum Press, New York.

Flower, R. J., Wood, J. N., and Parente, L., 1984, Macrocortin and the mechanism of action of glucocorticoids, *Adv. Inflammation Res.* **7:**61–69.

Glenney, J. R., 1986, Two related but distinct forms of the M_r 36,000 tyrosine kinase substrate (calpactin) that interact with phospholipid and actin in a Ca^{2+}-dependent manner, *Proc. Natl. Acad. Sci. U.S.A.* **83:**4258–4262.

Huang, K. S., Wallner, B. P., Mattaliano, R. J., Tizard, R., Burne, C., Frey, A., Hession, C., McGray, P., Sinclair, L. K., Chow, E. P., Browning, J. L., Ramachandran, K. L., Tang, J., Smart, J. E., and Pepinsky, R. B., 1986, Two human 35 kD inhibitors of phospholipase A_2 are related to substrates of pp60$^{v\text{-}src}$ and of the epidermal growth factor receptor/kinase, *Cell* **46:**191–199.

Jacobs, R. S., Culver, P., Langdon, R., and O'Brien, T., 1985, Some pharmacological observations on marine natural products, *Tetrahedron* **41:**981–984.

Kristensen, T., Saris, C. J. M., Hunter, T., Hicks, L. J., Noonan, D. J., Glenney, J. R., and Tack, B. F., 1986, Primary structure of bovine calpactin I heavy chain (p36), a major cellular substrate for retroviral protein–tyrosine kinases: Homology with the human phospholipase A_2 inhibitor lipocortin, *Biochemistry* **25:**4497–4503.

Lombardo, D., and Dennis, E. A., 1985, Cobra venom phospholipase A_2 inhibition by manoalide: A novel type of phospholipase inhibitor, *J. Biol. Chem.* **260:**7234–7240.

Roberts M. F., Deems, R. A., and Dennis, E. A., 1977, Dual role of interfacial phospholipid in phospholipase A_2 catalysis, *Proc. Natl. Acad. Sci. U.S.A.* **74:**1950–1954.

22

Molecular Mechanism of Regulation of Cellular Phospholipases

FUSAO HIRATA

1. Introduction

Many, if not all, cells release arachidonic acid when they are stimulated by hormones, neurotransmitters, antigens, and drugs. Arachidonic acid is mostly present as an esterified form in phosphatidylinositol and other alkenyl- and/or acylglycerophospholipids. The key enzymes involved in the arachidonate release are thus proposed to be phospholipase A_2 and phospholipase C (Van Den Bosh, 1980; Irvine, 1982). Phospholipase A_2 directly mobilizes arachidonic acid in the 2-position of the glycerol moiety of phospholipids (phosphatidylcholine, phosphatidylethanolamine, phosphatidylinositol, and alkenylglycerophospholipids), whereas phospholipase C converts phospholipids to diacylglycerol, a compound that releases arachidonic acid by the subsequent action of diacylglycerol or monoacylglycerol lipase. These phospholipases are inactive in unstimulated conditions (resting state), although they are surrounded by their lipid substrates. This raises important questions concerning the mechanism by which receptor stimulation can activate cellular phospholipases. This chapter discusses the molecular mechanisms that may possibly regulate cellular phospholipases.

2. Inhibition of Cellular Phospholipase(s) by Glucocorticoids

2.1. Inhibition of Arachidonic Acid Release

Glucocorticoids inhibit prostaglandin formation in a variety of cells. This action of glucocorticoids can be blocked by treatment with cyclohex-

FUSAO HIRATA • Department of Environmental Health Sciences, School of Hygiene and Public Health, The Johns Hopkins University, Baltimore, Maryland 21205–2179.

imide, actinomycin D, and other inhibitors of mRNA and protein synthesis (Danon and Assouline, 1978). Glucocorticoids themselves have no effect on the activities of enzymes involved in the *in vitro* synthesis of prostaglandins from arachidonic acid. Thus, glucocorticoids require intact cells for their action. Furthermore, the suppressive effects of glucocorticoids on prostaglandin formation depend on intracellular receptors for these steroids (Russo-Marie *et al.*, 1979). These observations suggest that the action of glucocorticoids on suppressing prostaglandin formation requires the synthesis of a new protein(s). Since inhibitors of protein synthesis can also diminish the anti-inflammatory action of glucocorticoids (Tsurufuji *et al.*, 1979), this interpretation fits well with the hypothetical mechanism proposed by Thompson and Lippman (1974): glucocorticoids first bind to cytosolic receptors, and the glucocorticoid–receptor complex is then transferred into nuclei, where this complex binds to the specific sites (enhancer region) of the genomic DNA and regulates its expression.

Glucocorticoids do not inhibit the transformation of arachidonic acid to prostaglandins when arachidonic acid is given to intact cells (Gryglewski *et al.*, 1975). Since the suppressive effect of glucocorticoids on prostaglandin formation is associated with inhibition of arachidonic acid release, Hong and Levine (1976) suggested that steroids induce (or reduce) the production of some proteins that regulate the cellular activities of phospholipases. Flower and Blackwell (1979) first suggested that a peptidic factor(s) that inhibits prostaglandin formation is produced in perfused rat lungs treated with glucocorticoids. Later, they partially purified this factor and reported that the glucocorticold-induced peptide has an apparent molecular weight of 16,000 and that it is synthesized mainly in macrophages (macrocortin) (Blackwell *et al.*, 1980). Since this protein inhibits the hydrolysis of exogenously added 2-[^{14}C-oleyl]-PtdCho (phosphatidylcholine) by the lung, they proposed that macrocortin is a phospholipase A_2 inhibitory protein.

We had worked independently on the mechanism of arachidonic acid release from rabbit neutrophils stimulated by *f*Met-Leu-Phe, a synthetic chemoattractant, and found that pretreating rabbit neutrophils with glucocorticoids reduced the amounts of arachidonic acid released (Hirata *et al.*, 1980). This glucocorticoid effect was also blocked by cycloheximide and actinomycin D. Under our experimental conditions, [1-^{14}C]-arachidonate was incorporated mainly into the phosphatidylcholine fraction and released from this fraction. We therefore proposed that a phospholipase A_2 inhibitory protein(s) was induced in the glucocorticoid-treated neutrophils. Using [^{14}C]-lysin, the synthesis of a protein with an apparent molecular weight of 40,000 was found to be induced in these cells after overnight incubation with glucocorticoids. Preparations of the partially purified protein accompanied an activity to inhibit porcine pancreas phospholipase A_2 *in vitro*. Hence, we labeled this protein "lipomodulin." Our group collaborated with Flower's to produce a comparative analysis of lipomodulin and macrocortin, which showed that these proteins are immunologically and biochemically related, if not identical (Hirata *et al.*, 1982). Several other groups also demonstrated the existence of such phospholipase inhibitory proteins in kidney interstitial cells (Cloix *et al.*, 1983), thymocytes (Gupta *et al.*, 1984), and lymphocytes (Iwata *et al.*, 1984). They designated them "renocortins," "phospholipase inhibitory protein (PLIP)," and "glycosylation inhibitory factor (GIF)," respectively. The collaboration between the groups of Russo-Marie and Flower, and between our group and Ishizaka's showed that these proteins were also immunologically related, if not identical. Thus, we proposed a common name: lipocortin(s) (DiRosa *et al.*, 1984).

2.2. *Properties of Lipocortins*

We first purified lipocortin from media in which rabbit neutrophils were cultured in the presence of fluocinolone acetonide, a synthetic glucocorticoid (Hirata *et al.*, 1980). Using a highly purified material, we described the general properties of lipocortins. The specific activity of lipocortin is about 5000 units, with 1 unit defined as the capacity to inhibit 0.14 μ of porcine pancreas phospholipase A_2 by 50% (Hirata *et al.*, 1982). Lipocortin has an apparent molecular weight of $38{,}000 \pm 2000$ and has a pI around 8.0 in the absence of Ca^{2+}, while Ca^{2+} shifts the pI of the protein to 9.0 (Hirata, 1983). This protein can bind Ca^{2+} (approximately 3.2 moles/mole protein) with an apparent K_m of 3 μM as measured by the equilibrium microdialysis method (F. Hirata, unpublished observations). The protein has approximately 10% of polysaccharides; thus, it binds to concanavilin A and protein A–Sepharose. Blackwell (1983) suggested that glycosylation might play a crucial role in the secretion of lipocortin from the glucocorticoid-treated cells, since tunicamycin inhibits both the glycosylation and secretion of lipocortin. Our recent results suggest that this protein is also acylated with palmitic acid and/or myristic acid (Hirata *et al.*, 1987). In addition, dexamethasone and other glucocorticoids can nonenzymatically make an adduct, probably with a lysine residue(s) of lipocortin as reported in the case of crystallin, a lens protein (Bucala *et al.*, 1985). Furthermore, this protein can be phosphorylated by various kinases, including protein kinase C and tyrosine kinase. The phosphorylated form of lipocortin loses its inhibitory activity toward procine pancreas phospholipase A_2 (Hirata, 1981).

Recently, the Biogen group has purified two phospholipase A_2 inhibitory proteins, which have properties similar to those described above (Huang *et al.*, 1986; Wallner *et al.*, 1986). Two clones of cDNA coding for these proteins have been isolated and the amino acid sequences of their primary structures have been predicted. These structures were found to have high homology to proteins structures previously reported to bind Ca^{2+}, and to substrate proteins for EGF receptor kinase or *src*-kinase (p35 and p36). They have sequences homologous to the tyrosine phosphorylating sites of the middle T antigen and the *src*-gene product, respectively. In addition, these teins share the consensus sequence of Lys-Gly-X-Gly-Thr-Asp-Glu-X-X-Leu-Ile-X-Ile-Ala-X-Arg (Kretsinger and Creutz, 1986). This sequence is now considered to play an important role in binding Ca^{2+} and acidic phospholipids such as PtdSer and PtdIns (Geisow *et al.*, 1986). We reported earlier that when lipocortin is added to cell culture media it can inhibit the release of arachidonic acid from intact cells stimulated by various ligands (Hirata *et al.*, 1981). In addition to its anti-inflammatory action, lipocortin can mimic other actions of glucocorticoids, such as their immunosuppressive effect (by inducing suppressor T cells), the arrest of cellular growth (and promotion of differentiation), and an antiedematic effect (Hirata, 1984). However, no biological action of recombinant lipocortins has been reported so far. Therefore, it remains unclear whether phospholipase inhibitory proteins encoded in cDNAs are identical to the natural lipocortin we previously reported.

2.3. *Inhibition of Phospholipases by Lipocortins*

The activity of lipocortin is generally assayed as an *in vitro* inhibitory activity toward porcine pancreas phospholipase A_2 with synthetic 1-stearyl-2-arachidonyl-PtdCho as substrate (Hirata *et al.*, 1982). Pancreatic phospholipase A_2 (soluble enzyme) has kinetic properties different from those of membrane phospholipase A_2 with regard to K_a for Ca^{2+} and substrate specificity (Jimeno-Abendano and Zahler, 1979): it requires higher concentrations of Ca^{2+} for maximal activity and hydrolyzes phosphatidylserine (PtdSer) more efficiently than phosphatidylcholine (PtdCho). Since lipocortin and related proteins such as p35 and p36 bind to PtdSer vesicles (Geisow *et al.*, 1986), we chose PtdCho as substrate. Under these conditions, the maximal inhibition was always obtained when the stoichiometric molar amount of lipocortin was given to porcine pancreas phospholipase A_2 in the reaction mixture. This suggests that 1 mole of lipocortin binds to 1 mole of phospholipase. Furthermore, this protein inhibited the lipase noncompetitively with respect to Ca^{2+}, an activator, and phospholipid substrates (Hirata, 1981). Lipocortin increased the K_a of lipase for Ca^{2+} but did not change the K_m of lipase for phospholipid substrates such as PtdCho and PtdEtn. These results suggest that lipocortin purified from the culture media of glucocorticoid-treated neutrophils inhibits phospholipase A_2 by forming a one-to-one complex with phospholipase A_2, rather than by binding phospholipid substrates. Lipocortin can bind to the phospholipase-A_2-coupled agarose (Hirata *et al.*, 1982; Parente and Flower, 1985). On the other hand, p35 and p36 bind phosphatidylserine (PtdSer) and other acidic phospholipids such as PtdIns and phosphatidic acid only in the presence of Ca^{2+}, a cation that is necessary for maximal activation of phospholipase A_2 (Creutz *et al.*, 1987). These proteins do not bind to PtdCho and PtdEtn, the better substrates for cellular phospholipase A_2, even in the presence of Ca^{2+}. From these observations, we proposed that lipocortin binds near Ca^{2+} binding sites(s) (active sites) of phospholipase A_2.

Recently, p35 and p36 (calpactins), which are proteins related to lipocortin, were reported to inhibit phospholipase A_2 by binding phospholipid substrates, rather than by making the complex with phospholipase A_2 (Davidson *et al.*, 1987). C-reactive protein and serum apolipoproteins, which bind PtdCho and other phospholipids, can also inhibit phospholipase A_2 *in vitro* (Miwa *et al.*, 1984; Vigo, 1985). These proteins inhibit porcine pancreas phospholipase A_2 at low substrate concentrations but stimulate it at high substrate concentrations (F. Hirata, unpublished observations). The activity of phospholipase A_2 is easily influenced by the physical states of phospholipid vesicles (Van Den Bosh, 1980). Alteration of the physical states of phospholipid vesicles by adding different classes of lipids, detergents, and hydrophobic proteins might enhance or inhibit phospholipase activities.

The C-MT peptide, which has 80% homology to the consensus sequence of endonexin and related calcium binding proteins, can inhibit various phospholipases, especially phospholipase C. In examining this peptide's inhibition of porcine pancreas phospholipase A_2, we found it competes with Ca^{2+} at the low-affinity site of Ca^{2+} in phospholipase A_2 (Notsu *et al.*, 1985). Porcine pancreas phospholipase A_2 appears to require 2 moles of Ca^{2+}; one at the low-affinity site (millimolar) and the other at the high-affinity site (micromolar) (Slotboom *et al.*, 1978). The Ca^{2+} at the low-affinity site appears to play a crucial role in bridging a phospholipid substrate molecule to the active (Ca^{2+} binding) site of phospholipase. Taken together, we concluded that the C-MT peptide binds to phospholipase A_2, interfering with the bridge formation between phospholipid substrates and Ca^{2+} in the active site of phospholipases, and that the physical state of phospholipid substrates influences the inhibitory activity of the C-MT peptide toward phospholipase A_2 (Notsu *et al.*, 1985). This mechanism of the inhibition of porcine pancreas phospholipase A_2 by lipocortin is essentially similar to that of the C-MT peptide. Davidson *et al.* (1987) used *E. coli* phospholipids, which are a mixture of phosphatidylserine (PtdSer) and phosphatidylethanolamine (PtdEtn). PtdSer is known to segregate in the presence of Ca^{2+}, and thus a mixture of PtdSer and PtdEtn forms a hexagonal structure rather than bilayer membranes (Tilcock *et al.*, 1982). Calpactins (p35 and p36) can bind to these vesicles in the presence of Ca^{2+} (Davidson *et al.*, 1987). Hence, it is feasible to assume that the kinetics of phospholipase A_2 inhibition by lipocortin and calpactins might differ, depending on phospholipid substrates and concentrations of Ca^{2+} in the reaction mixtures. Alternatively, calpactins might be proteins distinct from lipocortin.

Our initial results suggested that partially purified lipocortin can inhibit various phospholipases (Hirata, 1981). Our recent observations indicate that several isoforms of phospholipase inhibitory proteins may be present in the preparations. By using affinity chromatographies with the porcine pancreas phospholipase A_2-, *B. cerus* phospholipase C-, and human mononuclear phosphatidylinositol phospholipase C-coupled agarose, these isoforms of phospholipase inhibitory proteins were separated (Hirata, 1987). Each phospholipase inhibitory protein was specific for phospholipase A_2, phospholipase C, and PtdIns–phospholipase C, respectively, and the affinity of these inhibitory proteins for phospholipases belonging to other classes was two to three orders of magnitude lower (F. Hirata *et al.*, unpublished observations). Therefore, we proposed the names of members of the family of lipocortins as listed in Table I. However, this does not necessarily indicate that the synthesis of all lipocortins is induced by glucocorticoids. Recently, DeGeorge and associates (1987) reported that glucocorticoid treatment causes the inhibition of arachidonic acid release, but not of phosphatidylinositol turnover, in C-6 astrocytoma cells stimulated by muscarinic ligands. Bomalaski *et al.* (1986) also suggested that aspirin and related drugs might transiently induce the synthesis of phospholipase C inhibitory protein(s).

3. *Evidence for the Complex of Lipocortins with Phospholipases Inside Cells*

Phospholipases in membranes and even in cytosolic fractions are surrounded by phospholipid substrates, a major component of these membranes,

Table I. Various Phospholipase Inhibitory Proteins in Cells

Phospholipases	Lipocortins	Molecular weight
A_1	alpha	28,000[a]
A_2	beta	38,000 ± 2,000[b]
		11,000 ± 2,000[a]
C	gamma	36,000 ± 2,000[b]
		14,000 ± 2,000[a]
D	delta	—
Phosphatidylinositol phospholipase C	epsilon	30,000 ± 2,000[b]

[a]Bacterial origins
[b]Mammalian origins

which provide fluid matrix to membrane proteins. In the resting state, the release of arachidonic acid and the turnover of phospholipids is small, but these increase rapidly after various hormones, neurotransmitters, and other ligands bind to receptors on the cell surfaces. Phospholipase activities in homogenates and isolated membranes can be activated by increasing concentrations of Ca^{2+} (Van Den Bosh, 1980), adding GTP analogues such as GTPS (Litosh and Fain, 1986), adding ATP (plus kinases) (Moskowitz *et al.*, 1982; Wightman *et al.*, 1982), or adding trypsin and other proteases (Hirata *et al.*, 1980). All factors involved in these *in vitro* observations are closely associated with those in receptor functions. Since lipocortin can bind to phospholipase-A_2-coupled agarose (Hirata *et al.*, 1982; Parente and Flower, 1985), it is likely that lipocortin forms a complex with various members of phospholipases to make them inactive. If this is the case, it raises important questions about the mechanism by which phospholipases are activated in a receptor-mediated fashion.

3.1. Calcium and Phospholipases

Most, if not all, partially and highly purified phospholipases require Ca^{2+} for maximal activity. Most phospholipase A_2 requires a millimolar order of concentration of Ca^{2+} for maximal activity. Since intracellular concentrations of free Ca^{2+} range from 1 nM to 1 μM under physiological conditions, it was generally believed that phospholipase A_2 had no physiological role in the release of arachidonic acid. However, recent studies indicate that several isoforms of phospholipase A_2 exist inside cells, some with low K_a values (micromolar) for Ca^{2+} (Moskowitz *et al.*, 1982; Loeb and Gross, 1986). Even at low concentrations of Ca^{2+}, purified phospholipase A_2 can hydrolyze phospholipids at significant rates. It is likely that a protein(s) present in crude preparations might modify the affinity of these phospholipases for Ca^{2+}. Calmodulin, a calcium-binding protein, was once proposed to activate phospholipase A_2 (Wong and Cheung, 1979). Lipocortin can also change the affinity of phospholipase A_2 for Ca^{2+} *in vitro*. In the presence of lipocortin, phospholipase A_2 requires higher concentrations of Ca^{2+} for maximal activation; higher concentrations of Ca^{2+} can partially overcome the inhibitory effect of lipocortin against phospholipase A_2 (Hirata, 1983).

The turnover of phosphatidylinositol, particularly polyphosphoinositol, has been associated with the functions of many receptors regarding signal transduction, since these lipids serve as storage forms for a variety of messenger molecules released in response to specific extracellular signals (Berridge and Irvine, 1985). The first step in the metabolism of polyphosph-PtdIns is catalyzed by phospholipase C. This enzyme, which is distinct from those for other classes of phospholipids, catalyzes the degradation of PtdIns. A single enzyme from ram seminal vesicles utilizes all three phosphoinositide phospholipids. They produce diacylglycerol and inositolphosphates. Membrane-bound phospholipase C can be activated at low concentrations of Ca^{2+} (micromolar order), whereas the cytosolic enzyme, which represents the majority of detectable enzyme activity, requires high concentrations (millimolar order) of Ca^{2+} for maximal activity (Hofmann and Majerus, 1982). Although the role played by Ca^{2+} in the regulation of phospholipase A_2 activity might be applicable to the case of phospholipase C, further investigations are necessary to determine the role played by Ca^{2+} in regulating cellular phospholipases under physiological conditions.

3.2. Digestion of Lipocortins by Proteases

Thrombin is a serine protease that plays a central role in initiating procoagulant events (Glenn and Cunningham, 1979). This role includes the conversion of fibrinogen to fibrin, the stimulation of platelet aggregation and release reactions of serotonin and other biogenic amines, and the mobilization of arachidonate pathway metabolites from endothelial cells. In addition, this protein initiates the proliferation and prostaglandin production of fibroblasts. This mitogenic action of thrombin is closely associated with its protease activity. Similarly, trypsin and other proteases can stimulate the release of arachidonic acid (Henson *et al.*, 1976, 1980; Pickett, 1976), whereas various protease inhibitors can block it (Chang *et al.*, 1980). In addition, many ligands have been reported to initiate the activation of certain types of proteases after binding to their receptors, even though ligands themselves have no protease activity (Duque *et al.*, 1983; Peppin and Weiss, 1986). These include the early phase of cellular activation in neutrophils, macrophages, and endocrine cells. Various protease inhibitors, including tosyl-L-phenylalanylmethylketone (TPCK), are reported to inhibit the secretion of bioactive amines, O_2^- production, and prostaglandin formation in intact cells (Henson *et al.*, 1976; Duque *et al.*, 1983). In platelets, TPCK is reported to inhibit PtdIns–phospholipase C, but not phospholipase A_2 (Wallenga *et al.*, 1980). The treatment of fibroblasts and other cells with proteases including trypsin results in the release of prostaglandins and the initiation of proliferation. Interleukin-1 (IL-1), a growth factor produced in macrophages to act on T lymphocytes, is reported to release proteases from chondrocytes and to activate phospholipase A_2 in these cells (Chang *et al.*, 1986). All these findings implicate the involvement of certain types of proteases in the early stages of signal transduction and/or arachidonate release through various receptors.

When rabbit neutrophils were treated with pronase, the spontaneous release of arachidonic acid increased; this release was dependent on the time of the incubation and the dose of pronase employed (Hirata *et al.*, 1980). Under these conditions, the viability of neutrophils as measured by the trypan blue exclusion test did not change. Pronase treatment enhance the arachidonate release even from the glucocorticoid-treated cells, whose cellular level of lipocortin was increased. Thus, we assumed that pronase digests lipocortin in the plasma membranes. In keeping with this interpretation, the binding of monoclonal antilipocortin antibody decreases in the pronase-treated cells. The levels of lipocortin, as measured by antibody binding to intact neutrophils, were inversely related to the rates of arachidonic acid release. Essentially similar results were observed with human peripheral mononuclear cells (F. Hirata, unpublished observations).

3.3. Phosphorylation of Lipocortins

Many receptors, such as those for growth factors and insulin, have been reported to contain tyrosin kinase activities as an intrinsic enzyme activity (Hunter, 1986). The stimulation of these receptors results in increased phosphorylation of endogenous substrate proteins and their own receptors. This stimulation generally accompanies the enhanced release of arachidonic acid and/or the turnover of phospholipids, especially PtdIns, in a variety of cells. It is believed that the stimulation of many, if not all, receptors with various hormones, neurotransmitters, and drugs activates PtdIns–phospholipase C to form diacylglycerol and polyphosphoinositides (Berridge and Irvine, 1985). Diacylglycerol subsequently activates protein kinase C and inositol-polyphosphates mobilize intracellular calcium from the endoplasmic reticulum and microsomal fractions to the cytosols. Phorbol esters, activators of protein kinase C, have often been used as an activator of phospholipases. Hence, it is conceivable that the phosphorylation of lipocortin by these kinases might be a key event leading to the activation of cellular phospholipases and the subsequent release of arachidonic acid.

When purified lipocortin was incubated with the catalytic subunit of cyclic AMP-dependent kinase, a time- and dose-dependent incorporation of ^{32}P from ATP[$\gamma^{32}P$] to lipocortin was observed (Hirata, 1981). The inhibitory activity of lipocortin toward pancreas phospholipase A_2 was inversely related to the radioactivity of ^{32}P incorporated into lipocortin protein molecules. To confirm that this reaction was reversible, the phosphorylated lipocortin was treated with alkaline phosphatase. Such treatment caused the release of ^{32}P from lipocortin, and the antiphospholipase activity could be restored (Hirata *et al.*, 1982). These observations do not necessarily prove that the protein kinase responsible for the phosphorylation of lipocortin in intact cells is cyclic AMP dependent, although this protein kinase is suggested to be involved in brain synaptic vesicles (Moskowitz *et al.*, 1982). Partially purified tyrosine kinase and protein kinase C were able to phosphorylate lipocortin *in vitro* (Hirata *et al.*, 1984). When murine thymocytes were stimulated by various mitogens such as Con A, phorbol esters, and Ca^{2+}

ionophore, ^{32}P was primarily incorporated into a tyrosine residue rather than into serine or threonine residues (Hirata *et al.*, 1984). Certain amounts of phosphorylated lipocortin were also found in the media in which these thymocytes were stimulated. When the phosphorylated lipocortin was separated by Sephadex G-200 column chromatography, its inhibitory activity on phospholipase A_2 could be detected only after treatment with alkaline phosphatase, which cleaved ^{32}P from the lipocortin fraction (F. Hirata, unpublished observation).

Phospholipase inhibitory proteins, whose cDNA clones have recently been isolated by the Biogen group, have been identified as p36 and p35 substrate proteins for EGF receptors and *sre*-kinase, respectively (Brugge, 1986). These proteins contain sequences similar to those of the tyrosine phosphorylating sites of polyoma virus middle T antigen and of *sre*-gene products of sarcoma virus (Huang *et al.*, 1986; Wallner *et al.*, 1986). ^{26}Tyr and ^{23}Ser are potential candidates for the sites of phosphorylation by tyrosine kinase and protein kinase C, respectively (Brugge, 1986). Although it is still not certain that these proteins are identical to lipocortin, these results, taken together, support the idea that the phosphorylation–dephosphorylation process of lipocortin in intact cells might be a key mechanism in the release of arachidonic acid by the action of phospholipase A_2. Glucocorticoid treatment inhibits the release of arachidonic acid from carbamylcholine-stimulated C-6 glioma cells, but the turnover of PtdIns is not inhibited in these cells (DeGeorge *et al.*, 1987). Since glucocorticoids induce the synthesis of lipocortin specific for phospholipase A_2, but not for PtdIns–phospholipase C, these observations indicate that the release of arachidonic acid is mainly due to the activation of phospholipase A_2 through the phosphorylation of lipocortin by protein kinase C, whose activity is regulated by diacylglycerol, a product of PtdIns–phospholipase C.

3.4. N(G) Proteins and Lipocortins, GTP-Binding Proteins

Many receptors have been demonstrated to couple to N(GTP or nucleotide binding) proteins (Rodbell, 1985). Several kinds of N(G) proteins have been isolated; $N_s(G_s)$ is for activation of adenylate cyclase, $N_i(G_i)$ is for inhibition of adenylate cyclase, and $N_t(G_t)$ is for activation of cyclic GMP phosphodiesterase. These proteins are composed of heterotrimer peptides, α, β, and γ subunits. The α subunit is responsible for the binding as well as the hydrolysis of GTP and has distinct properties when each class of N(G) proteins is isolated. The function of N_i (G_i) and N_s (G_s) is modulated by the two bacterial toxins—pertussis and cholera. These toxins catalyze the incorporation of the adenosylribose (ADR) group from nicotinmaide adenine dinucleotide (NAD) into the α subunits of N_i (G_i), N_t (G_t), and N_s (G_s) and then block or facilitate the dissociation of the α subunit from the β-γ subunit complex. The dissociation is promoted by various activated receptors and is a prerequisite for functions of receptors. GTP and its nonhydrolyzable analogues such as GTPγS and GMPNP facilitate the dissociation and prevent the reassociation of the α subunit. Thus, these compounds modulate various receptor functions. Since the pertussis toxin treatment of cells can block the activation of phospholipase A_2 and PtdIns–phospholipase C (Burch *et al.*, 1986; Litosh and Fain, 1986), it is likely that the activation of phospholipases is a consequence of the interaction between N(G) proteins and phospholipases or between N(G) proteins and lipocortin.

In neutrophils permeated by detergents, GTP and its analogues stimulate the Ca^{2+}-dependent secretion of lysosomal enzymes (Gomperts *et al.*, 1986). These nucleotides also alter the affinity of chemotactic receptors for *f*Met-Leu-Phe and other synthetic chemoattractants. In addition, chemotaxis, O_2^- production, and lysozyme release of neutrophils can be blocked by treatment with pertussis toxin. Hence, N_i (G_i) and/or its related N(G) protein is proposed to couple *f*Met-Leu-Phe receptors. In addition, GTP and its analogues enhance the hydrolysis of polyphospho-PtdIns in the permeated neutrophils as well as in isolated plasma membranes (Smith *et al.*, 1986), suggesting that N(G) protein is responsible for the activation of PtdIns–phospholipase C in neutrophils. Receptor activation generally causes the reversible dissociation of the α subunit from the β-γ subunit complex; then the α-subunit–GTP complex activates the respective enzymes such as adenylate cyclase and cGMP phosphodiesterase (Rodbell, 1985). In the latter system, the α_T–GTP complex is proposed to bind to an inhibitory peptide of cGMP diesterase (Fung *et al.*, 1981). Therefore, we hypothesized that the α-subunit-GTP complex might stimulate phospholipases either by dissociating lipocortin from phospholipases or by activating phospholipases directly.

To examine our working hypothesis, we isolated plasma membranes from neutrophils and tested the effect of β- and ϵ lipocortin on phospholipase A_2 and PtdIns–phospholipase C. The maximal activation of phospholipase A_2 and PtdIns–phospholipase C with exogenously added 2-[1-^{14}C]arachidonyl-PtdCho and endogenously labeled Ptd-[^{3}H]inositol could be detected at 10^{-9} M and 3×10^{-8} M *f*Met-Leu-Phe, respectively (F. Hirata, unpublished observations). The β- and γ-lipocortins inhibited the activation of these phospholipases in the presence of both GTP and *f*Met-Leu-Phe. Therefore, we believe that the activity of both phospholipase A_2 and PtdIns–phospholipase C in neutrophil plasma membranes is regulated by lipocortin.

In order to demonstrate that the α-subunit–GTP complex interacts with lipocortin, we immunoprecipitated lipocortin from reaction mixtures in which neutrophil plasma membranes were incubated in the presence of GTP[γ-^{32}P] or GTP-γ[^{35}S]. Radioactivity was detected in the immunoprecipitates by the antilipocortin antibody only after stimulation with 10^{-9} M *f*Met-Leu-Phe. These results suggest that the α subunit of G protein is released from membranes after stimulation and that it binds to lipocortin. Essentially similar results were obtained with transducin prepared from bovine retina and purified lipocortin. To rule out the possibility that the β-γ subunit complex directly activates the phospholipases, the β-γ-subunit complex of transducin was isolated by chromatography with Blue Sephadex. However, it had virtually no effects on either phospholipase activity. Although lipocortin has been reported to be copurified with transducin (Huang *et al.*, 1986), a highly purified preparation of transducin did not inhibit porcine pancreas phospholipase A_2 *in vitro*, but rather contained phospholipase A_2 activity. Nevertheless, the inhibition of porcine pancreas phospholipase A_2 by lipocortin was reversed by the addition of transducin together with GTP and its nonhydrolyzable analogues, but not with GDP and GMP. These results indicate that the α-subunit–GTP complex reverses the anti phospholipase activity of lipocortin. Since lipocortin is a basic protein that binds to phosphatidylserine and other acidic phospholipids, the possibility cannot be ruled out that lipocortin might bind to the nucleotides bound to the α subunit. The amino acid sequence of p35, a lipocortin-related protein, has homology to that of c-K-ras protein, one of the G proteins considered to interact with PtdIns–phospholipase C (Munn and Mues, 1986). However, lipocortin failed to bind GTP. Details of how lipocortin interacts with G proteins will require further investigation.

4. Summary

Many cells release arachidonic acid when they are stimulated with various hormones, neurotransmitters, and drugs. The release of arachidonic acid is a consequence of events in which cellular phospholipases are activated after cell surface receptors are stimulated. Although these cellular phospholipases, regardless of their localization in the cytosol or membrane fractions, are surrounded by considerable amounts of phospholipid substrates, they remain inactive while in the resting (unstimulated) state. It is suggested that these lipases form a complex with inhibitory proteins called lipocortin.

The increase of Ca^{2+} concentration, phosphorylation by protein kinases, proteolysis by certain types of proteases, and GTP and its nonhydrolyzable analogues of nucleotides have been reported to activate phospholipases in intact cells, detergent-treated cells, cell-free homogenates, and plasma membranes and are proposed to be closely associated with signal transduction through various receptors. In this chapter we proposed a hypothetical mechanism by which cellular phospholipases are activated; lipocortin is dissociated from phospholipases, either by high concentrations of Ca^{2+}, by interaction with the GTP–α subunit of N(G) proteins, by phosphorylation with various kinases, or by digestion with proteases.

References

Berridge, M. J., and Irvine, R. F., 1985, Inositol trisphosphate, a novel second messenger in cellular signal transduction *Nature* **312:**315–321.

Blackwell, G. J., 1983, Specificity and inhibition of glucocorticoid-induced macrocortin secretion from rat peritoneal macrophages, *Br. J. Pharmacol.* **79:**578–594.

Blackwell, G. J., Carnuccio, R., DiRosa, M., Flower, R. J., Parente, L., and Persico, P., 1980, Macrocortin: A polypeptide causing the antiphospholipase effect of glucocorticoid, *Nature* **287:**147–149.

Bomalaski, J. S., Hirata, F., and Clark, M. A., 1986, Aspirin inhibits phospholipase C, *Biochem. Biophys. Res. Commun.* **139:**115–121.

Brugge, J. S., 1986, The p35/p36 substrates of protein-tyrosine kinases as inhibitors of phospholipase A_2, *Cell* **46:**149–150.

Bucala, R., Manabe, S., Urban, R. C., and Cerami, A., 1985, Nonenzymatic modifications of lens crystallins by prednisolone induces sulfhydryl oxidation and aggregate formation: *In vitro* and *in vivo* studies, *Exp. Eye Res.* **41:**353–363.

Burch, R. M., Luini, A., and Axelrod, J., 1986, Phospholipase A_2 and phospholipase C are activated by distinct GTP-binding proteins in response to α-adrenergic stimulation in FRTL 5 thyroid cells, *Proc. Natl. Acad. Sci. U.S.A.* **83:**7201–7205.

Chang, J., Wigley, F., and Newcombe, D., 1980, Neutral protease activation of peritoneal macrophage prostaglandin synthesis, *Proc. Natl. Acad. Sci. U.S.A.* **77:**4736–4740.

Chang, J., Gilman, S. C., and Lewis, A. J., 1986, Interleukin 1 activates phospholipase A_2 in rabbit chondrocytes: A possible signal for interleukin 1 action, *J. Immunol.* **136:**1283–1287.

Cloix, J. F., Colard, O., Rothhut, B., and Russo-Marie, F., 1983, Characterization and partial purification of "renocortins": Two polypeptides formed in renal cells causing the anti-phospholipase-like action of glucocorticoids, *Br. J. Pharmacol.* **79:**313–321.

Creutz, C. E., Zaks, W. J., Hamman, H. C., Crane, S., Martin, W. H., Gould, K. L., Oddie, K. M., and Parsons, S. J., 1987, Identification of chromaffin granule-binding proteins. Relationship of the chromobinding to calelectrin, synhibin, and the tyrosine kinase substrates p35 and p36, *J. Biol. Chem.* **262:**1860–1868.

Danon, A., and Assouline, G., 1978, Inhibition of prostaglandin biosynthesis by corticosteroids requires RNA and protein synthesis, *Nature* **273:**552–554.

Davidson, F. F., Dennis, E. A., Powell, M., and Glenny, J. R., Jr., 1987, Inhibition of phospholipase A_2 by "lipocortins" and calpactins, *J. Biol. Chem.* **262:**1698–1705.

DeGeorge, J. J., Ousley, A. H., McCarthy, K. D., Morell, P., and Lapetina, E. G., 1987, Gulcocorticoids inhibit the liberation of arachidonate but not the rapid production of phospholipase C dependent metabolites in acetylcoline-stimulated C62B glioma cells, *J. Biol. Chem.* (in press).

DiRosa, M., Flower, R. J., Hirata, F., Parente, L., and Russo-Marie, F., 1984, Letter to the editor; "Anti-phospholipase proteins, nonenclature announcement," *Prostaglandins* **28:**441–442.

Duque, R. E., Phan, S. H., Sulavik, M. C., and Ward, P. A., 1983, Inhibition by tosyl-L-phenylalanyl chloromethyl ketone of membrane potential changes in rat neutrophils. Correlation with the inhibition of biologial activity, *J. Biol. Chem.* **258:**8123–8128.

Flower, R. J., and Blackwell, C. J., 1979, Anti-inflammatory steroids induce biosynthesis of a phospholipase A_2 inhibitor which prevents prostaglandin generation, *Nature* **278:**456–459.

Fung, B. K.-K., Hurley, J. B., and Stryer, L., 1981, Flow of information in the light-triggered cyclic nucleotide cascade of vision, *Proc. Natl. Acad. Sci. U.S.A.* **78:**152–156.

Geisow, M. J., Fritsche, U., Hexham, J. M., Dash, B., and Johnson, T., 1986. A consensus amino-acid sequence repeat in torpedo and mammalian Ca^{2+}-dependent membrane-binding proteins, *Nature* **320:**636–638.

Glenn, K. C., and Cunningham, D. D., 1979, Thrombin-stimulated cell division involves proteolysis of its cell surface reeptor, *Nature* **278:**711–714.

Gomperts, B. D., Barrowman, M. M., and Cockcroft, S., 1986, Dual role for guanine nucleotides in stimulus–secretion coupling, *Fed. Proc.* **45:**2156–2161.

Gryglewski, R. J., Panczenko, B., Karbut, R., Grodzinska, L., and Ocetkiewicz, A., 1975, Corticosteroids inhibit prostaglandin release for perfused mesenteric blood vessels of rabbit and from perfused lungs of sensitized guinea pig, *Prostaglandins* **10:**343–355.

Gupta, C., Katsumata, M., Goldman, A. S., Herold, R., and Piddington, R., 1984, Glucocorticoid-induced phospholipase A_2-inhibitory proteins mediate glucocorticoid teratogenicity *in vitro, Proc. Natl. Acad. Sci. U.S.A.* **81:**1140–1143.

Henson, P. M., Gould, D., and Becker, E. L., 1976, Activation of stimulus-specific serine esterases (proteases) in the initiation of platelet secretion. I. Demonstration with organophosphorus inhibitors, *J. Exp. Med.* **144:**1657–1673.

Hirata, F., 1981, The regulation of lipomodulin, a phospholipase inhibitory protein in rabbit neutrophils by phosphorylation, *J. Biol. Chem.* **256:**7730–7733.

Hirata, F., 1983, A possible mediator of the action of glucocorticoids, *Adv. Prostaglandins Thromboxane Leukotriene Res.* **11:**73–78.

Hirata, F., 1984, Role of lipomodulin: A phospholipase inhibitory protein in immunoregulation, *Adv. Inflammation Res.* **7:**71–78.

Hirata, F., 1987, Role of lipocortins in cellular function as a second messenger of glucocorticoids, in: *Antiinflammatory Steroid Action: Basic and Clinical Aspects* (L. M. Lichtenstein, H. Claman, A. Oronsky, and R. Schleimer, eds.), Academic Press, Orlando (in press).

Hirata, F., Schiffmann, E., Venkatasubramanian, K., Salomon, D., and Axelrod, J., 1980, A phospholipase A_2 inhibitory protein in rabbit neutrophils induced by glucocorticoids, *Proc. Natl. Acad. Sci. U.S.A.* **77:**2533–2536.

Hirata, F., Del Carmine, R., Nelson, C. A., Axelrod, J., Schiffmann, E., Warabi, A., De Blas, A. L., Nirenberg, M., Manganiello, V., Vaughn, M., Kumagai, S., Green, I., Decker, J. L., and Steinberg, A. D., 1981, Presence of autoantibody for phospholipase inhibitory protein, lipomodulin, in patients with rheumatic diseases, *Proc. Natl. Acad. Sci. U.S.A.* **78:**3190–3194.

Hirata, F., Notsu, Y., Iwata, M., Parente, L., DiRosa, M., and Flower, R. J., 1982, Identification of several species of phospholipase inhibitory protein(s) by radioimmunoassay for lipomodulin, *Biochem. Biophys. Res. Commun.* **109:**223–230.

Hirata, F., Matsuda, K., Notsu, Y., Hattori, T., and Del Calmine, R., 1984, Phosphorylation at tyrosine residue of lipomodulin in mitogen-stimulated murine thymocytes. *Proc. Natl. Acad. Sci. U.S.A.* **81:**4717–4721.

Hirata, F., Stracke, M. L., and Schiffmann, E., 1987, Regulation of prostaglandin formation by glucocorticoid and their second messenger, lipocortins, in: *Proceedings of the 7th International Congress on Hormonal Steroids* (in press).

Hofmann, S. L., and Majerus, P. W., 1982, Identification and properties of two distinct phosphatidylinositol-specific phospholipase C enzymes from sheep seminal vesicular glands, *J. Biol. Chem.* **257:**6461–6469.

Hong, S.-C. L., and Levine, L., 1976, Inhibition of arachidonic acid release from cells as the biochemical action of anti-inflammatory corticosteroids, *Proc. Natl. Acad. Sci. U.S.A.* **73:**1730–1734.

Huang, K.-S., Wallner, B. P., Mattaliano, R. J., Tizard, R., Burne, C., Frey, A., Hession, C., McGray, P., Sinclair, L. K., Chow, E. P., Browning, J. L., Ramachandran, K. L., Tang, J., Smart, J. E., and Pepinsky, R. B., 1986, Two human 35 kD inhibitors of phospholipase A_2 are related to substrates of $pp60^{v-src}$ and of the epidermal growth factor/

kinase, *Cell* **46:**191–199.

Hunter, T., 1986, Cell growth control mechanisms, *Nature* **322:**14–16.

Irvine, R. F., 1982, How is the level of free arachidonic acid controlled in mammalian cells?, *Biochem. J.* **204:**3–32.

Iwata, M., Huff, T. F., and Ishizaka, K., 1984, Modulation of the biologic activities of IgE-binding factor. V. The role of glycosylation-enhancing factor and glycosylation-inhibiting factor in determining the nature of IgE-binding factor, *J. Immunol.* **132:**1286–1293.

Jimeno-Abendano, J., and Zahler, P., 1979, Purified phospholipase A_2 from sheep erythrocyte membrane: Preferential hydrolysis according to polar groups and 2-acyl chains, *Biochim. Biophys. Acta* **573:**266–275.

Kretsinger, R. H., and Creutz, C. E., 1986, Consensus in exocytosis, *Nature* **320:**573.

Litosch, I., and Fain, J. N., 1986, Regulation of phosphoinositide breakdown by guanine nucleotides, *Life Sci.* **39:**187–194.

Loeb, L. A., and Gross, R. W., 1986, Identification and purification of sheep platelet phospholipase A_2 isoforms. Activation by physiologic concentrations of calcium ion, *J. Biol. Chem.* **261:**10467–10470.

Miwa, M.,Kubota, I., Ichihashi, T., Motojima, H., and Matsumoto, M., 1984, Studies on phospholipase A inhibitor in blood plasma. I. Purification and characterization of phospholipase A inhibitor in bovine plasma, *J. Biochem. Tokyo* **96:**761–773.

Moskowitz, N., Schook, W., and Puszkin, S., 1982, Interaction of brain synaptic vesicles induced by endogenous Ca^{2+}-dependent phospholipase A_2, *Science* **216:**305–307.

Munn, T. Z., and Mues, G. I., 1986, Human lipocortin similar to *ras* gene products, *Nature* **322:**314–315.

Notsu, Y., Namiuchi, S., Hattori, T., Matsuda, K., and Hirata, F., 1985, Inhibition of phospholipase by Met-Leu-Phe-Ile-Leu-Ile-Lye-Arg-Ser-Arg-His-Phe, C terminus of middle-sized tumor antigen, *Arch. Biochem. Biophys.* **236:**195–204.

Parente, L., and Flower, R. J., 1985, Hydrocortisone and ''macrocortin'' inhibit the zymosan-induced release of lyso-PAF from rat peritoneal leucocytes, *Life Sci.* **36:**1225–1231.

Peppin, G. J., and Weiss, S. J., 1986, Activation of the endogenous metalloproteinase, gelatinase, by triggered human neutrophils, *Proc. Natl. Acad. Sci. U.S.A.* **83:**4322–4326.

Pickett, W. C., 1976, Trypsin-induced phospholipase activity in human platelets, *Biochem. J.* **160:**405–408.

Rodbell, M., 1985, Programmable messengers: A new theory of hormone action, *TIBS* **10:**461–464.

Russo-Marie, F., Paing, M., and Duval, D., 1979, Involvement of glucocorticoid receptors in steroid-induced inhibition of prostaglandin secretion, *J. Biol. Chem.* **254:**8498–8504.

Slotboom, A. J., Jansen, E. H. J. M., Vlijm, H., Pattus, F., Soares de Araujo, P., and De Haas, G. H., 1978. Ca^{2+} binding to porcine pancreatic phospholipase A_2 and its function in enzyme–lipid interaction *Biochemistry* **17:**4593–4600.

Smith, C. D., Cox, C. C. and Snyderman, R., 1986, Receptor-coupled activation of phosphoinositide-specific phospholipase C by an N protein, *Science* **232:**97–100.

Thompson, E. B., and Lippman, M. E., 1974, Progress in endocrinology and metabolism. Mechanism of action of glucocorticoids, *Metabolism* **32:**159–202.

Tilcock, C. P. S., Bally, M. B., Farren, S. B., and Cullis, P. R., 1982, Influence of cholesterol on the structural preferences of dioleoylphosphatidylethanolamine–dioleoylphosphatidylcholine systems: A phosphorus-31 and deuterium nuclear magnetic resonance study *Biochemistry* **21:**4596–4601.

Tsurufuji, S., Sugio, K., and Takemasa, T., 1979, The role of glucocorticoid receptor and gene expression in the anti-inflamatory action of dexamethasone, *Nature* **280:**408–410.

Van Den Bosh, H., 1980, Intracellular phospholipase A, *Biochim. Biophys. Acta* **604:**191–246.

Vigo, C., 1985, Effect of C-reactive protein on platelet-activating factor-induced platelet aggregation and membrane stabilization, *J. Biol. Chem.* **260:**3418–3422.

Wallenga, R., Vanderhoek, J. Y., and Feinstein, M. B., 1980, Serine esterase inhibitors block stimulus-induced mobilization of arachidonic acid and phospholipase C activity in platelets, *J. Biol. Chem.* **255:**6024–6027.

Wallner, B. P., Mattaliano, R. J., Hession, C., Cate, R. L., Tizard, R., Sinclair, L. K., Foeller, C., Chow, E. P., Browning, J. L., Ramachandran, K. L., and Pepinsky, R. B., 1986, Cloning and expression of human lipocortin, a phospholipase A_2 inhibitor with potential anti-inflammatory activity, *Nature* **320:**77–80.

Wong, P. Y.-K., and Cheung, W. Y., 1979, Calmodulin stimulates human platelet phospholipase A_2, *Biochem. Biophys. Res. Commun.* 90:473–480.

23

The Metabolism of Inositol Phosphates

PHILIP W. MAJERUS, THOMAS M. CONNOLLY, VINAY S. BANSAL, ROGER C. INHORN, and HANS DECKMYN

1. Introduction

Phosphatidylinositols are phospholipid precursors of a series of inositol phosphates. Several inositol phosphates function as messenger molecules that evoke responses in cells following stimulation by extracellular agonists. While many different compounds are formed, the functions of only a few are understood. The pathways of formation of the inositol phosphates are complex and only partially worked out. The very complexity of the system implies that many functions may be served by these molecules. *Myo*-inositol (Fig. 1) is a hexatol that can be substituted with various combinations of phosphate esters and 1,2 cyclic phosphate esters. Allowing for all possible

Abbreviations used in this chapter: Ins(1,4,5)P_3, inositol 1,4,5-trisphosphate; Ins(1,4)P_2, inositol 1,4-bisphosphate; cIns(1:2) P, inositol cyclic phosphate; cIns(1:2,4)P_2, cyclic inositol 1:2,4-bisphosphate; Ins(1,3,4,5)P_4, inositol 1,3,4,5 tetrakisphosphate; Ins(1,3,4)P_3, inositol 1,3,4-trisphosphate; Ins(3,4)P_2, inositol 3,4-bisphosphate; Ins(1,3)P_2, inositol 1,3-bisphosphate; Ins 5P, inositol 5-phosphate; Ins 4P, inositol 4-phosphate; Ins 3P, inositol 3-phosphate; Ins 2P, inositol 2-phosphate, Ins 1P, inositol 1-phosphate; Ins, inositol; PtdIns, phosphatidylinositol; PtdIns4P, phosphatidylinositol 4-monophosphate PtdIns(4,5)P_2, phosphatidylinositol 4,5-bisphosphate; EGTA, ethylene glycol bis(β-aminoethyl ether)-*N,N,N′N′*-tetraacetic acid.

PHILIP W. MAJERUS, THOMAS M. CONNOLLY, VINAY S. BANSAL, ROGER C. INHORN, and HANS DECKMYN • Division of Hematology–Oncology, Departments of Internal Medicine and Biological Chemistry, Washington University School of Medicine, St. Louis, Missouri 63110.

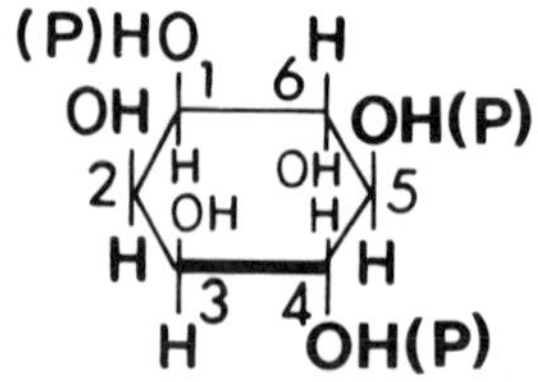

Figure 1. *Myo*-inositol. Numbering for the D isomer is shown with positions that contain phosphate esters in the precursor lipids indicated by (P).

combinations of phosphates, we could have as many as 63 distinct noncyclic inositol phosphates and at least 3 cyclic inositol phosphates. Each of these compounds could specify some function, and the amount of information that could be derived from this system is therefore immense. Thus far, 15 inositol phosphates have been described in tissues or extracts of tissues (listed in Table I).

Production of inositol phosphates from phosphatidylinositols occurs after stimulation of cells by several types of agonists. The initial step in this response involves an agonist binding to specific cell surface receptors. The many types of specific receptors that are coupled to phosphatidylinositol turnover have been summarized previously (Berridge and Irvine, 1984). The types of responses evoked vary widely, from those of short duration and immediate onset, such as secretion, aggregation, or contraction, to long-lasting effects evoked by mitogenic peptide growth factors. In all cases, the initial reaction involves activation of a phospholipase C enzyme that liberates inositol phosphates in the cell interior. As described below, this coupling between receptor and phospholipase C is probably achieved through the action of a yet undefined guanine nucleotide binding protein.

We have summarized the results from our own laboratory on the production and metabolism of inositol phosphates. Many general reviews on the physiology of these compounds have appeared to the past few years (Williamson *et al.*, 1985; Honkin, 1985; Litosch and Fain, 1986; Majerus *et al.*, 1986; Nishizuka, 1986) and will not be discussed here.

2. *Phospholipase C*

The phosphatidylinositols, phosphatidylinositol (PtdIns), phosphatidylinositol 4-phosphate (PtdIns4P), and phosphatidylinositol 4,5-bisphosphate (PtdIns(4,5)P_2), serve as storage forms for various water-soluble inositol phosphates that are produced when cells are stimulated. The phosphatidylinositols are degraded by phospholipase C to yield 1,2-diacylglycerol

Table I. Inositol Phosphates Found in Nature[a]

Compound	Source	Reference
S Ins 1-P	Numerous mammalian tissues	Eisenberg and Bolden, 1965; Ackerman *et al.*, 1987: Siess, 1985; Marjerus *et al.*, 1986; Morgan *et al.*, 1987.
D Ins 3-P or L Ins 1-P	Rat cerebral crotex, plants	Ackerman *et al.*, 1987; Ballou and Pizer, 1960
D Ins 4-P	Rat cerebral cortex, human platlets, calf brain, pituitary and glomerulosa cells	Ackerman *et al.*, 1987; Siess, 1985; Inhorn *et al.*, Morgan *et al.*, 1987.
DcIns(1:2)P	Pancreas, Kidney, platlets,SV40 transformed cells, Morris hepatoma cells	Majerus *et al.*, 1986; Graham *et al.*, 1987.
D Ins(1,3)P_2	Brain and liver extract, GH_4 cells	Bansal *et al.*, 1987; Irvine *et al.*, 1987; Hansen *et al.*, 1986.
D Ins (3,4)	Brain and liver extract, GH_4 cells	Inhorn *et al.*, 1987; Shears *et al.*, 1987; Irvine *et al.*, 1987
D Ins(1,4)P_2	Numerous mammian tissues	Majerus *et al.*, 1986; Berridge and Irvine, 1984.
D cIns(1:2,4)P_2	Pancreatic cells, sheep seminal vesicles	Majerus *et al.*, 1986; Graham *et al.*, 1987.
D Ins(1,4,5)P_3	Numerous mammalian tissues	Majerus *et al.*, 1986; Berridge and Irvine, 1984; Rittenhouse and Sasson, 1985; Beridge, 1986.
D Ins(1,3,4)P_3	Numerous tissues and cell lines	Majerus *et al.*, 1986; Berridge, 1986; Irvine *et al.*, 1984.
D cLns(1:2,4,5)P_3	Platelets and sheep seminal vesicles, pancreatic cells, erythrocytes	Majerus *et al.*, 1986; Chandra Sekar *et al.*, 1987; Irvine *et al.*, 1986a.
D Ins(1,3,4,5)P_4	Numerous tissues and cell lines	Majerus *et al.*, 1986; Berryridge 1986; Irvine *et al.*, 1986a; Batty *et al.*, 1985.
IP_4(?)	Plants, avian erythrocytes	Majumder *et al.*, 1972; Biswas *et al.*, 1978; Chakrabarti and Biswas, 1981; Isaaks *et al.*, 1977.
IP_5(?)	Plants, avian erythrocytes, Swiss 3T3	Majumder *et al.*, 1972; Boswas *et al.*, 1978; Chakrabarti and Biswas, 1981; Isaaks *et al.*, 1977; Heslop *et al.*, 1985, 1986 Guillemette *et al.*, 1987; Morgan *et al.*, 1987.
IP_6	Plants (predominant form), avian erythro-	Majumder *et al.*, 1972; Biswas *et al.*, 1978; Chakrabarti and Biswas, 1981; Isaaks *et al.*, 1977; Heslop *et al.*, 1985.

[a]The positions of the phosphates on plant and avian IP_4 are not defined, although, based on HPLC separation, they are different from Ins(1,3,4,5)P_4. Similarly, $InsP_5$ structures are uncertain in many cases, although Ins(1,3,4,5,6)P_5 is likely.

and inositol phosphates. The enzyme is present in most cell types and most of the activity is cytosolic when homogenates are prepared. Since the substrate is in a membrane bilayer, it is apparent that the enzyme must act in this location. It may be that in the intact cell the phospholipase C is loosely attached to the inner surface of the membrane (described below).

Two distinct soluble phospholipase C enzymes have been identified in ram seminal vesicles. The first enzyme, designated PLC-I, was purified to homogeneity from this tissue (Hofmann and Majerus, 1982). The pure enzyme has a specific activity of approximately 30 μmoles of PtdIns hydrolyzed per minute per milligram of protein. The molecular weight of the enzyme is 65,000 and it is a single polypeptide chain that is not glycosylated. Upon chromatography on ion exchange columns, such as aminohexyl agarose, the enzyme is resolved into two peaks of activity. The basis for this heterogeneity is unknown. The second enzyme from seminal vesicles, designated PLC-II, has a molecular weight of 85,000 and, although it has not ben purified to homogeneity, it appears to have an activity similar to PLC-I. Antibodies raised against each of these enzymes do not react with the other. The tissue distribution of the enzymes appears to differ: PLC-I is the predominant form in liver, and PLC-II is the major form in platelets and brain. Multiple forms of phospholipase C have been identified in partially purified preparations from many other tissues. The basis for the widely ranging molecular weights of phospholipase C enzymes reported by various investigators (from 65,000 to over 150,000) is unknown. The relation between these phospholipase C enzymes and those described above is presently unclear (Majerus *et al.*, 1986).

3. *Substrate Specificity*

The phospholipase C enzymes are specific for phosphatidylinositols. They do not hydrolyze other phospholipids, with the exception of phosphatidylglycerol, which is hydrolyzed at a rate 0.1% that of phosphatidylinositol (Hofmann and Majerus, 1982; Wilson *et al.*, 1984). The phospholipase C enzymes in *in vitro* assays require calcium ions to hydrolyze PtdIns. This property distinguishes the phospholipase C enzymes from lysosomal phospholipase C, which is neither specific for PtdIns nor inhibited by calcium chelation. The phospholipase C enzymes do not show specificity for particular fatty-acid-containing substrate molecules. The 2-position fatty acid is irrelevant, since 1-acyl-2-lyso-PtdIns is hydrolyzed at the same rate as PtdIns. However, in most cells the fatty acid composition of the phosphatidylinositols is highly uniform, since they contain predominantly a polyunsaturated fatty acid in the 2 position. In this way, the diacylglycerol product contains arachidonate or a related fatty acid in its 2 position and is thus a source for arachidonic acid, which is itself a messenger molecule after it is oxygenated to various icosanoid mediators.

Both PLC-I and PLC-II readily hydrolyze all three phosphatidylinositols (Wilson *et al.*, 1984). When the three lipids are incorporated into unilamellar vesicles, they compete with each other for enzyme. When present in equimolar concentrations, hydrolysis of phosphatidylinositol polyphosphates is favored. Increasing the portion of PtdIns or the calcium ion concentration favors hydrolysis of PtdIns. The activity of the enzyme is greatly affected by the lipid composition of the substrate bilayer, as reviewed previously (Majerus *et al.*, 1986). In fact, natural membranes are poor substrates for phospholipase C.

Calcium ion concentration determines both the rate of hydrolysis of phosphatidylinositols and the preferred substrate in *in vitro* assays. In the absence of calcium ions, PtdIns is not a substrate for phospholipase C. The K_m for PtdIns breakdown with respect to calcium ions is approximately 1 μM (Hofmann and Majerus, 1982). In contrast, phosphatidylinositol polyphosphates are hydrolyzed by phospholipase C even in the presence of EGTA, although calcium ions further stimulate hydrolysis (Wilson *et al.*, 1984). These findings are consistent with the idea that PtdIns(4,5)P_2 hydrolysis can occur at the low (basal) calcium ion concentration present in unstimulated cells. In this way, the Ins(1,4,5)P_3 product triggers calcium ion mobilization in the cell. PtdIns breakdown then follows in response to the elevated calcium ion concentration. The precise role of calcium ions in controlling PtdIns hydrolysis in cells is uncertain and awaits direct measurement in intact or permeabilized cells.

4. *Guanine Nucleotide Binding Proteins*

Over the past 10 years a great deal has been learned about a family of guanine nucleotide binding proteins referred to as G proteins (Gilman, 1987). These proteins function as intermediaries between receptors and intracellular effectors in transmembrane signaling pathways. The proteins consist of 3 subunits designated α, β, and γ in order of decreasing molecular weight. The α subunits differ among various G proteins. The β and γ subunits are similar in most G proteins, although threre may also be multiple forms of these subunits. The functions of G proteins in coupling receptors to adenylate cyclase and to cyclic GMP phosphodiesterase have been described in detail. Occupancy of receptors coupled to adenylate cyclase leads to formation of a complex between a G protein and the agonist-occupied receptor. The agonist-occupied receptor stimulates the exchange of GTP for GDP on the G protein.

Following this, the G protein is released from the receptor and dissociates into α and β/γ subunits. The α subunit then stimulates, if it comes from a stimulatory G protein (G_s), or inhibits (G_i) adenylate cyclase. The effect on adenylate cyclase is terminated by GTPase activity intrinsic to the G protein. Thus, nonhydrolyzable GTP analogues have a more pronounced and long-lasting effect.

A role for guanine nucleotide binding proteins in the activation of phospholipase C also appears likely. The first evidence suggesting the existence of such a protien was derived from studies showing that guanine nucleotides reduce the calcium ion requiremetn for secretion in permeabilized mast cells and platelets (Gomperts, 1980; Hasiam and Davidson, 1984). Later it was shown that the breakdown of phosphatidylinositols by phospholipase C in membranes from blowfly salivary glands was potentiated by guanine nucleotides (Litosch *et al.*, 1985). The ability of nonhydrolyzable analogues of guanine nucleotides to stimulate breakdown of endogenous phosphatidylinositols has now been demonstrated in many systems. These studies have led to the concept that the control of phospholipase C may occur in a manner analogous to that of adenylate cyclase. Studies showing guanine nucleotide stimulation of phospholipase C in membranes suggest that there may be membrane-bound phospholipase C enzymes that are distinct from the more prevalent soluble enzymes that have been purified and characterized. According to this idea, the membrane-bound enzyme is coupled to receptors and initiates phosphatidylinositol polyphosphate breakdown. It is possible, however, that the membrane-bound phospholipase C enzymes described in these systems are the same as the soluble enzymes that have previously been isolated. In support of this hypothesis, we recently found that guanine nucleotides stimulate cytosolic phospholipase C activity (Deckmyn *et al.*, 1986). In this way, the effects of guanine nucleotide binding proteins may be more similar to those of the rhodopsin phototransduction system (Kuhn, 1986; Stryer, 1987) than to the adenylate cyclase system described above.

Phototransduction is catalyzed by a guanine nucleotide binding protein named transducin. In this system, light photolyzes rhodopsin, resulting in an activated *agonist–receptor complex*. This complex catalyzes the exchange of GTP for GDP on transducin and thereby elutes the transducin molecule (mainly the α–GTP subunit) from the rod outer segment membrane into the cytosol. The α subunit then binds to cyclic GMP phosphodiesterase and activates this enzyme by displacing an inhibitory subunit. The duration of activation is determined by the intrinsic activity of the transducin GTPase. Once GTP is converted to GDP, the phosphodiesterase is no longer active. The transducin molecule must rebind to the rod outer segment membrane, join the β/γ subunits, and bind to an activated rhodopsin to allow exchange of GTP for GDP. The cyclic GMP phosphodiesterase is a loosely bound membrane protein. It can be eluted from membranes in light merely by washing in buffer solution (Baehr *et al.*, 1979). In this system, unlike the adenylate cyclase system, the guanine nucleotide binding protein or its α subunit appears to move from one part of the membrane to another, possibly eluting into the soluble fraction of the cell at some point. Whether the phosphodiesterase enzyme is firmly attached to the membrane at all times or whether it too is eluted at some point in the cycle is not clear.

Based on our recent studies, we propose that a similar system may operate in the case of phosphatidylinositol-specific phospholipase C. The phospholipase C enzyme in the unstimulated cell may be either in the cytosol, sequestered from its substrate, or loosely attached to the membrane in a relatively inactive state. Once an appropriate receptor is occupied by its agonist, guanine nucleotide binding protein may then be activated by GTP exchange. The G protein α subunit would then dissociate from the membrane to recruit phospholipase C in the cytosol or in another region of membrane, bringing it into an active conformation. In this state, the phospholipase C then initiates the phosphatidylinositol breakdown cycle.

We have recently obtained evidence for such a scheme using a crude enzyme from platelets (Deckmyn *et al.*, 1986). We assayed the enzyme in this preparation using unilamellar vesicles of total platelet lipids with ^{3}H-labeled PtdIns(4,5)P_2 and ^{32}P-labeled PtdIns. In this way, we could determine the effects of guanine nucleotides on the hydrolysis of both substrates in the same vesicles. These substrates retained a natural lipid composition but excluded all membrane proteins, including receptors and guanine nucleotide binding proteins. Vesicles thus constructed from whole platelet lipids are poor substrates for phospholipase C, mainly because they are rich in phosphatidylcholine. We found that the hydrolysis of PtdIns was minimally affected by guanine nucleotides. However, PtdIns(4,5)P_2 breakdown was sitmulated 10- to 20-fold by guanine nucleotides. GTPγS was the most postne guanine nucleotide studied. Similarly, sodium fluoride that mimics GTP action stimulated the hydrolysis of PtdIns(4,5)P_2. The effects of these substances on phospholipase C activity were not direct, since the nucleotides had no effect on the activity of isolated PLC-I or PLC-II. More recently, we succeeded in partially separating phospholipase C from a protein fraction that is required to demonstrate guanine nucleotide stimulation of phospholipase C activity. This implies that a guanine nucleotide binding protein is involved in phospholipase C activation. Whether this protein is one of the already isolated G protein or some yet undiscovered G protein remains to be elucidated. Studies with cholera toxin and pertussis toxin have been confusing, since pertussis toxin blocks the stimulation of phospholipase C by guanine nucleotides in some systems while it has no effect in others (Cockcroft, 1987). A firm answer concerning the nature of the guanine nucleotide binding protein that couples receptors to phospholipase C activation awaits isolation of the putative G protein. Further studies with toxins and crude systems are unlikely to resolve the uncertainties in this area.

5. Inositol Phosphates

Phospholipase C acts on all three phosphatidylinositols when the breakdown of these lipids is initiated. Although the hydrolysis of phosphatidylinositol polyphosphates precedes that of PtdIns in platelets and neutrophils, the hydrolysis of all three phosphoinositides occurs within a few seconds of stimulation (Wilson *et al.*, 1985; Lochner *et al.*, 1986). In these cells, the amount of each lipid degraded has been measured directly and phosphatidylinositol represents the major substrate. In some cells the hydrolysis of PtdIns follows that of the phosphatidylinositol polyphosphates by several minutes (Imai and Gershengorn, 1986).

Six different inositol phosphates are produced in this initial phospholipase C reaction (shown in Fig. 2). These products are the cyclic and noncyclic phosphate esters of inositol derived from each of the precursor lipids,

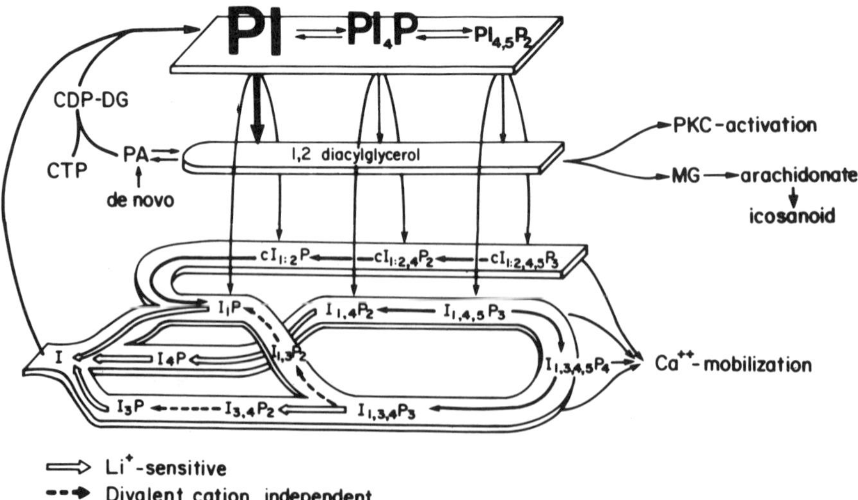

Figure 2. Pathway for inositol phosphate metabolism. I = inositol, P = phosphate, and the numbers refer to positions of phosphates on the inositol ring. Note that three different inositol bisphosphates and three different inositol monophosphates are formed during inositol phosphate metabolism.

namely, Ins 1P and cIns(1:2)P from PtdIns, Ins(1,4)P_2 and cIns(1:2,4)P_2 from PtdIns4P, and Ins(1,4,5)P_3 and cIns(1:2,4,5)P_3 from PtdIns(4,5)P_2. Both mammalian phospholipase C enzymes produce both cyclic and noncyclic products from each lipid (Majerus *et al.*, 1986). Unlike ribonuclease, which produces a cyclic phosphate as an intermediate in the hydrolysis of RNA, phospholipase C yields cyclic esters as final products rather than intermediates. Thus, phospholipase C does not catalyze the hydrolysis of the cyclic phosphate ester bond. The proportion of cyclic products ranges from one-third to two-thirds in various *in vitro* conditions. It has not been shown that physiologic events alter the proportion of cyclic versus noncyclic products, although this seems likely in view of the fact that the cyclic and noncyclic inositol phosphates are metabolized further by separate pathways utilizing different enzymes. Although cyclic phosphate esters of inositol have been found in a number of tissues, quantitation in intact cells has been difficult because of the acid lability of these compounds. cIns(1:2,4,5)P_3 has been demonstrated in thrombocin-stimulated platelets (Ishii *et al.*, 1986) and in carbamylcholine-stimulated pancreas minilobules (Chandra Sekar *et al.*, 1987). cIns(1:2)P has been shown in a variety of cells (Majerus *et al.*, 1986; Graham *et al.*, 1987).

There is currently no evidence for interconversion of cyclic and noncyclic inositol polyphosphates. When cIns(1:2,4,5)P_3 was added to crude extracts of platelets or rat kidney, no Ins(1,4,5)P_3 was formed, but cIns(1:2,4)P and Ins(1:2)P were formed (Connolly *et al.*, 1986a). cIns(1:2)P hydrolase converts the cyclic monophosphate to Ins 1P. This ubiquitous enzyme has been isolated from human placenta, has a molecular weight of 55,000, and is comprised of two subunits of 29,000 molecular weight (Ross and Majerus, 1986). The enzyme does not hydrolyze the cyclic phosphate esters in the inositol polyphosphates. The enzyme is not inhibited by lithium ions but is exquisitely sensitive to Ins 2P. The physiologic significance of this inhibition is uncertain, since Ins 2P levels have not been shown to change in cells under conditions of stimulation of phosphatidylinositol turnover.

In earlier studies of inositol phosphate metabolism, attempts were made to judge the importance of particular compounds by measuring the amount produced, usually by measuring changes in radioactively labeled compounds. Now that it is apparent that inositol phosphate metabolism is much more complex than initially envisioned, this strategy is of questionable value. Many different inositol phosphates are formed very rapidly when cells are stimulated, and the pathways of their interconversion are complex, as discussed below. Despite the experimental difficulties, it seems likely that differences in the proportion of various inositol phosphates and the temporal relations in thier production are important determinants of cellular responses to agonists.

6. *Ins(1,3,4)P₃ and Ins(1,3,4,5)P₄*

Efforts to find lipid precursors for inositol phosphates other than the three shown in Fig. 2 have been unsuccessful (Downes *et al.*, 1986); the many additional inositol phosphates are thus produced by further metabolism of the six initial products shown in Fig. 2. Recent studies indicate that the metabolism of inositol phosphates is considerably more complex than originally proposed. Irvine and co-workers (1984) isolated the inositol trisphosphates form carbachol-stimulated rat parotid glands (15-min stimulation) and found that the major inositol trisphosphate was not the expected Ins(1,4,5)P_3 isomer but rather Ins(1,3,4)P_3. The source of this isomer was elucidated by the discovery by Batty *et al.* (1985) that rat brain cortical slides stimulated with carbachol rapidly formed Ins(1,3,4,5)P_4. This suggested the presence of an Ins(1,4,5)P_3 3-kinase that converts Ins(1,4,5)P_3 to Ins(1,3,4,5)P_4. This kinase was initially demonstrated in the soluble fraction of homogenates of brain, liver, pancreas, and platelets (Irvine *et al.*, 1986b). The enzyme appears to have a very high affinity for Ins(1,4,5)P_3 ($K_m \approx 0.6\ \mu M$), much higher than the competing Ins(1,4,5)P_3 5-phosphomonoesterase described below. The tetrakisphosphate product of the kinase reaction is also a substrate for the same 5-phosphomonoesterase enzyme, which converts it to Ins(1,3,4)P_3, thereby explaining the presence of this isomer in stimulated tissues. Thus, a pathway for Ins(1,4,5)P_3 metabolism can be written

$$\text{Ins}(1,4,5)\text{P}_3 \xrightarrow[\text{ATP}]{\text{kinase}} \text{Ins}(1,3,4,5)\text{P}_4 \xrightarrow{\text{5-phosphotase}} \text{Ins}(1,3,4)\text{P}_3$$

In many tissues Ins(1,3,4)P_3 appears very rapidly after agonist stimulation, implying that the 3-kinase and 5-phosphomonoesterase rapidly metabolize Ins(1,4,5)P_3 via this pathway. The kinase does not phosphorylate cIns(1:2,4,5)P_3, thus, there does not appear to be a cyclic InsP_4 (Connolly *et al.*, 1987). The kinase enzyme is stimulated at low concentrations of Ca^{2+}, implying that diversion of Ins(1,4,5)P_3 via this pathway may be augmented after Ca^{2+} mobilization (Biden and Wallheim, 1986).

7. *Inositol Bisphosphate Metabolism*

We have recently discovered several additional enzymes that participate in inositol phosphate metabolism, including an enzyme, which we have named inositol polyphosphate 1-phosphatase, in crude extracts of calf brain and in platelets (Inhorn *et al.*, 1987). The enzyme has been purified 3600-fold from calf brain and is free of other phosphatase activities. It catalyzes the conversion of Ins(1,3,4)P_3 to Ins(3,4)P_2 and thus forms an inositol bisphosphate isomer that is different from the Ins(1,4)P_2 formed both directly from the lipid PtdIns4P and from Ins(1,4,5)P_3. The enzyme requires magnesium ions and is inhibited by lithium ions. Thc same enzyme also catalyzes the conversion of Ins(1,4)P_2 to Ins 4P. In brain and platelets, this appears to be the only enzyme that hydrolyzes Ins(1,4)P_2. Thus, most Ins(1,4,5)P_3 and all Ins(1,4)P_2 formed in these tissues is converted to Ins 4P and not to Ins 1P, as previously presumed.

We have also found another pathway for Ins(1,3,4)P_3 metabolism (Bansal *et al.*, 1987). This involves a 4-phosphatase enzyme that does not require magnesium ions, is fully active in the presence of EDTA, and is not inhibited by lithium ions. This enzyme converts Ins(1,3,4)P_3 to Ins(1,3)P_2, thus producing a third inositol bisphosphate. The three inositol biphosphates formed in calf brain homogenates are each produced and further metabolized by separate enzymes. This complexity suggests that one or more of these compounds may serve as messenger molecules. The enzymes that form and degrade these bisphosphates differ in magnesium ion requirement and inhibition by lithium. Therefore, changes in these ions could alter the relative proportions of the bisphosphates formed under different conditions or in different tissues. For example, lithium ions will cause Ins(1,4)P_2 to accumulate and block formation of Ins(3,4)P_2, thus increaisng the relative amount of Ins(1,3)P_2. A single chromatographic method does not resolve all the inositol monophosphates and bisphosphates clearly. Ins 1P and Ins 3P are stereoisomers that are difficult to resolve. Ins 4P is readily separated from these compounds, as is cIns(1:2)P. It is therefore necessary to reevaluate inositol bisphosphate and monophosphate production in various systems; preliminary work suggests that all three bisphosphates are found in cells (Irvine *et al.*, 1987).

8. *Inositol (1,4,5)P_3 5-Phosphomonoesterase*

Inositol (1,4,5)P_3 5-phosphomonoesterase is an enzyme that degrades Ins(1,4,5)P_3 to Ins(1,4)P_2. The enzyme was originally described as a particulate enzyme in erythrocyte membranes (Downes *et al.* 1982). Since then it has been found both in the membrane fraction and in the cytosol of most cells. The enzyme has been purified to homogeneity from the cytosol of human platelets (Connolly *et al.*, 1985).The isolated enzyme is a single polypeptide with an apparent molecular weight of about 45,000 as determined by SDS–polyacrylamide gel electrophoresis. The enzyme requires magnesium-ions and is present in large amounts in crude extracts of most tissues with the capacity to degrade within a few seconds any Ins(1,4,5)P_3 that is formed. This enzyme is actually a more general 5-phosphatase and hydrolyzes three substrates: Ins(1,4,5)P_3, Ins(1,3,4,5)P_4, and cIns(1:2,4,5)P_3. It removes the 5-phosphate from all three substrates. It does not hydrolyze other inositol phosphates, nor does it hydrolyze glycerol phosphoryl inositol derivatives such as Gro-PtdIns(4,5)P_2 or Gro-PtdIns 4P. The relative ability of the enzyme to utilize its three substrates varies considerably, as indicated in Table II. The catalytic efficiency of the 5-phosphomonoesterase in hydrolyzing Ins(1,3,4,5)P_4 is less than that observed with Ins(1,4,5)P_3 (Connolly *et al.*, 1987). The maximal velocity of hydrolysis of the Ins(1,3,4,5)P_4 substrate is approximately 1/30 of that observed with Ins(1,4,5)P_3. However, the affinity for Ins(1,3,4,5)P_4 is greater than that for Ins(1,4,5)P_3. These parameters suggest that at low substrate concentrations Ins(1,3,4,5)P_4 is hydrolyzed rapidly and will competitively inhibit hydrolysis of Ins(1,4,5)P_3. The capacity to cleave Ins(1,4,5)P_3 is much greater, especially at higher substrate concentrations. The 5-phosphomonoesterase also hydrolyzes cIns(1:2,4,5)P_3, although the capacity of the enzyme to utilize this substrate is less than 10% of its ability to hydrolyze Ins(1,4,5))P_3 at substrate concentrations that might occur in a stimulated cell. This finding, plus the fact that the Ins(1,4,5)P_3 3-kinase does not utilize cIns(1:2,,4,5)P_3 as a substrate, suggests that the cyclic inositol trisphosphate may be more stable in cells than its noncyclic counterpart and may thus have a more prolonged stimulatory effect.

The 5-phosphomonoesterase enzyme is regulated by protein kinase C (Connolly *et al.*, 1986b). Protein kinase C phosphorylates the 5-phosphomonoesterase at multiple sites, incorporating approximately 3–4 moles phosphate/mole enzyme. Phosphorylation of the enzyme by protien kinase C results in activation of the 5-phosphomonoesterase enzyme. We find two- to ten-fold stimulation of 5-phosphomonoesterase activity using various preparations of the enzyme. Presumably, the variable degree of stimulation results from different levels of basal phosphorylation of the enzyme used in the phosphorylation reaction. Protein kinase C also stimulates the activity of this enzyme with respect to its other substrates. The effect of protein kinase C on this enzyme could explain recent reports showing that treatment of platelets and other cells with phorbol esters (compounds that activate protein kinase C) decreases the subsequent calcium rise in response to agonists, decreases the levels of Ins(1,4,5)P_3, and causes rapid degradation of Ins(1,4,5)P_3 when the latter is added to permeabilized cells (Molina y Vedia and Lapetina, 1986).

Table II. Comparison of Substates for 5-Phosphomonoesterase

Substrate	1 K_m (μM)	2 V_m (moles/min/mole)	2/1 Relative efficiency[a]
Ins(1,4,5)P_3	10	240	1
cIns(1:2,4,5)P_3	250	240	0.04
Ins(1,3,4,5)P_4	1	8	0.3

[a]Catalytic efficiency is compared to Ins(1,4,5)P_3 set as 1.

In fact, when intact human platelets arc stimulated with thrombin after being labeled with [^{32}P]phosphate, there is a rapid increase in the phosphorylation of the IP_3 5-phosphomonoesterase. This phosphorylation is the major protein-kinase-C-stimulated phosphorylation that occurs in platelets.

Our hypothesis for the regulation of inositol trisphosphate production and subsequent calcium mobilization by changing activity of the 5-phosphomonoesterase can be summarized as follows: thrombin stimulation of platelets activates phospholipase C to cleave PtdIns(4,5)P_2, forming Ins(1,4,5)P_3, cIns(1:2,4,5)P_3, and 1:2-diacylglycerol. these inositol phosphates mobilize calcium from intracellular stores to elevate the cytosolic calcium concentration, which allows phospholipase C to utilize PtdIns as a substrate to form Ins 1P, cIns(1:2)P, and the major amount of 1:2-diacylglycerol. The 1:2-diacylglycerol derived from PtdIns, plus the elevated cytosolic calcium, activates protein kinase C, which phosphorylates the 5-phosphomonoesterase and thereby activates the phosphatase. Finally the phosphorylated 5-phosphomonoesterase hydrolyzes Ins(1,4,5)P_3, Ins(1,3,4,5)P_4, and cIns (1:2,4,5)P_3. The Ins(1,4)P_2 and cIns(1:2)4P_2 formed are inactive (at least with respect to calcium mobilization), and therefore the phosphoinositide-derived calcium-mobilizing signal is terminated.

9. Future Challenges

Future studies will be directed at resolving several issues. The enzymology of inositol phosphate metabolism must be further defined. Are there additional, as yet undiscovered, inositol phosphates formed in cells? Most importantly, functions for the various inositol phosphates need to be determined. Which ones are messenger molecules, and in what manner do they act?

ACKNOWLEDGEMENTS. This research was supported by Grants HLBI 14147 (Specialized Center for Research in Thrombosis), HL 16634, and Training Grant T32 HLBI 07088 from the National Institutes of Health, and by National Institutes of Health Research Service Award GM-07200, Medical Scientist, from the National Institute of General Medical Sciences. The authors thank Brian Whiteley, Theodora Ross, and Daniel Lips for many suggestions, and Teresa Bross, Susan Kaiser, and Cecil Buchanan for technical assistance.

References

Ackerman, K. E., Gish, B., G., Honchar, M. P., and Sherman, W. R., 1987, Evidence that inositol-1-phosphate in brain of lithium-treated rats results mainly from phosphatidylinositol metabolism, *Biochem. J.* **242:**517–514.

Baehr, W., Derbin, M. J., and Applebury, M. L., 1979, Isolation and characterization of cGMP phosphodiesterase from bovine rod outer segments, *J. Biol. Chem.* **254:**11669–11677.

Ballou, C. E., and Pizer, L. I., 1960, The absolute configuration of myoinositol 1-phosphate and a confirmation of bornesitol configurations, *J. Am. Chem. Soc.* **82:**333–335.

Bansal, V. S., Inhorn, R. C., and Majerus, P. W., 1987, The metabolism of inositol 1,3,4-trisphosphate to inositol 1,3-bisphosphate, *J. Biol. Chem.* (in press).

Batty, I. R., Nahorski, S. R., and Irvine, R. F., 1985, Rapid formation of inositol 1,3,4,5-tetrakisphosphate following muscarinic receptor stimulation of rat cerebral cortical slices, *Biochem. J.* **232:**211–215.

Berridge, M. J., 1986, Regulation of ion channels by inositol trisphosphate and diacylglycerol, *J. Exp. Biol.* **124:**323–335.

Berridge, M. J., and Irvine, R. F., 1984, Inositol trisphosphate, a novel second messenger in cellular signal transduction, *Nature* **312:**315–321.

Biden, T. J., and Wallheim, C. B., 1986, Ca^{++} regulates the inositol tris/tetrakisphosphate pathways in intact and broken preparations of insulin-secreting RINm5F cells, *J. Biol. Chem.* **261:**11931–11934.

Biswas, S., Maity, I. B., Chakrabarti, S., and Biswas, B. B., 1978, Purification and characterization of myo-inositol hexaphosphate-adenosine diphosphate phosphotransferase from *Phaseolus aureus, Arch. Biochem. Biophys.* **185:**557–566.

Chakrabarti, S., and Biswas, B. B., 1981, Evidence for the existence of phosphoinositol kinase in chicken erythrocytes, *Indian J. Biochem. Biophys.* **18:**398–401.

Chandra Sekar, M., Dixon, J. F., and Hokin, L. E., 1987, The formation of inositol 1,2–cyclic 4,5-trisphosphate and inositol 1,2-cyclic 4-bisphosphate on stimulation of mouse pancreatic minilobules with carbamylcholine, *J. Biol. Chem.* **262:**340–344.

Cockcroft, S., 1987, Polyphosphoinositide phosphodiesterase: Regulation by a novel guanine nucleotide binding protein Gp, *Trends Biochem. Sci.* **12:**75–78.

Connolly, T. M., Bross, T. E., and Majerus, P. W., 1985, Isolation of a phosphomonoesterase from human platelets that specifically hydrolizes the five-phosphate of inositol-(1,4,5)trisphosphate, *J. Biol. Chem.* **260:**7868–7874.

Connolly, T. M., Wilson, D. B., Bross, T. E., and Majerus, P. W., 1986, Isolation and characterization of the inositol cyclic phosphate products of phosphoinositide cleavage by phospholipase C. II. Metabolism in cell free extracts, *J. Biol. Chem.* **261:**122–126.

Connolly, T. M., Lawing, W. J., Jr., and Majerus, P. W., 1986b, Protein kinase C phosphorylates human platelet inositoltrisphosphate 5′-phosphomonoesterase increasing the phosphatase activity, *Cell* **46:**951–958.

Connolly, T. M., Bansal, V. S., Bross, T. E., Irvine, R. F., and Majerus, P. W., 1987, The metabolism of tris- and tetraphosphates of inositol by 5-phosphomonoesterase and 3-kinase enzymes. *J. Biol. Chem.* **262:**2146–2149.

Deckmyn, H., Tu, S.-M., and Majerus, P. W., 1986, Guanine nucleotides stimulate soluble phosphoinositide-specific phospholipase C in the absence of membranes, *J. Biol. Chem.* **261:**16553–16558.

Downes, C. P., Mussat, M. C., and Michell, R. H., 1982, The inositol trisphosphate phosphomonoesterase of the human erythrocyte membrane, *Biochem. J.* **203:**169–177.

Downes, C. P., Hawkins, P. T., and Irvine, R. F., 1986, Inositol 1,3,4,5-tetrakisphosphate and not phosphatidylinositol 3,4-bisphosphate is the probable precursor of the inositol 1,3,4-trisphosphate in agonist-stimulated parotid gland, *Biochem. J.* **238:**501–506.

Eisenberg, F., Jr., and Bolden, A. H., 1965, D-myo-inositol-1-phosphate, an intermediate in the biosynthesis of inositol in mammal, *Biochem. Biophys. Res. Commun.* **21:**100–105.

Gilman, A. J., 1987, G proteins: Transducers of receptor-generated signals, *Annu. Rev. Biochem.* **56** (in press).

Gomperts, B. D., 1980, Involvement of guanine nucleotide binding protein in the gating of CA^{2+} by receptors, *Nature* **306:**64–66.

Graham, R. A., Meyer, R. A., Szwergold, B. S., and Brown, T. R., 1987, Observation of myo-inositol 1,2-(cyclic)phosphate in a Morris hepatoma by ^{31}P NMR, *J. Biol. chem.* **262:**35–37.

Guillemette, G., Baukal, A. J., Balla, T., and Catt, K. J., 1987, Angiotensin-induced formation and metabolism of inositol polyphosphates in bovine adrenal glomerulosa cells, *Biochem. Biophys. Res. Commun.* **142:**15–22.

Hansen, C. A., Mah, S., and Williamson, J. R., 1986, Formation and metabolism of inositol 1,3,4,5-tetrakisphosphate in liver, *J. Biol. Chem.* **261:**8100–8103.

Haslam, R. J., and Davidson, M. M. L., 1984, Guanine nucleotides decrease the free [Ca^{2+}] required for secretion of serotonin from permeabilized blood platelets. Evidence for a role for a GTP-binding protein in platelet activation, *FEBS Lett.* **174:**90–95.

Heslop, J. P., Irvine, R. F., Tashjian, A. T., and Berridge, M. J., 1985, Inositol tetrakis- and pentakisphosphates in GH_4 cells, *J. Exp. Biol.* **119:**395–401.

Heslop, J. P., Blakeley, D. M., Brown, K. D., Irvine, R. F., and Berridge, M. J., 1986, Effects of bombesin and insulin on inositol(1,4,5)trisphosphate and inositol(1,3,4) trisphosphate formation in Swiss 3T3 cells, *Cell* **47:**703–709.

Hofmann, S. L., and Majerus, P. W., 1982, Identification and properties of two distinct phosphatidylinositol-specific phospholipase C enzymes from sheep seminal vesicular glands, *J. Biol. Chem.* **257:**6461–6469.

Hokin, L. E., 1985, Receptors and phosphoinositide-generated second messengers, *Annu. Rev. Biochem.* **54:**205–236.

Imai, A., and Gershengorn, M. C., 1986, Phosphatidylinositol 4,5-bisphosphate turnover is transient while phosphatidylinositol turnover is persistent in thyrotropin-releasing hormone-stimulated rat pituitary cells, *Proc. Natl. Acad. Sci. U.S.A.* **83:**8540–8544.

Inhorn, R. C., Bansal, V. S., and Majerus, P. W., 1987, Pathway for inositol 1,3,4-trisphosphate and 1,4-bisphosphate metabolism *Proc. Natl. Acad. Sci. U.S.A.* (in press).

Irvine, R. F., Letcher, A. J., Lander, D. J., and Downes, C. P., 1984, Inositol trisphosphates in carbachol-stimulated rat parotid glands, *Biochem. J.* **223:**237–243.

Irvine R. F., Letcher, A. J., Lander, D. J., and Berridge, M. J., 1986a, Specificity of inositol phosphate-stimulated Ca^{2+} mobilization from Swiss-mouse 3T3 cells, *Biochem. J.* **240:**301–304.

Irvine, R. F., Letcher, A. J., Heslop, J. P., and Berridge, M. J., 1986b, The inositol tris/tetrakisphosphate pathway—demonstration of Ins(1,4,5)P_3 3-kinase activity in animal tissues, *Nature* **320:**631–634.

Irvine, R. F., Letcher, A. J., Heslop, J. P., and Berridge, M. J., 1987, Inositol(3,4) bisphosphate and inositol(1,3)bisphosphate in GH_4 cells—Evidence for complex breakdown of inositol(1,3,4)trisphosphate, *Biochem. Biophys. Res. Commun.* **143:**353–359.

Isaacks, R., Harkness, D., Sampsell, R., Adler, J., Roth, S., Kim, C., and Goldman, P., 1977, Studies on avian erythrocyte metabolism. Inositol tetrakisphosphate: The major phosphate compound in the erythrocytes of the ostrich (*Struthio camelus camelus*), *Eur. J. Biochem.* **77:**567–574.

Ishii, H., Connolly, T. M., Bross, T. E., and Majerus, P. W., 1986, Inositol cyclic trisphosphate (inositol 1:2-cyclic,4,5-trisphosphate) is formed upon thrombin stimulation of human platelets, *Proc. Natl. Acad. Sci. U.S.A.* **83:**6397–6401.

Kühn, H., 1986, Proteins involved in the control of cGMP phosphodiesterase in retinal rod cells, *Fortschr. Zool.* **33:**289–297.

Litosch, I., and Fain, J. N., 1986, Mini review: Regulation of phosphoinositide breakdown by guanine nucleotides, *Life Sci.* **39:**187–194.

Litosch, I., Wallis, C., and Fain, J. N., 1985, 5-Hydroxytryptamine stimulates inositol phosphate production in a cell-free system from blowfly salivary glands. Evidence for a role of GTP in coupling receptor activation to phosphoinositide breakdown, *J. Biol. Chem.* **260:**5464–5471.

Lochner, J. E., Badway, J. A., Horn, W., and Karnovsky, M. L., 1986, All-*trans*-retinal stimulates superoxide release and phospholipase C activity in neutrophils without significantly blocking protein kinase C, *Proc. Natl. Acad. Sci. U.S.A.* **83:**7673–7677.

Majerus, P. W., Connolly, T. M., Deckmyn, H., Ross, T. S., Bross, T. E., Ishii, H. Bansal, V. S., and Wilson, D. B., 1986, The metabolism of phosphoinositide-derived messenger molecules, *Science* **234:**1519–1526.

Majumder, A. N. L., Mandal, N. C., and Biswas, B. B., 1972, Phosphoinositol kinase from germinating mung bean seeds, *Phytochemistry* **11:**503–508.

Molina y Vedia, L. M., and Lapetina, E. G., 1986, Phorbol 12,13-dibutyrate and 1-oleoyl-2-acetyldiacylglycerol stimulate inositol trisphosphate dephosphorylation in human platelets, *J. Biol. Chem.* **261:**10493–10495.

Morgan, R. O., Chang, J. P., and Catt, K. J., 1987, Novel aspects of gonadatropin-releasing hormone action on inositol polyphosphate metabolism in cultured pituitary gonadotrophs, *J. Biol. Chem.* **262:**1166–1171.

Nishizuka, Y., 1986, Studies and prospectives of protein kinase C *Science* **233:**305–312.

Rittenhouse, S. E., and Sasson, J. P., 1985, Mass changes in myo-inositol trisphosphate in human platelets stimulated by thrombin. Inhibitory effects of phorbol ester, *J. Biol. Chem.* **260:**8657–8660.

Ross, T. S., and Majerus, P. W., 1986, Isolation of D-*myo*-inositol 1:2 cyclic phosphate 2-inositolphosphohydrolase from human placenta, *J. Biol. Chem.* **261:**11119–11123.

Shears, S. B., Storey, D. J., Morris, A. J., Cubitt, A. B., Parry, J. B., Michell, R. H., and Kirk, C. J., 1987, Dephosphorylation of myo-inositol 1,4,5-trisphosphate and myo-inositol 1,3,4-trisphosphate, *Biochem. J.* **242:**393–402.

Siess, W., 1985, Evidence for the formation of inositol 4-monophosphate in stimulated human platelets, *FEBS Lett.* **185:**151–156.

Stryer, L., 1987, Visual transduction: Design and recurring motifs, in: *Membrane Proteins: Structure, Function, Assembly,* Nobel Symposium No. 66, Chemica Scripta, Stockholm (in press).

Williamson, J. R., Cooper, R. H., Joseph, S. K., and Thomas, A. P., 1985, Inositol trisphosphate and diacylglycerol as intracellular second messengers in liver, *Am. J. Physiol.* **248:**C203–C216.

Wilson, D. B., Bross, T. E., Hofmann, S. L., and Majerus, P. W., 1984, Hydrolysis of polyphosphoinositides by purified sheep seminal vesicle phospholipase C enzymes, *J. Biol. Chem.* **259:**11718–11724.

Wilson, D. B., Neufeld, E. J., and Majerus, P. W., 1985, Phosphoinositide interconversion in thrombin stimulated human platelets, *J. Biol. Chem.* **260:**1036–1051.

Index